D0935308

Organic Compounds of Sulphur, Selenium, and Tellurium

Volume 6

A Specialist Periodical Report

Organic Compounds of Sulphur, Selenium, and Tellurium

Volume 6

A Review of the Literature Published between
April 1978 and March 1980

Senior Reporter
D. R. Hogg *Department of Chemistry, University of Aberdeen*

Reporters
G. C. Barrett *Oxford Polytechnic*
E. Block *University of Missouri-St Louis, USA*
P. K. Claus *University of Wien, Austria*
D. L. J. Clive *University of Alberta, Canada*
M. Davis *La Trobe University, Victoria, Australia*
N. Furukawa *University of Tsukuba, Japan*
J. K. Landquist *formerly of ICI Ltd, Pharmaceuticals Division*
S. Oae *University of Tsukuba, Japan*
A. Ohno *University of Kyoto, Japan*
C. G. Venier *Iowa State University, USA*

The Royal Society of Chemistry
Burlington House, London W1V 0BN

British Library Cataloguing in Publication Data

Organic compounds of sulphur, selenium, and tellurium – Vol. 6.–
  (A Specialist periodical report)
  1. Organosulphur compounds – Periodicals
  2. Organoselenium compounds – Periodicals
  3. Organotellurium compounds – Periodicals
  I. Royal Society of Chemistry
  547'.06   QD305.S3

  ISBN 0-85186-299-3
  ISSN 0305-9812

Printed in Great Britain by
Spottiswoode Ballantyne Ltd., Colchester and London

Preface

These Reports cover the period April 1978 to March 1980 inclusive, except in the case of Chapter 2 Part III, where the period ends in February 1980.

The major changes in the format and contents of this volume have been necessitated by economic considerations. Sulphur-containing heteroaromatic compounds which previously were included in these Reports and in the Reports on 'Aromatic and Heteroaromatic Chemistry' are now covered in detail only in the new Reports, 'Heterocyclic Chemistry'. The literature on these compounds has been reviewed up to March 1978 in Volume 5 of this series and thereafter in Volume 1 of the new series. Highlights of the literature of these compounds are given as Chapter 6 of this volume. The organization of the remainder of this volume is essentially the same as that of Volume 5 except that the section on the ylides and carbanionic compounds of selenium and tellurium has been considerably extended and the coverage of Chapter 3, Part III has been extended to include dithiocarbamates, xanthates, and trithiocarbonates.

I would like to take this opportunity of expressing my appreciation to the contributors to this volume for all the care and effort they have put into their Reports. This has, of course, made my own task much less onerous than it could have been.

DRH

v

Contents

Abbreviations

The following abbreviations have been used:

c.d.	circular dichroism
c.i.d.n.p.	chemically induced dynamic nuclear polarization
c.i.m.s.	chemical ionization mass spectrometry
e.s.r.	electron spin resonance
i.r.	infrared
n.q.r.	nuclear quadrupole resonance
p.e.	photoelectron
u.v.	ultraviolet
FVP	flash vacuum pyrolysis
g.l.c.	gas–liquid chromatography
ASIS	aromatic-solvent-induced shift
LIS	lanthanide-induced shift
NOE	nuclear Overhauser effect
CNDO	complete neglect of differential overlap
HOMO	highest occupied molecular orbital
M.O.	molecular orbital
e.e.	enantiomeric excess
AIBN	azobis(isobutyronitrile)
DABCO	1,5-diazabicyclo[2.2.2.]octane
DCCI	dicyclohexylcarbodi-imide
DDQ	2,3-dichloro-5,6-dicyano-1,4-benzoquinone
DMF	NN-dimethylformamide
DMSO	dimethyl sulphoxide
HMPA	hexamethylphosphoric acid triamide
LDA	lithium di-isopropylamide
LAH	lithium aluminium hydride
NBS	N-bromosuccinimide
NCS	N-chlorosuccinimide
MCPBA	m-chloroperoxybenzoic acid
PPA	polyphosphoric acid
py	pyridine
TCNE	tetracyanoethylene
TCNQ	tetracyanoquinodimethane
TFA	trifluoroacetic acid
TFAA	trifluoroacetic anhydride

THF   tetrahydrofuran
TMEDA   tetramethylethylenediamine

Vol. 5,   Reference back to a preceding Volume of this series of
Vol. 4   'Specialist Periodical Reports'

1

Aliphatic Organosulphur Compounds, Compounds with Exocyclic Sulphur Functional Groups, and their Selenium and Tellurium Analogues

BY G. C. BARRETT

In spite of the large number of literature citations included in this Chapter, many of the routine papers on preparative chemistry have been excluded. The fact that so much work has been considered worthy of inclusion in this review on the basis of novelty and importance testifies as much to the intricacies and surprises still being found in the chemistry of sulphur- and selenium-containing functional groups as to the continuing growth in importance of these compounds in organic synthesis.

1 Textbooks and Reviews

Books published recently include several that are wholly devoted to organosulphur compounds[1a-d] and others[1e-i] in which specific chapters are relevant to the coverage in this chapter. Reviews[2] have appeared that deal with applications of organosulphur compounds in synthesis[2a-d] (including uses of β-keto-sulphides,[2b] vinyl sulphones,[2d] and vinyl sulphides[2e]) and the synthesis of organosulphur compounds,[2f] sulphenyl compounds,[2g] indolethiols,[2h] allenic sulphides[2i] and sulphenamides,[2j] organosilyl sulphides, selenides, and tellurides,[2k] optically active

---

[1] (a) 'Natural Sulphur Compounds', ed. D. Cavallini, G. E. Gaull, and V. Zappia, Plenum Press, New York and London, 1980; (b) H. Kwart and K. King, 'd-Orbitals in the Chemistry of S, Si, and P', Springer Verlag, Weinheim, 1977; (c) 'Sulfur Reports', ed. A. Senning, Horwood, Chichester, Vol. 1, 1980; (d) 'Topics in Organic Sulphur Chemistry: Proceedings of the 8th International Symposium on Organic Sulphur Chemistry, Portoroz, Yugoslavia', ed. M. Tisler, University Press Ljubljana, 1978; (e) 'Comprehensive Organic Chemistry', ed. Sir Derek H. R. Barton and W. D. Ollis, Pergamon Press, Oxford, 1979; (f) in 'Methodicum Chimicum', 1978, Vol. 7, Part B (ed. F. Korte, Academic Press, New York, 1978), R. M. Wilson and D. N. Buchanan, p. 585 (thiols and thiocarbonyl compounds), K. J. Wynne and I. Haiduc, p. 652 (sulphinyl compounds), and I. Haiduc and K. J. Wynne, p. 670 (sulphonyl compounds); (g) H. J. Reich, in 'Oxidations in Organic Chemistry, Part C', ed. W. S. Trahanovsky, Academic Press, New York, 1978; (h) Sir Derek H. R. Barton and S. V. Ley, Ciba Found Symp., 1978, No. 53, p. 53; (i) 'Chemistry of the Sulphonium Group', ed. S. Patai and C. J. M. Stirling, Wiley, Chichester, 1981.

[2] (a) B. M. Trost, Acc. Chem. Res., 1978, 11, 453; (b) B. M. Trost, Chem. Rev., 1978, 78, 363; (c) L. Field, Synthesis, 1978, 713; (d) M. Julia, D. Uguen, A. Righini, M. Launay, and J. N. Verpeaux, Int. Congr. Ser.–Excerpta Med., 1979, 457, 163; (e) S. Blechert, Nachr. Chem. Tech. Lab., 1979, 27, 634; (f) G. Tsuchihashi, Kagaku Sosetsu, 1978, 19, 75; (g) E. Kuehle, Chem. Labor. Betr., 1979, 30, 373; (h) T. F. Spande, Chem. Heterocycl. Compd., 1979, 25, 1; (i) W. Reichen, Chem. Rev., 1978, 78, 569; (j) F. A. Davis and U. K. Nadir, Org. Prep. Proced. Int., 1979, 11, 33; (k) D. Brandes, J.

sulphur compounds,[2l] sulphur-centred radicals,[2m] episulphonium intermediates,[2n] barriers to S—N bond rotation in sulphenamides,[2o] 1,2-phenylthio-shifts,[2p] thio-Claisen rearrangements,[2q] Pummerer reactions,[2r] and Ramberg–Bäcklund rearrangements.[2s] A review of nucleofugacity of different leaving groups in elimination reactions concerns several sulphur-containing functional groups.[2t]

Organoselenium chemistry has been reviewed from the point of view of applications in synthesis in general,[2u] and for mild oxidative transformations,[1g,h] as well as for its relevance in biochemistry.[2v] Coverage of the 1976–7 literature for organotellurium chemistry has been published.[2w]

2 Spectroscopic and Other Physical Properties of Organosulphur, Organoselenium, and Organotellurium Compounds

Molecular Orbital Calculations and Conformational Analysis.—*syn* Conformations are preferred for vinyl mercaptan,[3,4] for methyl vinyl sulphide,[4] and for methyl allenyl sulphide.[4] Similar M.O. studies of 1,1-bis(methylthio)ethylene,[5] simple dithioacetals,[6] and 1,2-dithiols and 1,2-bis(methylthio)ethane[6] relate to conformational aspects. Continuing themes are represented in alternative explanations for the role of the sulphur atom in determining which proton is most readily released from unsymmetrical sulphides $R^1CH_2SCH_2R^2$ (both hyperconjugation and C—S polarization are involved[7]). The stabilization of the α-carbanion is accounted for by the ability of the sulphur atom to allow excess negative charge to encroach into its sp-valence shell.[8] Activation of dioxygen by sulphonium salts arises through orbital interactions.[9]

Molecular Orbital (M.O.) studies of reaction intermediates have been reported for the rearrangement of 1-propenethiol,[10] for the addition of sulphenyl halides to ethylene (sulphurane rather than thiiranium salt as intermediate),[11] and for similar additions to alkynes (thiirenium ions or β-alkylthiovinyl cations, depending upon the substituents).[12] Electronic effects of substituents on energy levels of orbitals of

(*reference 2 continued*)

Organomet. Chem., Libr., 1979, **7**, 257; (*l*) A. Nudelman, *Phosphorus Sulfur*, 1976, **2**, 51; (*m*) K. D. Asmus, *Acc. Chem. Res.*, 1979, **12**, 436; (*n*) V. A. Smit, N. S. Zefirov, I. V. Bodrikov, and M. Z. Krimer, *ibid.*, 1979, **12**, 282; (*o*) G. Yamamoto, *Kagaku No Ryoiki*, 1978, **32**, 207; (*p*) S. Warren, *Acc. Chem. Res.*, 1978, **11**, 401; (*q*) L. Morin, J. Lebaud, D. Paquer, R. Chaussin, and D. Barillier, *Phosphorus Sulfur*, 1979, **7**, 69; (*r*) T. Numata, *Yuki Gosei Kagaku Kyokaishi*, 1978, **36**, 845; (*s*) L. A. Paquette, *Org. React.*, 1977, **25**, 1; (*t*) C. J. M. Stirling, *Acc. Chem. Res.*, 1979, **12**, 198; (*u*) H. J. Reich, *ibid.*, 1979, **12**, 22; J. V. Comasseto, J. T. B. Ferreira, and M. Marcuzzo do Canto, *Quim. Nova*, 1979, **2**, 58; K. Kondo and N. Sonoda, *Kagaku* (*Kyoto*), 1979, **34**, 652 and *Yuki Gosei Kagaku Kyokaishi*, 1979, **37**, 1047; (*v*) C. P. Downes, C. A. MacAuliffe, and M. R. C. Winter, *Inorg. Perspect. Biol. Med.*, 1979, **2**, 241; (*w*) K. J. Irgolic, *J. Organomet. Chem.*, 1978, **158**, 235.
[3] D. G. Lister and P. Palmieri, *J. Mol. Struct.*, 1978, **48**, 133.
[4] J. Kao, *J. Am. Chem. Soc.*, 1978, **100**, 4685.
[5] G. De Alti, P. Decleva, and A. Sgamellotti, *J. Mol. Struct.*, 1979, **53**, 129.
[6] M. Ohsaku, N. Bingo, W. Sugikawa, and H. Murata, *Bull. Chem. Soc. Jpn.*, 1979, **52**, 355.
[7] W. T. Borden, E. R. Davidson, N. H. Andersen, A. D. Denniston, and N. D. Epiotis, *J. Am. Chem. Soc.*, 1978, **100**, 1604.
[8] J. Fabian, P. Schoenfeld, and R. Mayer, *Phosphorus Sulfur*, 1976, **2**, 151.
[9] K. Ohkubo, Y. Azuma, and H. Sato, *Oxid. Commun.*, 1979, **1**, 49.
[10] V. I. Minkin and R. M. Minyaev, *Zh. Org. Khim.*, 1979, **15**, 1569.
[11] V. M. Csizmadia, *Prog. Theor. Org. Chem.*, 1977, **2**, 280.
[12] I. G. Csizmadia, V. Lucchini, and G. Modena, *Gazz. Chim. Ital.*, 1978, **108**, 543.

sulphur provide the motivation for M.O. studies of substituted phenyl methyl sulphides,[13–15] aryl trifluoromethyl sulphones,[16] and substituted styryl sulphones.[17] Force-field analysis assigns the *gauche* conformation to H_2S_4 and Me_2S_4.[18] Continuing studies (see Vol. 5, pp. 2, 7) indicate a preference for the equatorial orientation of the MeS substituent in 5-methylthio-1,3-dithians, although comparison of experiment with theory indicates the limited success of the calculations, since the equatorial preference is substantially larger than predicted.[19] The dangers of generalizations are implied in the results of conformational analysis of α-substituted cyclohexanones, since the order $PhSO_2$, PhS, PhSO emerges for the preferred adoption of the axial conformation,[20] in contrast to the order MeS, MeSO, $MeSO_2$ established earlier[21] for the corresponding cyclohexanones.

Ultraviolet Spectra, Circular Dichroism, and Optical Rotatory Dispersion.— Routine u.v. studies of the effects of sulphur-containing groups on the spectral absorption characteristics of substituted benzenes (references are collected at the end of this section) continue along familiar lines (see Vol. 5, p. 4). A model for the thioindigo chromophore, *i.e.* $MeCOC(SMe)=C(SMe)COMe$, has been shown by *X*-ray studies[22] to be non-planar, as a result of repulsion between methyl groups.

Studies of the circular dichroism of acyl derivatives of thioglycerols[23] and of 2- or 3-phenylalkyl isothiocyanates[24] constitute extensions of earlier work.

Additional Bibliography: B. Yu. Sultanov, B. A. Tagiev, and N. G. Radzhabov, *Azerb. Khim. Zh.*, 1979, 96 [thiophenols]; G. C. Chmutova, A. A. Karelov, and N. N. Vtyurina, *Zh. Obshch. Khim.*, 1979, **49**, 2275 [solvent effects; PhXMe (X = O, S, Se, or Te)]; V. Baliah and V. M. Kanagasabapathy, *Indian J. Chem., Sect. A*, 1978, **16**, 388 [*p*-dimethylaminophenyl sulphides]; G. V. Ratovskii, T. I. Rozova, M. A. Vasileva, and T. I. Bychkova, *Teor. Eksp. Khim.*, 1978, **14**, 245 [vinyl sulphones]; Yu. I. Naumov and V. A. Izmailskii, *Zh. Fiz. Khim.*, 1979, **53**, 2030 [*p*-hydroxyphenyl sulphones]; A. F. Moskvin, L. I. Doktorova, E. I. Kazankina, I. A. Reznikova, and V. A. Belyaev, *Zh. Obshch. Khim.*, 1979, **49**, 619 [anilinesulphonic acids].

Infrared, Raman, and Microwave Spectra.—Brief details of studies which provide information on conformations and electron distributions are summarized at the end of this section. A comparison of the polarized Raman spectra for $(C^2H_3)_2SO_2$ with far-i.r. data for Me_2SO_2[25] and polarized i.r. spectra for aryl thiocyanates and selenocyanates[26] represent less routine studies.

[13] G. A. Chmutova, N. N. Vtyurina, T. V. Komina, I. G. Gazizov, and H. Bock, *Zh. Obshch. Khim.*, 1979, **49**, 192.

[14] G. A. Chmutova, N. N. Vtyurina, and H. Bock, *Dokl. Akad. Nauk SSSR*, 1979, **244**, 1138.

[15] T. Matsushita, Y. Osamura, N. Misawa, K. Nishimoto, and Y. Tsuno, *Bull. Chem. Soc. Jpn.*, 1979, **52**, 2521.

[16] Yu. P. Egorov, Yu. Ya. Borovikov, S. I. Vdovenko, V. E. Didkovskii, V. N. Boiko, and G. M. Shchupak, *Teor. Eksp. Khim.*, 1978, **14**, 84 (*Chem. Abstr.*, 1979, **89**, 59 523).

[17] J. Sauer, I. Grohmann, R. Stoesser, and W. Wegener, *J. Prakt. Chem.*, 1979, **321**, 177.

[18] M. Askari, *Tetrahedron Lett.*, 1979, 3173.

[19] E. L. Eliel and E. Juaristi, *J. Am. Chem. Soc.*, 1978, **100**, 6114.

[20] H. Ozbal and W. W. Zajac, *Tetrahedron Lett.*, 1979, 4821.

[21] J. A. Hirsch, *Top. Stereochem.*, 1967, **1**, 199.

[22] H. L. Ammon and H. Hermann, *J. Org. Chem.*, 1978, **43**, 4581.

[23] S. Gronowitz, B. Herslof, P. Michelsen, and B. Akesson, *Chem. Phys. Lipids*, 1978, **22**, 307.

[24] V. M. Potapov, V. M. Demyanovich, L. D. Soloveva, and O. E. Vendrova, *Dokl. Akad. Nauk SSSR*, 1978, **241**, 592.

[25] K. Machida, Y. Kuroda, and K. Hanai, *Spectrochim. Acta, Part A*, 1979, **35**, 835.

[26] E. G. Yarkova, N. R. Safiullina, G. A. Chmutova, and I. V. Bautina, *Zh. Obshch. Khim.*, 1979, **49**, 2025.

Additional Bibliography: *Infrared spectra* P. Ruostesuo, *Finn. Chem. Lett.*, 1979, 202 [hydrogen bonding of phenols to Me$_2$SO or MeS(O)NMe$_2$]; I. Hargittai, *Z. Naturforsch., Teil A*, 1979, **34**, 755 [diaryl sulphones]; E. G. Yarkova, G. A. Chmutova, and N. N. Vtyurina, *Zh. Fiz. Khim.*, 1978, **52**, 498 [ArXR (X = O, S, Se, or Te; R = alkyl)]; Yu. P. Egorov, S. I. Vdovenko, Yu. Ya. Borovikov, V. A. Topchii, and V. I. Popov, *Teor. Eksp. Khim.*, 1978, **14**, 236 [arenesulphonyl chlorides]; K. Hanai, A. Noguchi, and T. Okuda, *Spectrochim. Acta, Part A*, 1978, **34**, 771 [MeSO$_2$NHPh]; E. G. Yarkova, N. R. Safiullina, G. A. Chmutova, and I. V. Bautina, *Zh. Strukt. Khim.*, 1979, **20**, 959 [aryl methyl sulphones]; D. Herrmann, *J. Prakt. Chem.*, 1978, **320**, 231 [areneselenyl bromides and aryl alkyl selenides]; *Infrared/Raman spectra* H. S. Randhawa and W. Walter, *Bull. Chem. Soc. Jpn.*, 1978, **51**, 1579 [BrCH$_2$COSH]; K. Ohno, A. Mitsui, and H. Murata, *Bull. Chem. Soc. Jpn.*, 1979, **52**, 2178; M. Ohsaku, H. Murata, and Y. Shiro, *J. Mol. Struct.*, 1977, **42**, 31; M. Sakakibara, I. Harada, H. Matsuura, and T. Shimanouchi, *ibid.*, 1978, **49**, 29; and H. Matsuura, N. Miyauchi, H. Murata, and M. Sakakibara, *Bull. Chem. Soc. Jpn.*, 1979, **52**, 344 [four studies of simple alkyl sulphides and selenides]; A. Bigotto, V. Galasso, G. C. Pappalardo, and G. Scarlata, *Spectrochim. Acta, Part A*, 1978, **34**, 435 [sulphides and sulphones]; B. Nagel and A. B. Remizov, *Zh. Obshch. Khim.*, 1978, **48**, 1189 [methyl vinyl sulphone and divinyl sulphoxide]; A. B. Remizov and G. G. Butenko, *Zh. Strukt. Khim.*, 1979, **20**, 63 [NN-dimethylsulphonamides]; N. S. Dance and C. H. W. Jones, *J. Organomet. Chem.*, 1978, **152**, 175 [dialkyl telluride–mercury(II) halide adducts]; P. Klaboe, C. J. Nielsen, R. Suchi, and O. Vikane, *Acta Chem. Scand., Ser. A*, 1978, **32**, 565 [phenyl dihalogenotellurides]; *Raman spectra* K. Hamada and H. Morishita, *J. Mol. Struct.*, 1978, **44**, 119 [MeXMe (X = O, S, Se, or Te)]; R. R. M. Brand, M. L. Halbensleben, H. K. Schenkel, and E. D. Schmid, *Z. Naturforsch., Teil B*, 1978, **33**, 197 [vinyl sulphides and RXC=CPh (R = Me or Ar; X = S, Se, or Te)]; *Microwave spectra* M. Tanimoto, V. Almond, S. W. Charles, J. N. Macdonald, and N. L. Owen, *J. Mol. Spectrosc.*, 1979, **78**, 95 and M. Tanimoto and J. N. Macdonald, *ibid.*, p. 106 [H$_2$C=CHSH]; R. Kewley, *Can. J. Chem.*, 1979, **56**, 772 [MeSCH$_2$CN]; C. J. Sylvia, N. S. True, and R. K. Bohn, *J. Mol. Struct.*, 1979, **51**, 163 [FCOSCHMe$_2$]; L. M. Boggia, R. R. Filgucira, J. Maranon, and O. M. Sorarrain, *Spectrosc. Lett.*, 1978, **11**, 143 [PhSO$_2$Cl].

Nuclear Magnetic Resonance Spectra.—Long-range coupling involving the thiol proton and ring protons in 2-methoxybenzenethiol,[27] halogenobenzenethiols,[28] and 2-hydroxybenzenethiol[29] has been suggested to indicate an out-of-plane conformational preference for the SH group in thiophenols (see also Vol. 5, p. 6), and hence reveals an important contrast with the analogous phenols. Other non-routine ^{1}H n.m.r. studies include those of lanthanide-induced shifts of sulphonium salts[30] and of alkylsulphinylmethyl alkyl sulphides[31] and assignment of the zwitterionic structure (1) to the Schiff bases derived from *o*-formylbenzenethiol.[32] Brief details of other ^{1}H n.m.r. work (but excluding data compilations) are collected together at the end of this section, where details of ^{13}C n.m.r. studies are also to be found. More complete coverage of the ^{13}C n.m.r. literature has been attempted, since missing details are still being located, with the objective of providing a full assessment of the influence of adjacent sulphur-containing functional groups on chemical shifts.

Nitrogen-15 n.m.r. spectra of alkanesulphonamides,[33] benzenesulphonamides,[34] and isothiocyanates[36] have been reported. Nitrogen-14 n.q.r. data on methanesulphonamide[36] and NN-disubstituted analogues[37] have established a valuable application of the technique, *i.e.* to assess the degree of delocalization of the lone

[27] T. Schaefer and T. A. Wildman, *Can. J. Chem.*, 1979, **57**, 450.
[28] T. Schaefer and W. J. E. Parr, *Can. J. Chem.*, 1979, **57**, 1421.
[29] T. Schaefer, T. A. Wildman, and S. R. Salman, *J. Am. Chem. Soc.*, 1980, **102**, 107.
[30] R. L. Caret and A. N. Vennos, *J. Org. Chem.*, 1980, **45**, 361.
[31] P. R. Jones, D. F. Sesin, and J. J. Uebel, *Phosphorus Sulfur*, 1978, **4**, 277.
[32] M. F. Corrigan, I. D. Rae, and B. O. West, *Aust. J. Chem.*, 1978, **31**, 587.
[33] H. R. Kricheldorf, *Angew. Chem.*, 1978, **90**, 489.
[34] I. I. Schuster, S. H. Doss, and J. D. Roberts, *J. Org. Chem.*, 1978, **43**, 4693.
[35] I. Yavari, J. S. Staral, and J. D. Roberts, *Org. Magn. Reson.*, 1979, **12**, 340.
[36] H. Negita, T. Kubo, M. Maekawa, A. Ueda, and T. Okuda, *Bull. Chem. Soc. Jpn.*, 1979, **52**, 1881.
[37] D. Ya. Osokin and G. G. Butenko, *Zh. Fiz. Khim.*, 1979, **53**, 130.

(1)

pair that is formally located on nitrogen in such compounds.[37] Chlorine-35 n.q.r. studies on α-chloroalkyl sulphides have been reported.[38]

Selenium-77 n.m.r. data for selenols, selenides, selenonium salts, diselenides, and selenoxy-acids[39] have been obtained under conditions (*i.e.* on the same spectrometer) which allow reliable comparisons of spectral parameters to be made. Benzeneseleninic acids[40] and selenolesters[41] have also been studied by this technique, and [77]Se spin–lattice relaxation times of organoselenium compounds have been collected in a pioneering study.[41a]

Tellurium-125 n.m.r. spectra of telluroesters have been determined.[41]

Additional Bibliography: *Proton n.m.r.* G. Llabres, M. Baiwir, L. Christiaens, and J. L. Piette, *Can. J. Chem.*, 1979, **57**, 2967 and G. Llabres, M. Baiwir, L. Christiaens, J. Denoel, L. Laitem, and J. L. Piette, *ibid.*, 1978, **56**, 2003 [PhXMe (X = O, S, Se, or Te) shown to adopt a screw conformation]; W. Walek, A. Priess, and S. Dietzel, *Z. Chem.*, 1978, **18**, 144 [Z/E isomerism of RSC(SMe)=NCN]; C. Fournier, B. Lemarie, B. Braillon, D. Paquer, and M. Vazeux, *Org. Magn. Reson.*, 1977, **10**, 20 [cyclobutane-1,1-dithiol]; F. Alcudia, J. L. Garcia Ruano, J. Rodriguez, and F. Sanchez, *Can. J. Chem.*, 1979, **57**, 2420 [β-hydroxyalkyl sulphides, sulphoxides, sulphones, and sulphonium salts, interpreted in terms of conformations]; Y. Takeda, T. Matsuda, and T. Tanaka, *Org. Magn. Reson.*, 1977, **10**, 98 [thiuram mono- and di-sulphides]; A. D. Pershin, I. Kende, V. Cholvad, and K. Mach, *Collect. Czech. Chem. Commun.*, 1978, **43**, 1349 [c.i.d.n.p. in radical recombination in reaction mixtures of azoisobutyronitrile and tetraethylthiuram disulphide]; I. D. Sadekov, A. Ya. Bushkov, and V. P. Metlushenko, *Zh. Obshch. Khim.*, 1978, **48**, 1658 [aryltellurium(IV) halides and isothiocyanates]; I. D. Sadekov, M. L. Cherkinskaya, V. L. Pavlova, V. A. Bren, and V. I. Minkin, *Zh. Obshch. Khim.*, 1978, **48**, 390 [diaryltellurium(IV) halides]; **Carbon-13 n.m.r.** G. Dauphin and A. Cuer, *Org. Magn. Reson.*, 1979, **12**, 557 [MeSH, Me₂S, Me₂S₂, Me₂S₃, and EtSSMe]; J. Sandstrom and I. Wennerbeck, *Acta Chem. Scand. Ser. B*, 1978, **32**, 421 [β-dimethylaminovinyl methyl sulphides]; G. A. Kalabin, D. F. Kushnarev, T. G. Mannafov, and A. A. Retinski, *Izv. Akad. Nauk SSSR, Ser. Khim.*, 1978, 2410; G. A. Kalabin, D. F. Kushnarev, V. M. Bzhezovskii, and G. A. Tschmutova, *Org. Magn. Reson.*, 1979, **12**, 598; G. A. Kalabin and D. F. Kushnarev, *Zh. Strukt. Khim.*, 1979, **20**, 617; and G. A. Kalabin, D. F. Kushnarev, G. A. Chmutova, and L. V. Kashurnikova, *Zh. Org. Khim.*, 1979, **15**, 24 [four studies of cyclopropyl and aryl selenides and aryl selenocyanates]; G. A. Kalabin, D. F. Kushnarev, L. M. Kataeva, L. V. Kashurnikova, and R. I. Vinokurova, *Zh. Org. Khim.*, 1978, **14**, 2478 [diaryl diselenides]; G. A. Kalabin, V. M. Bzhezovskii, B. A. Trofimov, D. F. Kushnarev, A. N. Volkov, A. G. Proidakov, R. N. Kudyakova, and A. N. Khudyakova, *Izv. Akad. Nauk SSSR, Ser. Khim.*, 1978, 1833 [conformations of (E)- and (Z)-RSCH=CHC≡CH]; K. H. Park, G. A. Gray, and G. D. Daves, *J. Am. Chem. Soc.*, 1978, **100**, 7475 [indolyldimethylsulphonium salts]; R. G. Petrova, I. I. Kandrov, V. I. Dostovalova, T. D. Churkina, and R. K. Friedlina, *Org. Magn. Reson.*, 1978, **11**, 406 [dithioacetals and 2-(alkylthio)alkyl sulphides]; A. H. Fawcett, K. J. Ivin, and C. D. Stewart, *Org. Magn. Reson.*, 1978, **11**, 360 [aliphatic sulphones and disulphones]; A. W. Douglas, *Can. J. Chem.*, 1978, **56**, 2129 [(Z)-5-fluoro-2-methyl-1-{[p-(methylsulphinyl)phenyl]methylene}-1H-inden-3-ylacetic acid ('Sulindac') and its methylthio-analogues]; S. S. McCrachren and S. A. Evans, *J. Org. Chem.*, 1979, **44**, 3551 and S. W. Bass and S. A. Evans, *ibid.*, 1980, **45**, 710 [sulphoxides, sulphones, thiolsulphinates, thiolsulphonates, and disulphides]; Y. Kosugi and T. Takeuchi, *Org. Magn. Reson.*, 1979, **12**, 435 [sodium sulphinates and sulphonates]; L. H. J. Lajunen and K. Raisanen, *Finn. Chem. Lett.*, 1978, 207 [hydroxynaphthalenesulphonic acids]; R. Radeglia and E. Fanghaenel, *J. Prakt. Chem.*, 1978, **320**,

[38] V. P. Feshin, M. G. Voronkov, P. A. Nikitin, G. M. Gavrilova, and V. B. Kobychev, *Zh. Obshch. Khim.*, 1979, **49**, 2496.
[39] J. D. Odom, W. H. Dawson, and P. D. Ellis, *J. Am. Chem. Soc.*, 1979, **101**, 5815.
[40] A. Fredga, S. Gronowitz, and A. B. Hornfeldt, *Chem. Scr.*, 1977, **11**, 37.
[41] (a) O. A. Gansow, W. D. Vernon, and J. J. Dechter, *J. Magn. Reson.*, 1978, **32**, 19; (b) B. Kohne, W. Lohner, K. Praefcke, H. J. Jakobsen, and B. Villadsen, *J. Organomet. Chem.*, 1978, **166**, 373.

339 [*N*-aryl arenesulphonamides]; T. Takata, Y. H. Kim, S. Oae, and K. T. Suzuki, *Tetrahedron Lett.*, 1978, 4303 [thiolsulphinates, thiolsulphonates, and disulphides]; V. V. Bairov, G. A. Kalabin, M. L. Alpert, V. M. Bzhezovskii, I. D. Sadekov, B. A. Trofimov, and V. I. Minkin, *Zh. Org. Khim.*, 1978, **14**, 671 [alkyl aryl tellurides].

Mass Spectra.—A continuing investigation into the finer details of the structures of secondary fragmentation products from thiols and sulphides (see Vol. 5, p. 7) involves several research groups. The ion $CH_2\overset{+}{=}SH$ that is formed from primary thiols by α-cleavage has been shown by collisional activation mass spectrometry to be more stable than CH_3S^+;[42] the same technique has been used[43] to show the occurrence of substantial C—S bond cleavage in sulphenyl cations (RS$^+$) after their formation by field desorption. Broer and Weringa have carried further their painstaking study of the alternative fragmentation course available to sulphide molecular ions, and have used ^{2}H- and ^{13}C-labelled sulphides to establish the formation of $MeS\overset{+}{=}CH_2$ and $MeCH\overset{+}{=}SH$ by loss of the methyl radical from the molecular ion of ethyl methyl sulphide.[44] They have also ascertained which of the isomeric $C_3H_7S^+$ ions fragments further into H_2S and ethylene.[45] The formation of methane[46] from EtS$^+$ and of hydrogen[47] from MeS$^+$ liberates the species [CHS]$^+$.

Ion cyclotron resonance studies of 3-methoxyalkanethiols[48] and homologous sulphides $MeO(CH_2)_nSMe$ ($n = 1$—3)[49] have provided information on the reactivity profile of the methoxymethyl cation towards alcohols, thiols, and amines,[48] and on the preferential abstraction of hydride ion from the carbon atoms that are alpha to sulphur (confirming that, in gas-phase reactions, S is more able than O to stabilize a neighbouring carbonium ion).[49] Gas-phase acidities of dimethyl sulphoxide and representative aliphatic sulphones reveal a lower degree of stabilization of the α-sulphonyl carbonium ions.[50]

Additional Bibliography: H. J. Moeckel, *Fresenius' Z. Anal. Chem.*, 1979, **295**, 241 [chemical ionization mass spectra of aliphatic thiols, sulphides, and disulphides]; R. Luedersdorf, K. Praefcke, and H. Schwartz, *Org. Mass Spectrom.*, 1978, **13**, 179 [aryl cyclopentanethiocarboxylates]; N. G. Foster, P. Chandraaurin, and R. W. Higgins, *Biomed. Mass Spectrom.*, 1979, **6**, 260 [alkyl 2-thienyl sulphides]; C. A. McAuliffe, F. P. McCullough, R. D. Sedgwick, and W. Levason, *Inorg. Chim. Acta*, 1978, **27**, 185 [dithioacetals, 1,*n*-bis(alkylthio)alkanes, 1,2-bis(alkylthio)ethylenes, *o*-(alkylthio)phenyl sulphides, and *o*-nitrophenyl sulphides]; B. A. Trofimov, V. Yu. Vitkovskii, G. K. Musorin, S. V. Amosova, and V. V. Keiko, *Zh. Obshch. Khim.*, 1979, **49**, 393 [butadienyl sulphides, selenides, and tellurides]; I. W. Jones and J. C. Tebby, *Phosphorus Sulfur*, 1978, **5**, 57 [aryl sulphides, disulphides, and sulphones]; M. Bihari, J. Tamas, I. Kapovits, and J. Rabai, *Adv. Mass Spectrom., Sect. B*, 1978, **7**, 1362 [*o*-(alkoxycarbonyl)aryl sulphides, sulphoxides, and sulphones]; L. I. Virin, A. R. Elman, Yu. A. Safin, Yu. V. Penin, L. I. Nekrasova, V. I. Zetkin, and R. V. Dzhagatspanyan, *Zh. Org. Khim.*, 1979, **15**, 991 [diaryl sulphides]; B. V. Rozynov, O. S. Reshetova, G. V. Golovkin, A. Yu. Krylova, and T. G. Perlova, *Khim. Sredstva Zashch. Rast.*, 1976, **7**, 41 (*Chem. Abstr.*, 1979, **91**, 90 744) [*p*-alkoxyphenyl sulphides, sulphoxides, sulphones, and disulphides]; D. N. B. Mallen and J. M. Smith, *Org. Mass Spectrom.*, 1979, **14**, 171 [di-(*o*-nitrophenyl) sulphides]; A. Robertiello, P. Bacchin, and V. Mancini, *Ann. Chim.* (*Rome*), 1977, **67**, 223 [2-azetidinonyl sulphides]; B. V. Rozynov, R. I.

[42] J. D. Dill and F. W. McLafferty, *J. Am. Chem. Soc.*, 1979, **101**, 6526.
[43] R. Weber, F. Borchers, K. Levsen, and P. W. Roellgen, *Z. Naturforsch., Teil A*, 1978, **33**, 540.
[44] W. J. Broer, W. D. Weringa, and W. C. Nieuwpoort, *Org. Mass Spectrom.*, 1979, **14**, 543.
[45] W. J. Broer and W. D. Weringa, *Org. Mass Spectrom.*, 1979, **14**, 36.
[46] W. J. Broer and W. D. Weringa, *Org. Mass Spectrom.*, 1978, **13**, 232.
[47] A. G. Harrison, *J. Am. Chem. Soc.*, 1978, **100**, 4911.
[48] J. K. Pau, J. K. Kim, and M. C. Caserio, *J. Am. Chem. Soc.*, 1978, **100**, 3838.
[49] J. K. Pau, M. B. Ruggera, J. K. Kim, and M. C. Caserio, *J. Am. Chem. Soc.*, 1978, **100**, 4242.
[50] J. B. Cumming and P. Kebarle, *J. Am. Chem. Soc.*, 1978, **100**, 1835.

Zhdanov, O. S. Reshetova, N. G. Kapitanova, Z. L. Gordon, and E. G. Rozantsev, *Izv. Akad. Nauk SSSR, Ser. Khim.*, 1978, 1838 [4,4-bis(ethanesulphonyl)piperidines and ethyl piperidein-3-yl sulphone]; L. K. Liu and C. T. Hong, *Org. Mass Spectrom.*, 1978, **13**, 675 [McLafferty rearrangement identified for (*E*)-PhCH=CHSO$_2$R]; A. Albasini, M. Melegari, G. Vampa, M. Rinaldi, and A. Rastelli, *Boll. Chim. Farm.*, 1978, **117**, 664 [arenesulphonamides]; C. Kascheres and R. van Fossen Bravo, *Org. Mass Spectrom.*, 1979, **14**, 293 [benzenesulphonylhydrazides]; A. D. Kossoy and J. L. Occolowitz, *Biomed. Mass Spectrom.*, 1978, **5**, 123 [*N*-methyl-*N*-toluene-*p*-sulphonyl-ureas]; V. K. Manzhos, G. E. Zhusupova, R. A. Muzychkina, and T. K. Chumbalov, *Zh. Prikl. Spektrosk.*, 1979, **30**, 739 [anthraquinonesulphonic acids]; N. G. Keats and L. A. Summers, *J. Heterocycl. Chem.*, 1979, **16**, 1369 [di-(2-pyridyl) selenide]; F. F. Knapp, *Org. Mass Spectrom.*, 1979, **14**, 341 [PhTeCH$_2$CH(OR)$_2$ → molecular ion → PhTeOR].

Photoelectron Spectra.—Organosulphur compounds are appropriately represented in the broad evaluation of this technique. The dihedral angle for the C–S–S–C grouping in di-t-adamantyl disulphide is 103°,[51] which is on the way to a fully *trans* configuration, favoured[52] for disulphides only when bulky substituents are involved.

Photoelectron spectral data have been collected for MeSCl,[53] MeSBr,[53] MeSCN,[54] and MeSeCN.[54] The trigonal-bipyramidal structure has been established for methylenesulphur tetrafluoride through p.e. spectroscopy, electron diffraction, and *X*-ray analysis.[55]

Electron Diffraction.—Among completed electron-diffraction studies are ethyl methyl sulphide,[56] chloromethyl methyl sulphide,[57] methyl phenyl sulphide,[58] di-(2-pyridyl) sulphide,[59] sulphones,[60] sulphoxides and sulphones,[61] and trifluoromethanesulphonyl chloride.[62] A number of precise analyses of gas-phase conformational equilibria have emerged. Methyl ethyl sulphide shows a preference for the *gauche* conformation rather than for the *trans* form.[56]

Dipole Moments and Studies of the Kerr Effect.—A number of references, together with brief details, have been collected from the recent literature:

S. G. Gagarin and R. Z. Zakharyan, *Izv. Akad. Nauk SSSR, Ser. Khim.*, 1979, 31 [MeSH, EtSH, Me$_2$S, and Et$_2$S]; D. M. Petkovic and J. S. Markovic, *Glas. Hem. Drus., Beograd*, 1979, **44**, 535 (*Chem. Abstr.*, 1980, **92**, 146 131) [thiols]; V. Baliah and V. M. Kanagasabapathy, *Indian J. Chem., Sect. B*, 1978, **16**, 810 [diaryl sulphides]; B. A. Arbuzov, A. M. Salikhova, S. M. Shostakovskii, A. N. Vereshchagin, and N. S. Nikolskii, *Izv. Akad. Nauk SSSR, Ser. Khim.*, 1979, 2786 [*gem*-dichlorocyclopropyl sulphides]; O. Exner and J. B. F. N. Engberts, *Collect. Czech. Chem. Commun.*, 1979, **44**, 3378 [halogenomethyl sulphones and *N*-nitro-*N*-methyl-arenesulphonamides]; M. S. R. Naidu, S. G. Peeran, and D. B. Reddy, *Indian J. Chem., Sect. B*, 1978, **16**, 1090 [unsymmetrically substituted 1,2-di-(arylsulphonyl)-1,2-diphenylethylenes]; P. Ruostesuo, *Finn. Chem. Lett.*, 1978, 159

[51] F. S. Joergensen and J. P. Snyder, *J. Org. Chem.*, 1980, **45**, 1015.

[52] F. S. Joergensen and J. P. Snyder, *Tetrahedron*, 1979, **35**, 1399.

[53] E. Nagy-Felsobuki and J. B. Peel, *Phosphorus Sulfur*, 1979, **7**, 157.

[54] M. V. Andreocci, M. Bossa, C. Furlani, M. N. Piancastelli, C. Cauletti, and T. Tarantelli, *J. Chem. Soc., Faraday Trans. 2*, 1979, **75**, 105.

[55] H. Bock, J. E. Boggs, G. Kleemann, D. Lentz, H. Oberhammer, E. M. Peters, K. Seppelt, A. Simon, and B. Solouki, *Angew. Chem.*, 1979, **91**, 1008.

[56] K. Ovanagi and K. Kuchitsu, *Bull. Chem. Soc. Jpn.*, 1978, **51**, 2243.

[57] V. A. Naumov, R. N. Garaeva, G. D. Geize, and F. S. Mailgoff, *Dokl. Akad. Nauk SSSR*, 1979, **245**, 650.

[58] N. M. Zaripov, A. V. Golubinskii, G. A. Chmutova, and L. V. Vilkov, *Zh. Strukt. Khim.*, 1978, **19**, 894.

[59] B. Rozsondai, I. Hargittai, and G. C. Pappalardo, *Z. Naturforsch., Teil A*, 1979, **34**, 752.

[60] I. Hargittai, *Kem. Kozl.*, 1978, **50**, 339, 457.

[61] B. Rozsondai, J. H. Moore, D. C. Gregory, and I. Hargittai, *J. Mol. Struct.*, 1979, **51**, 69.

[62] J. Brunvoll, I. Hargittai, and M. Kolonits, *Z. Naturforsch., Teil A*, 1978, **33**, 1236.

and 166 [benzenesulphenamides]; G. G. Butenko, A. N. Vereshchagin, A. B. Remizov, and R. N. Nurullina, *Izv. Akad. Nauk SSSR, Ser. Khim.*, 1979, 1763 and G. G. Butenko, A. N. Vereshchagin, and D. M. Nasyrov, *ibid.*, p. 2034 [methyl arenesulphonates and arenesulphonamides]; R. V. Sendega and T. A. Protsailo, *Ukr. Khim. Zh.*, 1978, **44**, 844 (*Chem. Abstr.*, 1979, **89**, 196 491) [alkyl and alkenyl benzenesulphonates]; A. M. Kamalyutdinova-Salikhova, T. G. Mannafov, R. B. Khismatova, S. G. Vulfson, and A. N. Vereshchagin, *Izv. Akad. Nauk SSSR, Ser. Khim.*, 1979, 1757 [*gem*-dichlorocyclopropyl aryl selenides].

X-Ray Crystal Analysis.—The host–guest partnership arrangements for hexakis-(alkylthiomethyl)benzenes[63] including chiral guests[64] and for the corresponding sulphones[65] with 1,4-dioxan[63] and squalene[64] as guests have been elucidated by *X*-ray analysis. Diethyl and diphenyl dithioacetals of D-ribose[66] and *N*-sulphinyl benzenesulphonamide[67] have also been studied, the latter compound being revealed as adopting the *cis* configuration and possessing an electron-rich S—N bond. A simple empirical relationship between the S—S bond lengths and the X—C—S—S torsion angles has emerged from a study of the crystal structures of 21 symmetrical disulphides.[68]

The mono- and di-hydrates of Ph_3SeCl are ionic compounds of five-co-ordinate selenium.[69]

Electron Spin Resonance Spectra.—Radicals arising from photolysis of diaryl disulphides (ArS·),[70] and of reaction mixtures of Bu^tSH and SCl_2 ($Bu^tSS·$)[71] have been studied by e.s.r. (a routine monitoring technique used in several of the studies of sulphur radicals mentioned later in this Chapter). Arenesulphonyl cation radicals, formed from 2-alkoxy-5-methylbenzenesulphonyl fluorides and sulphonic acids on treatment with PbO_2 and FSO_3H,[72] provide a further example of sulphur-centred radicals, while a variation on this aspect is provided by the report of the influence of the alkylthio-group in *p*-(alkylthio)nitrobenzenes on the hyperfine splitting in the spectra of derived radical anions and nitroxides.[73] The alkylthio-group has thus been shown to behave as a π-electron acceptor when a suitable angle exists between the vacant σ^* orbital of the S—R bond and the π-orbital of benzene.[73]

3 Thiols, Selenols, and Tellurols

Preparation.—An important addition to methods for the reduction of sulphonyl compounds to sulphenyl analogues, *i.e.* the direct reduction of sulphonic acids to thiols using trifluoroacetic anhydride and $Bu^n_4N^+I^-$,[74] has been reported. A mixture of thiol and the corresponding trifluorothiolacetate is obtained; this, on hydrolysis,

[63] A. D. U. Hardy, D. D. MacNicol, S. Swanson, and D. R. Wilson, *Tetrahedron Lett.*, 1978, 3579; D. D. MacNicol and S. Swanson, *J. Chem. Res. (S)*, 1979, 406.
[64] A. Freer, C. J. Gilmore, D. D. MacNicol, and D. R. Wilson, *Tetrahedron Lett.*, 1980, **21**, 1159.
[65] D. D. MacNicol and S. Swanson, *Tetrahedron Lett.*, 1980, **21**, 205.
[66] A. Ducruix, D. Horton, C. Pascard, J. D. Wander, and T. Prange, *J. Chem. Res. (S)*, 1978, 470.
[67] G. Deleris, C. Courseille, J. Kowalski, and J. Dunogues, *J. Chem. Res. (S)*, 1979, 122.
[68] L. S. Higashi, M. Lundeen, and K. Seff, *J. Am. Chem. Soc.*, 1978, **100**, 8101.
[69] R. V. Mitcham, B. Lee, K. B. Mertes, and R. F. Ziolo, *Inorg. Chem.*, 1979, **18**, 3498.
[70] W. Moerke, A. Jezierski, and H. Singer, *Z. Chem.*, 1979, **19**, 147.
[71] J. E. Bennett and G. Brunton, *J. Chem. Soc., Chem. Commun.*, 1979, 62.
[72] A. P. Rudenko, M. Ya. Zarubin, and A. M. Kutnevich, *Zh. Obshch. Khim.*, 1979, **49**, 954.
[73] A. Alberti, M. Guerra, G. Martelli, F. Bernardi, A. Mangini, and G. F. Pedulli, *J. Am. Chem. Soc.*, 1979, **101**, 4627.
[74] T. Numata, H. Awano, and S. Oae, *Tetrahedron Lett.*, 1980, **21**, 1235.

yields the thiol and easily separable elaboration products.[74] Most of the other recent papers describing preparations of thiols are collected together at the end of this section since, although often describing improved methodology, they are based on well-tried procedures.[75] Among methods using H_2S as reagent is an interesting study that has established a Meerwein–Ponndorf–Verley-type mechanism (Scheme 1) for the conversion of an alcohol into the corresponding thiol, using H_2S and Al_2O_3, with a ketone as co-catalyst.[76]

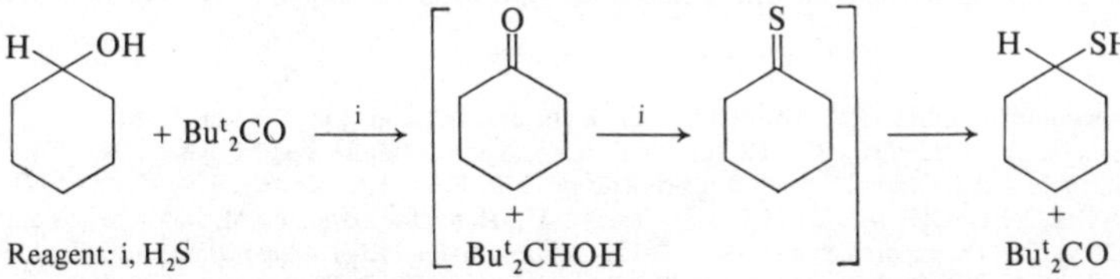

Reagent: i, H_2S

Scheme 1

Thiocarbonyl compounds offer reliable entries to thiols, and thiourea has been used for the conversion of chloromethylated polystyrene into the corresponding polymeric thiol.[77] *NN*-Dimethylthioformamide[78] and sodium *NN*-dimethyldithiocarbamate[79] have been used similarly, the latter leading to pyridine-2,4- and -2,5-dithiols from diazotized amino-chloropyridines. Another well-established route, *i.e.* conversion of phenols into benzenethiols *via* thionocarbamate intermediates and rearrangement at 200—280 °C, has been used[80] for the synthesis of 4-alkyl- and 4-cyano-derivatives. Reduction of di-t-butyl selenone with $LiAlH_4$ gives the corresponding selenol,[81] and the reduction by $NaBH_4$ of 2-aryl-3-piperidino-indane-1-thione gives the 2-arylindene-1-thiol; an unusual reductive cleavage of an enamine.[82] The thiol (2) is one of several trimers of dithioacetic acid.[83]

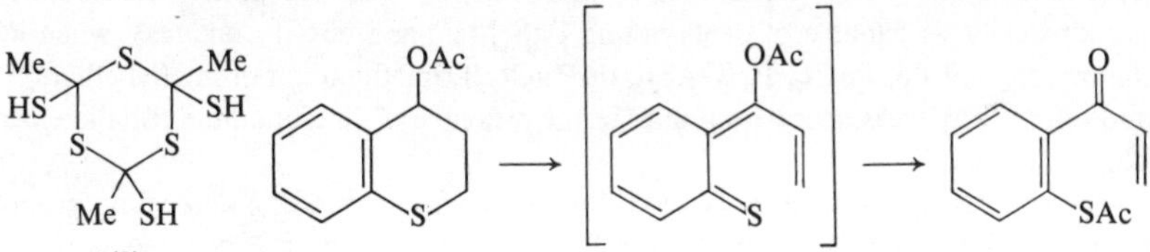

(2)   (3)

[75] G. C. Barrett, in 'Comprehensive Organic Chemistry', ed. Sir Derek H. R. Barton and W. D. Ollis, Pergamon Press, Oxford, 1979, Vol. 3 (ed. D. N. Jones), p.3.

[76] J. Barrault, M. Guisnet, J. Lucien, and R. Maurel, *J. Chem. Res. (S)*, 1978, 474; *Nouv. J. Chem.*, 1979, **3**, 15; *Ind. Eng. Chem., Prod. Res. Dev.*, 1978, **17**, 354.

[77] P. Kalck, R. Poilblanc, A. Gaset, A. Rovera, and R. P. Martin, *Tetrahedron Lett.*, 1980, **21**, 459.

[78] K. Hattori, T. Takido, and K. Habashi, *Nippon Kagaku Kaishi*, 1979, 101 (*Chem. Abstr.*, 1979, **90**, 137 380).

[79] K. Krowicki, *Pol. J. Chem.*, 1979, **53**, 701, 889.

[80] C. F. Shirley, *Mol. Cryst. Liq. Cryst.*, 1978, **44**, 193.

[81] B. J. McKinnon, P. De Mayo, N. C. Payne, and B. Ruge, *Nouv. J. Chim.*, 1978, **2**, 91.

[82] V. A. Usov, K. A. Petriashvili, and M. G. Voronkov, *Zh. Org. Khim.*, 1978, **14**, 2227.

[83] G. Levesque, A. Mahjoub, and A. Thuillier, *Tetrahedron Lett.*, 1978, 3847.

A conventional route from an alkanol to a thiol, *via* the trifluoromethane-sulphonate and treatment with $MeCOS^- Na^+$, has been used for the synthesis of the surfactant $C_{16}H_{33}\overset{+}{N}Me_2CH_2CH_2SH\ Cl^-$.[84]

Ring-opening reactions of sulphur-containing heterocyclic compounds which lead to thiols include Birch reduction of thiophen-2-carboxylic acid, to give a mixture of products that includes $HO_2CCH_2CH=CHCH_2SH$ and $HO_2CCH=CH-CHMeSH$;[85] photolysis of 4-thiochromanone enol acetate to give (3) by a novel 1,5-acyl migration in the intermediate *o*-thioquinone methylide;[86] and cleavage (by Na in NH_3) of *s*-trithians and related compounds $[(CH_2S)_3 \rightarrow MeSCH_2SCH_2S^- Na^+]$.[87]

Additional Bibliography: G. A. Tolstikov, F. Ya. Kanzafarov, Yu. A. Sangalov, and U. M. Dzhemilev, *Neftekhimiya*, 1979, **19**, 425 [thiols from alkenes, using H_2S plus $EtAlCl_2$]; H. Yamaguchi and S. Origuti, *Nippon Kagaku Kaishi*, 1979, 149 (*Chem. Abstr.*, 1979, **90**, 137 192) [$H_2NCH_2CH_2OSO_3H \rightarrow H_2NCH_2CH_2SH$, using NH_4SH, *via* the aziridine]; M. G. Voronkov and E. N. Deryagina, *Phosphorus Sulfur*, 1979, **7**, 123 [aryl bromides + H_2S at high temperatures → ArSH + ArSAr]; A. Etienne, J. C. Bore, G. Baills, G. Lonchambon, and B. Desmazieres, *C.R. Hebd. Seances Acad. Sci., Ser. C*, 1979, **288**, 49 [alkanethiols from $RNEt_2$ + HCl + H_2S]; J. E. Bittell and J. L. Speier, *J. Org. Chem.*, 1978, **43**, 1687 [alkyl halide + H_2S + NH_3 or an alkylamine]; M. A. Vasyanina, Yu. A. Efremov, and V. K. Khairullin, *Izv. Akad. Nauk SSSR, Ser. Khim.*, 1979, 1608 [$HO(CH_2)_3NH_2$ + S + red phosphorus → $HS(CH_2)_3NH_3^+\ S(CH_2)_2CH=NH$]; M. Mikolajczyk, S. Grzejszczak, A. Chefczynska, and A. Zatorski, *J. Org. Chem.*, 1979, **44**, 2967 [$(EtO)_2P(O)CHR + S_8 \rightarrow (EtO)_2P(O)CHRSH$]; H. Alper, J. K. Currie, and R. Sachdeva, *Angew. Chem.*, 1978, **90**, 722 [$H_2C=CR^1CONHR^2 + P_4S_{10}$ with $NaHCO_3 \rightarrow HSCH_2CHR^1CSNHR^2$]; A. M. Zeinalov, F. N. Mamedov, M. Morsum-Zade, and A. K. Ibad-Zade, *Zh. Org. Khim.*, 1979, **15**, 816 [1-naphthol + S_2Cl_2 followed by Zn plus HCl → 1-hydroxynaphthalene-2-thiol].

Reactions of Thiols, Selenols, and Tellurols.—Many of the reactions of thiols depend on the high nucleophilicity of the thiolate anion, and lead to sulphides or other sulphur-containing functional groups. Reactions of this type are therefore discussed in later sections of this Chapter.

Alkylthiolate anions are more reactive by a factor of 10^4—10^5 than OH^- towards tropylium cations and carbonyl compounds.[88] Furthermore, $PhSe^-$ is at least five times more nucleophilic than PhS^-,[89] but the latter nucleophile, paradoxically, is capable of dealkylating $PhSeMe$[89] and tertiary amines[90] when in the presence of Pd, $PdCl_2$, $Pd(OAc)_2$, or $RuCl_3$. Demethylation of methyl ethers[91a] and esters[91b] by treatment with a thiol in the presence of an aluminium trihalide has been reported.

The simplest method yet uncovered for the replacement of the diazonium group by H (or ^2H) is the use of PhSH (or PhS^2H);[92] so far, the reaction seems to be limited to arenediazonium cations. Further reductive processes that are mediated by thiols are the conversion of 1-nitro-alkenes into alkenes ($+ NaNO_2 + PhSSPh + S_8$)

[84] R. A. Moss, G. O. Bizzigotti, T. J. Lukas, and W. J. Sanders, *Tetrahedron Lett.*, 1978, 3661.
[85] W. G. Blenderman, M. M. Joullié, and G. Preti, *Tetrahedron Lett.*, 1979, 4985.
[86] I. W. J. Still and T. S. Leong, *Tetrahedron Lett.*, 1979, 3613.
[87] E. Weissflog and M. Schmidt, *Phosphorus Sulfur*, 1979, **6**, 453.
[88] C. D. Ritchie and J. Gandler, *J. Am. Chem. Soc.*, 1979, **101**, 7318.
[89] H. J. Reich and M. L. Cohen, *J. Org. Chem.*, 1979, **44**, 3148.
[90] S.-I. Murahashi and T. Yano, *J. Chem. Soc., Chem. Commun.*, 1979, 270.
[91] (*a*) M. Node, K. Nishide, M. Sai, K. Ichikawa, K. Fuji, and E. Fujita, *Chem. Lett.*, 1979, 97; (*b*) M. Node, K. Nishide, M. Sai, and E. Fujita, *Tetrahedron Lett.*, 1978, 5211.
[92] T. Shono, Y. Matsumura, and K. Tsubata, *Chem. Lett.*, 1979, 1051.

by PhSH and Na$_2$S,[93] corrole-catalysed photoreduction of benzaldehyde by PhSH,[94] reduction of benzyl halides by RS$^-$ Et$_3$NH$^+$ or RSe$^-$ Et$_3$NH$^+$,[95] and the use of propane-1,3-dithiol in the presence of triethylamine for the reduction of azides to amines.[96]

An 18-crown-6 ether that bears L-cysteinyl residues has been synthesized for use in studies of catalysed hydrolysis,[97] the thiol group being the effective functional group. Several papers have appeared that are based on the principles of addition to multiple bonds. Phenyl trimethylsilyl selenide has been advocated[98] as a convenient source of the phenylselenyl anion, for addition to aldehydes and ketones;[99] the anion is liberated by treatment of PhSeSiMe$_3$ with KF.[98] Asymmetric induction has been of continuing interest in this field, and addition of achiral thiols to (−)-menthyl crotonate[100] or to achiral cyclohexenols, catalysed by an alkaloid[101a] or by an (R)-alaninamide,[100b] leads to moderate enantiomeric excesses. The use of boron trifluoride diethyl etherate as a catalyst for accomplishing Markownikov addition of a thiol to an alkene[102] [PhCH$_2$SH → PhCH$_2$SCMe$_2$(CH$_2$)$_2$C(NHAc)(CO$_2$Et)$_2$] has not been established as a general method, since the thiolester, rather than the protected hexafluoropenicillamine, results from catalysed addition of PhCH$_2$SH to (CF$_3$)$_2$C=C(NHCOPh)CO$_2$H.[103]

Many more references to the recent literature, covering the additions of thiols to multiple bonds, are discussed in the later section 'Synthesis of Sulphides', but the use of Me$_2$AlSPh and Me$_2$AlSeMe (Me$_3$AlLi$^+$ PhS$^-$ is more reactive) for the formation of 1,4-adducts with 2-enones[104] and the preferential photoaddition of Bu^{n}SH to the C=C bond in the ester moiety of allyl and vinyl acrylates and to *trans*-crotonates[105] are items of unusual interest. Nucleophilic substitution reactions of thiols that lead to sulphides are also mostly covered in the later section. The generation of dianions from prop-2-ene-1-thiol[106] and from α-mercapto-γ-butyro-lactone[107] (using RLi with MgBr$_2$ and LiPr$^i_2$N with TMEDA, respectively) and their reactions with aldehydes and ketones provide valuable new uses in synthesis of sulphur nucleophiles, leading to oxirans and αβ-unsaturated esters, respectively, after desulphurization of the adducts.

Dehydrogenation and desulphurization of thiols, and related processes, are involved in reactions of simple alkanethiols with H[108] and O atoms,[109] and in

[93] N. Ono, S. Kawai, K. Tanaka, and A. Kaji, *Tetrahedron Lett.*, 1979, 1733.
[94] Y. Murakami, Y. Aoyama, and K. Tokunaga, *J. Chem. Soc., Chem. Commun.*, 1979, 1018.
[95] L. Hevesi, *Tetrahedron Lett.*, 1979, 3025.
[96] H. Bayley, D. N. Standring, and J. R. Knowles, *Tetrahedron Lett.*, 1978, 3633.
[97] J. M. Lehn and C. Sirlin, *J. Chem. Soc., Chem. Commun.*, 1978, 949.
[98] M. R. Detty, *Tetrahedron Lett.*, 1978, 5087.
[99] D. Liotta, P. B. Paty, J. Johnston, and G. Zima, *Tetrahedron Lett.*, 1978, 5091.
[100] M. Yoshihara, H. Fujihara, and T. Maeshima, *Chem. Lett.*, 1980, 195.
[101] (a) H. Pluim and H. Wynberg, *Tetrahedron Lett.*, 1979, 1251; (b) S. Inoue and Y. Kawano, *Makromol. Chem.*, 1979, **180**, 1405 (*Chem. Abstr.*, 1979, **91**, 108 216).
[102] G. A. Dilbeck, L. Field, A. A. Gallo, and R. J. Gargiulo, *J. Org. Chem.*, 1978, **43**, 4593.
[103] L. Field and A. A. Gallo, *J. Fluorine Chem.*, 1980, **15**, 29.
[104] A. Itoh, S. Ozawa, K. Oshima, and H. Nozaki, *Tetrahedron Lett.*, 1980, **21**, 361.
[105] L. I. Komarova, *Zh. Org. Khim.*, 1978, **14**, 737.
[106] M. Pohmakotr, K. H. Geiss, and D. Seebach, *Chem. Ber.*, 1979, **112**, 1420.
[107] K. Tanaka, N. Yamagishi, R. Tanikaga, and A. Kaji, *Bull. Chem. Soc. Jpn.*, 1979, **52**, 3619; K. Tanaka, H. Uneme, N. Yamagishi, N. Ono, and A. Kaji, *Chem. Lett.*, 1978, 653.
[108] O. Horie, J. Nishino, and A. Amano, *Int. J. Chem. Kinet.*, 1978, **10**, 1043.
[109] I. R. Seagle, F. Baiocchi, and D. Gutman, *J. Phys. Chem.*, 1978, **82**, 1333.

pyrolysis in the absence of a catalyst[110] or in the presence of poly(metal phthalocyanines).[111] Similarly, photolysis of alkanethiols in the liquid state[112] leads to loss of H_2S and hydrogen. A useful conversion of thiols into alkanes and thiolacetates employs $[Mo(CO)_6]$ that is adsorbed on alumina, this being suspended in AcOH.[113]

Straightforward exchange reactions $[RSH + HO^2H \rightleftharpoons RS^2H + H_2O;^{114}$ MeSH + H_2S *via* a cyclic complex[115]] depend on the acidities of the thiols, which are solvent-dependent.[116]

Excluding the extensive literature on metal complexes carrying bivalent sulphur ligands, some classes of compound formed between thiols and elements near sulphur in the Periodic Table are important in synthetic organic chemistry. White phosphorus gives a tris(alkylthio)phosphine with a mixture of a thiol and its sodium salt in CCl_4 ($CHCl_3$ is the other organic reaction product).[117] Arenethiolates do not react in the same way, owing to a competing reaction of the intermediate $ArS(P_4)Cl$ that involves the displacement of ArS^- (a good leaving group).[117] Desulphurization of thiols and sulphides with MeMgBr or ArMgBr in the presence of $[(Ph_3P)_2NiCl_2]$ leads to alkenes, toluenes, and biaryls in moderate to high yields.[118] Iron carbonyls give $[RSFe(CO)_3]_2$ with thiols, sulphides, and disulphides, and the corresponding disulphanes $[S_2Fe_2(CO)_6]$ and $[S_2Fe_3(CO)_9]$ with S_8.[119] Although the regeneration of the thiol can be accomplished, potential uses in synthesis do not appear to have been realized.

This Chapter has invariably included brief coverage of biochemically important reactions of thiols, since any insight which can be gained is based on a clear understanding of the fundamental chemistry of the SH group. Accurate determination of the relative numbers of SH groups in enzymes and other complex molecules is commonly achieved through the use of Ellman's reagent and then spectrophotometric assay of the resulting 5-mercapto-2-nitrobenzoic acid. Certain knowledge of the molar absorption coefficient of the dianion of this compound has been a limiting factor in this assay, but the parameters $\lambda_{max} = 412$ nm, $\varepsilon = 14\,150$ have been established for pure samples in dilute aqueous salt solutions.[120] One of the thiol-containing enzymes, glyceraldehyde 3-phosphate dehydrogenase, is efficiently deactivated by 2-nitro-3-(5-nitro-2-furyl)acrylates, and model reactions of this inhibitor with simple thiols have been reported.[121] Carbocyclic analogues of captopril (hetCOCHMeCH₂SH), a compound with inhibiting activity for angiotensin-converting enzyme, have been synthesized.[122] An intermediate (4) that was isolated during studies of the reduction of 1,3-dimethyl-5-(*p*-nitrophenyl-

[110] N. Barroeta, G. Martin, and J. Capasso, *React. Kinet. Catal. Lett.*, 1978, **9**, 177.
[111] N. Takamiya, T. Iwatsuki, T. Nagai and S. Murai, *Nippon Kagaku Kaishi*, 1979, 825.
[112] W. A. Pryor and E. G. Olsen, *J. Am. Chem. Soc.*, 1978, **100**, 2852.
[113] H. Alper and C. Blais, *J. Chem. Soc., Chem. Commun.*, 1980, 169.
[114] J. J. Khurma and D. V. Fenby, *Aust. J. Chem.*, 1979, **32**, 755.
[115] A. Wawer, *Pol. J. Chem.*, 1979, **53**, 1917.
[116] G. P. Sharnin, V. V. Nurgatiu, and B. M. Ginzburg, *Zh. Org. Khim.*, 1979, **15**, 1638.
[117] C. Brown, R. F. Hudson, and G. A. Wartew, *J. Chem. Soc., Perkin Trans. 1*, 1979, 1799.
[118] E. Wenkert, T. W. Ferreira, and E. L. Michelotti, *J. Chem. Soc., Chem. Commun.*, 1979, 637.
[119] N. S. Nametkin, V. D. Tyurin, and M. A. Kukina, *J. Organomet. Chem.*, 1978, **149**, 355.
[120] P. W. Riddles, R. L. Blakely, and B. Zerner, *Anal. Biochem.*, 1979, **94**, 75.
[121] E. Sturdik, L. Drobnica, S. Balaz, and V. Marko, *Biochem. Pharmacol.*, 1979, **28**, 2525.
[122] A. Sugie and J. Katsube, *Chem. Pharm. Bull.*, 1979, **27**, 1708.

(4)

imino)barbituric acid by thiols tends to confirm mechanisms previously advocated for the reduction of flavins by thiols.[123] Enhancement of the rates of hydrolysis of esters by nearby SH and $CONH_2$ groups has been explained on the basis of a mechanism that involves the formation of an intermediate cyclic imide.[124]

Thiyl and Related Selenium Radicals.—Direct generation of thiyl radicals (RS·) from thiols by γ-radiolysis (using a ^{60}Co source)[125] and indirect formation by the reaction of thiols with triarylmethyl radicals[126] or with selenocyanate anion radicals that are formed by pulse radiolysis of $(SeCN)_2$,[127] are accompanied by cleavage reactions of other bivalent sulphur compounds [photolysis of thionitrites[128] and cleavage of disulphides by $Pb(OAc)_4$[129]]. The first e.s.r. study of selenuranyl radicals, formed by attack on selenides by the CF_3· radical, has been reported.[130]

Oxidative decarboxylation of 4-carboxy-4-methylpentylthiyl radicals, leading to mixtures of 2,2-dimethyltetrahydrothiophens and 5,5-dimethyl-2,3-dihydrothiophens, has been proposed as a model for one of the stages in penicillin biosynthesis.[129] Comparative rate studies have shown that the addition of dioxygen to thiyl radicals is slower than to the corresponding carbon radicals.[131]

4 Thiolesters, Selenolesters, and Tellurolesters

Until recently, the relatively sparse contributions to the literature on thiolesters, R^1COSR^2, were easily accommodated as a small paragraph in the 'Thiols' section, but the substantial growth of interesting chemistry involving these compounds necessitates this separate section.

Preparation.—Modifications of the standard procedure ($R^1CO_2H + R^2SH$) employ 4-(dimethylamino)pyridine and dicyclohexylcarbodi-imide,[132] diphenylphosphinyl chloride,[133] or $Me_2NP(O)Cl_2$[134] as coupling reagents. Related methods (*e.g.* acyl

[123] J. M. Sayer, P. Coulon, J. Hupp, J. Fauchez, R. Belanger, and E. J. White, *J. Am. Chem. Soc.*, 1979, **101**, 1890.
[124] D. Petz, F. Schnieder, and H. G. Loeffler, *Z. Naturforsch., Teil C*, 1978, **33**, 151.
[125] D. J. Nelson, R. L. Petersen, and M. C. R. Symons, *J. Chem. Soc., Perkin Trans. 2*, 1977, 2005.
[126] T. H. Colle and E. S. Lewis, *J. Am. Chem. Soc.*, 1979, **101**, 1810.
[127] R. Badiello and M. Tamba, *Radiochem. Radioanal. Lett.*, 1979, **37**, 165.
[128] H. Chandra, B. G. Gowenlock, and J. Pfab, *J. Chem. Soc., Chem. Commun.*, 1979, 392.
[129] J. E. Baldwin and T. S. Wan, *J. Chem. Soc., Chem. Commun.*, 1979, 249.
[130] J. R. G. Giles, B. P. Roberts, M. J. Perkins, and E. S. Turner, *J. Chem. Soc., Chem. Commun.*, 1980, 504.
[131] K. Schaefer, M. Bonifacic, D. Bahnemann, and K. D. Asmus, *J. Phys. Chem.*, 1978, **82**, 2777.
[132] B. Nieses and W. Steglich, *Angew. Chem.*, 1978, **90**, 556.
[133] D. Scholz and D. Eigner, *Monatsh. Chem.*, 1979, **110**, 759.
[134] H.-J. Liu, S. P. Lee, and W. H. Chan, *Synth. Commun.*, 1979, **9**, 91.

chloride + Bu_3^nSnSR[135] and carboxylic ester + $Me_2AlSeMe$[136]) give excellent yields of thiolesters and selenolesters, respectively. The latter compounds can also be reached by alkylation of selenocarboxylates.[137] Addition of ArSeBr to ethyl vinyl ether gives the aryl selenide ArSeCH$_2$CHO rather than the selenolester ArSeCOMe.[138] Ketones yield β-keto-thiolesters by reaction with dithiolcarbonates and NaH.[139] Selenolesters have also been prepared from N-acylhydrazines and benzeneseleninic acid.[620]

Pummerer rearrangements leading to thiolesters are discussed in the later sections on sulphoxides. A superficially related Arbuzov-type reaction, with a 1,2-shift of sulphur, through the intermediate (5), has a similar outcome.[140] 2-Thiopyridyl chloroformate has been advocated as a reagent for the conversion of carboxylic acids into their pyridine-2-thiolesters under mild conditions.[141]

$$Ph_2P-C-Ph \longrightarrow Ph_2P=CPh \longrightarrow Ph_2P-CHPh \longrightarrow Ph_2P-CHPh$$

(5)

Reactions of Thiolacids and Thiolesters, Selenolesters, and Tellurolesters.—The well-known reactivity of thiolacids and thiolesters as acylating agents is shared with the selenolesters.[136] A novel acylating agent, 2-benzoylthio-1-methylpyridinium chloride, is effective in aqueous solutions.[142]

Uses in synthesis of thiolacids and esters are being explored for possible advantages *vis-à-vis* their oxygen analogues. The Dieckmann cyclization of ω-dicarboxylate esters is more easily effected in-the thiolester series,[143] and the equivalent in the thiolester series of the decarboxylation of a β-keto-ester is accomplished by using Raney nickel.[144] Dithiomalonate esters, $CH_2(COSEt)_2$, can be alkylated and reductively dethiocarboxylated by Raney nickel to yield hydroxyethyl compounds, thus reacting as masked ethanol carbanions.[145] Bis-metallated thiolacids, $RCH=C(OLi)SLi$, give β-hydroxy-thiolacids with aldehydes and ketones, and thiolactones by reaction with α-chloro-ketones.[146]

A photochemical Wolff rearrangement of a diazomonothiolmalonate[147] [$MeO_2C-C(N_2)COSR \rightarrow MeO_2CCH_2SR$] and the use of the same compound in a synthesis of an azetidinone by photolysis with an imine[148] follow established reactions of the

[135] D. N. Harpp, T. Aida, and T. H. Chan, *Tetrahedron Lett.*, 1979, 2853.
[136] A. P. Kozikowski and A. Ames, *J. Org. Chem.*, 1978, **43**, 2735.
[137] H. Ishihara and Y. Hirabayashi, *Chem. Lett.*, 1978, 1007.
[138] R. Baudar and M. Petrzilka, *Helv. Chim. Acta*, 1979, **62**, 1406.
[139] H.-J. Liu, S. K. Attah-Poku, and H. K. Lai, *Synth. Commun.*, 1979, **9**, 883.
[140] M. Yoshifuji, J. Tagawa, and N. Inamoto, *Tetrahedron Lett.*, 1979, 2415.
[141] E. J. Corey and D. A. Clark, *Tetrahedron Lett.*, 1979, 2875.
[142] M. Yamada, Y. Watabe, T. Sakakibara, and R. Sudoh, *J. Chem. Soc., Chem. Commun.*, 1979, 179.
[143] H.-J.Liu and H. K. Lai, *Tetrahedron Lett.*, 1979, 1193.
[144] H.-J. Liu, H. K. Lai, and S. K. Attah-Poku, *Tetrahedron Lett.*, 1979, 4121.
[145] H.-J. Liu and H. K. Lai, *Can. J. Chem.*, 1979, **57**, 2522.
[146] A. M. Sarpeshkar, G. J. Gossick, and J. Wemple, *Tetrahedron Lett.*, 1979, 703.
[147] V. Georgian, S. K. Boyer, and B. Edwards, *J. Org. Chem.*, 1980, **45**, 1686.
[148] V. Georgian, S. K. Boyer, and B. Edwards, *Heterocycles*, 1977, **7**, 1003.

oxygen analogue. Intramolecular Wittig alkene reactions of thiolesters, leading to fused azetidinone–pyrroline antibiotics, have been described.[149]

Substitution alpha to the carbonyl group of a thiolester is facilitated by prior conversion into silyl[150] or boranyl[151] enolates. Michael addition of cyclopentenone to $H_2C=C(SBu^t)OSiMe_3$, followed by alkylation with $BrCH_2C\equiv CCH_2CH_2OThp$ (Thp = tetrahydropyranyl) and conventional elaboration, leads to (±)-jasmine keto-lactone.[150] Self-condensation of alkyl arenethiolesters, using lithium *N*-hydroxy-2,2,6,6-tetramethylpiperidide, gives products which vary according to the reaction conditions [PhCOSEt → (PhCO)$_2$CHMe at ambient temperatures, but PhCOCHMeSCOPh at −98 °C].[152] A stable lithium enolate (6) has been discovered during this work. Macrocyclic lactones are accessible from heteroaryl ω-hydroxyalkanethiolates by treatment with $AgClO_4$.[153]

(6)

Alkylthioketenimines, formed from alkyl α-aminothiolacetates by treatment with $RP(O)Cl_2$, give back the original starting material rather than acetamides $R^1NHCOCH_2SR^2$ when they react with water.[154] Further work (see Vol. 5, p. 17) on the ozonolysis of thiolesters has implicated *S*-oxide intermediates as the initial reaction products, which undergo an *S,O*-acyl shift and further oxidation to mixed sulphonic–carboxylic anhydrides [$RCOS(O)R → RCO_2SR → RCO_2SO_2R$].[155]

A further example of ongoing studies (see Vol. 5, p. 17) is the establishment of competing photolysis and photo-Fries rearrangement processes with aryl areneselenol-[156] and arenetellurol-esters.[157]

Conventional structure–rate studies for the hydrolysis of thiolesters have been reported.[158]

5 Sulphides, Selenides and Tellurides

Preparation from Thiols, Selenols, and Tellurols.—The main methods for the synthesis of sulphides are substantially represented in the recent literature, and those papers whose topics can be clearly indicated by brief details are collected at the end of this section.

[149] R. J. Ponsford, P. M. Roberts, and R. Southgate, *J. Chem. Soc., Chem. Commun.*, 1979, 847; A. J. G. Baxter, R. J. Ponsford, and R. Southgate, *ibid.*, 1980, 429.
[150] H. Gerlach and R. Kuenzler, *Helv. Chim. Acta*, 1978, **61**, 2503.
[151] M. Hirama and S. Masamune, *Tetrahedron Lett.*, 1979, 2225; M. Hirama, D. S. Garvey, L. D. L. Lu, and S. Masamune, *ibid.*, p. 3937.
[152] D. B. Reitz, P. Beak, R. F. Farney, and L. S. Helmick, *J. Am. Chem. Soc.*, 1978, **100**, 5428.
[153] J. S. Nimitz and R. H. Wollenberg, *Tetrahedron Lett.*, 1978, 3523.
[154] L. I. Kruglik and Yu. G. Gololobov, *Zh. Org. Khim.*, 1978, **14**, 747.
[155] M. J. Sousa Lobo and H. J. Chaves das Neves, *Tetrahedron Lett.*, 1978, 2171.
[156] J. Martens, K. Praefcke, and H. Simon, *Chem.-Ztg.*, 1978, **102**, 108.
[157] W. Lohner, J. Martens, K. Praefcke, and H. Simon, *J. Organomet. Chem.*, 1978, **154**, 263.
[158] K. S. Venkatasubban, K. R. Davis, and J. L. Hogg, *J. Am. Chem. Soc.*, 1978, **100**, 6125; T. H. Fife and B. R. De Mark, *ibid.*, 1979, **101**, 7379.

The uses of thiols in the synthesis of sulphides can be discussed under the two main headings (*a*) additions to unsaturated systems and (*b*) substitution reactions. Several papers under the former heading have been discussed in an earlier section (p. 11). Addition of thiolate anions to 1-methoxy-[159] and 1-alkylthio-tropenylium fluorosulphonates[160] occurs at C-7. 1,4-Addition of PhSMgI to the enones $R^1CH=CR^2COR^3$, followed by addition of the resulting carbanion to a ketone R^4COR^5, provides a one-pot synthesis of 2-acyl-3-(phenylthio)alkan-1-ols $PhSCHR^1CR^2(COR^3)CR^4R^5OH$.[161] Several papers[162] deal with reactions of thiols with formaldehyde and HCl[162] (n-$C_8H_{17}SH$ → chloromethyl sulphide), with ketones (giving γ-keto-sulphides),[163] with sulphones,[164] with secondary amines,[165] and an interesting reaction with *N*-methylhydroxylamine hydrochloride, giving $(RSCH_2)_2CMeNOH$.[166] *NS*-Acetals are formed by the addition of thiols to imines.[167] An unusual conversion of perfluoro-2- and -4-methyl-2-pentenes into 1,3-dienes has been rationalized on the basis of elimination of a sulphenyl fluoride from the initial adduct that is formed with a thiol.[168]

Substitutions employing novel reagents include thallium thiolates and selenolates (PhSTl and PhSeTl from the thiol and selenol, using TlOEt or TlOPh) in reactions with acyl chlorides and α-halogenocarbonyl compounds,[169] and in the thiolysis of oxirans.[170] Trimethylsilyl phenyl selenide has been used for cleavage of esters (in conjunction with ZnI_2),[171] ring-opening of oxirans,[172] and Michael additions.[173] Interestingly, Me_2AlSR (from Me_3Al and RSH, at 0 °C) was utilized in the thiolysis of methanesulphonates.[174]

Important direct syntheses from alkanols (see Vol. 5, p. 19) have been supplemented by the high-yield synthesis of symmetrical sulphides, using $(CF_3CH_2O)_2PPh_3$ and a thiol.[175] Other less commonly used leaving groups in syntheses of sulphides from thiols are represented in the processes $ArCH_2NMe_2$ → $ArCH_2SR$.[176] dimedonyliodonium ylide → 2-phenylthio-5,5-dimethylcyclohexane-

[159] T. Komives, A. F. Marton, S. Holly, and F. Dutka, *React. Kinet. Catal. Lett.*, 1978, **8**, 19; J. P. Guthrie, *J. Am. Chem. Soc.*, 1978, **100**, 5892.
[160] M. Cavazza, G. Morganti, and F. Pietra, *J. Chem. Soc., Chem. Commun.*, 1978, 710.
[161] T. Shono, Y. Matsumura, S. Kashimura, and K. Hatanaka, *J. Am. Chem. Soc.*, 1979, **101**, 4752.
[162] J. Broniarz, J. Pernak, M. Pujanek, and W. Karminski, *Zesz. Nauk Politech. Slask. Chem.*, 1978, **81**, 85 (*Chem. Abstr.*, 1979, **90**, 167 989); J. Broniarz, J. Szymanowski, J. Pernak, and M. Pujanek, *Chem. Stosow.*, 1978, **22**, 201.
[163] V. I. Dronov, R. F. Nigmatullina, L. V. Spirikhin, and Yu. E. Nikitin, *Zh. Org. Khim.*, 1978, **14**, 2357; S. S. Shukurov, K. N. Negmatullaev, M. S. Danyarova, and S. S. Sabirov, *ibid.*, 1979, **15**, 1102.
[164] V. I. Dronov, R. F. Nigmatullina, E. E. Zaev, and Yu. E. Nikitin, *Zh. Org. Khim.*, 1979, **15**, 1706.
[165] A. M. Kuliev, K. Z. Guseinov, N. A. Aliev, A. K. Ibad-Zade, and N. Yu. Ibragimov, *Dokl. Akad. Nauk Az. SSR*, 1977, **33**, 53 (*Chem. Abstr.*, 1979, **89**, 75 228).
[166] K. Ito and M. Sekiya, *Chem. Pharm. Bull.*, 1979, **27**, 1691.
[167] H. Boehme and A. Ingendoh, *Liebigs Ann. Chem.*, 1978, 381.
[168] M. Maruta and N. Ishikawa, *J. Fluorine Chem.*, 1979, **13**, 111.
[169] M. R. Detty and G. P. Wood, *J. Org. Chem.*, 1980, **45**, 80.
[170] H. A. Klein, *Chem. Ber.*, 1979, **112**, 3037.
[171] N. Miyoshi, H. Ishii, S. Murai, and N. Sonoda, *Chem. Lett.*, 1979, 873.
[172] N. Miyoshi, K. Kondo, S. Murai, and N. Sonoda, *Chem. Lett.*, 1979, 909.
[173] N. Miyoshi, H. Ishii, K. Kondo, S. Murai, and N. Sonoda, *Synthesis*, 1979, 300.
[174] S. Ozawa, A. Itoh, K. Oshima, and H. Nozaki, *Tetrahedron Lett.*, 1979, 2909.
[175] T. Kubota, S. Miyashita, T. Kitazume, and N. Ishikawa, *Chem. Lett.*, 1979, 845.
[176] F. I. Gasanov, V. N. Mamedov, M. R. Kulibekov, and A. G. Kuliev, *Azerb. Khim. Zh.*, 1978, 40 (*Chem. Abstr.*, 1980, **91**, 5031).

1,3-dione,[177] and halogenonitrobenzenes → alkyl aryl sulphides with replacement of both halogen and nitro-groups.[178,179] The last-mentioned study is a further example of the role of HMPA in substitution reactions. While an *o*- or *p*-alkylthio-substituent activates an aromatic nitro-group towards nucleophilic substitution (by Pr^iSNa in HMPA),[178,180] the same reagent can bring about the substitution of the halogen atom of unactivated aryl halides.[179] In contrast, the reaction of 2,4-dichloronitrobenzene with PhSNa in DMF leads to only Cl-substitution products.[181] β-Aryl-nitroalkanes react with MeSNa to yield rearranged NO_2-replacement products; the reaction is suggested to involve a phenonium radical intermediate.[182]

A fifty-fold increase in the rate of thiolysis of *p*-nitrophenyl acetate by arylthiolate anions in the presence of $n\text{-}C_{16}H_{33}NMe_3^+$ Br^- micelles has been accounted for entirely on the basis of enhanced concentrations of reactants in the micelle phase.[183]

Additional Bibliography: M. Augustin, R. Werndl, M. Koehler, and H. H. Ruettinger, *Pharmazie*, 1978, **33**, 191 [Bu^nSH + *N*-arylmaleimides]; M. K. Gadzhiev, D. D. Gogoladze, and G. V. Maisuradze, *Azerb. Khim. Zh.*, 1979, 47 [Pr^nSH + allyl alcohol]; A. N. Mirskova, A. V. Martynov, I. D. Kalikhman, V. V. Keiko, Yu. V. Vitkovskii, and M. G. Voronkov, *Zh. Org. Khim.*, 1979, **15**, 1634 [Bu^nSH, *hv*, $H_2C{=}CCl_2$ and $ClCH{=}CHCl$]; M. Watanabe, K. Shirai, and T. Kumamoto, *Bull. Chem. Soc. Jpn.*, 1979, **52**, 3318 [PhSLi + 2-buten-4-olide]; H. Hertenstein, *Angew. Chem.*, 1980, **92**, 123 [R^1SH + diketen → anti-Markownikov adduct in Et_2O, but $R^1SCH_2CH{=}CHCO_2R^2$ in R^2OH]; K. K. Tokmurzin and N. Nurekeshova, *Izv. Akad. Nauk Kaz. SSR, Ser. Khim.*, 1978, **28**, 66 Bu^nSH + $H_2C{=}CHCOCH{=}CMe_2$ → $Bu^nSCH_2CH_2COCH{=}CMe_2$ in neutral conditions → $H_2C{=}CH\text{-}COCH_2CMe_2SBu^n$]; N. P. Petukhova, N. E. Dontsova and E. N. Prilezhaeva, *Izv. Akad. Nauk SSSR, Ser. Khim.*, 1979, 467 [tetrachloroethylene + RSNa → tris- and tetrakis-(alkylthio)ethylenes]; B. Czech, S. Quicic, and S. L. Regen, *Synthesis*, 1980, 113 [R^2SR^2 from R^1SH, R^2Br, and NaOH–Al_2O_3; R^2SR^2 from R^2Br or R^2Cl and Na_2S–Al_2O_3]; R. D. Little and J. R. Dawson, *J. Am. Chem. Soc.*, 1978, **100**, 4607 [$BrCH_2CH{=}CHCO_2Me$ + RSLi → (alkylthio)cyclopropanecarboxylic ester in a non-polar solvent, but the simple substitution product in polar solvents]; R. H. Everardus and L. Brandsma, *Synthesis*, 1978, 359 [$ClCH_2C{\equiv}CCH_2Cl$ → $RSCH_2C{\equiv}CCH_2SR$ → $RSC{\equiv}C\text{-}CH{=}CH_2$ with Bu^tOK]; S. Murahashi, M. Yamamura, K. Yanagisawa, N. Mita, and K Kondo, *J. Org. Chem.*, 1979, **44**, 2408 [alkenyl halides + RS^- and [$Pd(PPh_3)_4$] → alkenyl sulphides]; R. D. Miller and D. R. McKean, *Tetrahedron Lett.*, 1979, 1003 [cyclobutanones + γ-keto-sulphides with PhSH, $ZnCl_2$, and HCl]; R. W. Gray and A. S. Dreiding, *J. Indian Chem. Soc.*, 1978, **55**, 1224 [Cl substitution accompanied by dehydrochlorination and dehydrobromination by PhS^-, leading to 2-(phenylthio)-3,7-dehydrotropone]; B. C. Musial and M. E. Peach, *Phosphorus Sulfur*, 1977, **3**, 41; M. E. Peach and E. S. Rayner, *J. Fluorine Chem.*, 1979, **13**, 447; L. J. Johnston and M. E. Peach, *ibid.*, p. 41; and W. J. Frazee and M. E. Peach, *ibid.*, p. 225 [four studies of substitution of polybromo- and polyfluoro-benzenes with MeSCu and C_6F_5SCu]; H. Alsaidi, R. Gallo, and J. Metzger, *C.R. Hebd. Seances Acad. Sci., Ser. C*, 1979, **289**, 203 [2-chloro-5-nitropyridine + PhSH under phase-transfer conditions]; T. Ando and J. Yamawaki, *Chem. Lett.*, 1979, 45 [alkylation of PhSH, catalysed by KF on Celite]; A. B. Pierini and R. A. Rossi, *J. Org. Chem.*, 1979, **44**, 4667 and *J. Organomet. Chem.*, 1979, **168**, 163 [photostimulated substitution of halogenoarenes by $PhSe^-$ and $PhTe^-$]; N. V. Kondratenko, V. I. Popov, A. A. Kolomeitsev, I. D. Sadekov, V. I. Minkin, and L. M. Yagupolskii, *Zh. Org. Khim.*, 1979, **15**, 1561 [heptafluoro-n-propyl iodide + $ArTe^-$].

Preparation from Sulphenyl and Selenenyl Halides.—Most of the methods published in the period under review lead to β-halogeno-sulphides through additions

[177] G. F. Koser, S. M. Linden, and Y.-J. Shih, *J. Org. Chem.*, 1978, **43**, 2676.
[178] P. Gogolli, L. Testaferri, M. Tingoli, and M. Tiecco, *J. Org. Chem.*, 1979, **44**, 2636.
[179] J. R. Beck and J. A. Yahner, *J. Org. Chem.*, 1978, **43**, 2048, 2052.
[180] P. Cogolli, F. Maiolo, L. Testaferri, M. Tingoli, and M. Tiecco, *J. Org. Chem.*, 1979, **44**, 2642.
[181] R. F. Miller, J. C. Turley, G. E. Martin, L. Williams, and J. E. Hudson, *Synth. Commun.*, 1978, **8**, 371.
[182] N. Kornblum, J. Widmer, and S. C. Carlson, *J. Am. Chem. Soc.*, 1979, **101**, 658.
[183] I. M. Cuccovia, E. H. Schroeter, P. M. Monteiro, and H. Chaimovich, *J. Org. Chem.*, 1978, **43**, 2248.

to unsaturated systems, or to sulphides through nucleophilic attack at sulphur, but unusual variations on these general themes continue to be reported. Earlier work (see Vol. 5, p. 19), establishing the use of 6-diazopenicillinates for the synthesis of 6-chloro-6-phenylseleno-analogues, has been continued[184] and extended[185] to α-diazo-ketones, where the presence of the chlorine atom in the initial product was exploited by dehydrochlorination, leading to α-phenylselenyl-αβ-unsaturated ketones.

While the substantial output from the research groups led by Zefirov and by Schmid and Garratt is aimed at mechanistic details (thiiranium ion intermediates are preferred), other workers continue to seek outlets in general organic synthesis for these reactions. The fact that the initial reaction products carry a halogen atom frequently favours subsequent substitution reactions, and when these involve intramolecular processes, the outcome of reactions of sulphenyl halides can be useful 'cyclofunctionalization' procedures. An interesting example, involving participation by a carbonyl group (Scheme 2), provides a new route to 2,6-dideoxyglycosides.[186] Further studies of the analogous 'phenylsulpheno-(seleneno)lactonization' of unsaturated carboxylic acids,[187] and related cyclizations leading to 2,5-dihydrofurans (rather than vinyloxirans) from α-allenic alcohols,[188] functionalized tetrahydrofurans (from γ-hydroxy-β-keto-alkenes),[189] 1,2-disubstituted 2,3-dihydrocarbazoles (from o-alkenyl-anilines),[190] and cis-perhydroindanes (from cyclonona-1,5-dienes)[191] illustrate the principle further. The ease with which these reactions occur is notable, but conditions can be found[192] for reactions between PhSeCl and alkenes in which the ensuing Cl-substitution reactions can be avoided. In this context it is relevant that ω-hydroxyalkyl-alkynes also add PhSeCl without cyclization.[193] Intermolecular analogues of the cyclization reactions of PhSeCl–alkene adducts are also well established [hydroxyselenation of alkenes,[194]

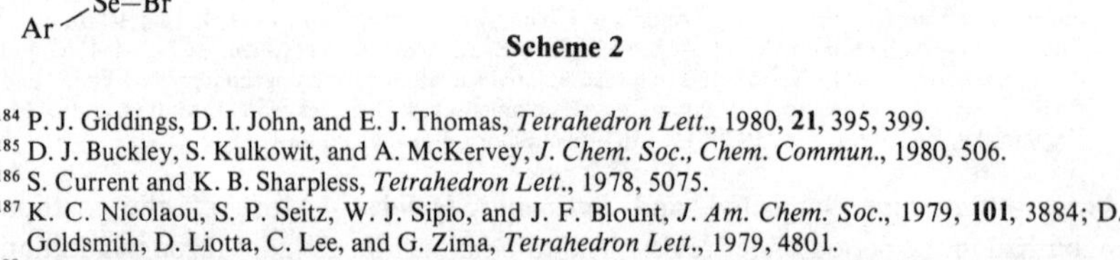

Scheme 2

[184] P. J. Giddings, D. I. John, and E. J. Thomas, *Tetrahedron Lett.*, 1980, **21**, 395, 399.
[185] D. J. Buckley, S. Kulkowit, and A. McKervey, *J. Chem. Soc., Chem. Commun.*, 1980, 506.
[186] S. Current and K. B. Sharpless, *Tetrahedron Lett.*, 1978, 5075.
[187] K. C. Nicolaou, S. P. Seitz, W. J. Sipio, and J. F. Blount, *J. Am. Chem. Soc.*, 1979, **101**, 3884; D. Goldsmith, D. Liotta, C. Lee, and G. Zima, *Tetrahedron Lett.*, 1979, 4801.
[188] P. L. Beaulieu, V. M. Morisset, and D. G. Garratt, *Tetrahedron Lett.*, 1980, **21**, 129.
[189] Z. Lysenko, F. Ricciardi, J. E. Semple, P. C. Wang, and M. M. Jouillié, *Tetrahedron Lett.*, 1978, 2679.
[190] D. L. J. Clive, C. K. Wong, W. A. Kiel, and S. M. Menchen, *J. Chem. Soc., Chem. Commun.*, 1978, 379.
[191] D. L. J. Clive, G. Chittattu, and C. K. Wong, *J. Chem. Soc., Chem. Commun.*, 1978, 441.
[192] D. Liotta and G. Zima, *Tetrahedron Lett.*, 1978, 4977.
[193] C. N. Filer, D. Ahern, R. Fazio, and E. J. Shelton, *J. Org. Chem.*, 1980, **45**, 1313.
[194] A. Toshimitsu, T. Aoai, H. Owada, S. Uemura, and M. Okano, *J. Chem. Soc., Chem. Commun.*, 1980, 412.

or oxoselenation ($RCH=CH_2 \rightarrow RCHBrCH_2SePh$, which with DMSO give $RCOCH_2SePh$)[195]].

1,2-Adducts result from the addition of arenesulphenyl chlorides to 1,3-butadienes[196] and O-silyl dienolates $Me_3SiOCH=CHCH=CH_2$,[197] the former rearranging to 1,4-adducts spontaneously (excepting the adduct with 1,3-butadiene itself), while the latter lose Me_3SiCl, resulting in overall γ-substitution of $\alpha\beta$-unsaturated aldehydes. The corresponding process occurs with $H_2C=CH$-$C(OSiMe_3)=CHMe$ and $PhSeCl$.[198]

Additional Bibliography: *Addition reactions* N. S. Zefirov, N. K. Sadovaya, R. S, Akhmedova, and J. V. Bodrikov, *Zh. Org. Khim.*, 1978, **14**, 662; N. S. Zefirov, V. A. Smit, I. V. Bodrikov, and M. Z. Krimer, *Dokl. Akad. Nauk SSSR*, 1978, **240**, 858; N. S. Zefirov, N. K. Sadovaya, L. A. Novgorodtseva, and I. V. Bodrikov, *Zh. Org. Khim.*, 1978, **14**, 1806; N. S. Zefirov, N. K. Sadovaya, R. S. Akhmedova, and I. V. Bodrikov, *ibid.*, 1979, **15**, 217; T. R. Cerksus, V. M. Csizmadia, G. H. Schmid, and T. T. Tidwell, *Can. J. Chem.*, 1978, **56**, 205; A. Modro, G. H. Schmid, and K. Yates, *J. Org. Chem.*, 1979, **44**, 4221; D. G. Garratt, *Can. J. Chem.*, 1978, **56**, 2184; D. G. Garratt and P. Beaulieu, *ibid.*, 1979, **57**, 119; G. H. Schmid, D. G. Garratt, and S. Yeroushalmi, *J. Org. Chem.*, 1978, **43**, 3764 [nine mechanistic studies of additions of RSCl and RSeCl to alkenes and alkynes]; A. S. Gybin, M. Z. Krimer, V. A. Smit, V. S. Bogdanov, and E. A. Vorobeva, *Izv. Akad. Nauk SSSR, Ser. Khim.*, 1979, 563; M. Z. Krimer, E. A. Vorobeva, and V. A. Smit, *Tezisy Dokl.-Vses. Konf. 'Stereokhim. Konform. Anal. Org. Neftekhim. Sint.' 3rd*, 1976, 110 (*Chem. Abstr.*, 1978, **88**, 189 899); and Yu. B. Kal'yan, M. Z. Krimer, and V. A. Smit, *Izv. Akad. Nauk SSSR, Ser. Khim.*, 1979, 2300 [ArSBr and ArS⁺ BF₄⁻ or ArS⁺ SbF₆⁻ giving thiiranium salts with alkenes]; F. Cooke, R. Moerck, J. Schwindeman, and P. Magnus, *J. Org. Chem.*, 1980, **45**, 1046 [Me₃SiCH=CH₂ + ArSCl → Me₃SiCH(SAr)CH₂Cl → vinyl sulphides]; V. R. Kartashev, I. V. Bodrikov, E. V. Skorobogatova, and N. S. Zefirov, *Phosphorus Sulfur*, 1977, **3**, 213 [allyl alcohols + PhSCl → 1,2-adducts + 1,2-rearrangement products]; Yu. V. Zeifman, L. T. Lantseva, and I. L. Knunyants, *Izv. Akad. Nauk SSSR, Ser. Khim.*, 1978, 946 [(CF₃)₂C=CF₂ → (CF₃)₂C(SR)CF₂Cl]; N. Pociute, D. Greiciute, and L. Rasteikiene, *Liet. TSR Mokslu Akad. Darb., Ser. B*, 1978, 75 (*Chem. Abstr.*, 1979, **89**, 179 518) [1,2-addition to unsaturated ketones]; K. Toyoshima, T. Okuyama, and T. Fueno, *J. Org. Chem.*, 1978, **43**, 2789 [addition of PhSCl to vinyl ethers and sulphides]; A. Chaudhuri, S. K. Bhattacharjee, and P. Sengupta, *Curr. Sci.*, 1978, **47**, 727 [ArSBr + alkene + EtOH → β-ethoxyalkyl sulphides]; F. R. Tantasheva, V. S. Savelev, E. A. Berdnikov, and E. G. Kataev, *Zh. Org. Khim.*, 1978, **14**, 478 [RSCl + vinyl sulphones]; E. V. Komissarova, N. N. Belyaev, and M. D. Stadnichuk, *Zh. Obshch. Khim.*, 1979, **49**, 938; M. D. Stadnichuk, V. A. Ryazantsev, and A. A. Petrov, *ibid.*, p. 956; and V. A. Ryazantsev and M. D. Stadnichuk, *ibid.*, p. 930 [whereas Me₃SiC≡CCH=CH₂ + MeSCl → Me₃SiC(SMe)=CClCH=CH₂, BuᵗC≡CCH=CH₂ gives C=C addition products and ArSeCl + Me₃SiC≡CCH=CH₂ gives a mixture of C≡C and C=C addition products]; G. Capozzi, V. Lucchini, G. Modena, and P. Scrimin. *Nouv. J. Chim.*, 1978, **2**, 95 [RSCl + alkynes in liquid SO₂]; *Substitution reactions of sulphenyl halides* J. F. Harris, *J. Org. Chem.*, 1979, **44**, 563 [free-radical substitution of toluene and saturated alkanes]; A. Chaudhuri and S. K. Bhattacharjee, *Indian J. Chem., Sect. B*, 1979, **18**, 279 [ArSBr + aromatic compounds]; T. S. Croft, *Phosphorus Sulfur*, 1976, **2**, 133 [phenols + CF₃SCl → *ortho*- and *para*-substitution products]; Yu. V. Zeifman, L. T. Lantseva, and I. L. Knunyants, *Izv. Akad. Nauk SSSR, Ser. Khim.*, 1978, 1229 and K. Hiroi, Y. Matsuda, and S. Sato, *Chem. Pharm. Bull.*, 1979, **27**, 2338 [sulphenylation of ketone anions and 1,3-dicarbonyl dianions, respectively; see also A. De Groot and B. J. M. Jansen, *Recl. Trav. Chim. Pays-Bas*, 1979, **98**, 487]; D. Caine and W. D. Samuels, *Tetrahedron Lett.*, 1979, 3609 [regioselectivity of α-phenylsulphenylation of ketone dienolates depends on structure of ketone and nature of reagent (PhSCl, PhSSPh, or PhSSO₂Ph)]; K. Anzai, *J. Heterocycl. Chem.*, 1979, **16**, 567 [RSCl + uracil → 5-(alkylthio)-derivative; indole → 3-substituted derivative]; M. Raban and L.-J. Cherm, *J. Org. Chem.*, 1980, **45**, 1688 [duplication of indole reaction described in the preceding citation]; Yu. V. Zeifman, L. T. Lantseva, and I. L. Knunyants, *Izv. Akad. Nauk SSSR, Ser. Khim.*, 1978, 2640 [bis(trifluoromethyl) carbanions + RSCl]; N. V. Kondratenko, V. I. Popov, L. G. Yurchenko, A. A. Kolomiitsev, and L. M. Yagupolskii, *Zh. Org. Khim.*, 1978, **14**, 1914 [ArSCl + (CF₃)₃CSH → ArSC(CF₃)₃].

[195] S. Raucher, *Tetrahedron Lett.*, 1978, 2261.
[196] G. H. Schmid, S. Yeroushalmi, and D. G. Garratt, *J. Org. Chem.*, 1980, **45**, 910.
[197] I. Fleming, J. Goldhill, and I. Peterson, *Tetrahedron Lett.*, 1979, 3205.
[198] S. Danishefsky and C. F. Yan, *Synth. Commun.*, 1978, **8**, 211.

Preparation from Disulphides, Diselenides, and Ditellurides.—The use of di-sulphides as sulphenylating agents towards alkenes and carbanions continue to offer a distinctive alternative to the use of sulphenyl halides for the same purpose. Although reaction conditions are more severe, cleaner results are usually obtained. The special case of the conversion of disulphides into corresponding sulphides is covered in the later section on Reactions of Disulphides.

Irradiation of a mixture of MeSSMe and $F_2C=CH_2$ with u.v. light gives $MeSCF_2CH_2SMe$, a compound which can be prepared less conveniently from $F_2C=CHBr$ with MeSH and MeSNa.[199] Anodic oxidation of disulphides in the presence of alkenes in MeCN gives β-acetamidoalkyl sulphides:[200]

$$RSSR \xrightarrow{-e^-} RSSR^{\cdot+} \xrightarrow{MeCN} RSN=\overset{+}{C}Me \xrightarrow{R'_2C=CR'_2} MeCONHCR'_2CR'_2SR$$

Related double-functionalization procedures for alkenes have been established for the preparation of 2-acetoxyalkyl phenyl selenides [PhSeSePh with $Cu(OAc)_2$ and AcOH],[201] β-hydroxyalkyl aryl sulphides [ArSSAr with $Pb(OAc)_4$ and TFA, for Markownikov 'trans-hydroxysulphenylation'],[202] and β-ketoalkyl selenides [PhSeSePh with Br_2 and $(Bu^n_3Sn)_2O$].[203]

α-Sulphenylation of carbonyl compounds is achieved by the use of a strong base (e.g. aldehyde + KH → potassium enolate → $PhSCR^1R^2CHO$ with PhSSPh[204]), but in the cyclohexanone series these conditions can be too severe.[205] An extraordinary discovery has been made[205] in a search for alternatives to the usual aprotic reaction media used for the sulphenylation of ketones, namely that in MeOH, with NaOMe as the base, dehydrogenative sulphenylation of cyclohexanones occurs, leading to o-(phenylthio)phenols (7).

(7)

2-Arylsulphenylation of 1,3-dicarbonyl compounds can be effected at room temperature in an $Et_4N^+ F^-$ emulsion,[206] but dianions $R^1\bar{C}HCO\bar{C}HCOR^2$, formed by successive treatment with NaH and Bu^nLi, are sulphenylated at the alternative site by PhSSPh.[207] β-Keto-esters, malonates, and α-cyanoalkanoates can be sulphenylated and decarbalkoxylated in a 'one-pot' process by reaction at 150—160 °C, in HMPA, with PhSSPh and NaI, providing α-(phenylthio)alkyl ketones, α-(phenylthio)alkanoates, and α-cyanoalkyl phenyl sulphides,

[199] D. W. A. Sharp and H. T. Miguel, *Isr. J. Chem.*, 1978, **17**, 144.
[200] A. Bewick, D. E. Coe, J. M. Mellor, and D. J. Walton, *J. Chem. Soc., Chem. Commun.*, 1980, 51.
[201] N. Miyoshi, Y. Ohno, K. Kondo, S. Murai, and N. Sonoda, *Chem. Lett.*, 1979, 1309.
[202] B. M. Trost, M. Ochiai, and P. G. McDougal, *J. Am. Chem. Soc.*, 1978, **100**, 7103.
[203] I. Kuwajima and M. Shimizu, *Tetrahedron Lett.*, 1978, 1277.
[204] P. Groenewegen, H. Kallenberg, and A. van der Gen, *Tetrahedron Lett.*, 1979, 2817.
[205] B. M. Trost and J. H. Rigby, *Tetrahedron Lett.*, 1978, 1667.
[206] J. H. Clark and J. M. Miller, *Can. J. Chem.*, 1978, **56**, 141.
[207] K. Hiroi, Y. Matsuda, and S. Sato, *Synthesis*, 1979, 621.

respectively.[208] The alkyl moiety of the displaced alkoxycarbonyl group appears in the alkyl phenyl sulphide which accompanies the main reaction product.[208] 1,3-Bis(methylthio)-2,2,4,4-tetramethylbicyclo[1.1.0]butane has been established[209] as the product from the parent hydrocarbon with MeSSMe and Bu^nLi, correcting earlier assignments. Sulphenylation of α-phenylalkyl cyanides with PhSSPh and KOH in THF gives PhCR(CN)SPh,[210] but acetonitrile gives the dithioacetal $(PhS)_2CHCN$. Similarly,[211] the ratio of disulphide to substrate must be carefully controlled to avoid *gem*-bis-phenylsulphenylation in the conversion of *o*-toluate esters into *o*-(phenylthiomethyl) analogues. [211] A more classical approach provides an *erythro/threo* mixture of MeSSCHMeSCHMeSMe (a constituent of roast pork) through treatment of 'but-2-ene thio-ozonide' (3,5-dimethyl-1,2,4-trithiolan) with MeLi, then with MeSSMe.[212]

A singlet benzyne intermediate is involved in the conversion of diaryliodonium 2-carboxylates into *o*-diselenides, *o*-ditellurides, and *o*-arylselenoaryl aryl tellurides by reaction with ArSeSeAr, ArTeTeAr, or ArSeTeAr respectively.[213]

Preparation from Thiocyanates, Selenocyanates, and Tellurocyanates.—Less commonly used, but nevertheless well-established, routes to sulphides, selenides, and tellurides are offered by the use of these reagents; more space is devoted to the derivatives of the heavier chalcogens.

Appropriate proportions of alkyl thiocyanate and oxalyl chloride condense to give $(RSCCl=NCO)_2$ or $RSCCl=NCOCOCl$.[214] A valuable route from alkanols to sulphides,[215] using a thiocyanate together with triphenylphosphine (or, better, tri-n-butylphosphine[216]), relies on the rearrangement of an intermediate phosphorane ether $R^1SPPh_3OR^2$, and consequently it is subject to variable behaviour, depending on the structural features in the alcohol. Whereas ArSCN + $HOCH_2CH_2OH$ (Ar = Ph) gives $ArSCH_2CH_2OH$, from which an unsymmetrical 1,2-bis(arylthio)ethane can be obtained by repetition of the process, both hydroxy-groups are substituted in the case of Ar = p-$NO_2C_6H_4$, and neither hydroxy-group is substituted when Ar is p-$MeOC_6H_4$.[216] The same reaction is suitable for the conversion of alk-2-enols into p-nitrophenyl selenides,[217] thus avoiding the 1,3-selenoallylic rearrangements which occur in other routes.

Copper(I or II)-catalysed addition of PhSeCN to an alkene, followed by reaction with an alcohol, gives *trans*-β-alkoxyalkyl phenyl selenides.[218] An extension of this reaction has been used[219] to convert cyclo-octa-1,5-diene into a mixture of the isomeric oxygen-bridged 1,4- and 1,5-bis(phenylseleno)cyclo-octanes. The first

[208] M. Asaoka, K. Miyake, and H. Takei, *Bull. Chem. Soc. Jpn.*, 1978, **51**, 3008.

[209] P. G. Gassmann and M. J. Mullins, *Tetrahedron Lett.*, 1979, 4457.

[210] E. Marchaud, G. Morel, and A. Foucaud, *Synthesis*, 1978, 360.

[211] F. M. Hauser, R. P. Rhee, S. Prasanna, S. M. Weinreb, and J. H. Dodd, *Synthesis*, 1980, 72.

[212] P. Dubs and M. Joho, *Helv. Chim. Acta*, 1978, **61**, 2809.

[213] J. B. S. Bouilha, N. Petragnani, and V. G. Toscano, *Chem. Ber.*, 1978, **111**, 2510.

[214] W. Boehmer and D. Herrmann, *Liebigs Ann. Chem.*, 1978, 1704.

[215] W. T. Flowers, G. Holt, F. Omogbai, and C. P. Poulos, *J. Chem. Soc., Perkin Trans. 1*, 1979, 1309.

[216] K. A. M. Walker, *Tetrahedron Lett.*, 1977, 4475.

[217] D. L. J. Clive, G. Chittattu, N. J. Curtis, and S. M. Menchen, *J. Chem. Soc., Chem. Commun.*, 1978, 770.

[218] A. Toshimitsu, T. Aoai, S. Uemura, and M. Okano, *J. Org. Chem.*, 1980, **45**, 1953.

[219] S. Uemura, A. Toshimitsu, T. Aoai, and M. Okano, *J. Chem. Soc., Chem. Commun.*, 1979, 610.

example of the use of PhCH$_2$TeCN as a *C*-benzyltellurenylation agent[220] gives full details of the synthesis of benzyl *o*-nitrophenyl telluride from *o*-NO$_2$C$_6$H$_4$Li.

Net *syn*-addition, with Markownikov orientation, has been established for the addition of PhSeSCN to (*E*)- and (*Z*)-1-phenylpropene under kinetic control, further interest arising in the fact that the (*Z*)-isomer leads to 2-(phenylselenyl)-1-phenylpropyl thiocyanate, while the (*E*)-isomer gives the corresponding iso-thiocyanate.[221]

Preparation of Sulphides, Selenides, and Tellurides using Other Sulphenylation, Selenenylation, and Tellurenylation Reagents.

—Sulphenamides are effective re-agents for α-heteroarylsulphenylation of β-keto-esters.[222] Sulphenamide inter-mediates are thought to be involved in the asymmetric 2-phenylsulphenylation of 4-alkylcyclohexanones in moderate optical yields, using a benzenesulphenyl halide with a chiral secondary alkylamine.[223]

Thiolsulphonates are effective in sulphenylation reactions, having been used recently for the stereoselective synthesis of allyl vinyl sulphides from RCH=CH-AlBu$^i_2$Bu^n.[224] Mild reaction conditions are called for in the sulphenylation of active-methylene compounds by (ArSO$_2$)$_2$CHSSO$_2$Ph.[225]

Alkyl potassium xanthates R^{1}OCSSK react with organic halides R^{2}X to give sulphides R^{1}SR^2 (R^1 = Et, R^2 = 4-trifluoromethyl-2,6-dinitrophenyl[226] and R^1,R^2 are various alkyl[227]). Under phase-transfer conditions, this is an efficient 'one-pot' synthesis of unsymmetrical sulphides.[227]

Important uses of selenenic and seleninic acids and anhydrides (see also the later sections) have been established, in the synthesis of β-silyloxy-α-(phenyl-seleno)aldehydes from allyl silyl ethers (Scheme 3)[228] and in the high-yield syntheses of 2-acetoxyalkyl phenyl selenides [using PhSe(O)OH and AcOH[229]] and 2-hydroxyalkyl selenides [using RSe(O)OH and H$_3$PO$_2$[230]] from alkenes. The effective reagent in the RSe(O)OH–H$_3$PO$_2$ system is assumed[230] to be the selenenic acid RSeOH, but this probably needs further consideration.

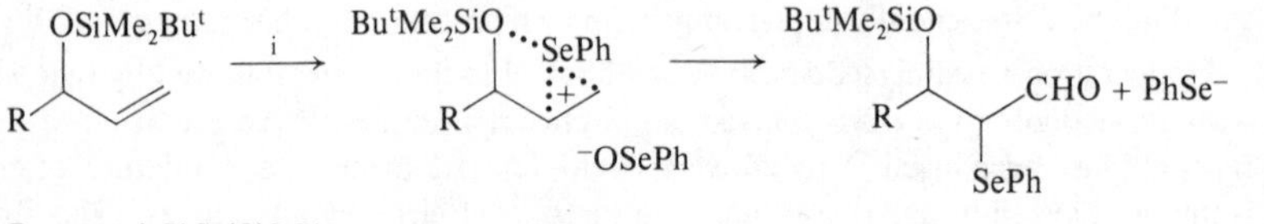

Reagents: i, PhSe(O)SePh

Scheme 3

[220] P. Wiriyachitra, S. J. Falcone, and M. P. Cava, *J. Org. Chem.*, 1979, **44**, 3957.
[221] D. G. Garratt, *Can J. Chem.*, 1979, **57**, 2180.
[222] S. Torii, H. Tanaka, and H. Okumoto, *Bull. Chem. Soc. Jpn.*, 1979, **52**, 267.
[223] K. Hiroi, M. Nishida, A. Nakayama, K. Nakazawa, E. Fujii, and S. Sato, *Chem. Lett.*, 1979, 969.
[224] A. P. Kozikowski, A. Ames, and H. Wetter, *J. Organomet. Chem.*, 1978, **164**, C33.
[225] A. Senning, *Synthesis*, 1980, 412.
[226] J. J. D'Amico, C. C. Tung, W. E. Dahl, and D. J. Dahm, *Phosphorus Sulfur*, 1978, **4**, 267.
[227] I. Degani, R. Fochi, and V. Regondi, *Synthesis*, 1979, 178.
[228] M. Shimizu, R. Takeda, and I. Kuwajima, *Tetrahedron Lett.*, 1979, 3461.
[229] N. Miyoshi, Y. Takai, S. Murai, and N. Sonoda, *Bull. Chem. Soc. Jpn.*, 1978, **51**, 1265.
[230] D. Labar, A. Krief, and L. Hevesi, *Tetrahedron Lett.*, 1978, 3967.

Preparation of Sulphides, Selenides, and Tellurides using the Elements Themselves, or Other Inorganic Sulphur, Selenium, and Tellurium Compounds.—Symmetrical sulphides can be obtained in 12—66% yields from an alkyl chloride, sulphur, and NaOH in DMSO.[231] The same system with a benzyl cyanide R^1PhCHCN produces the unsymmetrical sulphides R^1PhC(CN)SR^2 with an alkyl halide R^2X;[231,232] some disulphide is also formed in this reaction.[232] Lithium sulphide and persulphide, (which are formed *in situ* from S_8 and $LiEt_3BH$, and which remain dissolved in THF) give sulphides and disulphides, respectively, with alkyl halides.[233] Continuing studies are represented in the lithiation of 3,5-disubstituted thien-2-yl sulphides followed by reaction with S_8 and MeI[234] and in the synthesis of sulphides[235] from alkynes $[ArSO_2CH_2CH=CMeC\equiv CH + Na_2S \rightarrow (ArSO_2CH_2CH=CMeC\equiv C)_2S]$. Routine kinetic studies of the reactions of Bu^nCl with Na_2S and with Bu^nSNa have been reported.[236] Sulphides can be prepared from alkenes and SCl_2.[625]

Selenides and tellurides are formed through the reaction sequences $PhCH_2MgCl + Se_8 \rightarrow PhCH_2SeMgCl$, which with $RN_2^+ Cl^- \rightarrow PhCH_2SeR$,[237] and thiophens $+ Te_8 + MeI \rightarrow$ thienyl methyl tellurides.[238] Unexpected selenide by-products have, paradoxically, become common in oxidations by SeO_2, and the selenides (9) and (10) accompany the expected 1,2-diketone in the oxidation of (8) with SeO_2 in dioxan.[239]

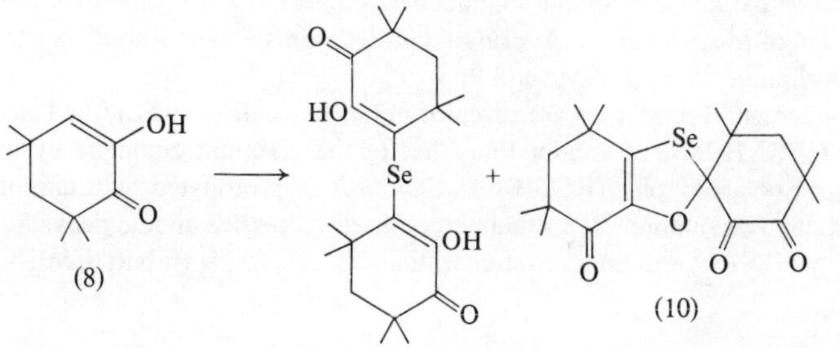

Miscellaneous Methods of Preparation.—This Chapter is interwoven with accounts of the reactions of sulphur, selenium, and tellurium compounds which amount to the conversion of one sulphur-containing functional group into another. This small section is included to collect together a variety of methods for the synthesis of sulphides from sulphur compounds other than those mentioned in the preceding sections.

Thiones participate in an ene-reaction with alkylidenecyclohexanes and methyl-

[231] A. Jonczyk, *Angew. Chem.*, 1979, **91**, 228.

[232] E. Marchaud, G. Morel, and A. Foucaud, *C.R. Hebd. Seances Acad. Sci., Ser. C*, 1979, **289**, 57.

[233] J. A. Gladysz, V. K. Wong, and B. S. Jick, *J. Chem. Soc., Chem. Commun.*, 1978, 838.

[234] Ya. L. Gol'dfarb and M. A. Kalik, *Zh. Org. Khim.*, 1978, **14**, 2603.

[235] A. D. Bulat, M. A. Antipov, and B. V. Passet, *Zh. Org. Khim.*, 1979, **15**, 2225.

[236] S. R. Rafikov, R. S. Aleev, A. L. Dobrikov, S. N. Salazkin, and V. D. Komissarov, *Izv. Akad. Nauk SSSR, Ser. Khim.*, 1978, 2812.

[237] V. P. Krasnov, V. I. Naddaka, V. P. Garkin, and V. I. Minkin, *Zh. Org. Khim.*, 1978, **14**, 2620.

[238] N. Dereu and J. L. Piette, *Bull. Soc. Chim. Fr., Part 2*, 1979, 623.

[239] T. Laitalainen, T. Simonen, and R. Kivekas, *Tetrahedron Lett.*, 1978, 3079.

cyclohexane to give allyl sulphides.[240] An unusual reaction has been described[241] in which pyridine-4-thione gives di-(4-pyridyl) sulphide in refluxing decalin; pyridine-2-thione reacts similarly. Addition of pyridine-3-thiol to the reaction mixtures gives crossed products. Studies of 1,2- and 1,3-bis(dimethylamino)cyclobutene-dithiones (see Vol. 4, p. 9) have continued, and reactions with MeSH to give various products, including methyl sulphides, have been described.[242] Diphenylphosphino-dithioate esters $Ph_2P(S)SR$ react with alkyl-lithiums at $-78\,°C$ to establish a general synthesis of unsymmetrical sulphides.[243]

Examples of previously described methods are the ring-opening of thiirans [ethylene sulphide with Pr^nONa and $H_2C=CHCH_2Cl \rightarrow Pr^nO(CH_2CH_2S)_n$-$CH_2CH=CH_2$ ($n = 1$ or 2)],[244] the use of a chlorosulphonium salt ($RCO_2^-Et_3NH^+$ + $Me_2\overset{+}{S}Cl\ X^- \rightarrow RCO_2CH_2SMe$),[245] C–Se bond cleavage of a diseleno-acetal [$(PhSe)_2CRSiMe_3 + Bu^nLi \rightarrow PhSeCRLi(SiMe_3)$],[246] cleavage of sulphimides [$R^1R^2S=NR^3 + P_4S_{10} \rightarrow R^1SR^2$],[247] and methylthiomethylation of phenols (phenols + $Me_2S=NR \rightarrow o$-methylthiomethyl-phenols).[248]

Reactions of Sulphides: Simple Reactions and Fundamental Properties.—This title is used as a collecting point for papers describing the pyrolysis of Me_2S or thiiran (giving $H_2C=CHSH$ and thiophen),[249] hydrogen abstraction from Me_2S by $CF_3\cdot$ radical,[250] oxidation of simple sulphides by singlet oxygen[251] or ozone,[252] C–S bond cleavage of Me_2S by $H\cdot$[253] or using recoil ^3H atoms,[254] and a study of the exchange equilibrium involving PhSMe and PhS$^-$.[255]

Fundamental properties are revealed in kinetic acidity studies (the kinetic acidity of m-$CF_3C_6H_4SMe$ is greater than that of the selenium analogue by a factor of 1.38),[256] polarography ($RSCH_2CH_2COCO_2R$ is protonated at a carbonyl group rather than at sulphur),[257] voltammetry (three successive anodic peaks for Ph_2Se at Pt in MeCN),[258] and complexation enthalpies of $1,\omega$-bis-(n-butylthio)alkanes with

[240] Y. Inoue and D. J. Burton, *J. Fluorine Chem.*, 1979, **14**, 89.

[241] K. Krowicki, *Pol. J. Chem.*, 1978, **52**, 2349.

[242] G. Seitz, R. Schmiedel, and K. Mann, *Arch. Pharm.* (*Weinheim, Ger.*), 1977, **310**, 991; G. Seitz, K. Mann, R. Schmiedel, and R. Matusch, *Chem. Ber.*, 1979, **112**, 990; R. Matusch, R. Schmiedel, and G. Seitz, *Liebigs Ann. Chem.*, 1979, 595.

[243] M. Yoshifuji, F. Hanafusa, and N. Inamoto, *Chem. Lett.*, 1979, 723; K. Goda, F. Hanafusa, and N. Inamoto, *Bull. Chem. Soc. Jpn.*, 1978, **51**, 818.

[244] A. A. Rodin, K. A. Vyunov, A. I. Ginak, and E. G. Sochilin, *Zh. Org. Khim.*, 1979, **15**, 2252.

[245] T.-L. Ho, *Synth. Commun.*, 1979, **9**, 267.

[246] D. Van Ende, W. Dumont, and A. Krief, *J. Organomet. Chem.*, 1978, **149**, C10.

[247] I. W. J. Still and K. Turnbull, *Synthesis*, 1978, 540.

[248] T. Yunamoto and M. Okawara, *Bull. Chem. Soc. Jpn.*, 1978, **51**, 2443.

[249] E. N. Deryagina, E. N. Sukhomazova, O. B. Bannikova, and M. G. Voronkov, *Izv. Akad. Nauk SSSR, Ser. Khim.*, 1979, 2103.

[250] N. L. Arthur and K. S. Yeo, *Aust. J. Chem.*, 1979, **32**, 2077.

[251] B. M. Monroe, *Photochem. Photobiol.*, 1979, **29**, 761; M. L. Kacher and C. S. Foote, *ibid.*, p. 765.

[252] R. I. Martinez and J. T. Herron, *Int. J. Chem. Kinet.*, 1978, **10**, 433.

[253] T. Yokata and O. P. Strausz, *J. Phys. Chem.*, 1979, **83**, 3196.

[254] M. Casiglioni and P. Volpe, *Gazz. Chim. Ital.*, 1979, **109**, 187.

[255] E. S. Lewis and S. Kukes, *J. Am. Chem. Soc.*, 1979, **101**, 417.

[256] H. J. Reich, F. Chow, and S. K. Shah, *J. Am. Chem. Soc.*, 1979, **101**, 6638.

[257] J. Moiroux, *Electrochim. Acta*, 1978, **23**, 571.

[258] R. Seeber, A. Cinquantini, P. Zanello, and G. A. Mazzocchin, *J. Electroanal. Chem. Interfacial Electrochem.*, 1978, **88**, 137.

BBr$_3$, AlBr$_3$, and GaCl$_3$[259] (a topic distantly related to an analysis of g.l.c. data for 39 n-alkyl isoalkyl sulphides[260]).

Reactions of Sulphides, Selenides, and Tellurides: Rearrangements, and C—S, C—Se, and C—Te Bond Cleavage Reactions.—There is some overlap between this and later sections, since many uses of sulphides, selenides, and tellurides in synthesis involve rearrangements, and most uses in synthesis depend on a clean cleavage of the sulphide, selenide, or telluride functional group.

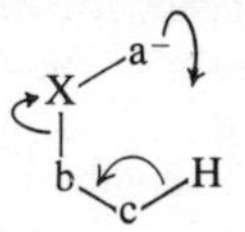

(11)

An important approach to alkene synthesis, using the so-called αβ-elimination (11) of α-sulphenyl carbanions, has been studied for the synthesis of *cis*-cyclo-octene from allyl, benzyl, and methyl cyclo-octyl sulphides.[261] Photorearrangement of benzyl phenyl sulphides[262] and the corresponding selenides,[263] followed by methylation, gives o-arylphenyl methyl sulphides and selenides (Scheme 4). Further studies of the Claisen rearrangements of allyl 2- and 3-thienyl sulphides, leading to 3-(2-)allylthiophen-2-(3-)thiols, respectively, and their cyclization products, have been described.[264] 1,3-Shifts of alkylthio- and arylthio-groups which occur following the creation of a carbonium ion at the α-carbon atom (or an equivalent process) in alkyl benzyl or phenyl sulphides continue to be studied in several laboratories. Acetolysis of [^{13}C]diazomethyl 1-(phenylthio)alkyl ketones gives products resulting from a 1,3-shift of PhS (PhSCH$_2$CO^{13}CHN$_2$ → PhSCH$_2$CO^{13}CH$_2$OAc + PhS^{13}CH$_2$COCH$_2$OAc). The corresponding shift does not occur in the reaction

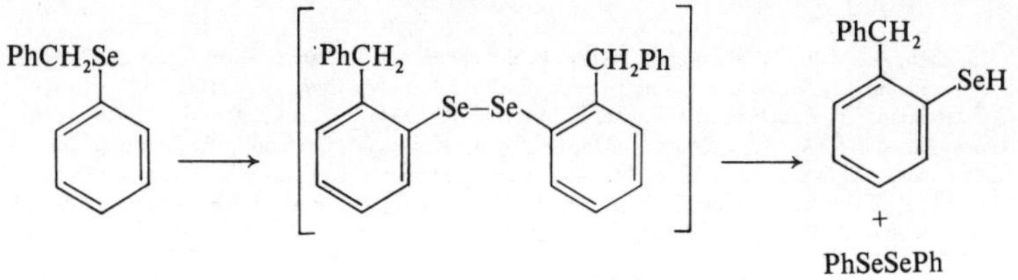

Scheme 4

[259] L. A. Ganyushin, E. N. Guryanova, and I. P. Romm, *Zh. Obshch. Khim.*, 1978, **48**, 2478.
[260] R. V. Golovnya, V. G. Garbuzov, and T. A. Misharina, *Izv. Akad. Nauk SSSR, Ser. Khim.*, 1978, 387.
[261] J. F. Biellmann, H. d'Orchymont, and J. L. Schmitt, *J. Am. Chem. Soc.*, 1979, **101**, 3283.
[262] J.-L. Fourrey and P. Jouin, *J. Org. Chem.*, 1979, **44**, 1892.
[263] J.-L. Fourrey, G. Henry, and P. Jouin, *Tetrahedron Lett.*, 1980, **21**, 455.
[264] A. V. Anisimov, V. Fionova, and E. A. Viktorova, *Khim. Geterotsikl. Soedin.*, 1978, 186; A. V. Anisimov, V. F. Ionova, V. K. Govorek, V. S. Babaitsev, and E. A. Viktorova, *Dokl. Akad. Nauk SSSR*, 1979, **244**, 362; A. V. Anisimov, V. F. Ionova, V. S. Babaitsev, and E. A. Viktorova, *Zh. Org. Khim.*, 1979, **15**, 882; A. V. Anisimov, V. F. Ionova, and E. A. Viktorova, *ibid.*, p. 1970.

with the oxygen analogue.[265] Extensive studies[266] of equivalent aminolysis and solvolysis studies of *erythro/threo*-β-chloroalkyl phenyl sulphides, in which 1,3-shifts are accompanied by E1cb elimination, leading to vinyl sulphides, indicate generally lower rates for the *erythro*-isomers. Detailed assessment of the effect of structure has been reported[267] for the methanolysis of benzyl γ-toluene-*p*-sulphonyloxyalkyl sulphides, $PhCH_2SCR^1R^2CH_2CHR^3OTs$, indicating that at least two of the three groups R^1, R^2, and R^3 should be an alkyl group if migration of $PhCH_2S$ is to occur (a good illustration of the Thorpe–Ingold effect). There is clearly considerable scope for rationalization, with respect to structure and reaction conditions, of the results which have accumulated on the tendency for 1,3-shifts to occur; the corresponding rearrangements of allylic sulphides are discussed in the later section dealing with unsaturated sulphides.

Shifts from S to N in 1-alkyl-2-alkylthio- or -arylthio-imidazoles have been studied.[268]

2-(Alkylthio)cycloalkanones give C–S cleavage products on photolysis in MeOH, in interesting contrast with the behaviour of 2-cyano-analogues, which yield ω-cyanoalkanoate esters.[269] Photolysis of 4-alkenyl sulphides, leading to mixtures of thians and thiolans, involves reversible C–S bond homolysis,[270] while insertion into the C–S bond by carbenes through photolysis of ethyl α-(trimethyl-stannyl)diazoacetate in Me_2S leads predominantly (35%) to Me_3SnCMe-(SMe)CO_2Et, with 12% of $MeSCHMeCO_2Et$.[271] Electrochemical cleavage processes are represented in the formation of thiols and alkanes,[272, 273] and in the oxidative decarboxylation and cleavage of α-(phenylthio)alkanoic acids in alcoholic solutions, leading to aldehydes and acetals.[274]

Reductive cleavage of sulphides, selenides, and tellurides on Raney nickel[275] and cleavage by lithium in THF, leading to secondary or tertiary alkyl-lithium compounds,[276] are useful synthetic operations; novel variations are involved in the cleavage of dialkenyl sulphides with Et_3SiH (giving low yields of silyl sulphides)[277] and of alkenyl, aryl, and allyl sulphides and selenides with PPh_3 and

[265] S. Gladiali, A. Pusino, V. Rosnati, A. Saba, F. Soccolini, and A. Selva, *Gazz. Chim. Ital.*, 1977, **107**, 535; A. Pusino, V. Rosnati, A. Saba, F. Soccolini, and A. Selva, *ibid.*, 1978, **108**, 557.

[266] L. Rasteikiene, V. Zabelaite-Miskiniene, E. Stumbreviciute, M. G. Linkova, and I. L. Knunyants, *Izv. Akad. Nauk SSSR, Ser. Khim.*, 1978, 1099; J. Kulis, D. Greiciute, S. Jonusauskas, and L. Rasteikiene, *Zh. Org. Khim.*, 1978, **14**, 1492; L. Rasteikiene, J. Kulis, and V. Vidugiriene, *ibid.*, 1979, **15**, 531; Z. Talaikyte, V. Vidugiriene, and L. Rasteikiene, *Zh. Vses. Khim., O-va.*, 1979, **24**, 102 (*Chem. Abstr.*, 1979, **90**, 185 962); D. Greiciute, J. Kulis, and L. Rasteikiene, *Phosphorus Sulfur*, 1977, **3**, 261.

[267] E. L. Eliel, W. H. Pearson, L. M. Jewell, A. G. Abatjoglou, and W. R. Kenan, *Tetrahedron Lett.*, 1980, **21**, 331.

[268] J. Kister, G. Assef, G. Mille, and J. Metzger, *Can. J. Chem.*, 1979, **57**, 822.

[269] M. Tokuda, Y. Watanabe, and M. Itoh, *Bull. Chem. Soc. Jpn.*, 1978, **51**, 905.

[270] G. Bastien and J. M. Surzur, *Bull. Soc. Chim. Fr., Part 2*, 1979, 601; G. Bastien, M. P. Crozet, E. Flesia, and J. M. Surzur, *ibid.*, p. 606.

[271] W. Ando, M. Takata, and A. Sekiguchi, *J. Chem. Soc., Chem. Commun.*, 1979, 1121.

[272] M. Miyake, Y. Nakayama, M. Nomura, and S. Kikkawa, *Bull. Chem. Soc. Jpn.*, 1979, **52**, 559.

[273] G. Farnia, M. G. Severin, G. Capobianco, and E. Vianello, *J. Chem. Soc., Perkin Trans. 2*, 1978, 1.

[274] J. Nokami, M. Kawada, R. Okawara, S. Torii, and H. Tanaka, *Tetrahedron Lett.*, 1979, 1045.

[275] L. Horner and G. Doms, *Phosphorus Sulfur*, 1978, **4**, 259.

[276] C. G. Screttas and M. Micha-Screttas, *J. Org. Chem.*, 1979, **44**, 713.

[277] M. G. Voronkov, S. A. Bolshakova, N. N. Vlasova, S. V. Amosova, and B. A. Trofimov, *Zh. Obshch. Khim.*, 1979, **49**, 1914.

NiCl$_2$, leading to alkenes and arenes.[278] In a comparative assessment of the PPh$_3$–NiCl$_2$ system, the order of bond cleavage PhSeMe $\gg$ PhCl > PhSMe was established.[278] Several papers describe the conversion of β-hydroxyalkyl sulphides, readily obtained from α-(alkylthio)alkyl-lithium compounds and carbonyl compounds, into alkenes, using P$_2$I$_4$, PI$_3$, or SOCl$_2$,[279] electrochemical reduction,[280] or 1-ethyl-2-fluoropyridinium tetrafluoroborate and LiI.[281] The analogous process with β-hydroxyalkyl selenides, using MeSO$_2$Cl and Et$_3$N, has also been described.[256] Mild catalytic oxidation, employing physiological conditions {O$_2$ with [R$_2$CO^{II}], a cytochrome system, or [Fe(ClO$_4$)$_2$] with ascorbic acid}, of alkyl phenyl sulphides in which the alkyl group carries an α-hydrogen atom leads to products from C–S bond cleavage (PhSCHR^{1}R^2 → PhSSPh + R^{1}COR^2) and sulphoxides.[282] Whereas β-keto-alkyl sulphides are converted into α-keto-aldehyde dimethyl acetals by [Tl(NO$_3$)$_3$] in MeOH, the same treatment of β-keto-alkyl selenides gives $\alpha\beta$-unsaturated ketones,[283] as does the oxidation of β-hydroxyalkyl selenides with Jones' reagent.[284] Oxidation without C–Se bond cleavage occurs, however, when β-hydroxyalkyl selenides are treated with N-succinimidyldimethylsulphonium chloride and Et$_3$N, providing a useful synthesis of β-keto-alkyl selenides.[284]

Free-radical intermediates are involved in the pyrolytic cleavage of C–S bonds of 5-(methylthio)imidazoles, *i.e.* (12) → (13).[285]

(12)   (13)

Reactions of Sulphides, Selenides, and Tellurides: Effects of Neighbouring Functional Groups.

—Kinetic studies of dehydrochlorination reactions in the gas phase reveal participation of neighbouring groups in aliphatic 2-chloroethyl sulphides.[286] Physical studies[287] indicate interactions of functional groups in β-keto-sulphides and in the corresponding nitriles. A more unusual example of participation by sulphur is the enhanced rate of reduction by I$^-$ of γ-(methylthio)alkyl methyl sulphoxides.[288]

Activation of the adjacent proton at α-sp^3 centres has been exploited in uses of sulphides in synthesis (see later section), and must be the basis for a novel acetoxyl-transfer reaction, (14) → (15), that is observed in refluxing benzene

[278] H. Okamura, M. Miura, K. Kosugi, and K. Takei, *Tetrahedron Lett.*, 1980, **21**, 87.
[279] J. N. Denis, W. Dumont, and A. Krief, *Tetrahedron Lett.*, 1979, 4111.
[280] T. Shono, Y. Matsumura, S. Kashimura, and H. Kyutoku, *Tetrahedron Lett.*, 1978, 2807.
[281] T. Mukaiyama and M. Imaoka, *Chem. Lett.*, 1978, 413.
[282] T. Numata, Y. Watanabe, and S. Oae, *Tetrahedron Lett.*, 1979, 1411; 1978, 4933.
[283] Y. Nagao, K. Kaneko, and E. Fujita, *Tetrahedron Lett.*, 1978, 4115.
[284] J. Lucchetti and A. Krief, *C.R. Hebd. Seances Acad. Sci., Ser. C*, 1979, **288**, 537.
[285] D. Gyorgy, J. Nyitrai, K. Lempert, W. Voelter, and H. Horn, *Chem. Ber.*, 1978, **111**, 1464.
[286] G. Chuchani, I. Martin, and D. B. Bigley, *Int. J. Chem. Kinet.*, 1978, **10**, 649.
[287] K. C. Kole, C. Sandorfy, M. T. Fabi, P. R. Olivato, R. Rittner, C. Trufem, H. Viertler, and B. Wladislaw, *J. Chem. Soc., Perkin Trans. 2*, 1977, 2025.
[288] J. T. Doi and W. K. Musker, *J. Am. Chem. Soc.*, 1978, **100**, 3533.

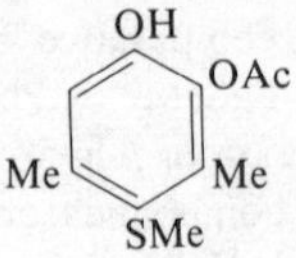

(14) 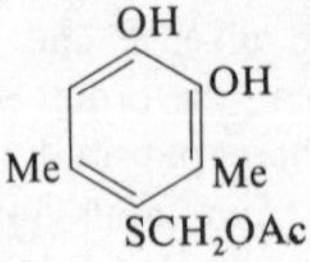   (15)

solution in the presence of Et_3N.[289] Simple examples of reactions employing sulphenyl-stabilized carbanions are Michael reactions with PhSCHLiCN[290] and Knoevenagel condensation of $MeSCH_2CN$, using 1-(methylthio)-1-alkenyl cyanides.[291] These compounds are also susceptible to reduction by $NaBH_4$ $[\rightarrow RCH_2CH(SMe)CN]$ and desulphurization with Raney nickel ($\rightarrow RCH=CHCN$).

Arylthio-carbenes yield arylthiocyclopropanes with alkenes,[292,293] and add to alkynes[293] and to allyl sulphides.[294] Scheme 5 illustrates a convenient route to vinyl(arylthio)carbenes and their uses in synthesis.

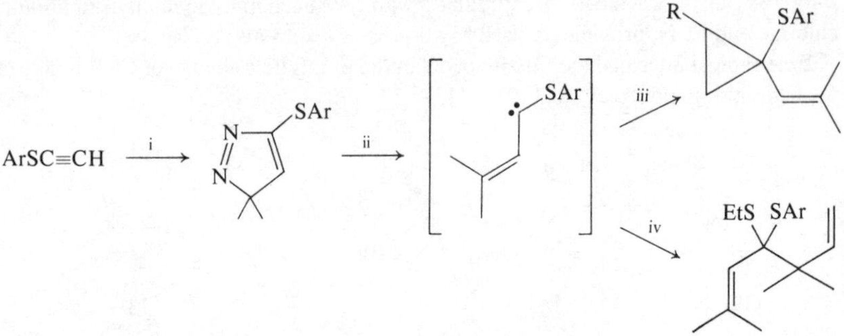

Reagents: i, Me_2CN_2, ether, at 0 °C; ii, $h\nu$; iii, $RCH=CH_2$; iv, $EtSCH_2CH=CMe_2$

Scheme 5

Evidence for the existence in solution of free bis(phenylthio)carbene, formed by the decomposition of $(PhS)_3CLi$, has been reported.[295]

Uses of Saturated Sulphides, Selenides, and Tellurides in Synthesis.—These applications are mostly based on the C—C bond-forming reactions of sulphenyl carbanions; the extensive uses of vinyl sulphides and their allyl and propargyl analogues are discussed in a later section. Some methods for C—S or C—Se bond cleavage, which are used to terminate a use in synthesis, have been discussed in an earlier section; other methods, notably the conversion of the sulphide into a sulphoxide followed by pyrolytic elimination of the sulphenic acid, are covered in later sections.

[289] R. R. King, *J. Org. Chem.*, 1979, **44**, 4194.
[290] N. Wang, S. Su, and L. Tsai, *Tetrahedron Lett.*, 1979, 1121.
[291] S. Kano, T. Yokomatsu, T. Ono, S. Hibino, and S. Shibuya, *Chem. Pharm. Bull.*, 1978, **26**, 1874.
[292] T. Balaji and D. B. Reddy, *Bull. Chem. Soc. Jpn.*, 1979, **52**, 3434.
[293] M. Franck-Neumann and J. J. Lohmann, *Tetrahedron Lett.*, 1979, 2075.
[294] M. Franck-Neumann and J. J. Lohmann, *Tetrahedron Lett.*, 1978, 3729.
[295] M. Nitsche, D. Seebach, and A. K. Beck, *Chem. Ber.*, 1978, **111**, 3644.

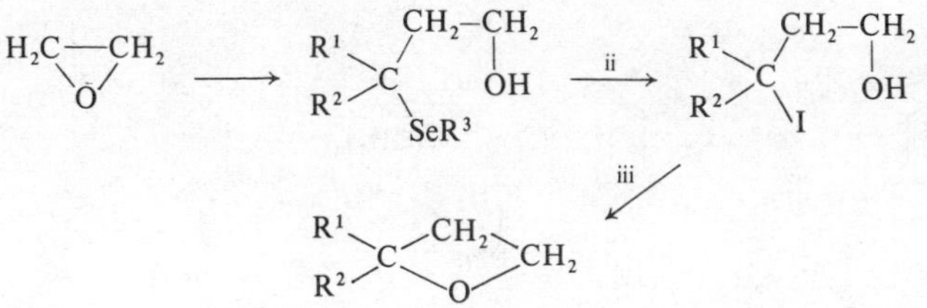

Reagents: i, $R^1R^2C(SeR^3)Li$; ii, MeI, NaI, DMF, $CaCO_3$, at 80 °C; iii, Bu^tOK, DMSO, at 20 °C

Scheme 6

'One-carbon' homologations of unusual types are illustrated by the conversion of an oxiran into an oxetan (Scheme 6) through the use of α-lithioalkyl selenides,[296] a method which can, in principle, involve higher members of the saturated cyclic ether series. Further examples of uses of α-lithioalkyl selenides[297] have been described: 1,2-addition of $RSeCHLiCO_2Et$ to cyclohexenone;[297] the use of this reagent for the synthesis of $\alpha\beta$-unsaturated esters by condensation with carbonyl compounds;[298] the formation of alkylidenecyclopropanes from α-lithiocyclopropyl phenyl selenide and primary alkyl halides;[299] the use of $RSeCH_2Li$ for the conversion of alkyl halides into the next higher homologous 1-alkene;[299] and the use of the dianion of $PhSeCHMeCO_2H$ for the conversion of an oxiran into an α-methylene-γ-lactone.[300] Corresponding uses of sulphides have also been reported. α-Lithiocyclopropyl phenyl sulphide reacts with formaldehyde to give α-(hydroxymethyl)cyclopropyl phenyl sulphide, which gives cyclobutanone with TsOH and $HgCl_2$, in 56% overall yield.[301] The same reagent adds to ketones analogously, and subsequent cleavage of the cyclopropane ring with HCl and $ZnCl_2$ gives γ-keto-sulphides ($R^1COR^2 \rightarrow R^1CHR^2COCH_2CH_2SPh$).[302] It may also be added to $\alpha\beta$-unsaturated ketones, in which case subsequent cleavage with HBr and $ZnBr_2$ gives 1-(phenylthio)buta-1,3-dienes $[R^1R^2C=CHCOR^3 \rightarrow R^1R^2C=CHCR^3=C(SPh)CH_2CH_2Br]$.[303] α-Sulphenyl carbanions ($R\bar{C}HSPh$) and $BrCH_2CH=CHCO_2Et$ have been utilized in a synthesis of ethyl 2-(phenylthiomethyl)cyclopropanecarboxylate.[304] The use of α-chloroalkyl sulphides can lead to corresponding products; $ClCH_2SPh$ brings about $TiCl_4$-promoted phenylthiomethylation of silyl enol ethers and enolates, subsequent alkene-forming desulphurization resulting in overall introduction of a methylene group into the α-position of a ketone.[305,306] A more routine use of an α-chloromethyl sulphide, *i.e.* the conversion of alcohols into methylthiomethyl ethers,[307] is facilitated by having $AgNO_3$ and Et_3N in the reaction mixture.

[296] M. Sevrin and A. Krief, *Tetrahedron Lett.*, 1980, **21**, 585.
[297] J. Lucchetti and A. Krief, *Tetrahedron Lett.*, 1978, 2697.
[298] J. Lucchetti and A. Krief, *Tetrahedron Lett.*, 1978, 2693.
[299] S. Halazy and A. Krief, *Tetrahedron Lett.*, 1979, 4233.
[300] N. Petragnani and H. M. C. Ferraz, *Synthesis*, 1978, 476.
[301] B. M. Trost and W. C. Venduchick, *Synthesis*, 1978, 821.
[302] R. D. Miller and D. R. McKean, *Tetrahedron Lett.*, 1979, 583.
[303] R. D. Miller, D. R. McKean, and D. Kaufmann, *Tetrahedron Lett.*, 1979, 587.
[304] E. Ghera and Y. Ben-David, *Tetrahedron Lett.*, 1979, 4603.
[305] I. Paterson and I. Fleming, *Tetrahedron Lett.*, 1979, 2179.
[306] I. Paterson and I. Fleming, *Tetrahedron Lett.*, 1979, 993, 995.
[307] K. Suzuki, J. Inanaga, and M. Yamaguchi, *Chem. Lett.*, 1979, 1277.

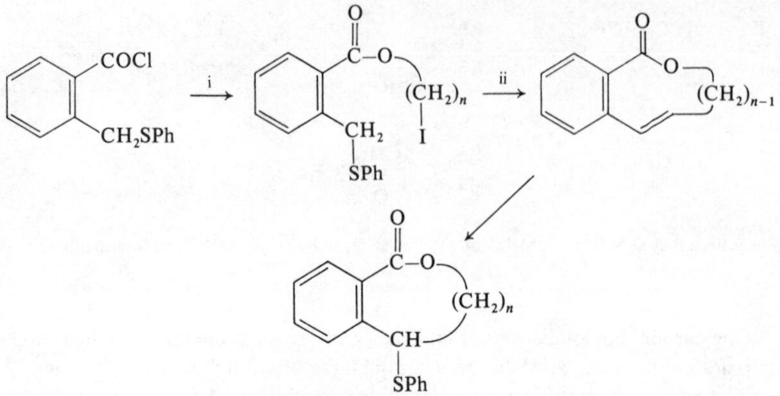

Reagents: i, HO(CH$_2$)$_n$I; ii, K(Me$_3$SiN)$_2$, THF

Scheme 7

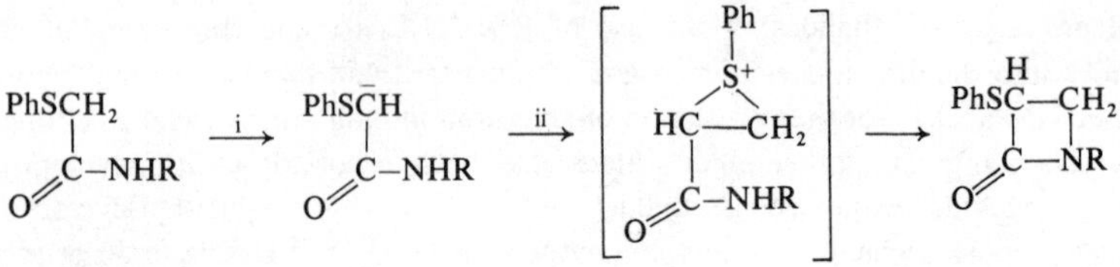

Reagents: i, NaH; ii, CH$_2$I$_2$

Scheme 8

An application of sulphides in the synthesis of natural products, reminiscent of an earlier era of organic chemistry, is the functional group conversion RCO$_2$Et → RCH$_2$OTs → RCH$_2$SPh → RCH$_3$ (the last step being brought about with Li and EtNH$_2$), used in a synthesis of (+)-(3R,4R)-4-methyl-3-heptanol, a pheromone of the European elm bark beetle (*Scolytus multistriatus*).[308] Cyclization methods based on α-sulphenyl carbanions include a new macrolide synthesis (Scheme 7),[309] cyclization of (2E,6Z)-10,11-epoxyfarnesyl phenyl sulphide with Bu^{n}Li (in DABCO, at −78 °C) to form (6Z)-hedycaryols,[310] and a convenient [3 + 1]-cyclization (Scheme 8) that leads to β-lactams.[311] Cyclopropanes are formed by the electrolysis of β-arylthioalkyl methanesulphonates, themselves formed in straightforward ways from αβ-unsaturated ketones *via* β-keto-sulphides.[312] The opposite approach, *i.e.* ring-opening of γ-phenylthiomethyl- or γ-phenylseleno-methyl-γ-lactones by sodium in liquid NH$_3$, as well as the cleavage of ω-phenyl-selenoalkyl ethers and α-silyloxyalkyl aryl selenides, leads to alkenes in good

[308] G. Frater, *Helv. Chim. Acta*, 1979, **62**, 2829.
[309] T. Takahashi, K. Kasuga, and J. Tsuji, *Tetrahedron Lett.*, 1978, 4917.
[310] M. Kodama, S. Yokoo, H. Yamada, and S. Ito, *Tetrahedron Lett.*, 1978, 3121.
[311] K. Hirai and Y. Iwano, *Tetrahedron Lett.*, 1979, 2031.
[312] T. Shono, Y. Matsumura, S. Kashimura, and H. Kyutoku, *Tetrahedron Lett.*, 1978, 1205.

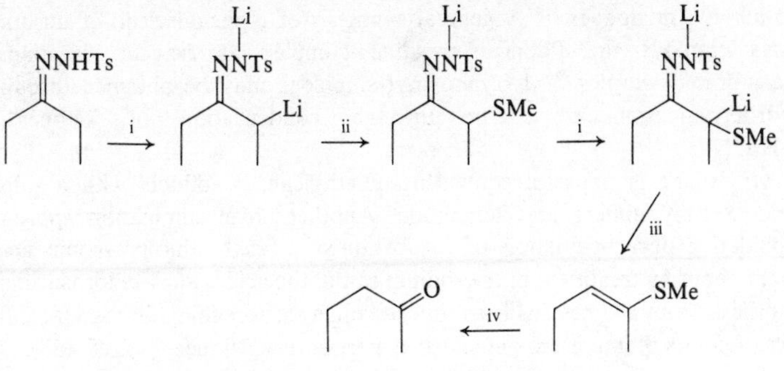

Li   Li   Li   Li
| | | |
NNHTs NNTs NNTs NNTs

$$\xrightarrow{i} \quad \xrightarrow{ii} \quad \xrightarrow{i}$$

(structures bearing Li, SMe, Li/SMe substituents)

$$\downarrow iii$$

(ketone, O) $\xleftarrow{iv}$ (vinyl sulphide, SMe)

Reagents: i, Bu^{n}Li, hexane, TMEDA, THF at -50 °C; ii, MeSSMe; iii, NH$_4$Cl,H$_2$O; iv, HgCl$_2$, hot aq. MeCN

Scheme 9

yields;[313] this is effectively a reversal of the 'cyclofunctionalization' process discussed in the earlier section 'Preparations . . . from Sulphenyl Halides'.

A sequence amounting to 1,2-transposition of aliphatic ketones (Scheme 9) illustrates another field of organic synthesis to which sulphides make a contribution.[314] PhSeCH$_2$CH$_2$C(OMe)$_3$ is a useful synthon for 2-substituted acrylates or for α-methylene-γ-butyrolactones after a Claisen orthoester rearrangement with allyl alcohols [R^{1}R^{2}C=CHCH$_2$OH → PhSeCH$_2$CH-(CO$_2$Me)CR^{1}R^{2}CH=CH$_2$ → H$_2$C=C(CO$_2$Me)CR^{1}R^{2}CH=CH$_2$ after elimination of selenoxide].[315]

Vinyl Sulphides, Selenides, and Tellurides.—This account is divided into sections covering the synthesis of vinyl sulphides and selenides, their reactions, and their uses in synthesis.

Most main routes to vinyl sulphides and selenides have been represented in recent papers, together with some novel variations. Halogenation–dehydrohalogenation has been applied as a convenient method for introducing a vinylthio-group into organic compounds; a novel procedure was used, involving the reaction of a Grignard reagent or an organolithium compound with ClCH$_2$CH$_2$SCN and treatment of the resulting chloroethyl sulphide with Bu^{t}OK.[316] This is an alternative to the direct synthesis from alkenes (addition of sulphenyl halide followed by elimination with Bu^{t}OK),[317,318] in which *cis*-isomers result from kinetic control (at -78 °C) and *trans*-isomers from thermodynamic control (at $+77$ °C).[317] Stereo-specific synthesis of (*E*)-1-alkenyl selenides is achieved by the dehydrobromination of 2-bromoalkyl selenides, and (*Z*)-isomers are prepared by the hydroboration of

[313] K. C. Nicolaou, W. J. Sipio, R. L. Magolda, and D. A. Claremon, *J. Chem. Soc., Chem. Commun.*, 1979, 83.
[314] T. Nakai and T. Mimura, *Tetrahedron Lett.*, 1979, 531.
[315] S. Raucher, K.-J. Hwang, and J. E. Macdonald, *Tetrahedron Lett.*, 1979, 3057.
[316] W. Verboom, J. Meijer, and L. Brandsma, *Synthesis*, 1978, 577.
[317] B. Geise and S. Lachheim, *Chem. Ber.*, 1979, **112**, 2503.
[318] Y. Masaki, K. Sakuma, and K. Kaji, *Chem. Lett.*, 1979, 1235.

the 1-alkynyl analogues.[319] A general synthesis of α-phenylseleno-$\alpha\beta$-unsaturated ketones employs the PhSeCl—pyridine complex as reagent, reacting with $\alpha\beta$-unsaturated ketones.[320] 1-Cyanovinyl sulphides may be obtained through the dehydrobromination route, or through condensation of ketones with $RSCH_2CN$.[321,291]

A vinylsilane is a useful equivalent of ethylene,[322] adducts with a sulphenyl chloride readily undergoing elimination. Another novel elimination approach,[323] *i.e.* oxidative decarboxylation of 2-substituted 2-(methylthio)propionic acids, is brought about by treatment of the sodium salt of the acid with *N*-chlorosuccinimide.

Synthesis from alkynes has been touched on in the foregoing discussion, and brief details of points of interest are included as part of ref. 319 (see also ref. 365).

Further examples of syntheses of vinyl sulphides from alkene derivatives include unusual examples, such as condensation of indole-3-thiol with ethylene at 180 °C,[324] the reaction of vinylidene chloride with thiourea and benzyl chloride to give $PhCH_2SCH{=}CHSCH_2Ph$,[325] cathodic reduction of α-keto-keten dithioacetals $[PhCOCH{=}C(SMe)_2 \rightarrow PhCOCH{=}CHSMe]$,[326] and the use of divinyl sulphoxide as a vinylating agent $[R_3Si(CH_2)_nSH + (H_2C{=}CH)_2SO \rightarrow R_3Si(CH_2)_nSCH{=}CH_2$; whereas $Bu^tSH \rightarrow Bu^tSCH{=}CH_2$ (15%) $+ Bu^tSCH_2CH_2SOCH{=}CH_2]$.[327]

An unusual reaction in which vinyl sulphides are formed[328] (Scheme 10) is of further interest because a quite different product $(MeO_2CCHR^1CHR^2COR^3)$ is formed with ^-OMe.

Syntheses of vinyl sulphides from dithioesters include the conversions $MeCSSMe \rightarrow (E/Z){-}MeSCMe{=}CHCSSMe$[329] and $Me_2CBrCSSMe \rightarrow (RO)_2P{-}(O)SC(SMe){=}CMe_2$[330] when a trialkyl phosphite was used as the reagent. The conversion $RSC(S)CN \rightarrow (E/Z){-}RSC(CN){=}C(CN)SR$ occurred during attempted hydrolysis.[331]

[319] S. Raucher, M. R. Hansen, and M. A. Cotter, *J. Org. Chem.*, 1978, **43**, 4885; R. N. Kudyakova, A. N. Volkov, and B. A. Trofimov, *Izv. Akad. Nauk SSSR, Ser. Khim.*, 1979, 213; A. N. Mirskova, N. V. Lutskaya, D. I. Kalikhman, B. A. Shainyan, and M. G. Voronkov, *ibid*, p. 572; I. A. Aslanov, E. F. Dzhafarov, A. A. Dzhafarov, K. B. Kurbanov, and A. G. Abdullaev, *Azerb. Khim. Zh.*, 1978, 45; G. G. Skvortsova, D. G. Kim, and L. V. Andriyankova, *Khim. Geterotsikl. Soedin.*, 1978, 364; S. V. Amosova, N. N. Skatova, O. A. Tarasova, and B. A. Trofimov, *Zh. Org. Khim.*, 1979, **15**, 2038; A. N. Volkov, R. N. Kudyakova, and B. A. Trofimov, *ibid.*, p. 1554; V. E. Statsyuk, S. P. Korshunov, N. V. Korzhova, and I. V. Bodrikov, *ibid.*, p. 1998; E. Larsson, *J. Prakt. Chem.*, 1978, **320**, 353 and 1979, **321**, 267; D. Schorstein, C. J. Suckling, and R. Wrigglesworth, *J. Chem. Soc., Chem. Commun.*, 1978, 795; B. A. Trofimov, S. V. Amosova, G. K. Musorin, D. F. Kushnarev, and G. A. Kalabin, *Zh. Org. Khim.*, 1979, **15**, 619; M. Hojo, R. Masuda, and S. Takagi, *Synthesis*, 1978, 284.
[320] G. Zima and D. Liotta, *Synth. Commun.*, 1979, **9**, 697.
[321] F. Pochat, *Tetrahedron Lett.*, 1978, 2683.
[322] F. Cooke, R. Moerck, J. Schwindeman, and P. Magnus, *J. Org. Chem.*, 1980, **45**, 1046.
[323] B. M. Trost, M. J. Crimmin, and D. Butler, *J. Org. Chem.*, 1978, **43**, 4549.
[324] G. G. Skvortsova, L. F. Teterina, B. V. Trzhitsinskaya, and V. K. Voronov, *Khim. Geterotsikl. Soedin.*, 1979, 352.
[325] P. Cassoux, R. Lahana, and J. F. Normant, *Bull. Soc. Chim. Fr., Part 2*, 1979, 427.
[326] H. H. Ruettinger, W. D. Rudorf, and H. Matschiner, *J. Prakt. Chem.*, 1979, **321**, 443.
[327] M. G. Voronkov, F. P. Kletsko, N. N. Vlasova, N. K. Gusarova, G. G. Efremova, V. V. Keiko, and B. A. Trofimov, *Izv. Akad. Nauk SSSR, Ser. Khim.*, 1978, 1690.
[328] K. Kobayashi, T. Taguchi, T. Morikawa, T. Takase, and H. Takanashi, *Tetrahedron Lett.*, 1980, **21**, 1047.
[329] Z. Yoshida, S. Yoneda, T. Kawase, and M. Inaba, *Tetrahedron Lett.*, 1978, 1285.
[330] Yu. G. Gololobov and M. N. Danchenko, *Zh. Obshch. Khim.*, 1979, **49**, 231.
[331] R. Mayer, W. Thiel, and H. Viola, *Z. Chem.*, 1979, **19**, 56.

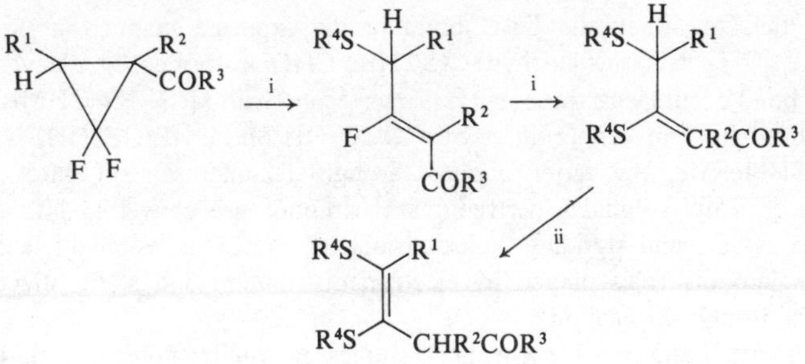

Reagents: i, R^4S^-; ii, KOH, MeOH, THF

Scheme 10

Desulphurization and deselenization of vinyl sulphides and selenides has been the feature of several studies recently; in many cases, C—S or C—Se bond cleavage was unexpected, but useful alkene syntheses have been sought successfully, employing $[Pb(OAc)_4]^{332}$ or a Grignard reagent with $NiCl_2$ and PPh_3.[333] Keto-vinyl sulphides $R^1COCH=CR^2SR^3$ undergo desulphurization and concomitant reduction of C=C with $NaBH_4$ and $CoCl_2$ or $NiCl_2$,[334] but with $LiAlH_4$ and $R^2COCH=C(R^2)SR^3$ a mixture of reduction product and $R^1CH=CHCOR^2$ is formed; this arises through allylic rearrangement of the first-formed product, in which the carbonyl group is reduced.[335] Whereas $E \rightarrow Z$ photoisomerization of these keto-vinyl sulphides occurs without cleavage,[336] easy displacement of the sulphide group is brought about with a secondary amine in benzene solution, giving the corresponding keto-vinylamine.[337] Vinyl sulphides give alkynes with $H_2N(CH_2)_3NHK$.[338]

A confusing variety of results is accumulating for the competition between deprotonation and C—Se bond cleavage of phenyl vinyl selenide. Butyl-lithium causes cleavage,[338,339] while $LiNPr^i_2$[339] brings about both metallation and elimination. In other laboratories, $KNPr^i_2$ with $LiOBu^t$,[340a] or an alkyl-lithium,[340b] is advocated for deprotonation without cleavage (see also refs. 341–345). The conventional procedure for the conversion of a vinyl sulphide or selenide into a ketone, involving acid hydrolysis, has not always been reliable, and further studies (see Vol. 5, p. 26) establish an order of rates vinyl ether > vinyl sulphide > vinyl selenide for this reaction, consistent with the generation of a heteroatom-stabilized carbonium ion in the rate-determining step.[346]

[332] B. M. Trost and Y. Tanigawa, *J. Am. Chem. Soc.*, 1979, **101**, 4413, 4743.
[333] H. Okamura, M. Miura, and H. Takei, *Tetrahedron Lett.*, 1979, 43.
[334] T. Nishio and Y. Omote, *Chem. Lett.*, 1979, 1223.
[335] T. Nishio and Y. Omote, *Chem. Lett.*, 1979, 365.
[336] N. Furukawa, M. Fukumura, T. Nishio, and S. Oae, *Phosphorus Sulfur*, 1978, **5**, 191.
[337] T. Nishio and Y. Omote, *Synthesis*, 1980, 390.
[338] C. A. Brown, *J. Org. Chem.*, 1978, **43**, 3083.
[339] M. Sevrin, J. N. Denis, and A. Krief, *Angew. Chem.*, 1978, **90**, 550.
[340] S. Raucher and G. A. Koolpe, (a) *J. Org. Chem.*, 1978, **43**, 3794; (b) *ibid.*, p. 4252.
[341] R. C. Cookson and P. J. Parsons, *J. Chem. Soc., Chem. Commun.*, 1978, 821.
[342] R. H. Everhardus, R. Grafing, and L. Brandsma, *Recl. Trav. Chim. Pays-Bas*, 1978, **97**, 69.
[343] J. J. Fitt and H. W. Gschwend, *J. Org. Chem.*, 1979, **44**, 303.
[344] R. R. Schmidt, H. Speer, and B. Schmid, *Tetrahedron Lett.*, 1979, 4277.
[345] K. Isobe, M. Fuse, H. Kosugi, H. Hagiwara, and H. Uda, *Chem. Lett.*, 1979, 785.
[346] R. A. McLelland and M. Leung, *J. Org. Chem.*, 1980, **45**, 187.

Ozone fails to attack the C=C bond in the expected manner in the case of RSCBu^t=CH$_2$, and the aldehyde RSCHBu^{t}CHO is formed.[347] Cleavage of the double bond occurs with the isobutyl isomer[347] and with MeSCEt=CHMe,[348] giving thiolesters, but further products EtCOCH(OH)Me or EtCH(OH)COMe and EtCOCHMeSMe, the latter arising through 1,2-migration of MeS, are also formed.[348] Vinyl sulphides derived from camphor are converted into camphorquinone by O$_3$ and pyridine.[347] Extensions of these studies might lead to new insights into the roles played by electron distribution and steric effects on the classical ozonolysis reaction.

Mechanistic and routine chemical studies of vinyl sulphides include stereochemical evidence that cycloaddition to TCNE involves a zwitterionic intermediate[349,350] (vinyl sulphides are more reactive than vinyl ethers[350]); the demonstration of [1,7]-σ-shifts in 1,7-bis(alkylthio)cycloheptatrienes which account for the rapid tautomeric interconversions;[351] laser-flash photocyclization of aryl vinyl sulphides;[352] and dye-sensitized photo-oxygenation leading to sulphoxides, sulphones, and products from C=C bond cleavage.[353] Hydrolysis of RCBr=C(CN)SEt gives 2-keto-1-cyanoalkyl ethyl sulphides,[354] while reaction with hydrazine gives 3-amino-4-ethylthio-pyrazoles.[355] Cl$_2$C=CHSR^1 can be converted into (E/Z)-R^{2}SCCl=CHSR^1, (R^{2}S)$_2$C=CHSR^1, ClC≡CSR^1, and R^{2}SC≡CSR^1 by treatment with R^{2}SH and KOH,[356] and the addition of R$_3$SiH to H$_2$C=CHSEt in the presence of H$_2$PtCl$_6$ or [(Ph$_3$P)$_3$RhCl] gives mainly 1-(trialkylsilyl)alkyl sulphides.[357]

Alkylation and other electrophilic reactions of vinyl sulphides and selenides following α-metallation[339-345] include a number of useful synthetic procedures. Addition of an alkyl-lithium to phenyl vinyl selenide gives the α-lithio-β-alkylated selenide, which, on further alkylation and oxidative deselenation, gives the 1,2-disubstituted ethylene (H$_2$C=CHSePh → R^{1}CH$_2$CHLiSePh → R^{1}CH$_2$CHR^{2}SePh → R^{1}CH=CHR^2), thus permitting the vinyl selenide to be described as a $\overset{+}{C}H=\overset{-}{C}H$ synthon.[340b] A similar overall result (RCHO → RCOCH=CH$_2$) is based on acid-catalysed isomerization of 1-(phenylthio)vinyllithium–aldehyde adducts, followed by oxidative desulphurization.[341] Metallation of buta-1,3-dienyl ethers and sulphides with Bu^{n}Li and Bu^{t}OK occurs at C-1 and can be followed by alkylation with MeI, MeSCN, or MeCHO;[342] similar reactions have been described for 3-(dimethylamino)prop-1-enyl phenyl sulphide.[343] Electrophilic attack on (E)-β-(arylsulphinyl)vinyl sulphides (Scheme 11) occurs at the sulphinyl-substituted carbon atom, mainly with retention of configuration.[344]

[347] R. Chaussin, P. Leriverend, and D. Paquer, *J. Chem. Soc., Chem. Commun.*, 1978, 1032.
[348] M.-P. Strobel, L. Morin, and D. Paquer, *Tetrahedron Lett.*, 1980, **21**, 523.
[349] H. Graf and R. Huisgen, *J. Org. Chem.*, 1979, **44**, 2594, 2595.
[350] T. Okuyama, M. Nakada, K. Toyoshima, and T. Fueno, *J. Org. Chem.*, 1978, **43**, 4546.
[351] M. Cavazza, G. Morganti, and F. Pietra, *Recl. Trav. Chim. Pays-Bas*, 1979, **98**, 165.
[352] T. Wolff, *J. Photochem.*, 1979, **11**, 215 and *J. Am. Chem. Soc.*, 1978, **100**, 6157.
[353] W. Ando, T. Nagashima, K. Saito, and S. Kohmoto, *J. Chem. Soc., Chem. Commun.*, 1979, 154.
[354] F. Pochat, *Tetrahedron Lett.*, 1979, 19.
[355] F. Pochat, *Tetrahedron Lett.*, 1979, 2991.
[356] A. N. Mirskova, T. L. Usova, N. V. Lutskaya, I. D. Kalikhman, and M. G. Voronkov, *Izv. Akad. Nauk SSSR, Ser. Khim.*, 1978, 426.
[357] M. G. Voronkov, N. N. Vlazova, S. V. Kirpichenko, S. A. Bolshakova, V. V. Keiko, E. O. Tsetlina, S. V. Amosova, B. A. Trofimov, and V. Chvalovsky, *Izv. Akad. Nauk SSSR, Ser. Khim.*, 1979, 422.

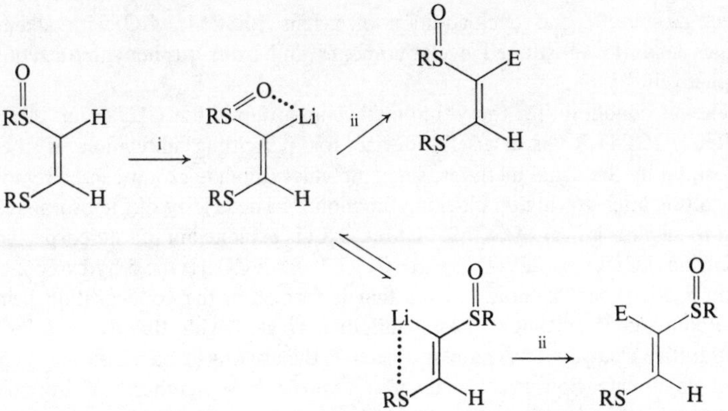

Reagents: i, Bu^{t}Li; ii, E$^+$

Scheme 11

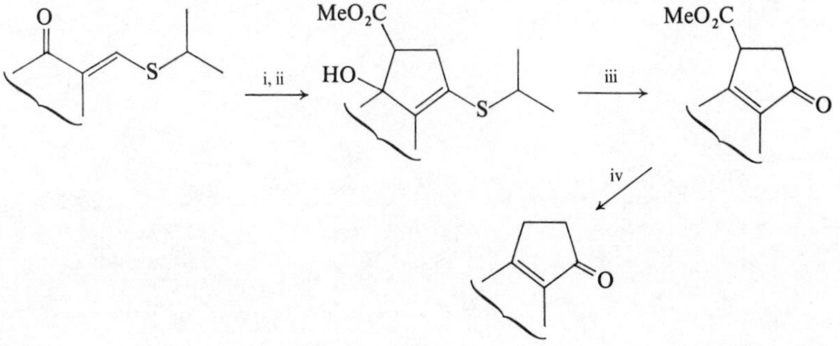

Reagents: i, lithium 2,2,6,6-tetramethyl-*N*-hydroxypiperidide, at -78 °C; ii, H$_2$C=CHCO$_2$Me; iii, 5% HCl in THF; iv, Me$_3$SiI, at 180 °C, then HCl in refluxing toluene

Scheme 12

Michael addition of 1-lithio-2-methoxycarbonylprop-2-enyl phenyl sulphide to acrylates gives 3-(phenylthio)cyclopentadienones, while addition to aldehydes is followed similarly by cyclization to furan-2-ones.[345] Further uses of vinyl sulphides in syntheses of cyclic compounds include an efficient pentannelation procedure (Scheme 12),[358] cyclization of γ-(3-phenylthio-2-methylprop-2-enyl)-β-keto-esters to 3-(phenylthio)cyclohexanones, using TFA,[359] and an aromatic substitution reaction [HOCPh$_2$C(SEt)=CH$_2$ → 3-phenyl-2-(ethylthio)indene] that involves attack at the β-carbon atom of the vinyl sulphide moiety, promoted by acid-catalysed dehydration.[360] 1-(Phenylthio)vinylphosphonium salts PhSC(=CH$_2$)PPh$_3$ X$^-$, have been advocated for use in the synthesis of cyclopentanone by condensation with

[358] J. P. Marino and L. C. Ketterman, *J. Chem. Soc., Chem. Commun.*, 1979, 946.
[359] A. S. Kende and J. A. Schneider, *Synth. Commun.*, 1979, **9**, 419.
[360] M. Braun, *Chem. Ber.*, 1979, **112**, 1495.

γ-keto-diesters,[361] and cycloaddition of ketens PhSCMe=CO with alkenes and imines leads to substituted cyclobutanones and *trans*-α-(phenylthio)azetidinones, respectively.[362]

The introduction of a (butylthio)methylene group in the CH_2 group of a ketone $R^1R^2CHCOCH_2R^3$ is a useful device for permitting alkylation at the other α-position by the usual methods, since the vinyl sulphide group can be removed by hydrolysis after serving its blocking function.[363] The acidity of the protons that are beta to sulphur in (*E*)-MeSCMe=CMeCO$_2$Et is exploited in the stereospecific chain extension RCHO $\rightarrow$ (2*Z*,4*E*)-RCH=CHCH=CMeCO$_2$H, the 5-hydroxy-2-methyl-3-(methylthio)-pent-2-enoic lactone that is formed in the condensation being first desulphurized (by Raney nickel) and then treated with Bu^{t}OK to bring about dehydrative cleavage.[364] A similar objective, the use of 4-(t-butylthio)but-3-en-2-one as a chain-extension agent, has been realized in a synthesis of isorenieratene (Scheme 13).[365,366]

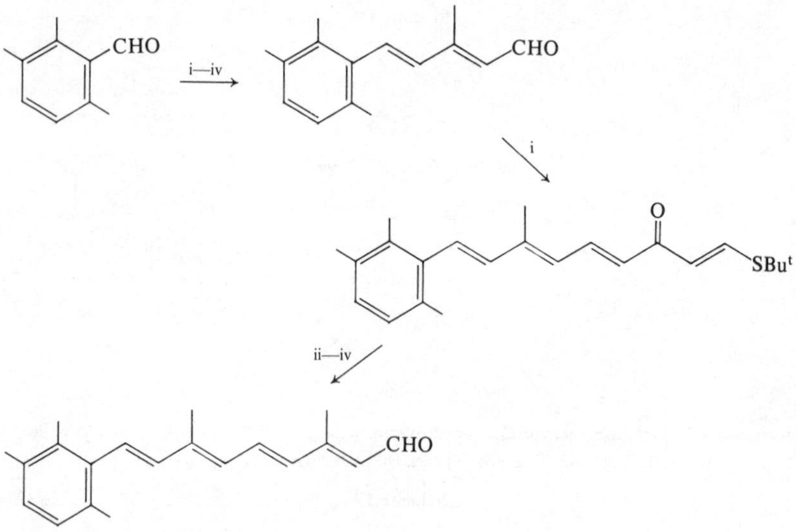

Reagents: i, MeCOCH=CHSBu^t; ii, MeLi; iii, H$^+$; iv, I$_2$

Scheme 13

Allenic Sulphides.—This short section is mainly concerned with alka-1,2-dienyl sulphides and related compounds.[367] The reaction of Cl$_2$C=CClC≡CPh or of PhCCl=CClC≡CCl with an alkanethiolate gives (RS)$_2$C=C=C=CPhSR, while

[361] A. T. Hewson, *Tetrahedron Lett.*, 1978, 3267.
[362] M. Ishida, T. Minami, and T. Agawa, *J. Org. Chem.*, 1979, **44**, 2067.
[363] S. Campbell, N. Petragnani, T. Brocksom, and L. Tsuchiya, *Farmaco, Ed. Sci.*, 1978, **33**, 421.
[364] E. J. Corey and G. Schmidt, *Tetrahedron Lett.*, 1979, 2317.
[365] S. Akiyama, S. Nakatsuji, T. Hamamura, M. Kataoka, and M. Nakagawa, *Tetrahedron Lett.*, 1979, 2809.
[366] S. Akiyama, S. Nakatsuji, S. Eda, M. Kataoka, and M. Nakagawa, *Tetrahedron Lett.*, 1979, 2813.
[367] A. Roedig and G. Zaby, *Liebigs Ann. Chem.*, 1979, 1606, 1614, 1626.

$Cl_2C=C=C=CClMe$ gives $RSC \equiv CC(SR)=CClMe$. Similarly, $Cl_2C=CClC \equiv CCl$ with five equivalents of $NaSR^1$ gives $(R^1S)_2C=C(SR^1)C(SR^1)=CHSR^1$, while $Cl_2C=CClC \equiv CSR^2$ gives $(R^1S)_2C=C=C(SR^1)SR^2$ *via* $R^1SC \equiv CC(SR^1)=C-(SR^1)SR^2$.

Allyl Sulphides and Selenides.—Preparative methods described for saturated sulphides may be used for the allyl analogues, and 1,2-phenylthio-shifts can be exploited.[2p] Scheme 14 illustrates some of the methods used, and applications for the separate synthesis of 1- and 2-(phenylthio)buta-1,3-dienes.[368] Dehydrochlorination of adducts formed from trisubstituted alkenes and PhSCl yields allyl sulphides.[369]

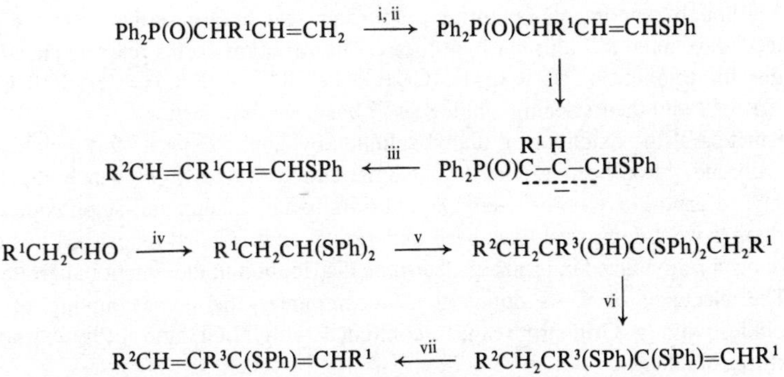

$$Ph_2P(O)CHR^1CH=CH_2 \xrightarrow{\text{i, ii}} Ph_2P(O)CHR^1CH=CHSPh$$

$$\downarrow \text{i}$$

$$R^2CH=CR^1CH=CHSPh \xleftarrow{\text{iii}} Ph_2P(O)\overset{R^1}{\underset{}{C}}-\overset{H}{\underset{}{C}}-CHSPh$$

$$R^1CH_2CHO \xrightarrow{\text{iv}} R^1CH_2CH(SPh)_2 \xrightarrow{\text{v}} R^2CH_2CR^3(OH)C(SPh)_2CH_2R^1$$

$$\downarrow \text{vi}$$

$$R^2CH=CR^3C(SPh)=CHR^1 \xleftarrow{\text{vii}} R^2CH_2CR^3(SPh)C(SPh)=CHR^1$$

Reagents: I, Bu^nLi; ii, $PhSSPh$; iii, R^2CHO; iv, $PhSH$, H^+; v, Bu^nLi, $R^2CH_2COR^3$; vi, $SOCl_2$, Et_3N; vii, $NaIO_4$; then heat

Scheme 14

The delocalized anion shown in Scheme 14 is representative of the crucial intermediate in many applications of allyl sulphides in synthesis. The discovery[370] that HMPA induces 1,4-addition of thioallyl anions to $\alpha\beta$-unsaturated ketones through the α-carbon atom of the allyl sulphide makes the use of these compounds more attractive in synthesis, so that they rival organocopper compounds in versatility (see also Chap. 2, Pt. I, p. 97). A special advantage is the fact that conversion of the alkylated allyl sulphide into the corresponding sulphoxide, followed by rearrangement (see later section on 'Sulphoxide–Sulphenate Rearrangements'), places an oxygen function on the γ-carbon atom of the original allyl grouping $(R^1SCHR^2CH=CHR^3 \rightarrow R^2CH=CHCHR^3OH)$.[369,370] The same regioselectivity is seen in the preferred γ-alkylation of alkylthioallylboronyl anions Li^+ $Me_2CHSCH=CHCH_2\bar{B}Et_3$, which are formed from an alkylthioallyl-lithium and a trialkylborane at $-78\ ^\circ C$. An interesting application is the γ-alkylation by allyl halides, leading to head-to-tail 1,5-dienes in good yields.[371] Conversion of geraniol

[368] P. Baldwin, J. I. Grayson, and S. Warren, *J. Chem. Soc., Chem. Commun.*, 1978, 657.
[369] Y. Masaki, K. Hashimoto, and K. Kaji, *Tetrahedron Lett.*, 1978, 4539.
[370] M. R. Binns, R. K. Haynes, T. L. Houston, and W. R. Jackson, *Tetrahedron Lett.*, 1980, **21**, 573.
[371] Y. Yamamoto, H. Yatagai, and K. Maruyama, *J. Chem. Soc., Chem. Commun.*, 1979, 157; *Chem. Lett.*, 1979, 385.

(16)   (17)

into fragranol and grandisol skeletons is assisted by a more passive use of a sulphide group, introduced to stabilize an allyl-lithium intermediate.[372]

Thio-Claisen rearrangement of allyl sulphides, $Me_2CHC(=CMe_2)SCH_2CH=CH_2 \rightleftharpoons Me_2CHC(=S)CMe_2CH_2CH=CH_2$,[373] is a reversible [3,3]-sigmatropic process which may involve a biradicaloid (16) or an aromatic (17) transition state, and although experiments favouring the former have been described,[374] more work is needed to make this a firmer hypothesis. The variation of this rearrangement that leads to thioketens $(R^1R^2C=CHCHR^3SC\equiv CR^4 \rightarrow R^3CH=CHCR^1R^2CR^4=C=S)$[375,376] and their selenium analogues[376] has been described.

Kinetics of the oxidation of diallyl sulphide by singlet oxygen, and its inhibition by rubrene, have been reported.[377] Isomerization studies[378] in which the base-catalysed prop-2-enyl–prop-1-enyl equilibrium for allyl sulphides, sulphoxides, and sulphones were compared have been undertaken, and isolated reports on this topic have been rationalized in terms of electronic distribution in the functional groups.

The cleavage of C—S bonds that accompanies the cross-coupling of allyl sulphides with a Grignard reagent combined with $NiCl_2$ and PPh_3[379] leads to mixtures of alkenes.

Acetylenic Sulphides.—Ethynyl phenyl sulphide, formed from bromoacetaldehyde diethyl acetal with PhSH by condensation followed by elimination by base, reacts with allyl alcohols to give allyl 2-(phenylthio)vinyl ethers; these, on oxidative desulphurization, lead to 2,4-dienals.[380] A simple new synthesis of 1-alkynyl sulphides $(CF_3CH_2SR^1 + R^2Li \rightarrow R^2C\equiv CSR^1)$[381] and another route to propargyl sulphides $[R^1SH + (Me_2N)_2P(O)OCH_2C\equiv CR^2 + NaOH \rightarrow R^1SCH_2C\equiv CR^2]$[382] should stimulate new research into applications of these compounds in synthesis. Some possibilities are shown in a reductive cyclization of N-substituted suc-cinimides (Scheme 15)[383] and a use of $PhSeCHLiC\equiv CLi$ in a synthesis of 7-hydroxymyoporone (18), revealing the propargyl selenide to be the synthetic equivalent of the synthon $\overset{+}{C}H=CHC\overset{-}{H}=O$.[384]

[372] V. Rautenstrauch, *J. Chem. Soc., Chem. Commun.*, 1978, 519.

[373] P. Metzner, N. P. Thi, and J. Vialle, *Nouv. J. Chim.*, 1978, **2**, 179.

[374] P. Metzner, T. N. Pham, and J. Vialle, *J. Chem. Res. (S)*, 1978, 478.

[375] E. Schaumann and F. F. Grabley, *Liebigs Ann. Chem.*, 1979, 1746.

[376] R. S. Sukhai and L. Brandsma, *Recl. Trav. Chim. Pays-Bas*, 1979, **98**, 55.

[377] B. M. Monroe, *Photochem. Photobiol.*, 1979, **29**, 761.

[378] V. Svata, M. Prochazka, and V. Bakos, *Collect. Czech. Chem. Commun.*, 1978, **43**, 2619.

[379] H. Okamura and H. Takei, *Tetrahedron Lett.*, 1979, 3425.

[380] R. C. Cookson and R. Gopalan, *J. Chem. Soc., Chem. Commun.*, 1978, 924.

[381] K. Tanaka, S. Shiraishi, T. Nakai, and N. Ishikawa, *Tetrahedron Lett.*, 1978, 3103.

[382] B. Corbel and J. P. Pangam, *Synthesis*, 1979, 882.

[383] P. M. M. Nossin and W. N. Speckamp, *Tetrahedron Lett.*, 1979, 4411.

[384] H. J. Reich, P. M. Gold, and F. Chow, *Tetrahedron Lett.*, 1979, 4433.

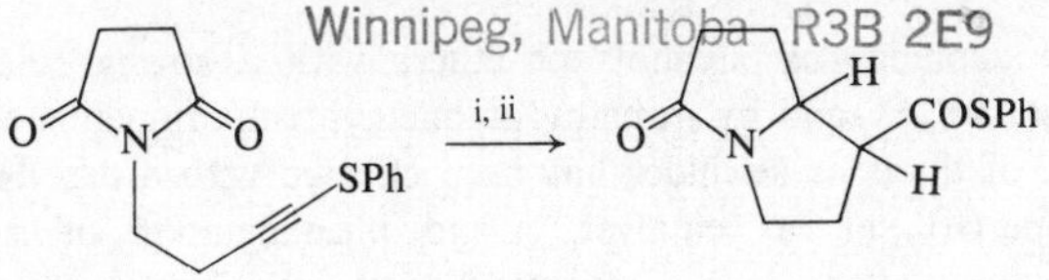

Reagents: i, HCO$_2$H; ii, H$^+$

Scheme 15

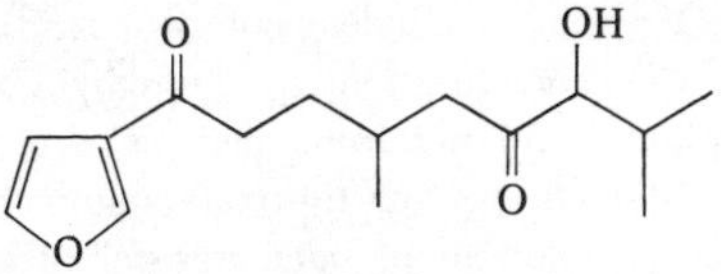

(18)

6 Naturally Occurring Organosulphur Compounds

This short section is located immediately before coverage of sulphonium salts, a group of compounds most importantly represented in naturally occurring compounds. Possible roles for S-adenosyl-L-methionine have stimulated studies with model compounds;[385] similar studies may follow from the fact that Bleomycin A$_2$ is a sulphonium salt, carrying the grouping $-CONH(CH_2)_3\overset{+}{S}Me_2$.[386] 5-(Dimethylsulphonio)pentanoic acid has been identified as a constituent of *Diplotaxis tenuifolia*.[387]

Block's book[388] lists a number of simple organosulphur compounds which occur naturally, and further examples include S-methylthiomethyl 2-methylbutanethioate, MeCH$_2$CHMeC(O)SCH$_2$SMe (a new addition to the 24 organosulphur compounds already isolated from the essential oil of hops),[389] and methyl 1-(1-methylthioethyl)-thioethyl disulphide, a constituent of roast pork[212] (see p. 21).

7 Sulphonium Salts

The opportunity has been taken to bring together recent papers on trialkyl- and triaryl-sulphonium salts with those covering S-heteroatom-sulphonium salts, R^{1}R$^2\overset{+}{S}$X Y$^-$ (X = halogen, NR^{3}R^4, OR^3, or SR^3), the latter class having been mostly covered in previous volumes of this series in the sections on sulphides and sulphoxides.

Trialkyl- and Triaryl-sulphonium, -selenonium, and -telluronium Salts.—Some modifications of known preparative methods are summarized, with references, at the end of this section. Interesting practical procedures have been described for the

[385] T. Irie and H. Tanida, *J. Org. Chem.*, 1979, **44**, 325.
[386] N. J. Oppenheimer, L. O. Rodriguez, and S. M. Hecht, *Biochemistry*, 1979, **18**, 3439.
[387] F. Larker and J. Hamelin, *Phytochemistry*, 1979, **18**, 1396.
[388] E. Block, 'Reactions of Organosulfur Compounds', Academic Press, New York, 1978.
[389] M. Moir, I. M. Gallacher, J. Hobkirk, J. C. Seaton, and A. Suggett, *Tetrahedron Lett.*, 1980, **21**, 1085.

alkylation of sulphides by alcohols or ethers with a strong acid,[390] by alkyl methanesulphonates,[391] and by trimethylsilylmethyl trifluoromethanesulphonate.[392] The arylation of thiols or selenides has been effected with a diaryliodonium salt, using a copper(II) salt as catalyst,[393] and phenacylation of sulphides with diazoacetophenone has been reported.[394] A similar type of product, $R^1R^2\overset{+}{S}CHArCOCN$, has been formed from a sulphide with a *gem*-dicyano-oxiran.[395] Unusual routes to selenonium salts (SeOCl$_2$ + ArMgX → Ar$_3$Se$^+$ X$^-$;[396] Ph$_2$SeO + RCH$_2$NO$_2$ with Ac$_2$O and HX → Ph$_2\overset{+}{S}$eCHRNO$_2$ X$^-$ [397]) and telluronium salts (Ar$_2$TeO + active-methylene compounds[398]) have been reported.

Assignment of the (*S*) configuration to (+)-ethylmethylpropylsulphonium 2,4,6-trinitrobenzenesulphonate follows from chemical correlation with (*S*)-lactic acid *via* a (4*S*,6*S*)-4-ethyl-6-methyl-2-oxo-1,4-oxathianium salt intermediate.[399]

An important reaction of sulphonium salts, transfer of an *S*-alkyl group to a nucleophilic species, is also shared with selenonium salts. Both trimethyl-sulphonium[400] and -selenonium[401] hydroxides are effective methylating agents towards carboxylic acids, thiols, phenols, and other acidic substrates (*e.g.* heteroaromatic NH groups), and the more reactive trimethylselenonium salt has been advocated as a useful alternative to diazomethane which avoids known carcinogenic materials.[401] The (*pro-R*)-methyl group of *S*-methyl-L-methionine is transferred to L-homocystine when catalysed by a jack bean transferase.[402] Further references to papers on this topic, and on other reactions and uses of sulphonium salts in synthesis, which extend well-established chemistry, are summarized at the end of this section. The main topic areas covered in these papers are the synthesis of cyclopropanes and oxirans, and substitution reactions. An interesting combination of reactions is seen in the 'epoxyannelation' of aldehyde and ketone enolates (Scheme 16).[403]

Additional Bibliography: N. V. Kondratenko, G. N. Timofeeva, V. I. Popov, and L. M. Yagupolskii, *Zh. Org. Khim.*, 1979, **15**, 2432 [sulphides, selenides, or tellurides + MeI and AgBF$_4$]; Y. L. Chow, B. H. Bakker, and K. Iwai, *J. Chem. Soc., Chem. Commun.*, 1980, 521 [styrene, Br$_2$, and Me$_2$S + base → PhC(=CH$_2$)$\overset{+}{S}$Me$_2$ Br$^-$]; M. J. Farrall, T. Durst, and J. M. J. Frechet, *Tetrahedron Lett.*, 1979, 203 [sulphonium-salt-substituted polymeric reagents]; M. Ando and S. Emoto, *Bull. Chem. Soc. Jpn.*, 1978, **51**, 2435 [substituted phenyldimethylsulphonium fluorosulphonates]; J. F. King and G. T. Y. Tsang, *J. Chem. Soc., Chem. Commun.*, 1979, 1131 [kinetics of substitution at methylene groups of benzyl in (PhCH$_2$)$_2\overset{+}{S}$Et BF$_4$$^-$]; M. Cavazza, G. Morganti, C. A. Veracini, and F. Pietra, *J. Chem.*

[390] B. Bader and M. Julia, *Tetrahedron Lett.*, 1979, 1101.
[391] S. Kozuka, S. Taniyasu, A. Kikuchi, and K. Ogino, *Chem. Lett.*, 1979, 129.
[392] E. Vedejs and G. R. Martinez, *J. Am. Chem. Soc.*, 1979, **101**, 6452.
[393] J. V. Crivello and J. H. W. Lam, *J. Org. Chem.*, 1978, **43**, 3055; *Synth. Commun.*, 1979, **9**, 151.
[394] W. T. Flowers, G. Holt, and P. McCleery, *J. Chem. Soc., Perkin Trans. 1*, 1979, 1485.
[395] A. Robert, M. T. Thomas, and A. Foucaud, *J. Chem. Soc., Chem. Commun.*, 1979, 1048.
[396] Y. Ishii, Y. Iwama, and M. Ogawa, *Synth. Commun.*, 1978, **8**, 93.
[397] S. A. Shevelev, V. V. Semenov, and A. A. Fainzilberg, *Izv. Akad. Nauk SSSR, Ser. Khim.*, 1978, 1091.
[398] I. D. Sadekov, A. I. Usachev, A. A. Maksimenko, and V. I. Minkin, *Zh. Obshch. Khim.*, 1978, **48**, 934.
[399] E. Kelstrup, *J. Chem. Soc., Perkin Trans. 1*, 1979, 1037.
[400] K. Yamauchi, T. Tanabe, and M. Kinoshita, *J. Org. Chem.*, 1979, **44**, 638.
[401] K. Yamauchi, K. Nakamura, and M. Kinoshita, *Tetrahedron Lett.*, 1979, 1787.
[402] G. Grue-Sorensen, E. Kelstrup, A. Kjaer, and J. O. Madsen, *J. Chem., Soc., Chem. Commun.*, 1980, 19.
[403] M. E. Garst, *J. Org. Chem.*, 1979, **44**, 1578.

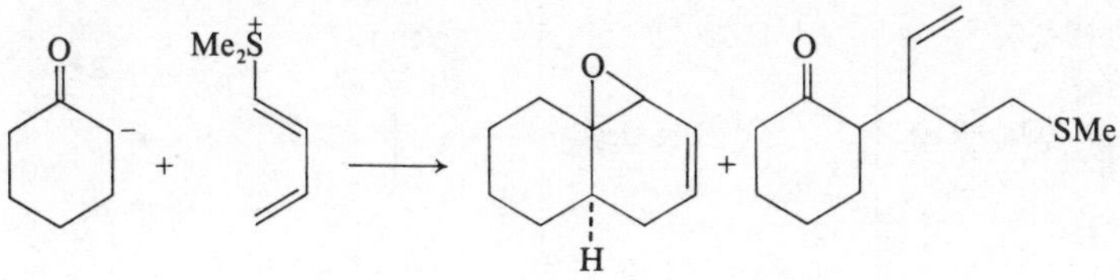

Scheme 16

Soc., Chem. Commun., 1979, 912 [competition between 3- and 7-substitution in 2-(dimethylsulphonio)cyclohepta-2,4,6-trienone]; D. M. Hedstrand, W. H. Kruizinga, and R. M. Kellogg, *Tetrahedron Lett.*, 1978, 1255 [dye-sensitized photoreduction of phenacylsulphonium salts by transfer of hydride ion from 1,4-dihydropyridines]; T. Minami, I. Niki, and T. Agawa, *Phosphorus Sulfur*, 1977, **3**, 55 [demethylation of vinyldimethylsulphonium salts with PhO⁻]; B. A. Arbuzov, Yu. V. Belkin, N. A. Polezhaeva, and G. E. Buslaeva, *Izv. Akad. Nauk SSSR, Ser. Khim.*, 1979, 1625 [demethylation of methylphenylselenonium salts with PPh₃]; P. G. Gassman, T. Miura, and A. Mossman, *J. Chem. Soc., Chem. Commun.*, 1980, 558 [dealkylation of benzylalkylselenonium salts by base; NaNH₂ leads to a [2,3]-sigmatropic rearrangement, giving 2-alkylbenzyl selenides]; H. J. Reich and M. L. Cohen, *J. Am. Chem. Soc.*, 1979, **101**, 1307 [competition between *S*- or *Se*-allylation and α-alkylation of phenacyl sulphides and selenides; [2,3]-sigmatropic rearrangements of allyl-sulphonium and -selenonium salts]; S. Kano, T. Yokomatsu, and S. Shibuya, *Tetrahedron Lett.*, 1978, 4125 [the [2,3]-sigmatropic rearrangements of β-hydroxy-β-phenylalkyl-sulphonium ylides to 3-phenylalk-3-enyl sulphides]; H. Matsuyama, M. Matsumoto, M. Kobayashi, and H. Minato, *Bull. Chem. Soc. Jpn.*, 1979, **52**, 1139 [catalysis of sulphide–sulphonium salt equilibrium by ArSCN, ArSCF₃, or ArSOMe]; S. Kano, T. Yokomatsu, and S. Shibuya, *J. Chem. Soc., Chem. Commun.*, 1978, 785 [general synthesis of oxirans, using β-hydroxyalkyl-dimethylsulphonium salts]; D. H. Rich and J. P. Tam, *Synthesis*, 1978, 46 [cyclization of derivatives of *S*-methylmethionine to cyclopropanes with NaH]; S. Brandt and P. Helquist, *J. Am. Chem. Soc.*, 1979, **101**, 6473 {conversion of alkenes into cyclopropanes with retention of configuration, using [(C₅H₅)Fe-(CO)₂(CH₂S⁺Me₂)]BF₄⁻}; K. Takaki, K. Negoro, and T. Agawa, *J. Chem. Soc., Perkin Trans. 1*, 1979, 1490 [addition of carbanions to vinyldimethylsulphonium salts, leading to cyclopropanes, alkenes, or 1-pyrrolines]; T. Severin and I. Braentigam, *Chem. Ber.*, 1979, **112**, 3007, and E. Vilsmaier and W. Tröger, *Angew. Chem.*, 1979, **91**, 860 and *Synthesis*, 1980, 466 [enamino-sulphonium salts].

Halogeno-sulphonium Salts.—Uses for *S*-chlorodimethylsulphonium chloride and the corresponding bromine compound have been extended to a conversion of oxirans into α-halogeno-ketones (Scheme 17),[404] to oxidative dimerization of malonates,[405] and to the conversion of sulphonamides into sulphimides.[405]

The formation of *S*-halogeno-sulphonium salts from the halogen and a sulphide is in competition with an α-halogenation–oxidation reaction which leads to sulphonyl chlorides when water is present.[406] The initial attack at sulphur may be followed by a Pummerer rearrangement to the adjacent carbon atom, and in the case of unsymmetrical sulphides the regioselectivity is determined by the relative electron density at the α-carbon atoms.[407]

A mixture of ClCH₂SMe and SbCl₅ contains the chloro-sulphonium salts ClCH₂S⁺ClMe SbCl₃⁻ and Cl₂CHS⁺ClMe SbCl₆⁻, and not the carbonium salts, as previously claimed.[408] However, MeS⁺Cl₂ SbCl₆⁻ is formed from Cl₃CSMe.

[404] G. A. Olah, Y. D. Vankar, and M. Arvanaghi, *Tetrahedron Lett.*, 1979, 3653.
[405] N. Furukawa, T. Inoue, T. Aida, T. Akasaka, and S. Oae, *Phosphorus Sulfur*, 1978, **4**, 15.
[406] R. F. Langler, Z. A. Marini, and E. S. Spalding, *Can. J. Chem.*, 1979, **57**, 3193.
[407] T. P. Ahern, D. G. Kay, and R. F. Langler, *Can. J. Chem.*, 1978, **56**, 2422.
[408] K. Hartke and E. Akgun, *Chem. Ber.*, 1979, **112**, 2436.

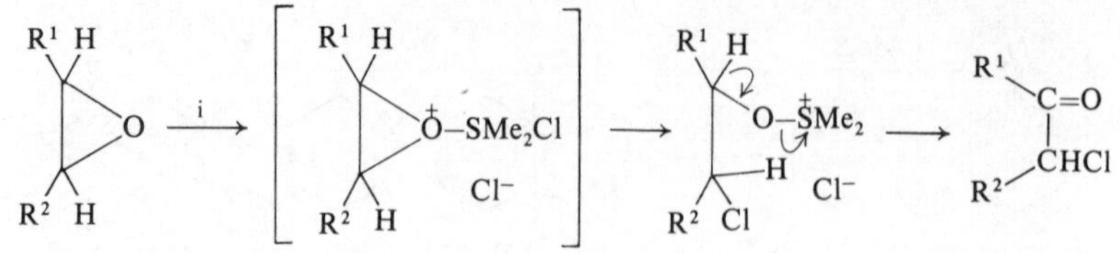

Reagents; i, Me$_2$SCl$^+$ Cl$^-$

Scheme 17

Amino-sulphonium Salts.—*N*-Chlorosuccinimide reacts with sulphides in the presence of SbCl$_5$ to give *S*-succinimidosulphonium salts, which can undergo Stevens rearrangement to 5-alkylthiomethoxypyrrolin-2-ones.[409] Similar results have been obtained with hydantoins,[410] and analogous reactions are being exploited to effect the *ortho*-substitution of aromatic amines.[411]

N-Chlorosuccinimide plus Me$_2$S is a convenient reagent for introducing a dimethylsulphonio-group, and it forms Me$_2$NCH=C($\overset{+}{S}$Me$_2$)CHO Cl$^-$ from the corresponding aldehyde.[412] The reagent can be used with triethylamine to couple carboxylic acids to amines, but its merits relative to other reagents in amide synthesis have not yet been determined, so that its potential for peptide synthesis is as yet unknown.[413]

Dialkylamino-alkoxysulphonium salts R$_2$N$\overset{+}{S}$(OR)Ph BF$_4$$^-$ are formed from sulphenamides and an alcohol with Bu^{t}OCl and AgBF$_4$;[414] these are conditions which are much milder than those used in the alternative methods of alkylation of sulphinamides.

Alkoxy-sulphonium Salts.—Alkylation of sulphoxides by a trialkyloxonium tetrafluoroborate gives these salts, which are relatively stable to air and moisture.[415] The compounds, which are intermediates in one form of the Pummerer rearrangement,[416] are efficient alkylating agents.[415] The role of methyl phenyl sulphide as the catalytic electron-carrier in an electrochemical oxidation of alcohols is proposed to involve alkoxysulphonium salt formation, followed by a hydride shift.[417] The fact that some sulphoxide is also formed in this reaction appears to have encouraged the authors in their proposal: PhSMe $-$ 2e$^-$ $\rightarrow$ Ph$\overset{++}{S}$Me $\rightarrow$ Ph$\overset{+}{S}$MeOCHR^{1}R^2 $\rightarrow$ Ph$\overset{+}{S}$($\bar{C}$H$_2$)OCHR^{1}R^2 $\rightarrow$ PhSMe + R^{1}COR^2. However, there are inconsistencies in the adoption of this mechanism, which has been established for dimethyl sulphoxide in the presence of a proton acceptor, particularly since alcohols are readily oxidized in the same electrochemical system in the absence of a

[409] E. Vilsmaier, J. Schuetz, and S. Zimmerer, *Chem. Ber.*, 1979, **112**, 2231.

[410] E. Vilsmaier, R. Bayer, I. Laengenfelder, and U. Wetz, *Chem. Ber.*, 1978, **111**, 1136; E. Vilsmaier, R. Bayer, U. Wetz, and K.-H. Dittrich, *ibid.*, p. 1147.

[411] P. G. Gassmann and H. R. Drewes, *J. Am. Chem. Soc.*, 1978, **100**, 7600.

[412] V. Kral and Z. Arnold, *Collect. Czech. Chem. Commun.*, 1978, **43**, 1248.

[413] K. Takeda, T. Kobayashi, and H. Ogura, *Chem. Pharm. Bull.*, 1979, **27**, 536.

[414] M. Haake and H. Gebbing, *Synthesis*, 1979, 98.

[415] H. U. Wagner and A. Judelbaum, *Angew. Chem.*, 1978, **90**, 487.

[416] S. Wolfe and P. M. Kazmaier, *Can. J. Chem.*, 1979, **57**, 2388, 2397; S. Wolfe, P. M. Kazmaier, and H. Anksi, *ibid*, p. 2404.

[417] T. Shono, Y. Matsumura, M. Mizoguchi, and J. Hayashi, *Tetrahedron Lett.*, 1979, 3861.

sulphide. The presence of other side-products which would be consistent with the alkoxysulphonium ion intermediate must now be established to give support to the proposed mechanism.

Anomalous results in the conversion of allylic alcohols into halides *via* alkenoxysulphonium salts (formed using dimethyl sulphide and NCS) have been ascribed to participation by remote functional groups in the rearrangement.[418]

Alkylthio-sulphonium Salts.—Admirably thorough studies have been made by Caserio and her co-workers on the preparation and reactions of alkylthio-sulphonium salts, which are now being recognized as intermediates in a number of organosulphur reactions.

Alkylation of a disulphide with Me_3O^+ BF_4^-,[419] or the reaction of a trifluoroacetoxysulphonium salt with a thiol $[Me_2SO + (CF_3CO)_2O \rightarrow Me_2\overset{+}{S}OCOCF_3\ CF_3CO_2^- \rightarrow Me_2\overset{+}{S}SR\ CF_3CO_2^-)$[420] are direct methods of synthesis. A bis(arylthio)sulphonium salt, probably better represented as the symmetrical cation (19), is formed either from a disulphide or from an arenesulphenyl chloride with $SbCl_5$.[421]

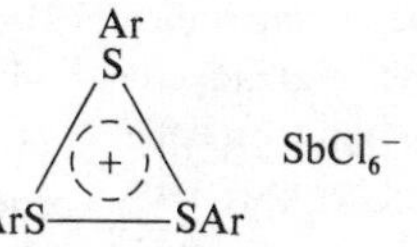

(19)

Alkylthio-sulphonium salts are efficient alkylthiolating agents, and several papers describing comparisons between $Me_2\overset{+}{S}SMe\ BF_4^-$ and MeSCl have appeared.[422–424] Labelling studies[422] indicate that the reaction of this salt with diastereoisomeric forms of MeSCHMeCHMeSMe, which involves C—S bond cleavage in the latter, is not simply a matter of S_N2 displacement, or *S*-methylation by the salt. The initial *S*-methylthiolation liberates Me_2S, while a further equilibrium, forming MeSSMe and the *S*-methylthiiranium salt, provides all the intermediates needed to account for the formation of MeSCHMeCHMe$\overset{+}{S}Me_2$ and MeSCHMeCHMe$\overset{+}{S}$(Me)SMe.[422] Exchange of MeS between MeSX (X = Cl or $-\overset{+}{S}Me_2\ BF_4^-$) and $MeSCH_2CH=CH_2$ occurs with the methylthio-sulphonium salt, but not with the sulphenyl chloride.[423] The overall reaction leads to $MeSCH_2CH(SMe)CH_2X$ with scrambling of the ^2H, which indicates a reversible degenerate rearrangement of the intermediate *S*-methylthio-sulphonium ion. Cleavage of both C—S bonds of a dithioacetal by

[418] F. Bellesia, R. Grandi, U. M. Pagnoni, and R. Trave, *J. Chem. Soc., Perkin Trans. 1*, 1979, 851.

[419] J. K. Kim, E. Lingman, and M. C. Caserio, *J. Org. Chem.*, 1978, **43**, 4545.

[420] R. Tanikaga, K. Tanaka, and A. Kaji, *J. Chem. Soc., Chem. Commun.*, 1978, 865.

[421] A. S. Gybin, W. A. Smit, V. S. Bogdanov, M. Z. Krimer, and J. B. Kalyan, *Tetrahedron Lett.*, 1980, **21**, 383.

[422] S. A. Anderson, J. K. Kim, and M. C. Caserio, *J. Org. Chem.*, 1978, **43**, 4822.

[423] J. K. Kim, M. L. Kline, and M. C. Caserio, *J. Am. Chem. Soc.*, 1978, **100**, 6243; J. K. Kim and M. C. Caserio, *J. Org. Chem.*, 1979, **44**, 1897; M. L. Kline, N. Bentov, J. K. Kim, and M. C. Caserio, *ibid.*, p. 1904.

$Me_2\overset{+}{S}SMe$ BF_4^- proceeds *via* methylthio-sulphonium ions, which rearrange by cleavage and re-formation of C—S and S—S bonds by way of an α-(methylthio)carbonium ion.[424]

Conversions of thiols into disulphides and thiolsulphonates through their reactions with DMSO and TFAA involve the alkylthio-sulphonium salts as intermediates, the thiolsulphonate arising when water is present.[420] Bis-(arylthio)sulphonium salts are effective reagents for the conversion of alkenes into thiirans.[421]

8 Thioacetals [1-(Alkylthio)alkyl Sulphides] and their Selenium Analogues

Saturated Thioacetals and Selenoacetals.—Standard preparations of dithioacetals and diselenoacetals have been developed further. Preparations from carbonyl compounds summarized in ref. 425 indicate current approaches, but some examples show unusual features [MeSeH and $ZnCl_2$ → $R^1R^2C(SeMe)_2$ + $R^1R^2CHSeMe$,[426] and R^1R^2C=CHCHO + PhSH → $R^1R^2C(SPh)CH_2CH(SPh)_2$[427]]. Syntheses from halides [CH_2Br_2 + ArSNa + 18-crown-6 ether → $CH_2(SAr)_2$;[428] H_2C=CHCl + Me_2S (at 450 °C) → $MeCH(SMe)_2$ + H_2C=CHSMe + 7% of H_2C=CHSH[429]]; by α-bis(sulphenylation) of ketones, using a disulphide and NaH;[430,431] by trapping the intermediate in the Pummerer rearrangement of a sulphoxide [$R^1SOCH_2R^2$ + TFAA → $R^1S\overset{+}{C}HR^2$ → $R^1SCHR^2(SR^3)$];[432] and from thiocarbonyl compounds [RCSSEt + EtMgI → $RC(SEt)_2MgI$;[433] a process used[434] in a synthesis of isoartemesia ketone] have also been recorded. Conversion of an acetal into a dithioacetal [2,5-dimethoxytetrahydrofuran → $(PhS)_2CH(CH_2)_2CH(SPh)_2$[435] with PhSH and HCl] and conversion of one dithioacetal [$PhSCR^1R^2CH_2CH$-$(SPh)_2$][427,436] into another [1,1-bis(phenylthio)cyclopropanes] by MeLi and TMEDA, illustrate further possibilities, the last-mentioned reaction involving intramolecular nucleophilic displacement of PhS⁻ by a dithioacetal carbanion.

The synthesis of the mono-*S*-oxo-dithioacetal moiety of the antibiotic sparso-mycin has been accomplished, starting from *N*-benzyloxycarbonyl-L-cysteine.[437]

O,S-Acetals can be prepared by treatment of α-(phenylthio)alkylboranes with

[424] J. K. Kim, J. K. Pau, and M. C. Caserio, *J. Org. Chem.*, 1979, **44**, 1544.
[425] A. Cravador and A. Krief, *C.R. Hebd. Seances Acad. Sci., Ser. C*, 1979, **289**, 267; M. Tazaki and M. Takagi, *Chem. Lett.*, 1979, 767; D. L. J. Clive and S. M. Menchen, *J. Chem. Soc., Chem. Commun.*, 1978, 356 and *J. Org. Chem.*, 1979, **44**, 1883; H. A. Klein, *Z. Naturforsch., Teil B*, 1979, **34**, 999.
[426] A. Cravador, A. Krief, and L. Hevesi, *J. Chem. Soc., Chem. Commun.*, 1980, 451.
[427] T. Cohen, W. M. Daniewski, and R. B. Weisenfeld, *Tetrahedron Lett.*, 1978, 4665.
[428] S. Tanimoto, T. Imanishi, S. Jo, and M. Okano, *Bull. Inst. Chem. Res., Kyoto Univ.*, 1978, **56**, 297 (*Chem. Abstr.*, 1979, **90**, 168 204).
[429] M. A. Kuznetsova, E. N. Deryagina, and M. G. Voronkov, *Zh. Org. Khim.*, 1979, **15**, 1996.
[430] M. Braun, *Tetrahedron Lett.*, 1978, 3695.
[431] Y. Nagao, K. Kaneko, K. Kawabata, and E. Fujita, *Tetrahedron Lett.*, 1978, 5021.
[432] R. Tanikaga, Y. Hiraki, N. Ono, and A. Kaji, *J. Chem. Soc., Chem. Commun.*, 1980, 41.
[433] A. I. Meyers, T. A. Tait, and D. L. Comins, *Tetrahedron Lett.*, 1978, 4657.
[434] P. Gosselin, S. Masson, and A. Thuillier, *Tetrahedron Lett.*, 1978, 2717.
[435] T. Cohen, D. Ouellette, and W. M. Daniewski, *Tetrahedron Lett.*, 1978, 5063.
[436] T. Cohen and J. R. Matz, *J. Org. Chem.*, 1979, **44**, 4816; T. Cohen and W. M. Daniewski, *Tetrahedron Lett.*, 1978, 2991.
[437] H. C. J. Ottenheijm and R. M. J. Liskamp, *Tetrahedron Lett.*, 1978, 2437.

N-chlorosuccinimide in MeOH (PhSCHRBR$_2$ → PhSCHROMe),[438] methanolysis of *S*-monomethyl dithioacetal fluorosulphonates,[439] and addition of an alkyl-lithium to thiono-esters.[440] Quantitative studies have been reported for an interesting variation on the simplest preparation, in which hemithioacetals exist in equilibrium with an aldehyde and a mixture of two thiols.[441]

Reactions of thioacetals and selenoacetals are overwhelmingly directed towards applications in synthesis, most of which rely on the formation of stabilized carbanions. The HMPA-induced addition of (PhS)$_2$CRLi to $\alpha\beta$-unsaturated ketones is almost entirely 1,4-oriented (see also ref. 370), providing a characteristic example of a dithioacetal carbanion as an acyl-ion equivalent[442] (see also Chap. 2, Pt. I, p. 95). Similar uses for the THF-soluble potassium dianion of bis-(ethylthio)acetic acid[443] (a new α-keto-acid synthon) and alkylselenylcyclopropyl carbanions[444] have been described. Full details have been published[445] of the uses of bis(phenylthio)-carbanions in C—C bond formation with carbonyl compounds which create several synthetic possibilities [(PhS)$_2$C(CH$_2$R^1)CHR^{2}OH + SOCl$_2$ + Et$_3$N → PhSC(=CHR^1)CHR^{2}SPh → sulphoxide → PhSC-(CHR^{1}OH)=CHR^2 by sulphoxide–sulphenate rearrangement].[446] The C—Se bond cleavage of diselenoacetals,[444] *S*,*Se*-acetals,[442] and seleno-orthoesters[447] is brought about by an alkyl-lithium, but deprotonation without C—Se bond cleavage is carried out with lithium tetramethylpiperidide.[447] This partial deselenation approach allows vinyl selenides and vinylsilanes to be prepared from β-hydroxy-α-trimethylsilyl selenides.[448]

The conclusion to uses of dithioacetals and selenium analogues in synthesis is usually a form of hydrolysis to give an aldehyde or a ketone. New 'dethio-acetalization' studies describe uses of NO$^+$ HSO$_4^-$, of either NaNO$_2$ or NaNO$_3$ with TFA, or of NO$_2^+$ BF$_4^-$,[449] of isoamyl nitrite,[450] or of Me$_2$$\overset{+}{S}$Br Br$^-$.[451] A comparison of four methods commonly used for dethioacetalization [HgCl$_2$ with CaCO$_3$; HgCl$_2$ with CuCl$_2$; H$_2$O$_2$; and (PhSe)$_2$O] has been made for the corresponding deselenoacetalization.[452] It appears that selenoacetals are more easily cleaved than thioacetals but more prone to elimination to give vinyl selenides.[452] Methyl selenoacetals are more easily cleaved than the phenyl analogues.[452]

[438] A. Mendoza and D. S. Matteson, *J. Organomet. Chem.*, 1978, **156**, 149; *J. Chem. Soc., Chem. Commun.*, 1978, 357.

[439] T. A. Hase and R. Kivikari, *Synth. Commun.*, 1979, **9**, 107.

[440] L. Narasimhan, R. Sanitra, and J. S. Swenton, *J. Chem. Soc., Chem. Commun.*, 1978, 719.

[441] M. S. Kanchuger and L. D. Byers, *J. Am. Chem. Soc.*, 1979, **101**, 3005.

[442] J. Lucchetti, W. Dumont, and A. Krief, *Tetrahedron Lett.*, 1979, 2695; L. Wartski, M. El Bouz, J. Seyden-Penne, W. Dumont, and A. Krief, *ibid.*, p. 1543.

[443] G. S. Bates, *J. Chem. Soc., Chem. Commun.*, 1979, 161.

[444] S. Halazy and A. Krief, *J. Chem. Soc., Chem. Commun.*, 1979, 1136; S. Halazy, J. Lucchetti, and A. Krief, *Tetrahedron Lett.*, 1978, 3971.

[445] P. Blatcher and S. Warren, *J. Chem. Soc., Perkin Trans. 1*, 1979, 1074.

[446] P. Blatcher and S. Warren, *Tetrahedron Lett.*, 1979, 1247.

[447] D. van Ende, A. Cravador, and A. Krief, *J. Organomet. Chem.*, 1979, **177**, 1.

[448] W. Dumont, D. van Ende, and A. Krief, *Tetrahedron Lett.*, 1979, 485.

[449] G. A. Olah, S. C. Narang, G. F. Salem, and B. G. B. Gupta, *Synthesis*, 1979, 273.

[450] K. Fuji, K. Ichikawa, and E. Fujita, *Tetrahedron Lett.*, 1978, 3561.

[451] G. A. Olah, Y. D. Vankar, M. Arvanaghi, and G. K. S. Prakash, *Synthesis*, 1979, 720.

[452] A. Burton, L. Hevesi, W. Dumont, A. Cravador, and A. Krief, *Synthesis*, 1979, 877.

Protection of carboxyl groups using methylthiomethyl esters can be reversed either by oxidation with H_2O_2 or methylation, followed by mild hydrolysis.[453]

A small number of papers that cover mechanistic aspects deal with the factors that stabilize the developing carbonium ion in the hydrolysis of O,S-thioacetals.[454] An electron-transfer, mild reduction system (iron polyphthalocyanine) has been shown[455] to reduce benzil dithioacetal to the β-keto-sulphide [PhCOCPh(SPh)$_2$ → PhCOCHPhSPh]. Homolysis of dithioacetals on heating with Bu^tOOBu^t in PhCl followed by a 1,2-shift of an alkylthio-group leads to 1,2-bis-(alkylthio)alkyl compounds.[456]

Synthesis of α-alkylthioalkyl sulphones by oxidation of dithioacetals with $KMnO_4$ has been reported.[457]

Keten Dithioacetals.—The structural relationship of these compounds with saturated dithioacetals is shown by their formation from dithioacetal carbanions *via* α-di(phenylthio)alkylboronate esters and reaction with carbonyl compounds and LiNPr$^i_2$ [(PhS)$_2$CHLi → (PhS)$_2$CHB(OR)$_2$ → (PhS)$_2$C=CR^{1}R^2].[458] Several alternative routes are available, including one involving the reaction of a carboxylic acid or ester with Al(SPh)$_3$, which has been used with $\alpha\beta$-unsaturated esters to give β-(phenylthio)keten dithioacetals [R^{1}R^{2}C=CR^{3}CO$_2$R^4 → PhSCR^{1}R^{2}C=C-(SPh)$_2$].[459] Synthesis from thiocarbonyl compounds, including CS$_2$, is a standard route [R^{1}R^{2}CHCS$_2$H → dianion with LiNPr$^i_2$ → R^{1}R^{2}C=C(SR^3)$_2$ with R^{3}X;[460] alk-2-enedithioate esters → conjugated keten dithioacetals;[461] α-(methylthio)methylenation of ketones with CS$_2$, base, and MeI,[462, 463] and similar treatment of phenyl cyanomethyl sulphones also gives the keten dithioacetals: PhSO$_2$CH$_2$CN → PhSO$_2$C(CN)=C(SMe)$_2$].[464] Difluoroketen dithioacetals F$_2$C= C(SEt)$_2$ are available from CF$_3$CH(SEt)$_2$ by reaction with Bu^{n}Li or LiNPr$^i_2$.[465] The further replacement of halogen by organometallic compounds yields the homologues RCF=C(SEt)$_2$, and the overall process amounts to the conversion of CF$_3$CO$_2$H into RCHFCO$_2$H, taking into account the other established transformations of functional groups.[465] [2,3]-Sigmatropic rearrangement of PhṠ-(ĊHSPh)CH$_2$C≡CMe gives (PhS)$_2$C=CMeCH=CH$_2$ in high yield,[466] and another

[453] J. M. Gerdes and L. G. Wade, *Tetrahedron Lett.*, 1979, 689.

[454] J. L. Jensen and W. P. Jencks, *J. Am. Chem. Soc.*, 1979, **101**, 1476; J. P. Ferraz and E. H. Cordes, *ibid.*, p. 1488; T. H. Fife and T. J. Przystas, *ibid.*, 1980, **102**, 292.

[455] H. Inoue, T. Nagata, H. Hata, and E. Imoto, *Bull. Chem. Soc. Jpn.*, 1979, **52**, 469.

[456] R. G. Petrova, T. D. Churkina, and V. I. Dostoralova, *Izv. Akad. Nauk SSSR, Ser. Khim.*, 1979, 1818.

[457] M. Poje and K. Balenovic, *Tetrahedron Lett.*, 1978, 1231.

[458] A. Mendoza and D. S. Matteson, *J. Org. Chem.*, 1979, **44**, 1352.

[459] T. Cohen and R. B. Weisenfeld, *J. Org. Chem.*, 1979, **44**, 3601; T. Cohen, R. E. Gapinski, and R. R. Hutchins, *ibid.*, p. 3599.

[460] F. E. Ziegler and C. M. Chan, *J. Org. Chem.*, 1978, **43**, 3065.

[461] M. Pohmakotr and D. Seebach, *Tetrahedron Lett.*, 1979, 2271.

[462] M. Augustin and C. Groth, *J. Prakt. Chem.*, 1979, **321**, 215.

[463] M. Augustin, C. Groth, H. Kristen, K. Peseke, and C. Wiechmann, *J. Prakt. Chem.*, 1979, **321**, 205.

[464] Y. Tominaga, S. Hidaki, Y. Matsuda, G. Kobayashi, and K. Sakemi, *Yakugaku Zasshi*, 1979, **99**, 540 (*Chem. Abstr.*, 1980, **91**, 157 539); S. Hidaki, Y. Tominaga, Y. Matsuda, G. Kobayashi, and K. Sakemi, *ibid.*, p. 1234.

[465] K. Tanaka, T. Nakai, and N. Ishikawa, *Chem. Lett.*, 1979, 175.

[466] V. Ratovelomanana and S. Julia, *Synth. Commun.*, 1978, **8**, 87; B. Cazes and S. Julia, *Tetrahedron Lett.*, 1978, 4065.

route to conjugated keten dithioacetals is based on carbene insertion into a disulphide [$ClCMe_2C \equiv CH$ + MeLi → $Me_2C=C=C$: → $Me_2C=C=C(SR)_2$ → $H_2C=CMeCH=C(SR)_2$].[467] An alternative rearrangement approach has been used[468] in which the carbene $(RS)_2C$: (generated from the sodium salt of *S*-allyl-*S'*-methyldithiocarbonate tosylhydrazone by pyrolysis) undergoes a [2,3]-sigmatropic rearrangement and is trapped with MeI to give $(MeS)_2C=CR^1CR^2=CH_2$.[468]

Applications of keten dithioacetals in synthesis that have been reported recently are mainly developments of existing possibilities, *i.e.* cleavage with lithium naphthalenide at −78 °C to give S-stabilized vinyl-lithium compounds for use in known routes to allyl alcohols or *αβ*-unsaturated carboxylic acids,[459] and nucleophilic substitution of an RS group in a keto-keten dithioacetal or related compounds,[463,464] giving pyridin-2(1*H*)-ones and indolizines,[464] or other heterocyclic compounds.[469] Conjugated keten dithioacetals are valuable synthons,[461,466] and the extension of a carbon skeleton through addition of an alkyl-lithium has been illustrated in a synthesis of (−)-(*E*)-lanceol.[466] 1,1-Bis-(phenylthio)cyclopropanes have been prepared from $H_2C=C(SPh)_2$ with an allyl sulphide carbanion.[470] Regioselectivity of the alkylation of lithiated $MeCH=C$-$(SR)_2$ is influenced by the structure of the group R, a greater degree of *γ*-alkylation being associated with greater bulk. Irrespective of this factor, only *γ*-alkylation is observed in the presence of a copper(I) salt.[471] Addition of dimethyl phosphonate to $PhCH=C(SMe)_2$, followed by catalytic reduction, yields secondary alkyl phosphonates $PhCHMeP(O)(OMe)_2$ in an unusual example of a standard use of keten dithioacetals.[472]

The rotational barrier about the double bond in $(MeS)_2C=CRCN$ (R = NO_2) is greater than that in the analogue (R = H).[473]

9 Trithio-orthoesters [1,1-Bis(alkylthio)alkyl Sulphides]

Limited but important uses of these compounds as equivalents of the formyl anion include interesting desulphurization techniques. 1,4-Additions to *αβ*-unsaturated carbonyl compounds[474,475] were followed by reduction with $CrCl_2$ [$(PhS)_3CR$ → $(PhS)_2CHR$] and subsequent Ag^+-catalysed dethioacetalization with trichloroisocyanuric acid, yielding *γ*-keto-aldehydes.[474]

Three sulphenyl substituents stabilize a carbon radical to a sufficient extent that 1,1,1,2,2,2-hexakis(trifluoromethylthio)ethane undergoes reversible homolysis in benzene solution.[476]

[467] J.-C. Clinet and S. Julia, *J. Chem. Res. (S)*, 1978, 12.
[468] T. Nakai and K. Mikami, *Chem. Lett.*, 1978, 1243.
[469] A. Kumar, H. Ila, and H. Junjappa, *J. Chem. Res. (S)*, 1979, 268; J. P. Marino and J. L. Kostusyk, *Tetrahedron Lett.*, 1979, 2489, 2493.
[470] T. Cohen, R. B. Weisenfeld, and R. E. Gapinski, *J. Org. Chem.*, 1979, **44**, 4744.
[471] F. E. Ziegler and C. C. Tam, *J. Org. Chem.*, 1979, **44**, 3428.
[472] M. Yamashita, T. Miyano, T. Watabe, H. Inokawa, H. Yoshida, T. Ogata, and S. Inokawa, *Bull. Chem. Soc. Jpn.*, 1979, **52**, 466.
[473] C. Dreier, L. Herricksen, S. Karlsson, and J. Sandstrom, *Acta Chem. Scand., Ser. B*, 1978, **32**, 281.
[474] T. Cohen and S. M. Nolan, *Tetrahedron Lett.*, 1978, 3533.
[475] R. A. J. Smith and A. R. Lal, *Aust. J. Chem.*, 1979, **32**, 353.
[476] A. Haas, K. Schlosser, and S. Steenkan, *J. Am. Chem. Soc.*, 1979, **101**, 6282.

10 Sulphuranes and Hypervalent Sulphur, Selenium, and Tellurium Compounds

The particular interest under this heading lies in current studies of sulphuranes, both as synthetic objectives and as reaction intermediates. While the numerous papers dealing with compounds that are multisubstituted at Se and Te atoms are summarized only briefly in this section, the growing interest in Se and Te analogues of reactions of organosulphur compounds will depend more and more on this work.

Extensions of recently established preparations of sulphuranes include the stable thiosulphurane $CF_3S(SCF_3)[OC(CF_3)_3]_2$, formed by oxidative addition of perfluoro-t-butyl hypochlorite to bis(trifluoromethyl) disulphide, and its C_6F_5 analogues.[477] A member of a further new family, $(CF_3)_2S(OCF_3)_2$, was formed by photolysis of CF_3OCl with bis(trifluoromethyl) sulphide, and the corresponding sulphurane oxide was similarly obtained from the bis(trifluoromethyl) sulphoxide.[478] Hydrolysis of these products gives bis(trifluoromethyl) sulphoxide and sulphone, respectively. The sulphurane oxide gives a new type of stable sulphurane oxide, $(CF_3)_2S(O)[N=C(CF_3)_2]$, with $LiN=C(CF_3)_2$. (+)-Bis-(o-hydroxymethylphenyl) sulphoxide gives a chiral spiro-sulphurane with MeCOCl at $-78\ °C$, which gives a racemic sulphurane oxide with ozone;[479] these are applications of methods introduced earlier by Martin's group (see Vol. 4, p. 30), but the stereochemical implications may need to be accepted with caution until studies of the racemization of chiral sulphuranes have been undertaken.

An anomalous reaction $[Me_2\overset{+}{S}CR(CN)CO_2Me \rightarrow MeO_2CCR(CN)CR-(CN)CO_2Me]$, which has been observed during an intended nucleophilic substitution, has been ascribed to homolysis of the sulphurane formed by addition of the nucleophile at sulphur.[480]

Direct fluorination by F_2,[481] or by XeF_2,[482] of sulphides, selenides, and tellurides gives the difluorochalcogen(IV) compounds,[481] further fluorination giving '*cis*' (21) and '*trans*' (20) Ph_2SF_4 in the case of the sulphide.[483] The additional bibliography is concerned with preparations and reactions of halogeno-chalcogens.

(20)   (21)

Additional Bibliography: G. Kleeman and K. Seppelt, *Angew. Chem.*, 1978, **90**, 547 and *Chem. Ber.*, 1979, **112**, 1140 [MeSF$_5$; from SClF$_5$ + CH$_2$C=O → F$_5$SCH$_2$COCl and conventional elaboration *via* F$_5$SCH$_2$Br, from which methylenesulphur tetrafluoride (H$_2$C=SF$_4$), which shows some of the reactions expected of a carbene, can be prepared, using Bu^{n}Li at $-110\ °C$]; R. A. DeMarco and W. B. Fox, *J. Fluorine Chem.*, 1978, **12**, 137 [addition of HF (from KF and formamide) to F$_5$SCH=CH$_2$]; A. H. Cowley, D. J. Pagel, and M. L. Walker, *J. Am. Chem. Soc.*, 1978, **100**, 7065

[477] Q. C. Mir, D. P. Babb, and J. M. Shreeve, *J. Am. Chem. Soc.*, 1979, **101**, 3961.
[478] T. Kitazume and J. M. Shreeve, *Inorg. Chem.*, 1978, **17**, 2173.
[479] P. Huszthy, I. Kapovits, A. Kucsman, and L. Radics, *Tetrahedron Lett.*, 1978, 1853.
[480] G. Morel, E. Marchand, and A. Foucaud, *Tetrahedron Lett.*, 1978, 3719.
[481] I. Ruppert, *Chem. Ber.*, 1979, **112**, 3023.
[482] Yu. L. Yagupolskii and T. I. Savina, *Zh. Org. Khim.*, 1979, **15**, 438.
[483] I. Ruppert, *J. Fluorine Chem.*, 1979, **14**, 81.

[(Me$_2$N)$_x$$\overset{+}{S}F_{3-x}$ from (Me$_2$N)$_x$SF$_{4-x}$ with BF$_3$, PF$_5$, or AsF$_5$ in SO$_2$]; D. G. Garratt, M. Ujjainwalla, and G. H. Schmid, *J. Org. Chem.*, 1980, **45**, 1206 [stepwise *anti*-addition of alkene to SeCl$_4$ to give (2*RS*,3*RS*)-(MeCHClCHMe)$_2$SeCl$_2$]; D. G. Garratt and G. H. Schmid, *Chem. Scr.*, 1977, **11**, 170 [reactions of MeSeCl$_3$ with alkynes are neither stereospecific nor regiospecific]; E. Lehmann, *J. Chem. Res.* (*S*), 1978, 42 [CF$_3$SeSeCF$_3$ + BrF$_3$ → CF$_3$SeF$_3$]; N. Ya. Derkach, N. P. Tishchenko, and V. G. Voloshchuk, *Zh. Org. Khim.*, 1978, **14**, 958 [R^{1}SeR^2 + Ph$_2$SeCl$_2$ + Bu^{t}OCl → R^{1}R^{2}Se(OBu^t)Cl]; E. S. Mamelov, R. S. Salakhova, and T. M. Gadzhily, *Azerb. Khim. Zh.*, 1979, 107 [R$_2^1$SeCl$_2$ + R^{2}OH → R$_2^1$Se(OR^2)$_2$]; I. D. Sadekov, A. A. Maksimenko, and B. B. Rivkin, *Zh. Org. Khim.*, 1978, **14**, 874 [silyl enol ethers + RTeCl$_3$ or TeCl$_4$ → RCOCH$_2$TeCl$_2$R or (RCOCH$_2$)$_2$TeCl$_2$]; S. Uemura, H. Miyoshi, and M. Okano, *Chem. Lett.*, 1979, 1357 [PhC≡CR + TeCl$_4$ → (*Z*)-PhCCl=CRTeCl$_3$]; J. Bergmann and L. Engman, *J. Organomet. Chem.*, 1979, **181**, 335 [alkene + TeCl$_4$ → β-chloroalkyltellurium(IV) chloride, while alkene + TeCl$_4$ + EtOH → β-ethoxyalkyl analogue]; J. Bergmann and L. Engman, *J. Organomet. Chem.*, 1979, **175**, 233 [R$_n$TeCl$_{4-n}$ + Ni(CO)$_4$ in DMF → RCO$_2$H]; V. Kumar, P. H. Bird, and B. C. Pant, *Synth. React. Inorg. Met.-Org. Chem.*, 1979, **9**, 203 [R$^1_4$Pb or R$^1_6$Pb$_2$ + R$^2_2$TeCl → R^{1}R^{2}TeCl$_2$]; I. D. Sadekov, A. I. Usachev, and V. I. Minkin, *Zh. Obshch. Khim.*, 1978, **48**, 475 [telluronium ylide + halogen → Ar$_2$TeHal$_2$]; T. N. Srivastava, R. C. Srivastava, and M. Singh, *Indian J. Chem., Sect. A*, 1979, **18**, 527 [R$^1_2$TeCl$_2$ + R$^2_4$M$^+$ X$^-$ → (R$^2_4$M$^+$)$_2$ [R$^1_2$TeCl$_2$X$_2$]$^{2-}$ (M = N, P, As, or Sb; X = Cl or Br)]; T. N. Srivastava, R. C. Srivastava, and M. Singh, *J. Organomet. Chem.*, 1978, **160**, 449 [R$_2$Te + ICl or (SCN)$_2$ → R$_2$TeICl or R$_2$Te(SCN)$_2$]; T. N. Srivastava, R. C. Srivastava, and M. Singh, *Inorg. Chim. Acta*, 1979, **33**, 299 [R^{1}R^{2}Te + R^{3}CON(Hal)COR^3 → R^{1}R^{2}Te(Hal)N(COR^3)$_2$]; V. I. Naddaka, V. P. Garkin, I. D. Sadekov, V. P. Krasnov, and V. I. Minkin, *Zh. Org. Khim.*, 1979, **15**, 896 [R$^1_2$TeCl(OBu^t) → R$^1_2$Te=NSO$_2$R^2]; F. H. Musa and W. R. McWhinnie, *J. Organomet. Chem.*, 1978, **159**, 37 [R$_2$Te(NCS)$_2$ and Ph$_2$MeTe(NCS), the latter showing telluronium salt character in DMSO but a covalent structure in C^{2}HCl$_3$]; I. D. Tseimakh, A. Ya. Bushkov, L. E. Nivorozhkin, and I. D. Sadekov, *Zh. Obshch. Khim.*, 1978, **48**, 1658 [Ar$^1_2$Te + Ar$^2_2$TeCl$_2$ ⇌ Ar$^1_2$TeCl$_2$ + Ar$^2_2$Te, studied as a function of structure and solvent].

11 Sulphoxides, Selenoxides, and Telluroxides

Preparation.—Methods involving the oxidation of sulphides and selenides are either well established, and a subsidiary part of recent papers, or are the main feature in papers describing novel modifications of standard methods. Points of interest arising from references on this topic, collected at the end of this section, include the use of standard oxidants [NaIO$_4$ and Tl(NO$_3$)$_3$], but adsorbed on silica gel or alumina, and the asymmetric induction, at high levels, that accompanies microbiological or protein-mediated oxidation. 2-Arenesulphonyl-3-aryl-oxaziridines have emerged as suitable reagents for the oxidation of sulphides to sulphoxides,[484] the (−)-camphor-10-sulphonyl analogue showing a low level of enantioselectivity in this process.[485]

A study has been made[486] of the conversion of selenides and tellurides into their mono-oxides, using *N*-chlorosuccinimide or Bu^{t}OCl followed by alkaline hydrolysis; some fragmentation of aliphatic selenoxides occurs at this stage.

Other routes to sulphoxides include the first preparation from sulphones (though the direct partial reduction has yet to be discovered). Conversion of a sulphone into the corresponding aryloxy-salt, R$^1\overset{+}{S}$(O)(OR^2)R^3 X$^-$, can be performed surprisingly easily,[487,488] using an arenediazonium tetrafluoroborate. This salt is then open to reduction (with RSH[487] or NaBH$_4$ with Al$_2$O$_3$[488]), reduction with α-substitution

[484] F. A. Davis, R. Jenkins, and S. G. Yocklovich, *Tetrahedron Lett.*, 1978, 5171.

[485] F. A. Davis, R. Jenkins, S. Q. A. Rizvi, and T. W. Panunto, *J. Chem. Soc., Chem. Commun.*, 1979, 600.

[486] M. R. Detty, *J. Org. Chem.*, 1980, **45**, 274.

[487] M. Shimagaki, H. Tsuchiya, Y. Ban, and T. Oishi, *Tetrahedron Lett.*, 1978, 3435.

[488] I. W. J. Still and S. Szilagyi, *Synth. Commun.*, 1979, **9**, 923.

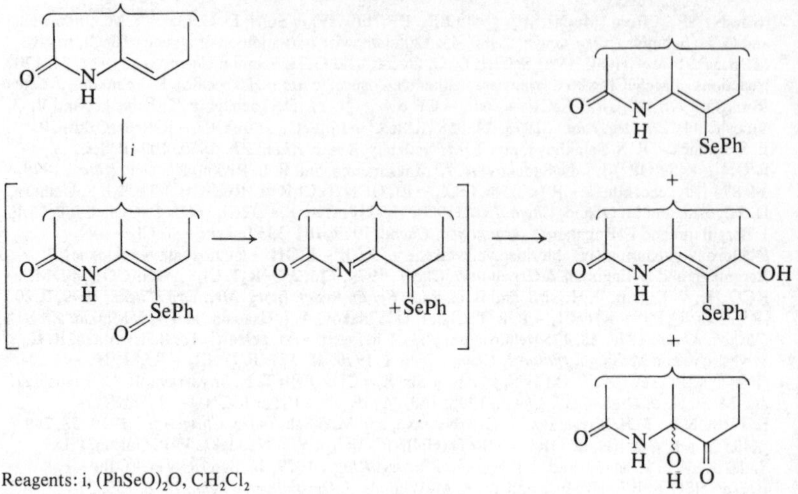

Reagents: i, (PhSeO)$_2$O, CH$_2$Cl$_2$

Scheme 18

[Ph$\overset{+}{\text{S}}$(O)(OPh)Me + RLi and LiX → PhSOCH$_2$X + PhSO$_2$Me],[487] or reduction with homologation (with R$_2$CuLi and LiX).[487] Standard sulphinylation procedures reported include electrophilic substitution of phenols with ArSOCl and AlCl$_3$[489] and the preparation of optically active sulphoxides from resolved (−)-menthyl sulphinates with (E)-1-alkenylmagnesium bromides[490] or arylthiomethyl-[491] or arylsulphonylmethyl-lithium compounds.[492] A practical route to 1-alkenyl sulphoxides uses non-activated alkynes and sulphenic acids, generated *in situ* from the thermolysis of 2-cyanoethyl phenyl sulphoxides.[493]

The wider use of benzeneseleninic anhydride as a mild oxidant can be expected to generate reports of side-reactions, and the formation of a selenoxide during oxidation of 4-azapregn-5-ene-3,20-dione (Scheme 18), followed by Pummerer-type reactions, accounts for the observed products.[494]

Additional Bibliography: T. Nishio and Y. Omote, *Synthesis*, 1980, 392 [NaIO$_4$ on silica gel for R^{1}COCH$_2$CR^{2}R^{3}SR^4 → R^{1}COCH=CR^{2}R^3 *via* the sulphoxide]; K.-T. Liu and Y.-C. Chien, *J. Org. Chem.*, 1978, **43**, 2717 [NaIO$_4$ on Al$_2$O$_3$]; K.-T. Liu and Y.-C. Tong, *J. Chem. Res.*(*S*), 1979, 276 [NaIO$_4$ on Al$_2$O$_3$, with (−)-2-(*S*)-methylbutanol for conversion of sulphide → sulphoxide, but only with 0.6—8.9% enantiomeric excess]; K.-T. Liu, *J. Chin. Chem. Soc.* (*Taipei*), 1977, **24**, 217 [Tl(NO$_3$)$_3$ on Al$_2$O$_3$ for sulphides → sulphoxides, but with 15% sulphone in the case of Bu^{t}S]; E. Abushanab, D. Reed, F. Suzuki, and C. J. Sih, *Tetrahedron Lett.*, 1978, 3415 [*p*-tolyl methyl sulphide gives (+)-(*R*)-sulphoxide with *Mortierella isabellina*, but gives (−)-(*S*)-sulphoxide with *Helminthosporium* sp. NRRL 4671]; T. Sugimoto, T. Kokubo, J. Miyazaki, S. Tanimoto, and M. Okano, *J. Chem. Soc., Chem. Commun.*, 1979, 1052 [diaryl sulphides → sulphoxides and sulphones with H$_2$O$_2$ and bovine serum albumin, the high (>90%) optical purity of the sulphoxides being a cumulative result of stereoselective oxidation of the sulphide and enantioselectivity in the further oxidation to the sulphone]; F. Alcudia, E. Brunet, and J. L. Garcia Ruano, *An. Quim.*, 1979, **75**, 162

[489] D. W. Chasar and T. M. Pratt, *Phosphorus Sulfur*, 1978, **5**, 35.
[490] G. H. Posner and P.-W. Tang, *J. Org. Chem.*, 1978, **43**, 4131.
[491] L. Colombo, C. Gennari, and E. Narisano, *Tetrahedron Lett.*, 1978, 3861.
[492] R. Annunziata, M. Cinquini, and F. Cozzi, *Synthesis*, 1979, 535.
[493] D. N. Jones, P. D. Cottam, and J. Davies, *Tetrahedron Lett.*, 1979, 4977.
[494] T. G. Back and N. Ibrahim, *Tetrahedron Lett.*, 1979, 4931.

[formation of diastereoisomers from (+)-MeSCHPhCH$_2$OR → sulphoxide by using NaIO$_4$, *m*-chloroperoxybenzoic acid, or H$_2$O$_2$, but with a higher (80%) enantioselectivity with *m*-chloroperoxybenzoic acid]; C. Srinivasan, P. Kuthalingam, and N. Arumuyan, *Can. J. Chem.*, 1978, **56**, 3043 and *Indian J. Chem., Sect. A*, 1978, **16**, 478 [S$_2$O$_8$$^{2-}$]; J. Drabowicz and M. Mikolajczyk, *Synthesis*, 1978, 758 [H$_2$O$_2$ and SeO$_2$ in MeOH]; J. Drabowicz, W. Midura, and M. Mikolajczyk, *Synthesis*, 1979, 39 [Br$_2$ and KHCO$_3$, suitable for the synthesis of ^{18}O-enriched sulphoxides in high yields]; N. Kunieda, H. Mochizuki, and M. Kinoshita, *Mem. Fac. Eng., Osaka City Univ.*, 1978, **19**, 207 (*Chem. Abstr.*, 1980, **92**, 5664) [Br$_2$ and Tröger's base is a system that is potentially suitable for asymmetric synthesis]; L. G. Faehl and J. L. Kice, *J. Org. Chem.*, 1979, **44**, 2357 [*p*-chlorobenzeneselenonic acid (ArSeO$_3$H), which, on oxidation of a sulphide to a sulphoxide, is itself reduced to ArSeO$_2$H; this is an even more effective oxidant for synthesis of sulphoxides in the presence of acid]; V. A. Likholobov, A. V. Mashkina, M. G. Volkhonskii, and V. N. Yakovleva, *Kinet. Katal.*, 1979, **20**, 1152 [O$_2$ and inorganic nitrate]; E. Fujita, M. Ochiai, and K. Watanabe, *J. Indian Chem. Soc.*, 1978, **55**, 1226 [Tl(NO$_3$)$_3$, for selenides → selenoxides]; L. Jirackova, T. Jelinkova, J. Rotschova, and J. Pospisil, *Chem. Ind.* (*London*), 1979, 384 [Bu^{t}OOH]; H. C. Brown and A. K. Mandal, *J. Org. Chem.*, 1980, **45**, 916 [whereas the use of H$_2$O$_2$ to terminate a hydroboronation that is conducted in the presence of Me$_2$S and excess NaOH does not oxidize the sulphide, the use of NaOCl is unsatisfactory because it oxidizes the sulphide and not the organoborane]; G. P. Dareshwar and B. D. Hosangadi, *Indian J. Chem., Sect. B*, 1978, **16**, 143 [intramolecular oxidation of *o*-NH$_2$OCOC$_6$H$_4$SPh → *o*-NH$_2$COC$_6$H$_4$SOPh]; H. S. Laver and J. R. MacCallum, *Photochem. Photobiol.*, 1978, **28**, 91 [photosensitized oxidation]; W. Ando, S. Komoto, and K. Nishizawa, *J. Chem. Soc., Chem. Commun.*, 1978, 894 [methylene-blue-sensitized photo-oxidation of Ph$_2$S + Me$_2$S̈–C̄(CO$_2$Me)$_2$ → Ph$_2$SO *via* the oxenoid intermediate rather than a 1,2-dioxetan]; S. Iriuchijima and G. Tsuchihashi, *Yuki Gosei Kagaku Kyokaishi*, 1978, **36**, 1050 (*Chem. Abstr.*, 1979, **91**, 20 039) [photoaddition with concurrent photo-oxygenation; PhSH + H$_2$C=CMeCN → PhSOCH$_2$CMe(CN)OH → PhSOCH$_2$COMe]; T. Tezuka, H. Miyazaki, and H. Suzuki, *Tetrahedron Lett.*, 1978, 1959 [Sinnreich mechanism for photo-oxygenation of sulphides *via* persulphoxides, RSCH(OOH)R, (see Vol. 5, pp. 22, 34) was rejected in favour of direct attack of singlet oxygen on sulphur].

Deoxygenation of Sulphoxides, Selenoxides, and Telluroxides.—References and brief details from recent papers on this topic include:

G. A. Olah, Y. D. Vankar, and M. Arvanaghi, *Synthesis* 1979, 984 [pyridine or R$_3$N with SO$_3$, I$_2$, and NaI]; G. A. Olah, B. G. B. Gupta, and S. C. Narang, *J. Org. Chem.*, 1978, **43**, 4503 [(Me$_2$N)$_3$P with I$_2$]; G. A. Olah, R. Malhotra, and S. C. Narang, *Synthesis*, 1979, 58 [(COCl)$_2$ with NaI, or NaI with SOCl$_2$, POCl$_3$, or PCl$_5$]; G. A. Olah, S. C. Narang, B. G. B. Gupta, and R. Malhotra, *Synthesis*, 1979, 61 [Me$_3$SiCl with NaX (X = I, CN, SMe, or SCN)]; T. Numata, H. Togo, and S. Oae, *Chem. Lett.*, 1979, 329 [PhSSiMe$_3$]; M. R. Detty, *J. Org. Chem.*, 1979, **44**, 4528 [PhSeSiMe$_3$ for the deoxygenation of sulphoxides, selenoxides, and telluroxides]; H. S. D. Soysa and W. P. Weber, *Tetrahedron Lett.*, 1978, 235 [disilathianes, *e.g.* (Me$_3$Si)$_2$S]; M. Sekine, H. Yamagata, and T. Hata, *Tetrahedron Lett.*, 1979, 375 [P(O)(SiMe$_3$)$_3$]; I. W. J. Still, J. N. Reed, and K. Turnbull, *Tetrahedron Lett.*, 1979, 1481 [PSBr$_3$]; I. W. J. Still, S. K. Hasan, and K. Turnbull, *Can. J. Chem.*, 1978, **56**, 1423 [P$_4$S$_{10}$ for deoxygenation of sulphoxides and selenoxides]; J. N. Denis and A. Krief, *Tetrahedron Lett.*, 1979, 3995 [P$_2$I$_4$]; D. L. J. Clive and S. M. Menchen, *J. Chem. Soc., Chem. Commun.*, 1979, 168 [B(SePh)$_3$ or B(SeMe)$_3$]; J. Drabowicz and M. Mikolajczyk, *Synthesis*, 1978, 542 [H$_2$NC(=NH)SO$_2$H with I$_2$ as catalyst]; V. Ruffato and U. Miotti, *Gazz. Chim. Ital.*, 1978, **108**, 91 [diallyl sulphide and HClO$_4$]; I. D. Sadekov and A. A. Maksimenko, *Zh. Org. Khim.*, 1978, **14**, 2621 [tellurides from telluroxides in HCONH$_2$ at 120–150 °C].

Reactions of Saturated Sulphoxides and Selenoxides.—The main reactions of saturated sulphoxides and selenoxides are associated with the nucleophilic properties of the sulphinyl and seleninyl groups, and their activating effect on the adjacent saturated carbon atoms. The Pummerer rearrangement (as well as rearrangements of unsaturated sulphoxides discussed in a later section) and alkene-forming eliminations are instances of the former property; the main consequence of the latter property, the formation of sulphinyl-stabilized carbanions, is covered in Chapter 2.

An extraordinary example of a Pummerer rearrangement[495] is that in which

[495] R. R. King, *J. Org. Chem.*, 1978, **43**, 3784.

(22)

4-methylsulphinyl-3,5-xylenol gives the 4-methylthio-2-acetoxy analogue and (22) on treatment with Ac_2O, and the 4-chloro- and 4-methylthio-2-chloro-analogues on treatment with AcCl. This has been accounted for on the basis of acetoxy-sulphonium ion intermediates, as in more typical examples of the reaction. Continuing studies of asymmetric induction accompanying the rearrangement have established an unusually high level (70% e.e.) for optically active β-oxoalkyl sulphoxides when DCCI is used with Ac_2O,[496] but lower levels with phosphonates $(R^1O)_2P(O)CH_2SOR^2$.[497] The sulphoxide of S-benzylcysteinylglycine undergoes a Pummerer rearrangement with Ac_2O, to give a 1,3-thiazin-4-one.[498]

Racemization of sulphoxides in non-aqueous solvents containing HCl involves an S-chlorosulphonium ion, since products from cleavage of the C—S bond and α-chloro-sulphides are formed.[499-501] Chlorination of sulphoxides leads to α-chloro-sulphides[501, 502] in non-aqueous solvents, while a 'sulphohaloform' reaction is brought about by chlorinolysis in aqueous media.[503] This reaction, leading to a sulphonyl chloride, also applies to aliphatic sulphides, which undergo S-oxidation.[503] A chiral N-halogeno-caprolactam oxidizes the (S)-enantiomer of $(\pm)$-p-$MeC_6H_4SOCH_2Ph$ to the sulphone at a faster rate than the (R)-isomer.[504] An α-chloro-sulphoxide yields a thiolester by a process that has been wrongly described[505] as a modified Pummerer reaction [$RCHClS(O)Ph + LiNPr^i_2 + Me_3SiCl \rightarrow RCOSPh$]. The conditions are quite different. Mention has been made in discussing the preceding example of 'halogeno-Pummerer reactions' of C—S bond cleavage; this is a process which occurs in aqueous acid media[506] or in TFAA followed by hydrolysis,[507] and which converts benzyl sulphoxides into aromatic aldehydes.

Acid-catalysed exchange between sulphoxides and sulphides occurs *via* sulphonium intermediates,[508] whereas less certainty can be asserted for the mechanistic details of the photo-metathesis $PhSMe + PhSe(=O)Ph \rightarrow PhS(=O)Me + Ph_2Se$.[509]

[496] T. Numata, O. Itoh, and S. Oae, *Tetrahedron Lett.*, 1979, 1869.
[497] M. Mikolajczyk, A. Zatorski, S. Grzejszczak, B. Costisella, and W. Midura, *J. Org. Chem.*, 1978, **43**, 2518.
[498] S. Wolfe, P. M. Kazmaier, and H. Auksi, *Can. J. Chem.*, 1979, **57**, 2412.
[499] E. Ciuffarin and S. Gambarotta, *J. Chem. Res. (S)*, 1978, 274; M. Cioni, E. Ciuffarin, S. Gambarotta, M. Isola, and L. Senatore, *ibid.*, p. 270.
[500] N . Kunieda, T. Numata, and S. Oae, *Phosphorus Sulfur*, 1977, **3**, 1.
[501] J. Klein, *Chem. Lett.*, 1979, 359.
[502] T. Masuda, N. Furukawa, and S. Oae, *Bull. Chem. Soc. Jpn.*, 1978, **51**, 2659.
[503] D. G. Kay, R. F. Langler, and J. E. Trenholm, *Can. J. Chem.*, 1979, **57**, 2185.
[504] Y. Sato, N. Kunieda, and M. Kinoshita, *Mem. Fac. Eng., Osaka City Univ.*, 1977, **18**, 435 (*Chem. Abstr.*, 1979, **90**, 22 190).
[505] K. M. More and J. Wemple, *J. Org. Chem.*, 1978, **43**, 2713.
[506] G. A. Russell and J. M. Pecoraro, *J. Org. Chem.*, 1979, **44**, 3990.
[507] H. Sugihara, R. Tanikaga, and A. Kaji, *Synthesis*, 1978, 881.
[508] E. Ciuffarin, S. Gambarotta, M. Isola, and L. Senatore, *J. Chem. Res. (S)*, 1978, 272.
[509] T. Tezuka, H. Suzuki, and H. Miyazaki, *Tetrahedron Lett.*, 1978, 4885.

The c.i.d.n.p. associated with the photolysis of aryl alkyl sulphoxides is consistent with C–S bond homolysis (ArSOMe → ArSO· + Me·).[510] An optically active N-chloro-sulphomimide reacts with a sulphoxide to give the corresponding α-chloroalkyl sulphoxide with a small degree of enantiomeric excess involving S as the chiral centre.[511] Labelling studies with ^{18}O would seem to be necessary to establish the oxygen-exchange process which might be involved in this reaction. Exchange of an alkyl group between a sulphoxide and a Grignard reagent has been reported,[512] but this is not the only reaction occurring in this system, since sulphides (including products corresponding to the α-alkylation of the initial sulphoxide) are the reaction products.[512] Fluorination of sulphoxides gives the sexivalent sulphur compounds $R_2S(O)F_2$.[513]

Papers describing reactions of simple sulphoxides and uses of dimethyl sulphoxide as a reagent in organic synthesis are collected together here. The thermal decomposition of dialkyl sulphoxides[514] involves the formation of an alkene, a breakdown reaction not available to dimethyl sulphoxide, although other pathways account for side-products in reactions of this compound. Other reactions of DMSO include those with Se at 110 °C in aqueous KOH, which give Me_2S and MeSSMe, but only traces of MeSeSeMe, MeSeSMe, and $MeSCH_2SeMe$;[515] ion–molecule reactions in the gas phase;[516] oxidation by aqueous Chloramine-T;[517] and γ-radiolysis, which gives methane, ethane, propane, HCHO, CO_2, and methyl sulphides, disulphides, thiolsulphonates and Me_3S^+ $MeSO_3^-$.[518] DMSO has also been used as a reagent for the photo-oxidation of Bu^n_2S;[519] as a reagent with Bu^tBr for the conversion of carboxylic acids into methylthiomethyl esters, the effective species being the Pummerer rearrangement intermediate;[520] and to effect the related reaction Cl_3CCN → $Cl_3CCONHCH_2SMe$.[521] Its use as an oxidizing agent encompasses the oxidation of chloro-thiolformates (PhSCOCl → CO_2 + Me_2S + $MeSCH_2Cl$);[522] the oxidation of α-bromo-ketones ($RCOCH_2Br$ → $RCOCH_2\overset{+}{S}Me_2$ Br^- → RCOCHO + RCOCOSMe + $RCOCO_2H$ + RCO_2H);[523] the oxidation of alcohols in combination with $(COCl)_2$,[524,525] or $SOCl_2$,[524] or TFAA;[526] and a

[510] K. A. Muszkat, K. Praefcke, I. Khait, and R. Lüdersdorf, *J. Chem. Soc., Chem. Commun.*, 1979, 898.

[511] H. Morita, H. Itoh, N. Furukawa, and S. Oae, *Chem. Lett.*, 1978, 817.

[512] N. A. Nesmeyanov, V. A. Kalyavin, and O. A. Reutov, *Izv. Akad. Nauk SSSR, Ser. Khim.*, 1978, 2633.

[513] I. Ruppert, *Angew. Chem.*, 1979, **91**, 941.

[514] Yu. E. Nikitin, E. M. Baranovskaya, and G. G. Bikbaeva, *Zh. Prikl. Khim.*, (*Leningrad*), 1979, **52**, 430 (*Chem. Abstr.*, 1979, **91**, 38 636).

[515] G. K. Musovin, B. A. Trofimov, S. V. Amosova, V. Yu. Vitkovskii, and G. A. Kalabin, *Zh. Obshch. Khim.*, 1979, **49**, 831.

[516] J. E. Fulford, J. W. Dupuis, and R. E. March, *Can. J. Chem.*, 1978, **56**, 2324.

[517] D. S. Mahadevappa, M. B. Jadhav, and H. M. K. Naidu, *Int. J. Chem. Kinet.*, 1979, **11**, 261.

[518] M. C. Gutierrez, R. Barrera, H. Benz, and R. Parellada, *An. Quim.*, 1977, **73**, 520.

[519] M. A. Fox, P. K. Miller, and M. D. Reiner, *J. Org. Chem.*, 1979, **44**, 1103.

[520] A. Dossena, R. Marchelli, and G. Casnati, *J. Chem. Soc., Chem. Commun.*, 1979, 370.

[521] I. V. Bodrikov, A. A. Michurin, and N. N. Bochkareva, *Zh. Org. Khim.*, 1979, **15**, 436.

[522] A. Queen, A. E. Lemire, A. F. Janzen, and M. N. Paddon-Row, *Can. J. Chem.*, 1978, **56**, 2884.

[523] N. Saldabols and A. Cimanis, *Zh. Org. Khim.*, 1978, **14**, 1910.

[524] K. Omura and D. Swern, *Tetrahedron*, 1978, **34**, 1651.

[525] A. J. Mancuso, S.-L. Huang, and D. Swern, *J. Org. Chem.*, 1978, **43**, 2480; A. J. Mancuso, D. S. Brownfain, and D. Swern, *ibid.*, 1979, **44**, 4148.

[526] S. L. Huang, K. Omura, and D. Swern, *Synthesis*, 1978, 297.

curious observation that, since some labelled Me$_2$S is formed in the reduction of PhCHO by NaBH$_4$ in [^{14}C]DMSO, the aldehyde must act as a relay in the reaction through the formation of an alkoxysulphonium salt.[527]

Diphenyl selenoxide has been advocated for the oxidation of catechols to *ortho*-quinones in the synthesis of benzylisoquinoline alkaloids.[528]

Elaboration of other functional groups in aliphatic sulphoxides is illustrated in the preparation of α-chloro-β-hydroxyalkyl sulphoxides by the phase-transfer-catalysed reaction of α-chloroalkyl sulphoxides with carbonyl compounds, and their further conversion into $\alpha\beta$-epoxyalkyl sulphoxides.[529] Optically active β-oxoalkyl sulphoxides are reduced by hydride reagents to the β-hydroxyalkyl analogues with 60—70% optical yields.[530] Phenylsulphinylcarbene (PhSOCH:) is a relatively stable species, thus requiring the sulphinyl group to act as a good electron donor in spite of the conventional view that it is an electron-withdrawing group.[531] This carbene undergoes the expected addition to alkenes, and structure–rate relationships have been reported.[531] Vinylsulphinylcarbenes, formed from 5-alkylsulphinyl-3,3-dimethyl-3H-pyrazoles by photolysis, or from sulphinylcyclopropenes, readily undergo sigmatropic [1,2]-rearrangements to vinylsulphines.[532]

The *syn*-elimination of aliphatic sulphoxides and selenoxides has become a widely used method for the introduction of a double bond;[533-542] recent papers referring to elimination reactions of sulphenic acids have involved the preparation of vinyl chlorides[534] and bromomethyl ketones[535] from halogenomethyl phenyl sulphoxides (PhSOCHClLi + RCH$_2$X → PhSOCHClCH$_2$R → RCH=CHCl); the use of γ-keto-sulphoxides as equivalents of vinyl ketones [EtSO(CH$_2$)$_2$COR^1 → EtSOCHR^{2}CH$_2$COR^1 → R^{2}CH=CHCOR^1];[536] and the generation of inden-1-one, which was trapped as the Diels–Alder adduct.[537] Similar uses of selenoxides are the conversion of 4,4-disubstituted cyclohex-2-en-1-ones into cyclohexa-2,5-dienones;[538] the sequence PhSeBr + H$_2$C=CHOEt, then an allyl alcohol, → PhSeCH$_2$CH(OEt)OCR^{1}R^{2}CR^3=CR^{4}R^5 → R^{1}R^{2}C=CR^{3}CR^{4}R^{5}CH$_2$CO$_2$Et after elimination of selenoxide and Claisen rearrangement;[539] and the preparation of allylic alcohols and oxirans from β-hydroxy-selenoxides.[540, 541] In the case of alkyl aryl selenoxides, the areneselenenic acid formed by the *syn*-elimination adds rather readily to the alkene, leading to a β-hydroxy-selenide;[542] this can be avoided by

[527] D. C. Wigfield and R. T. Pon, *J. Chem. Soc., Chem. Commun.*, 1979, 910.

[528] J. P. Marino and A. Schwartz, *Tetrahedron Lett.*, 1979, 3253.

[529] F. Durst, K.-C. Tin, F. De Reinach-Hirtzbach, J. M. Decesare, and D. M. Ryan, *Can. J. Chem.*, 1979, **57**, 258.

[530] R. Annunziata, M. Cinquini, and F. Cozzi, *J. Chem. Soc., Perkin Trans. 1*, 1979, 1687.

[531] C. G. Venier and M. A. Ward, *Tetrahedron Lett.*, 1978, 3215.

[532] M. Franck-Neumann and J. J. Lohmann, *Tetrahedron Lett.*, 1979, 2397.

[533] T. E. Boothe, J. L. Greene, P. B. Shevlin, M. R. Willcott, R. R. Inness, and A. Cornelis, *J. Am. Chem. Soc.*, 1978, **100**, 3874.

[534] V. Reutrakul and P. Thamnusan, *Tetrahedron Lett.*, 1979, 617.

[535] V. Reutrakul, A. Tiensripojamam, K. Kusamran, and S. Nimgirawath, *Chem. Lett.*, 1979, 209.

[536] Y. Nagao, K. Seno, and E. Fujita, *Tetrahedron Lett.*, 1979, 3167.

[537] H. H. Szmant and R. Nanjundiah, *J. Org. Chem.*, 1978, **43**, 1835.

[538] H. Plieninger and W. Gramlich, *Chem. Ber.*, 1978, **111**, 1944; see also T. Kametani, K. Suzuki, H. Kurobe, and H. Nemoto, *J. Chem. Soc., Chem. Commun.*, 1979, 1128.

[539] R. Pitteloud and M. Petrzilka, *Helv. Chim. Acta*, 1979, **62**, 1319.

[540] T. Hori and K. B. Sharpless, *J. Org. Chem.*, 1978, **43**, 1689.

[541] H. J. Reich, S. K. Shah, and F. Chow, *J. Am. Chem. Soc.*, 1979, **101**, 6648.

[542] H. J. Reich, S. Wollowitz, J. E. Trend, F. Chow, and D. F. Wendelborn, *J. Org. Chem.*, 1978, **43**, 1697.

adding an alkylamine to the reaction mixture,[542] or exploited[541] if an overall conversion of the type (E)-EtCH$_2$CH=CHPr^n → (E)-EtCH=CHCH(OH)Pr^n is needed, simply by oxidizing the β-hydroxy-selenide to the selenoxide and repeating the *syn*-elimination step. Further examples of these alkene-forming methods are discussed in the next section, covering unsaturated sulphoxides.

The planar pericyclic mechanism which is accepted for the sulphoxide-elimination reaction is supported by the temperature-dependence of the relative reaction rates for n-C$_7$H$_{15}$SOPh and its β-^{2}H-analogue.[543]

Reactions of Unsaturated Sulphoxides and Selenoxides.—Deferring details of the chemistry of carbanions derived from alkenyl sulphoxides to Chapter 2, the main groupings into which recent work can be separated are preparative studies and rearrangements. The main methods for introducing a sulphinyl group are suitable in one form or another for preparing unsaturated sulphoxides, although special strategies are needed in particular cases, *e.g.* in the preparation of allenyl phenyl sulphides from propargyl alcohols (HOCR^{1}R^{2}C≡CH + PhSCl → R^{1}R^{2}C= C=CHSOPh).[544] This example of a sulphenate–sulphoxide rearrangement is one of several in the recent literature: *e.g.* ThpO(CH$_2$)$_8$CH(OH)CH=CHMe → sulphenate → sulphoxide, which on elimination yields (E)-dodeca-9,10-dien-1-ol, which gives the pheromone of the red bollworm moth on acetylation;[545] a similarly organized conversion of a *cis*-allylic alcohol into a *trans*-isomer in a total synthesis of (±)-PGF$_{2\alpha}$;[546] regio- and stereo-specific allylic oxidation in terpene synthesis [RCH$_2$CH=CMe$_2$ → RCH$_2$CH(SPh)CClMe$_2$ → RCH$_2$CH(SOPh)CMe=CH$_2$ → RCH$_2$CH=CMeCH$_2$OH];[547] preparation of β-acetoxy-α-methylene-γ-butyrolactones from β-phenylsulphenyl-α-methylenecarboxylic esters;[548] and the synthesis of dienes by regiospecific dehydration of allyl alcohols *via* the selenate–selenoxide–elimination sequence.[549]

Establishment of the absolute configuration of the sulphenate (23) that is formed from one enantiomer of a sulphoxide (24) gives a clear picture of the conformation of the reactant.[550]

The *syn*-elimination of sulphenic and selenenic acids from unsaturated sulphoxides and selenoxides introduces a further site of unsaturation, examples being the addition of allyl alcohols to allenic sulphoxides to give β-(allyloxy)vinyl sulphoxides

(24)   (23)

[543] H. Kwart, T. J. George, R. Louw, and W. Ultee, *J. Am. Chem. Soc.*, 1978, **100**, 3927.
[544] R. C. Cookson and P. J. Parsons, *J. Chem. Soc., Chem. Commun.*, 1978, 822.
[545] J. H. Babler and B. J. Invergo, *J. Org. Chem.*, 1979, **44**, 3723.
[546] R. Davis and K. G. Untch, *J. Org. Chem.*, 1979, **44**, 3755.
[547] Y. Masaki, K. Hashimoto, K. Sakuma, and K. Kaji, *J. Chem. Soc., Chem. Commun.*, 1979, 855.
[548] J.-P. Corbet and C. Benezra, *Tetrahedron Lett.*, 1979, 4003.
[549] H. J. Reich, I. L. Reich, and S. Wollowitz, *J. Am. Chem. Soc.*, 1978, **100**, 5981.
[550] R. W. Hoffmann, R. Gerlach, and S. Goldmann, *Tetrahedron Lett.*, 1978, 2599.

(which, after Claisen rearrangement and elimination, give $\bar{2}$,4-dienones)[551] and the conversion of α-alkylidene-γ-lactones into dienes $RCH=C(CO_2Me)CH=CH_2$ *via* α-(2-phenylseleninylethyl)acrylates.[552]

α-Phenylsulphinylacrylates $(RCH_2)_2C=C(CO_2Me)SOPh$ have two rearrangement pathways open to them; the vinylogous Pummerer rearrangement, leading to γ-hydroxy- or -acetoxy-alkyl sulphides, or a prototropic shift in aqueous pyridine, followed by rearrangement of the resulting allyl sulphoxides to the sulphenates, which give the γ-hydroxy-acrylates on hydrolysis.[553] In an 'additive Pummerer rearrangement' of these compounds, using AcCl, TFAA, or $SOCl_2$ as reagents, γ-substituted sulphides are formed.[553] C—S Bond cleavage of 1,3-dienyl,[554] allenyl,[554] and alkenyl[555] sulphoxides follows reactions with alkyl-lithium compounds, leading to terminal allenes in the last named series. Organocopper reagents R_2CuLi react with allenyl sulphoxides to give the 2-alkylated 1-lithio-allyl compounds[556] and with allyl sulphoxides to cause γ-substitution.[557] After desulphurization of the γ-substituted products, this constitutes a synthesis of trisubstituted alkenes. The *cis*-addition of organocopper(I) compounds to alkynyl sulphoxides[558] and a simpler equivalent alkylation of ethynyl sulphoxides, using a malonate,[559] have been described. The reaction of trimethylhydroquinone with the allyl alcohol $H_2C=CHCMe(OH)CH_2SOPh$, catalysed by $BF_3 \cdot Et_2O$ or $SnCl_4$, gives (25), which by Pummerer rearrangement and hydrolysis yields the chroman-2-aldehyde (26); this is an intermediate in a new synthesis of α-tocopherol.[560]

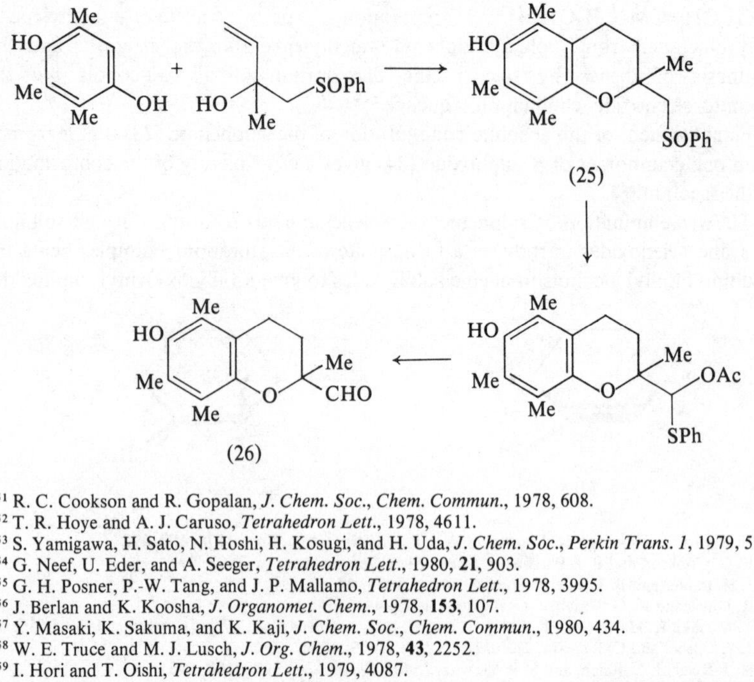

(25)

(26)

[551] R. C. Cookson and R. Gopalan, *J. Chem. Soc., Chem. Commun.*, 1978, 608.
[552] T. R. Hoye and A. J. Caruso, *Tetrahedron Lett.*, 1978, 4611.
[553] S. Yamigawa, H. Sato, N. Hoshi, H. Kosugi, and H. Uda, *J. Chem. Soc., Perkin Trans. 1*, 1979, 570.
[554] G. Neef, U. Eder, and A. Seeger, *Tetrahedron Lett.*, 1980, **21**, 903.
[555] G. H. Posner, P.-W. Tang, and J. P. Mallamo, *Tetrahedron Lett.*, 1978, 3995.
[556] J. Berlan and K. Koosha, *J. Organomet. Chem.*, 1978, **153**, 107.
[557] Y. Masaki, K. Sakuma, and K. Kaji, *J. Chem. Soc., Chem. Commun.*, 1980, 434.
[558] W. E. Truce and M. J. Lusch, *J. Org. Chem.*, 1978, **43**, 2252.
[559] I. Hori and T. Oishi, *Tetrahedron Lett.*, 1979, 4087.
[560] J. M. Akkermann, H. de Koning, and H. O. Huisman, *J. Chem. Soc., Perkin Trans. 1*, 1979, 2124.

Further Diels–Alder syntheses using β-phenylsulphinyl-αβ-unsaturated carbonyl compounds (see Vol. 5, p. 39) have established these compounds as useful αβ-ethynylcarbonyl synthons,[561] taking into account the scope for *syn*-elimination that is offered by the products.

Physical Properties and Stereochemistry of Sulphoxides.—Resolution of sulphoxides, sulphinates, and thiolsulphinates *via* β-cyclodextrin inclusion complexes is not efficient ($< 15\%$) with alkyl, aryl, or benzyl sulphoxides, but reaches 68% in the case of $MeS(O)OPr^i$.[562] Photoracemization of 1-naphthyl methyl sulphoxide in a cholesteric phase leads to less than 2% e.e. at equilibrium.[563]

Benzyl and 1-phenylethyl t-butyl sulphoxides have been shown, by measurement of dipole moments, to adopt the same conformation in solution (involving close approach of bulky groups) as has been established for the solid state through X-ray crystal analysis.[564]

12 Sulphones

Preparation of Saturated Sulphones.—Oxidation of sulphides and sulphoxides, using H_2O_2,[565,566] illustrates a standard method, though in the case of keten dithioacetals some C=C bond cleavage occurs,[566] and trifluoroperoxyacetic acid is more satisfactory.[566] Oxidation of sulphides to sulphones with hypochlorous acid can be stopped at the sulphoxide stage.[567] The same reagent is involved in the oxidative chlorinolysis of n-butyl glycidyl sulphoxide to $Bu^nSO_2CH_2CHCl$-CH_2Cl.[568] A combination of PhSeSePh with H_2O_2, presumably $PhSe(O)OOH$, has been used as a mild, selective method of oxidation.[569]

Preparations from sulphinic acids, SO_2, other simple sulphur compounds, and sulphonyl halides are reported in the papers that are listed (with brief details) at the end of this section. The $AlCl_3$-catalysed rearrangement of an aryl arenesulphonate gives the corresponding diaryl sulphone.[570]

Additional Bibliography: R. A. Hancock and S. T. Orszulik, *Tetrahedron Lett.*, 1979, 3789 [PhSO₂Cu + ArTl(OCOCF₃)₂]; P. Messinger and H. Greve, *Arch. Pharm. (Weinheim, Ger.)*, 1978, **311**, 827, 280 [RSO₂H → RSO₂CH₂OH; Michael additions and Mannich reactions involving RSO₂H]; F. Manescalchi, M. Orena, and D. Savoia, *Synthesis*, 1979, 445 [benzenesulphinate salts of polymer-bound cations, with alkyl halides]; U. M. Dzhemilev, R. V. Kunakova, R. L. Gaisin, and G. A. Tolstikov, *Izv. Akad, Nauk SSSR, Ser. Khim.*, 1979, 2702; U. M. Dzhemilev, R. V. Kunakova, R. L. Gaisin, E. V. Vasileva, and G. A. Tolstikov, *Zh. Org. Khim.*, 1978, **14**, 2223; and R. V. Kunakova, G. A. Tolstikov, U. M. Dzhemilev, F. V. Sharipova, and D. L. Sazikova, *Izv. Akad. Nauk SSSR, Ser. Khim.*, 1978, 931 [telomerization of butadiene with RSO₂H, catalysed by Pd, Ph₃P, and

561 S. Danishefsky, T. Harayama, and R. K. Singh, *J. Am. Chem. Soc.*, 1979, **101**, 7008.
562 M. Mikolajczyk and J. Drabowicz, *J. Am. Chem. Soc.*, 1978, **100**, 2510.
563 C. Eskenazi, J. F. Nicoud, and H. B. Kagan, *J. Org. Chem.*, 1979, **44**, 995.
564 M. Hirota, Y. Takahashi, M. Nishio, and K. Nishihata, *Bull. Chem. Soc. Jpn.*, 1978, **51**, 2358.
565 A. Bergman and C. A. Wachtmeister, *Chemosphere*, 1978, **7**, 949.
566 N. P. Petukhova, N. E. Dontsova, O. M. Sazonova, and E. N. Prilezhaeva, *Izv. Akad. Nauk SSSR, Ser. Khim.*, 1978, 2654.
567 A. R. Derzhinskii, L. D. Konyushkin, and E. N. Prilezhaeva, *Izv. Akad. Nauk SSSR, Ser. Khim.*, 1978, 2070.
568 A. R. Derzhinskii, L. D. Konyushkin, and E. N. Prilezhaeva, *Izv. Akad. Nauk SSSR, Ser. Khim.*, 1979, 2650.
569 K. C. Nicolau, R. L. Magolda, and W. E. Barnette, *J. Chem. Soc., Chem. Commun.*, 1978, 375.
570 N. K. Undavia, M. L. Dhanani, and K. A. Thaker, *J. Inst. Chem. (India)*, 1978, **50**, 38 (*Chem. Abstr.*, 1979, **89**, 108 476).

Et$_3$Al]; K. B. Rall, A. I. Vildavskaya, and A. A. Pterov, *Zh. Org. Khim.*, 1978, **14**, 1107 [ArSO$_2$Na + vinylacetylenes]; J. Wildeman and A. M. Van Leusen, *Synthesis*, 1979, 733; A. D. Bulat, M. A. Antipov, and B. V. Passet, *Zh. Org. Khim.*, 1979, **15**, 651; R. Moreau, Y. Adam, and P. Loiseau, *C.R. Hebd. Seances Acad. Sci., Ser. C*, 1978, **287**, 39; and V. N. Mikhailova, A. D. Bulat, and V. P. Yurevich, *Zh. Obshch. Khim.*, 1978, **48**, 217 [RSO$_2$Na + alkyl and alkenyl halides]; V. M. Naidan, G. D. Naidan, and G. V. Litvin, *Zh. Org. Khim.*, 1978, **14**, 2622 [styrene, PhN$_2$$^+$ Cl$^-$, and SO$_2$]; Yu. A. Moskvichev, M. I. Farberov, I. M. Rubleva, and G. V. Samchenko, *Khim. Prom-st. (Moscow)*, 1978, 262 (*Chem. Abstr.*, 1979, **89**, 23 886) [PhCl + SOCl$_2$ + SO$_3$ → (4-ClC$_6$H$_4$)$_2$SO$_2$]; L. K. Liu, Y. Chi, and K. Y. Jan, *J. Org. Chem.*, 1980, **45**, 406 [β-iodo-sulphones from RSO$_2$I with CuCl$_2$ + alkene]; M. M. Tanaskov and M. D. Stadnichuk, *Zh. Obshch. Khim.*, 1978, **48**, 1140 [CuCl-catalysed addition of RSO$_2$Cl to alkenes and alkynes *via* RSO$_2$·]; A. V. Kalabina, M. A. Vasileva, and T. I. Bychkova, *Zh. Org. Khim.*, 1979, **15**, 268 [u.v.–AIBN-initiated addition of RSO$_2$Cl to vinyl ethers]; A. I. Khodair, A. A. Abdel-Wahab, and A. M. El-Khawaga, *Z. Naturforsch., Teil B*, 1978, **33**, 403 [R^{1}SO$_2$F + R^{2}MgBr → sulphones, but no reaction occurs with organocadmium reagents]; Yu. A. Moskvichev, I. M. Rubleva, B. N. Bychkov, G. S. Mironov, and M. I. Farberov, *Kinet. Katal.*, 1978, **19**, 895 [aromatic sulphonylation with ArSO$_2$Cl and FeCl$_3$].

Reactions of Saturated Sulphones.—Applications of sulphones in synthesis are discussed in a later section, and sulphonyl-stabilized carbanions are dealt with elsewhere in this Volume. This section is therefore concerned with reactions of saturated alkyl and aryl sulphones, particularly C—S bond cleavage reactions, on which many applications of sulphones in synthesis depend.

Phase-transfer catalysis is a useful adjunct to the alkylation of benzyl[571] and iodomethyl[572] aryl sulphones. In the latter case, an α-proton is substituted, rather than the halogen, and this same initial deprotonation has been proposed[573] for the novel, quaintly-named, 'vicarious' nucleophilic substitution of nitrobenzene by ClCH$_2$SO$_2$Ph in the presence of NaOH (Scheme 19). The novel α-keto-sulphones ArCOSO$_2$Ar are obtained by ozonolysis of the α-diazo- or α-alkoxymethylene analogues.[574]

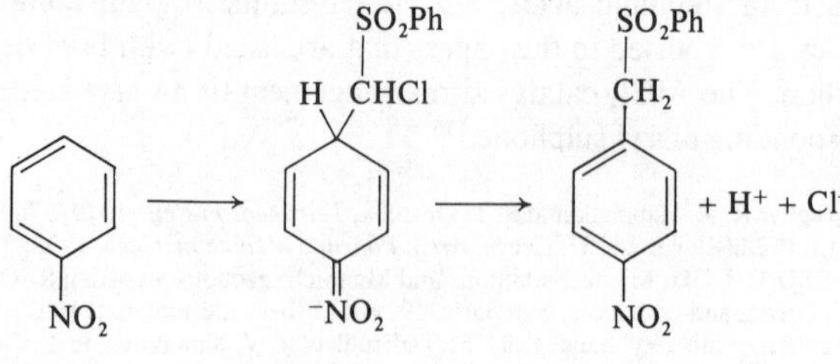

Scheme 19

An intramolecular electron-transfer/radical anion mechanism has been advocated[575] for the Truce–Smiles-type rearrangement of t-butyl aryl sulphones by Bu^{n}Li at −78 °C, *ortho*-lithiation of the aryl moiety being followed by migration of the t-butyl group to the metallated site. Other methods of C—S bond cleavage that are covered in recent papers include photolysis[576] and electrochemical reduction,[577]

[571] J. Golinski, A. Jonczyk, and M. Makosza, *Synthesis*, 1979, 461.
[572] A. Jonczyk and T. Pytlewski, *Synthesis*, 1978, 883.
[573] J. Golinski and M. Makosza, *Tetrahedron Lett.*, 1978, 3495.
[574] K. Schank and F. Werner, *Liebigs Ann. Chem.*, 1979, 1977.
[575] D. M. Snyder and W. E. Truce, *J. Am. Chem. Soc.*, 1979, **101**, 5432.
[576] R. F. Langler, Z. A. Marini, and J. A. Pincock, *Can. J. Chem.*, 1978, **56**, 903.
[577] B. Lamm and K. Ankner, *Acta Chem. Scand., Ser. B*, 1978, **32**, 264; E. A. Berdnikov, S. B. Fedorov, and Yu. M. Kargin, *Zh. Obshch. Khim.*, 1978, **48**, 875; S. Gambino, P. Martigny, G. Mousset, and J. Simonet, *J. Electroanal. Chem. Interfacial Electrochem.*, 1978, **90**, 105.

the former method leading to sulphinic acids and sulphonyl chlorides in hydrochloric acid,[576] and the latter yielding sulphinic acids and alkenes with alkyl sulphones that carry an electronegative β-substituent.[577] Analogously, β-trimethylsilylalkyl phenyl sulphones undergo fluoride-ion-induced elimination ($PhSO_2$-$CHRCH_2SiMe_3 \rightarrow RCH{=}CH_2$).[578] Diastereoisomers of 1-methyl-2-bromopropyl phenyl sulphone yield *cis-* and *trans*-but-2-enes with tri-n-butylstannyl radicals, the non-stereospecific nature of the process indicating rapid rotation within the intermediate 2-phenylsulphonyl-1-methylpropyl radicals.[579] Fluorinolysis of dimethyl sulphone yields CF_3SOF, CF_3OSOF, and CHF_2OSO_2F.[580]

Unsaturated Sulphones.—In addition to modifed versions of methods used for the synthesis of saturated sulphones [*e.g.* $ArSO_2H$ + $H_2C{=}CHCR^1{=}CHR^2 \rightarrow$ $H_2C{=}CHCR^1(CH_2R^2)SO_2Ar$;[581] peroxide-initiated addition of RSO_2Br to alkynes;[582] formation of rearranged allyl sulphones from RSO_2Cl + allylcobaloxime;[583] RSO_2Na + alk-1-enylmercury halides;[584] RSO_2Na + 1,3-diene with $PdCl_2$;[585] Michael addition of oxazolin-5-ones[586] and organocuprates[587] to ethynyl phenyl sulphones], standard methods of introducing double bonds have been applied to saturated sulphones and are summarized at the end of this section.

Rearrangements of allyl sulphones during electrolysis compete with reduction,[588] while a synthesis of allenyl sulphones by rearrangement of propargylic sulphinates[589] illustrates a different aspect of organosulphur rearrangements. Deprotonation of δ-oxo-alkenyl sulphones $MeCOCR^1{=}CHCHR^2SO_2Ph$ provides an interesting example in which comparisons between carbonyl and sulphonyl groups can be made; alkylation of the derived carbanion occurs preferentially γ to $C{=}O$ because sulphonyl groups tend to deconjugate to greater extents than carbonyl groups.[590]

Grignard reagents, in the presence of copper(II) acetylacetonate, effect nucleophilic substitution of the sulphonyl group and provide homologated alkenes.[591] 1-Alkynyl sulphones behave similarly with Grignard reagents and with alkyl-lithium compounds,[592] although some isomerization also takes place. The organolithium compounds effect α-lithiation of vinyl sulphides at $-95\,^{\circ}C$,[593] providing a useful synthesis of *trans*-2-substituted cyclopropyl sulphones from *trans*-$BrCH_2CH{=}CHSO_2Ph$.[594] The α-position in allyl sulphides is also lithiated,

[578] P. J. Kocienski, *Tetrahedron Lett.*, 1979, 2649.

[579] T. E. Boothe, J. L. Greene, and P. B. Shevlin, *J. Org. Chem.*, 1980, **45**, 794.

[580] L. A. Harmon and R. J. Lagow, *J. Chem. Soc., Perkin Trans. 1*, 1979, 2675.

[581] M. Julia, M. Nel, and L. Saussine, *J. Organomet. Chem.*, 1979, **181**, C17.

[582] W. Boell, *Liebigs Ann. Chem.*, 1979, 1665.

[583] A. E. Crease, B. D. Gupta, M. D. Johnson, E. Bialkowska, K. N. V. Duong, and A. Gaudemer, *J. Chem. Soc., Perkin Trans. 1*, 1979, 2611.

[584] J. Hershberger and G. A. Russell, *Synthesis*, 1980, 475.

[585] Y. Tamaru, M. Kagotani, and Z. Yoshida, *J. Chem. Soc., Chem. Commun.*, 1978, 367.

[586] W. Steglich and H. Wegmann, *Synthesis*, 1980, 481.

[587] V. Fiandanese, G. Marchese, and F. Naso, *Tetrahedron Lett.*, 1978, 5131.

[588] J. Y. Pape and J. Simonet, *Electrochim. Acta*, 1978, **23**, 445.

[589] M. Cinquini, F. Cozzi, and M. Pelosi, *J. Chem. Soc., Perkin Trans. 1*, 1979, 1430.

[590] P. T. Lansbury and R. W. Erwin, *Tetrahedron Lett.*, 1978, 2675.

[591] M. Julia, A. Righini, and J.-N. Verpeaux, *Tetrahedron Lett.*, 1979, 2393.

[592] R. L. Smorada and W. E. Truce, *J. Org. Chem.*, 1979, **44**, 3444.

[593] J. J. Eisch and J. E. Galle, *J. Org. Chem.*, 1979, **44**, 3279.

[594] J. J. Eisch and J. E. Galle, *J. Org. Chem.*, 1979, **44**, 3277.

(27)   (28)

and the resulting carbanion can participate in Michael reactions.[595] Electro-chemical reduction of methyl styryl sulphone in the presence of Bu^tCl gives sulphone-replacement products $PhCH=CHBu^t$ and $4\text{-}Bu^tC_6H_4CH=CH_2$, as well as the reduction products styryl methyl sulphide and $PhCHBu^tCH_2SO_2Me$ and the corresponding sulphoxide.[596] The addition process in this example is also seen in the addition of MeLi to the enantiomer (27); the extraordinarily high level (>99%) of asymmetric induction (especially remarkable in an acyclic case), leading to the *threo*-isomer (28), has been taken[597] to indicate efficient co-ordination of the lithium cation.

Addition reactions of vinyl sulphones (Diels–Alder addition,[598] [2 + 2]-photocycloaddition,[599] and conventional functionalization processes, *e.g.* $IN_3 \rightarrow$ 2-azido-1-iodoalkyl sulphones[600]) and substitution reactions of β-halogeno-vinyl sulphones[601] have been reported. A point of interest in the cycloaddition study is the effect of (E)–(Z) equilibration in hindering the participation of simple vinyl sulphones in photocycloaddition reactions.[599] Elimination of SO_2 has been observed in the photolysis of 3-oxo-alk-1-enyl sulphones in benzene, giving radicals which attack the solvent.[602]

Additional Bibliography: P. Messinger and H. Greve, *Arch. Pharm.* (*Weinheim, Ger.*), 1979, **312**, 792 [bromination–dehydrobromination of alkyl aryl sulphones]; M. Yamamoto, *Nippon Kagaku Kaishi*, 1978, 771 (*Chem. Abstr.*, 1979, **89**, 108 474) [methyl phenyl sulphone + RCHO and micelles]; K. Laping and M. Hanack, *Tetrahedron Lett.*, 1979, 1309 [$C_4F_9SO_2MgX + R^1COR^2$, followed by elimination of H_2O]; P. O. Ellingsen and K. Undheim, *Acta Chem. Scand.*, *Sect. B*, 1979, **33**, 528 [Horner–Wittig reaction with $(EtO)_2P(O)CH_2SO_2Ph$]; Y. Ueno, H. Setoi, and M. Okawara, *Chem. Lett.*, 1979, 47 and H. Kotake, K. Inomata, and M. Sumita, *ibid.*, 1978, 717 [de-acylative methylenation of arylsulphonylmethyl ketones with HCHO].

Sulphones Used in Synthesis.—The range of examples of reactions of sulphones discussed in the preceding section makes clear the scope for applications in synthesis. Most applications are based on sulphonyl carbanions, and are covered in the next Chapter, but ring syntheses are illustrated by a novel synthesis of naphthalene that involves the condensation of *o*-formylbenzyl *p*-tolyl sulphones

[595] G. A. Kraus and K. Frazier, *Synth. Commun.*, 1978, **8**, 483.

[596] H. Lund and C. Degrand, *C.R. Hebd. Seances Acad. Sci.*, *Ser. C*, 1978, **287**, 535.

[597] M. Isobe, M. Kitamura, and T. Goto, *Tetrahedron Lett.*, 1979, 3465.

[598] R. V. C. Carr and L. A. Paquette, *J. Am. Chem. Soc.*, 1980, **102**, 853.

[599] M. A. A. M. El Tabei, N. V. Kirby, and S. T. Reid, *Tetrahedron Lett.*, 1980, **21**, 565.

[600] Y. Tamura, J. Haruta, S. M. Bayomi, M. W. Chun, S. Kwon, and M. Ikeda, *Chem. Pharm. Bull.*, 1978, **26**, 784.

[601] M. S. R. Naidu and S. G. Peeran, *Indian J. Chem.*, *Sect. B*, 1978, **16**, 1086; A. V. Martynov, A. N. Mirskova, I. D. Kalikhman, P. V. Makerov, and M. G. Voronkov, *Zh. Org. Khim.*, 1979, **15**, 427.

[602] Y. Tamura, H. Kiyokawa, and Y. Kita, *Chem. Pharm. Bull.*, 1979, **27**, 676.

$$PhSO_2(CH_2)_nCH \overset{O}{\overbrace{\qquad}} CH_2$$

(29)

with $\alpha\beta$-unsaturated ketones,[603] the cyclization of epoxyalkyl phenyl sulphones (29) with MeMgI to give cyclobutanols ($n = 2$) and cyclopentanols ($n = 3$),[604] and the thermal extrusion of SO_2 from $Me(CH_2)_5C(=CH_2)CH_2SO_2CH_2CO_2Me$ to give a disubstituted γ-lactone.[605] Pyrolysis of sulphones has been reviewed.[606] The synthesis of alkynes, $R^1C{\equiv}CR^2$, from carboxylic acids, R^1CO_2H, *via* β-keto-alkyl sulphones and their enol phosphates[607,608] has been used in a synthesis of cyclododecyne.[607] Effective procedures for the cleavage of C—S bonds (reduction with sodium amalgam continues to be frequently used,[598] and Bu^n_3SnH gives good yields of terminal alkenes from $ArSO_2CHRCH{=}CH_2$[609]) support syntheses of homologated allyl alcohols[610] and $\alpha\beta$-unsaturated ketones[611] *via* $\beta\gamma$-epoxy-alkyl sulphones [a disadvantage being the formation of $(E)/(Z)$ mixtures of allyl alcohols[610]]. Similarly, phenyl *trans*-2-chlorovinyl sulphone[612] and ethynyl phenyl sulphone[586] serve as equivalents of a vinyl cation in the synthesis of α-vinyl-α-amino-acids.

13 Sulphenic and Selenenic Acids, and their Derivatives

Sulphenic and Selenenic Acids.—Flash vacuum pyrolysis of alkyl methyl sulphoxides or methyl methanethiolsulphinate, MeS(O)SMe, gives methane-sulphenic acid, characterized by microwave spectrometry as a dico-ordinate sulphur species, MeSOH, in the gas phase, with geometry similar to that of H_2S_2. This important study establishes the viability of separate molecules of the simplest sulphenic acid in the gas phase and also provides further examples of the self-condensation reactions used earlier to infer the transient generation of sulphenic acids in organosulphur reactions. Thus,[613] MeS(O)SMe is formed on cooling the reaction mixture, but thioformaldehyde is formed at 750 °C. Extension of this work[614] to the formation of ethenesulphenic and (E)-1-propenesulphenic acids has been stimulated by the establishment of the structure of the lachrymatory factor in onion (*Allium cepa*) as (Z)-propanethial S-oxide, formed from the propene-sulphenic acid that is liberated by the action of alliinase on *trans*-(+)-S-(1-propenyl)-L-cysteine sulphoxide *in vivo*. Pyrolysis of N-substituted azetidin-

[603] J. Wildeman, P. C. Borgen, H. Phim, P. H. F. M. Rouwette, and A. M. van Leusen, *Tetrahedron Lett.*, 1978, 2213.
[604] B. C. Corbel, J. M. Decesare, and T. Durst, *Can. J. Chem.*, 1978, **56**, 505.
[605] R. D. Little, S. Wolf, T. Smestad, S. C. Seike, L. W. Lindner, and L. Patron, *Synth. Commun.*, 1979, **9**, 545.
[606] F. Vögtle and L. Rossa, *Angew. Chem.*, 1979, **91**, 534.
[607] P. A. Bartlett, F. R. Green, and E. H. Rose, *J. Am. Chem. Soc.*, 1978, **100**, 4852.
[608] B. Lythgoe and I. Waterhouse, *Tetrahedron Lett.*, 1978, 2625.
[609] Y. Ueno, S. Aoki, and M. Okawara, *J. Am. Chem. Soc.*, 1979, **101**, 5414.
[610] P. J. Kocienski, *Tetrahedron Lett.*, 1979, 441.
[611] J. C. Saddler, P. C. Conrad, and P.L. Fuchs, *Tetrahedron Lett.*, 1978, 5079.
[612] B. W. Metcalf and E. Bonilavri, *J. Chem. Soc., Chem. Commun.*, 1978, 914.
[613] R. E. Penn, E. Block, and L. K. Revelle, *J. Am. Chem. Soc.*, 1978, **100**, 3622.
[614] E. Block, R. E. Penn, and L. K. Revelle, *J. Am. Chem. Soc.*, 1979, **101**, 2200.

2-on-yl sulphoxides generates azetidinonesulphenic acids, which may be trapped intramolecularly if the *N*-substituent is an alk-2-enyl group, leading to penam sulphoxides,[615a] but which yield 4-thionoazetidin-2-ones *via* thiolsulphinates when the *N*-substituent is an alk-2-enyl group.[615b] Details of other recent papers dealing with pyrolysis of sulphoxides can be found in the 'Sulphoxides' section. Pyrolysis of sulphinimides ArS(O)N=CHR gives arenesulphenic acids.[616]

Generation of sulphenic acid anions from disulphides by chlorinolysis and hydrolysis in aqueous media competes with oxidation to sulphonic acids.[617] Full details have been published[618] of the propensity for attack by electrophiles at both S and O in the sulphenate anions that are generated by the hydrolysis of disulphides and sulphenic acid derivatives.

Simple selenenic acids are also generated by pyrolysis of selenoxides,[315] and show some distinctive properties compared with sulphenic acids; thus,[541,542,619] addition to alkenes gives β-hydroxy-selenides. In the case of cyclo-octa-1,5-dienes, the cyclic ether corresponding to the introduction of phenylseleno-groups at C-1 and C-6 is the reaction product.[619] The oxidative reactivity of 'PhSeOH' has been illustrated in a previous section,[228-230] and further examples are the synthesis of selenolesters from *N*-acyl-hydrazines[620] (RCONHNH$_2$ → [RCON=NH] → RCON=NSePh → RCOSePh) and the PhSeSePh-catalysed oxidation of benzylic and primary allyl alcohols with Bu^{t}OOH,[621] a reaction which is considered to involve the areneselenenic acid or its anhydride. The reaction is also applicable to saturated alcohols if bis-(2,4,6-trimethylphenyl) diselenide is used.[621] The general picture presented in these reactions suggests that the structure 'RSeOH' can only be an approximation of the true structure of the organoselenium reagents involved.

Sulphenate Esters.—The essence of sulphenate ester chemistry is contained in an account of continuing mechanistic studies,[622] showing that S—O bond cleavage of CF$_3$SOCH$_2$Ar occurs on ethanolysis, while its CCl$_3$ analogue undergoes cleavage of the C—O bond.[622] References to allyl sulphenates are discussed with unsaturated sulphoxides in an earlier section.

Sulphenyl and Selenenyl Halides.—The preparation of these compounds by halogenolysis of disulphides and diselenides, a particular example being the preparation of the stable 3-nitropyridine-2-sulphenyl chloride and bromide,[623] has not proved suitable for sulphenyl fluorides. Evidence for the existence of PhSF as an intermediate in the conversion of diphenyl disulphide into thianthrene has been provided by [19]F n.m.r. studies.[624] Addition of SCl$_2$ to alkenes[625] is rarely used, since

[615] (a) A. C. Kaura, C. D. Maycock, and R. J. Stoodley, *J. Chem. Soc., Chem. Commun.*, 1980, 34; (b) M. D. Bachi, O. Goldberg, A. Gross, and J. Vaya, *J. Org. Chem.*, 1980, **45**, 1477, 1481.

[616] F. A. Davis, A. J. Friedman, and U. K. Nadir, *J. Am. Chem. Soc.*, 1978, **100**, 2844.

[617] V. N. R. Pillai and P. V. Padmanabhan, *J. Indian Chem. Soc.*, 1978, **55**, 279 (*Chem. Abstr.*, 1979, **89**, 163 195).

[618] D. R. Hogg and A. Robertson, *J. Chem. Soc. Perkin Trans. 1*, 1979, 1125.

[619] R. M. Scarbrough, A. B. Smith, W. E. Barnette, and K. C. Nicolaou, *J. Org. Chem.*, 1979, **44**, 1742.

[620] T. G. Back and S. Collins, *Tetrahedron Lett.*, 1979, 2661.

[621] M. Shimizu and I. Kuwajima, *Tetrahedron Lett.*, 1979, 2801.

[622] S. Braverman and H. Manor, *Phosphorus Sulfur*, 1976, **2**, 215.

[623] R. Matsueda and K. Aiba, *Chem. Lett.*, 1978, 951.

[624] F. Seel, R. Budenz, R. D. Flaccus, and R. Stark, *J. Fluorine Chem.*, 1978, **12**, 437.

[625] M. Mühlstädt, N. Stransky, E. Kleinpeter, and H. Meinhold, *J. Prakt. Chem.*, 1978, **320**, 107.

the β-chloro-alkanesulphenyl chloride that is formed adds to the alkene to give the symmetrical sulphide.

References to addition reactions of sulphenyl halides with alkenes and alkynes are given in the 'Sulphides' section. An incidental bonus from a study of the addition of PhSCl to adamantylideneadamantane[626] is the formation of the 4e-chloro-derivative and PhSSPh, *via* an intermediate thiiranium salt. Sulphenyl and selenenyl chlorides have been used for the dehydration of aldoximes to nitriles.[627] Whereas sulphinyl chlorides yield sulphinate esters with alkoxytrimethylsilanes, no reaction occurs with sulphenyl chlorides.[628]

Sulphenamides and Selenenamides.—Most of the methods of preparation reported in the recent literature are based on the straightforward route from a sulphenyl or selenenyl halide by reaction with an amine, and these papers are listed at the end of this section. Oxidative condensation of a thiol with an amine, using $K_2S_2O_8$,[629] cleavage (in liquid NH_3) of 4,4'-bis(nitrophenyl) disulphide[630] or of methyl benzenesulphenate,[631] and the reaction of an *N*-chloro-amide with selenium, to give $(RCONH)_2SeCl_2$, or with a selenenyl chloride, to give $RCONHSeRCl_2$,[632] are more unusual routes.

Reactions involving cleavage of an S—N bond include anodic oxidation[633, 634] or oxidation with PbO_2,[635] giving sulphonic acids[633] or phenazines[634, 635] *via* nitrenes; reaction with a thiol, giving a disulphide;[636] ammonolysis, giving thiols, amines, disulphides, and thiosulphenamides $ArSSNH_2$ *via* sulphonium salt intermediates;[637] and a reaction with an isocyanide, giving an *S*-substituted isothiourea by insertion into the S—N bond.[638] Rearrangements [$RSNMePh \rightarrow o$- and p-RSC_6H_4NHMe[639] and $Ar^1N(SR^1)CR^2=NAr^2 \rightarrow Ar^1N=CR^2N(SR^1)Ar^2$ [640]] also involve S—N bond cleavage. Radicals $Ar^1\dot{N}SAr^2$, formed by abstraction of H from sulphenanilides, have been studied by e.s.r. spectroscopy,[641] and the sulphenylnitrenes that are formed by oxidation of 2,4-dinitrobenzenesulphenamide with $Pb(OAc)_4$[642] yield aziridines with alkenes and $Ar\overline{SN}-\overset{+}{N}H_2SAr$ with the sulphenamide, from which ArNHSAr is formed by two successive Smiles rearrangements.

Rotational barriers in arenesulphenamides, established by dynamic ^1H n.m.r. spectroscopy, are considered to arise from a balance of steric and electronic

[626] J. Bolster, R. M. Kellogg, E. W. Meijer, and H. Wynberg, *Tetrahedron Lett.*, 1979, 285.

[627] G. Sosnovsky and J. A. Krogh, *Z. Naturforsch., Teil B*, 1979, **34**, 511.

[628] D. N. Harpp. B. T. Friedlander, C. Larsen, K. Steliou, and A. Stockton, *J. Org. Chem.*, 1978, **43**, 3481.

[629] V. A. Ignatov, P. G. Valkov, S. V. Sukhanov, and A. N. Lazovenko, *Zh. Obshch. Khim.*, 1979, **49**, 841.

[630] R. Sato, S. Takizawa, and S. Oae, *Phosphorus Sulfur*, 1979, **7**, 185.

[631] D. A. Armitage and M. J. Clark, *Phosphorus Sulfur*, 1978, **5**, 41.

[632] N. Ya. Derkach, T. V. Lyapina, and E. S. Levchenko, *Zh. Org. Khim.*, 1978, **14**, 280.

[633] H. Sayo, K. Mori, and T. Michida, *Chem. Pharm. Bull.*, 1979, **27**, 2093.

[634] H. Sayo, K. Mori, A. Ueda, and T. Michida, *Chem. Pharm. Bull.*, 1978, **26**, 1682.

[635] H. Sayo, K. Mori, and T. Michida, *Chem. Pharm. Bull.*, 1979, **27**, 2316.

[636] V. A. Ignatov, *Zh. Obshch. Khim.*, 1979, **49**, 1122.

[637] R. Sato, K. Araya, Y. Takikawa, S. Takizawa, and S. Oae, *Phosphorus Sulfur*, 1978, **5**, 245.

[638] J. P. Chupp, J. J. D'Amico, and K. L. Leschinsky, *J. Org. Chem.*, 1978, **43**, 3553.

[639] P. Ainpour and N. E. Heimer, *J. Org. Chem.*, 1978, **43**, 2061.

[640] L. P. Olekhnovich, I. E. Mikhailov, N. M. Ivanchenko, Yu. A. Zhdanov, and V. I. Minkin, *Zh. Org. Khim.*, 1979, **15**, 1355.

[641] Y. Miura, Y. Katsura, and M. Kinoshita, *Bull. Chem. Soc. Jpn.*, 1979, **52**, 1121.

[642] R. S. Atkinson and B. D. Judkins, *J. Chem. Soc., Chem. Commun.*, 1979, 832, 833.

effects,[643] but probing into the details to try to secure a less vague interpretation has led to this being a controversial subject over recent years.[644] Rotational barriers for hydroxylamines cannot be accounted for by any of the proposed mechanisms for sulphenamides.[644] The high barriers to rotation that lead to there being separable diastereoisomers of *N*-heteroarylamino-sulphenamides are due to restricted rotation about the N—N bond, not the S—N bond.[645]

Sulphenimines ArSN=CR^{1}CO$_2$R^2 are formed from an arenesulphenyl chloride with an α-amino-acid ester in the presence of an acid scavenger, and they yield α-keto-esters by reaction with PPh$_3$ and silica gel.[646] Sulphenimines PhSN= CR^{1}CH$_2$R^2 are formed by the alkylation of carbanions formed from a thiolate and a nitrile.[647]

Additional Bibliography: K. B. Kurbanov, Z. A. Mamedov, and S. T. Akhmedov, *Azerb. Khim. Zh.*, 1978, 50 (*Chem. Abstr.*, 1980, **91**, 5006) [ArSNR$_2$ from ArSCl]; I. V. Koval, T. G. Oleinik, and M. M. Kremlev, *Zh. Org. Khim.*, 1979, **15**, 2319 [Ar^{1}SCl + Ar^{2}SO$_2$NH$_2$]; ref. 720 [RSCl + AcNHSiMe$_3$]; A. Darmadi and M. Kaschani-Motlagh, *Z. Anorg. Allg. Chem.*, 1979, **448**, 35 [CF$_3$SeBr + NH$_3$ → (CF$_3$Se)$_n$NH$_{3-n}$].

Thionitrites and Thionitrates.—Treatment of a thiol with N$_2$O$_4$ gives the corresponding thionitrite RSNO,[648] further reaction with N$_2$O$_4$ leading to the sulphonic acid or the thiolsulphonate. t-Butyl thionitrite or the corresponding thionitrate Bu^{t}SNO$_2$ (formed from a thiol and N$_2$O$_4$), can be used with a copper(II) halide to convert anilines into halogenobenzenes.[649] Thionitrates also convert *p*-amino-phenols into *p*-benzoquinone-t-butylthioimines.[650]

14 Thiocyanates and Isothiocyanates, and their Selenium and Tellurium Analogues

Thiocyanates.—Relatively little new work has appeared in the period under review that concerns reactions of these compounds, but modifications of known preparative methods continue to be studied. These papers are mostly to be found at the end of this section. Several sections of a recent book[651] deal with these compounds in some depth.

A novel method for the synthesis of thiocyanates amounts to the overall conversion RNH$_2$ → RSCN.[652] The amine is allowed to react with 2,4,6-triphenylpyrylium thiocyanate and the resulting pyridinium thiocyanate is pyrolysed at 180—200 °C. The point is made[652] that this method is safer than the classical diazotization–substitution sequence for bringing about this conversion (illustrated recently for 3-aminothiophens, to give 3-thienyl thiocyanates and selenocyanates[653]).

[643] M. Raban and G. Yamamoto, *J. Am. Chem. Soc.*, 1979, **101**, 5890.
[644] D. Kost and E. Berman, *Tetrahedron Lett.*, 1980, **21**, 1065.
[645] R. S. Atkinson, B. D. Judkins, and B. H. Patwardhan, *Tetrahedron Lett.*, 1978, 3137.
[646] E. M. Gordon and J. Pluscec, *J. Org. Chem.*, 1979, **44**, 1218.
[647] F. A. Davis and P. A. Mancinelli, *J. Org. Chem.*, 1978, **43**, 1797.
[648] Y. H. Kim, K. Shinhama, D. Fukushima, and S. Oae, *Tetrahedron Lett.*, 1978, 1211.
[649] Y. H. Kim, K. Shinhama, and S. Oae, *Tetrahedron Lett.*, 1978, 4519.
[650] S. Oae, K. Shinhama, and Y. H. Kim, *Chem. Lett.*, 1979, 1077.
[651] 'Chemistry of Cyanates and their Thio Derivatives', ed. S. Patai, Wiley, Chichester, 1977.
[652] A. R. Katritzky and S. S. Thind, *J. Chem. Soc., Chem. Commun.*, 1979, 138; A. R. Katritzky, U. Gruntz, N. Mongelli, and M. C. Rezende, *J. Chem. Soc., Perkin Trans. 1*, 1979, 1953.
[653] C. Paulmier, *Bull. Soc. Chim. Fr., Part 2*, 1979, 237.

Many examples of isomerization into isothiocyanates are found in papers describing the uses of thiocyanates in synthesis, one of which is based on their sulphenylation reactivity [$H_2C=C(CH_2SCN)_2 \rightarrow H_2C=C(CH_2NCS)_2$ + 1,2,6,7-tetrathia-4,9-dimethylenecyclodecane].[654] This rearrangement is brought about for poly(alkylbenzyl) thiocyanates and benzyl selenocyanates by u.v. irradiation.[655] Allyl thiocyanates undergo [3,3]-sigmatropic rearrangement during distillation [$R^1O_2CCH=CR^2CH_2SCN \rightarrow R^1O_2CC(NCS)=CR^2Me$].[656] Thiocyanates are also useful in thiazolinone synthesis,[657] and an interesting variation of the Japp–Klingemann reaction [$MeCOCH(COMe)SCN + ArN_2^+ Cl^-$, giving 2-acetyl-4-aryl-5-imino-Δ^2-1,3,4-thiadiazoline] has been reported.[658]

Conversion of a thiocyanate into a nitrile [$(EtO_2C)_2CHSCN \rightarrow (EtO_2C)_2CHCN$] with aqueous $MeNH_2$ or NH_3 is suggested to proceed *via* formation of a thiiran, followed by extrusion of sulphur,[659] but other mechanisms must also be considered. 2-Nitrophenyl thiocyanate with Bu^n_3P is an effective reagent for coupling carboxylic acids with amines,[660] and it may have considerable value in peptide synthesis.

Methods of preparation of selenocyanates are entirely analogous to those of thiocyanates, and the papers listed at the end of this section include examples of methods. The previously unknown compound PhTeCN is formed in only low yield from PhTeTePh and KCN, but in high yield through reductive cyanation of $ArTeCl_3$ and by the general route from ArTeBr with KCN.[661]

Additional Bibliography: K. Fujuki, T. Nishio, and Y. Omote, *Bull. Chem. Soc. Jpn.*, 1979, **52**, 614 [photosubstitution of halogenobenzenes with SCN$^-$]; G. Cainelli, F. Manescalchi, and M. Panunzio, *Synthesis*, 1979, 141 and C. R. Harrison and P. Hodge, *ibid.*, 1980, 299 [alkyl halide + SCN$^-$ that is bound to an ion-exchange resin]; M. S. Korobov, L. E. Nivorozhkin, V. I. Minkin, M. M. Levkovich, and S. I. Testoedeva, *Zh. Org. Khim.*, 1978, **14**, 788 [MeCCl=CMeCHO + SCN$^-$]; D. G. Garratt, M. D. Ryan, and M. Ujjainwalla, *Can. J. Chem.*, 1979, **57**, 2145 [alkenes + PhSeSCN]; K. Falkenstein, C. Wolff, and W. Tochtermann, *Liebigs Ann. Chem.*, 1979, 1483 [ring opening of oxanorbornadiene by SCN$^-$]; V. R. Kartashov, N. F. Akimkina, E. V. Skorobogatova, and N. L. Sanina, *Zh. Org. Khim.*, 1979, **15**, 545 and *Kinet. Katal.*, 1978, **19**, 785 [alkenes + (SCN)$_2$]; R. J. Maxwell and L. S. Silbert, *Tetrahedron Lett.*, 1978, 4991 and *J. Am. Oil Chem. Soc.*, 1978, **55**, 583 [alkenes + (SCN)$_2$ or ISCN which is in equilibrium with I_2 + (SCN)$_2$]; P. D. Woodgate, H. H. Lee, P. S. Rutledge, and R. C. Cambie, *Synthesis*, 1978, 152 [alkenes + I_2 + KSeCN]; R. C. Cambie, H. H. Lee, P. S. Rutledge, and P. D. Woodgate, *J. Chem. Soc., Perkin Trans. 1*, 1979, 757, 765 [alkenes + ISCN]; I. V. Bodrikov, L. I. Kovaleva, L. V. Chumakov, and N. S. Zefirov, *Zh. Org. Khim.*, 1978, **14**, 2457 [alkene + RSSCN]; K. Tamao, T. Kakui, and M. Kumada, *Tetrahedron Lett.*, 1980, 111 [(*E*)-alkenyl pentafluorosilicates + copper(II) thiocyanate]; Y. Tamura, T. Kawasaki, N. Golida, and Y. Kita, *Tetrahedron Lett.*, 1979, 1129 [epoxy-ketones + Ph$_3$P(SCN)$_2$]; Y. Tamura, M. Adachi, T. Kawasaki, and Y. Kita, *Tetrahedron Lett.*, 1979, 2251 [primary alcohols + Ph$_3$P(SeCN)$_2$ at −60 °C]; J. Burski, J. Kieszkowski, J. Michalski, M. Pakulski, and A. Skowronska, *J. Chem. Soc., Chem. Commun.*, 1978, 940 [alcohols + chlorophosphoranes + lead thiocyanate]; C. Paulmier, *Bull. Soc. Chim. Fr.*, Part 2, 1979, 592 [thiophens + Br$_2$ and SCN$^-$]; D. N. Harpp, B. T. Friedlander, and R. A. Smith, *Synthesis*, 1979, 181 [Me$_3$SiCN + RSCl]; K. Fujiki, *Bull. Chem. Soc. Jpn.*, 1977, **50**, 3065 [1-alkoxy-naphthalenes + Cu(SCN)$_2$ → 4-thiocyanates]; V. M. Naidan, G. D. Naidan, and A. V. Dombrovskii, *Zh. Obshch. Khim.*, 1979, **49**, 1829 [H$_2$C=CR^{1}R^2 with ArN$_2$$^+$ HSO$_4$$^-$ and NH$_4$SCN → ArCH$_2$CR^{1}R^{2}SCN, while with SO$_2$ + ArN$_2$$^+HSO_4$$^-$ + NH$_4$SCN → ArSO$_2$CH$_2$CR^{1}R^2-

[654] M. Mühlstädt, G. Winkler, A. Koennecke, and B. Schulze, *Z. Chem.*, 1978, **18**, 89; B. Schulze, M. Mühlstädt, and I. Schubert, *ibid.*, 1979, **19**, 41.
[655] H. Suzuki, M. Usuki, and T. Hanafusa, *Synthesis*, 1979, 705 and *Bull. Chem. Soc. Jpn.*, 1979, **52**, 836.
[656] R. A. Volkmann, *Synth. Commun.*, 1978, **8**, 541.
[657] G. C. Barrett, *Tetrahedron*, 1980, **36**, 2123.
[658] N. F. Eweiss and A. Osman, *Tetrahedron Lett.*, 1979, 1169.
[659] S. Kambe and T. Hayashi, *Chem. Ind. (London)*, 1979, 479.
[660] P. A. Grieco, D. S. Clark, and G. P. Withers, *J. Org. Chem.*, 1979, **44**, 2945.
[661] S. J. Falcone and M. P. Cava, *J. Org. Chem.*, 1980, **45**, 1044.

SCN]; A. Toshimitsu, Y. Kozawa, S. Uemura, and M. Okano, *J. Chem. Soc., Perkin Trans. 1*, 1978, 1273 [alkene + KSeCN and CuCl₂ in EtOH → β-alkoxy-selenocyanates]; A. Toshimitsu, S. Uemura, and M. Okano, *J. Chem. Soc., Perkin Trans. 1*, 1978, 1206 [diene + KSeCN and CuCl₂ → saturated cyclic selenides rather than selenocyanates]; S. Rajappa, T. G. Rajagopalan, R. Sreenivasan, and S. Kanal, *J. Chem. Soc., Perkin Trans. 1*, 1979, 2001 [pyrolysis of *N*-aryl-*N'*-benzoyl-thioureas → ArSCN].

Isothiocyanates.—Reference has been made in the preceding section to the formation of mixtures of thiocyanates and isothiocyanates in syntheses employing metal thiocyanates. Specific methods of synthesis for isothiocyanates are illustrated in $RNH_2 + CS_2 + As_2O_3 \rightarrow RNCS$;[662] $Ph_2CHC(S)NHR + BuLi \rightarrow Ph_2C=C-(S^-)NHR \rightarrow Ph_2C=C=NR$, which, with CS_2, → RNCS;[663] and RNC + (ArCOS)₂ → RNCS.[664] Various ring-opening reactions lead to unsaturated isothiocyanates [5,5-disubstituted thiazoline-2-thiones → $R^1R^2C=CR^3NCS$ with $(MeO)_3P$;[665] 2-methyl-oxazoles with CS_2 and base → 2-isothiocyanatovinyl acetates[666]]. Dehydrobromination of α-bromoalkyl isothiocyanates, formed from the alkyl analogues and NBS,[667] yields vinyl isothiocyanates.[668] Competition between substitution of bromine and addition to the carbon atom of the isothiocyanate group is reported in the formation of α-bromoalkyl isothiocyanates.[668] The latter route is adopted preferentially by softer nucleophiles, and subsequent hydrolysis gives aldehydes and ketones, and therefore provides an overall transformation $R^1R^2CHNH_2 \rightarrow R^1COR^2$. Addition to the carbon atom of isothiocyanate is the characteristic reaction of this class, illustrated in the ring-opening of 3-dimethylamino-2-phenyl-2*H*-azirines[669] with MeNCS → $RCH=CPhC(NMe_2)=NC(S)NHMe$, while other isothiocyanates give products from cleavage of the 1,3-bond. Other examples include the synthesis of thio-oximes (PhNCS + $R^1R^2C=NONa \rightarrow R^1R^2C=NOC(S)NHPh \rightarrow R^1R^2C=NSR^3$),[670] the formation of bis(imino)thietans with isocyanides,[671] and the production of imino-1,3-oxathietans from cyclobutene-3,4-diones and EtO_2CNCS[672] through [2 + 2] cycloadditions to the C=S moiety.

Photolysis of isothiocyanates yields isocyanides.[673] Iodine forms charge-transfer complexes, comparative studies indicating that isothiocyanates are more powerful donors than the corresponding thiocyanates.[674]

Enzymic hydrolysates of glucosinolates from seeds of nine species of *Tropaeolum* contain volatile isothiocyanates.[675]

[662] I. Okuda and H. Sugiyama, *Yuki Gosei Kagaku Kyokaishi*, 1978, **36**, 1090 (*Chem. Abstr.*, 1979, **90**, 137 196).

[663] T. Fujinami, N. Otani, and S. Sakai, *Nippon Kagaku Kaishi*, 1978, 265 (*Chem. Abstr.*, 1978, **88**, 169 697).

[664] S. Tanaka, K. Masue, and M. Okano, *Bull. Chem. Soc. Jpn.*, 1978, **51**, 659.

[665] A. Q. Hussein, A. Abu-Taha, and J. C. Jochims, *Chem. Ber.*, 1978, **111**, 3750.

[666] A. W. Faull and R. Hull, *J. Chem. Res. (S)*, 1979, **240**, 148.

[667] A. Q. Hussein and J. C. Jochims, *Chem. Ber.*, 1979, **112**, 1948.

[668] A. Q. Hussein and J. C. Jochims, *Chem. Ber.*, 1979, **112**, 1956.

[669] E. Schaumann and S. Grabley, *Liebigs Ann. Chem.*, 1978, 1568.

[670] R. F. Hudson and H. Dj-Forudian, *J. Chem. Soc., Perkin Trans. 1*, 1978, 12; *Phosphorus Sulfur*, 1978, **4**, 9.

[671] G. L'abbè, L. Huybrechts, J.-P. Declercq, G. Germain, and M. Van Meerssche, *J. Chem. Soc., Chem. Commun.*, 1979, 160.

[672] G. Seitz and R. Sutrisno, *Synthesis*, 1978, 831.

[673] R. Jahn and U. Schmidt, *Monatsh. Chem.*, 1978, **109**, 161.

[674] C. Raby and J. Buxeraud, *Bull. Soc. Chim. Fr., Part 2*, 1978, 439.

[675] A. Kjaer, J. Ogaard Madsen, and Y. Maeda, *Phytochemistry*, 1978, **17**, 1285.

15 Sulphinic and Seleninic Acids, and their Derivatives

Sulphinic and Seleninic Acids.—An improved method of preparation of simple sulphinic acids, using sulphur dioxide[676] with a Grignard reagent or an organolithium compound (also applied for the preparation of $F_2C=CFSO_2H$ from $F_2C=CFLi^{677}$), and a promising use of organoaluminium compounds for the same purpose[678] offer convenient access to these compounds. Camphene has been converted into the *endo*-sulphinic acid *via* the tris(camphanyl)aluminium reagent.[678] Other organometallic compounds for which SO_2-insertion studies have been reported[679] generally tend to form *O*-sulphinates (*i.e.* C—O bond formation).

Certain reactions of alkenes in liquid SO_2 can be accounted for on the basis of addition of SO_2 followed by elimination of sulphoxylic acid, H_2SO_2, through an ene reaction (*e.g.* aromatization of cyclohexa-1,4-diene[680] and the easy exchange of allylic protons in solutions of alkenes in SO_2^{681}). The rationalization of allylic oxidation by SeO_2 on the basis of allylseleninic acid intermediates in a 1973 paper[682] has been largely supported,[683] though with more complexities arising from ionic intermediates than originally proposed.

Routes to sulphinic acids from the cleavage of sulphones, discussed in an earlier section, involve more drastic reaction conditions than are needed for the cleavage of *N*-(alkylsulphonylmethyl)imides, themselves formed from thiols through standard procedures [$RCON(COR)CH_2Br + RSH \rightarrow RCON(COR)CH_2SR \rightarrow RCON(COR)CH_2SO_2R \rightarrow RSO_2Na$ with NaOEt].[684] Sulphides, disulphides, or their oxides undergo oxidative cleavage to sulphinic and sulphonic acids with superoxide anion.[685] Oxidative chlorinolysis of disulphides gives sulphinyl chlorides,[686] and a route to perfluoroalkanesulphinic acids depends on the addition of R_2NSF_3 to a perfluoroalkene [$R^1R^2CFSF_2NR_2 \rightarrow R^1R^2CFSONR_2 \rightarrow R^1R^2CFSO_2H$, the last step being effected with conc. H_2SO_4].[687]

In a modified standard route, arenesulphonyl chlorides are easily reduced to sulphinic acids, using a saturated hydrocarbon and $AlCl_3$.[688]

Allyl sulphinates, formed by ring-opening of certain cyclic sulphinamides in aqueous alkali, readily undergo loss of SO_2 through a pericyclic route [(*Z*)-$ArSO_2NHCHMeCH=CHCHMeSO_2^- \rightarrow$ (*E*)-$ArSO_2NHCHMeCH_2CH=$

[676] H. W. Pinnick and M. A. Reynolds, *J. Org. Chem.*, 1979, **44**, 160.
[677] R. Sarwetre, D. Masure, C. Chuit, and J. F. Normant, *C.R. Hebd. Seances Acad. Sci., Ser. C*, 1979, **288**, 335.
[678] A. V. Kuchin, L. I. Akhmetov, V. P. Yurev, and G. A. Tolstikov, *Zhur. Obshch. Khim.*, 1978, **48**, 469; *ibid.*, 1979, **49**, 401.
[679] A. Dormond, C. Moise, A. Dahchour, and J. Tirouflet, *J. Organomet. Chem.*, 1979, **177**, 181; R. G. Severson and A. Wojcicki, *J. Am. Chem. Soc.*, 1979, **101**, 877; A. E. Crease and M. D. Johnson, *ibid.*, 1978, **100**, 8013.
[680] D. Masilamani and M. M. Rogic, *Tetrahedron Lett.*, 1978, 3785.
[681] D. Masilamani and M. M. Rogic, *J. Am. Chem. Soc.*, 1978, **100**, 4634.
[682] D. Arigoni, A. Vasella, K. B. Sharpless, and H. P. Jensen, *J. Am. Chem. Soc.*, 1973, **95**, 7917.
[683] L. M. Stephenson and D. R. Speth, *J. Org. Chem.*, 1979, **44**, 4683; see also W.-D. Woggon, F. Ruther, and H. Egli, *J. Chem. Soc., Chem. Commun.*, 1980, 706.
[684] M. Uchino, K. Suzuki, and M. Sekiya, *Chem. Pharm. Bull.*, 1978, **26**, 1837.
[685] T. Takata, Y. H. Kim, and S. Oae, *Tetrahedron Lett.*, 1979, 821.
[686] A. Etienne, G. Baills, G. Lonchambon, and B. Desmazieres, *C.R. Hebd. Seances Acad. Sci., Ser. C*, 1979, **288**, 493.
[687] O. A. Radchenko, A. Ya. Ilchenko, L. N. Markovskii, and L. M. Yagupolskii, *Zh. Org. Khim.*, 1978, **14**, 275.
[688] R. Fields, G. Holt, M. O. A. Orabi, and B. Naseeri-Noori, *J. Chem. Soc., Perkin Trans. 1*, 1979, 233.

CHMe],[689] thus supporting the interpretation of rearrangements of alkenes in liquid SO_2 that is based on the instability of allyl sulphinic acids. Sodium salts of sulphinic acids take the role of co-catalyst in $PdCl_2$-catalysed dimerization of butadiene in an alcohol solvent, leading to octa-2,7-dienyl alkyl ethers,[690] and act as arylating agents towards cyclo-octa-1,5-diene in the presence of $PdCl_2$.[691]

Oxidation of sulphinic acids by air (under u.v. irradiation) gives sulphonic acids,[692] but N_2O_4 leads to unstable sulphonyl nitrites RSO_2NO rather than to oxidized products.[693]

Seleninic acids are effective catalysts for the conversion of alkenes into oxirans and of sulphides into sulphones by H_2O_2.[694] They can be used for the partial oxidation of thiols, three equivalents of the thiol with PhSe(O)OH giving equimolar amounts of PhSeSR and RSSR.[695] Benzeneseleninic anhydride (ref. 1h), formed by oxidation of PhSeSePh with HNO_3[696-698] or with t-butyl hydroperoxide,[699] continues to show promise as an oxidizing agent. It oxidizes alcohols under essentially neutral conditions,[696] converts benzylic hydrocarbons into aldehydes or ketones,[697] introduces conjugated double bonds into lactones,[697] converts thiocarbonyl compounds into their carbonyl analogues,[698] and produces α-phenylselenoalkyl ketones from alkenes[699] (see also Vol. 5, p. 54 and ref. 494 for further uses of benzeneseleninic anhydride).

Sulphinyl and Seleninyl Halides.—Chlorinolysis of *N*-(alkylsulphinylmethyl)-phthalimides [PhtCH$_2$S(O)R → RS(O)Cl][700] and the formation of α-chloro-alkane-sulphinyl chlorides from reactive methylene compounds with $SOCl_2$ and pyridine[701] illustrate new examples of known methods of preparation. The latter compounds have been shown to yield α-chloro-sulphines on reaction with a tertiary amine (MeCHClSOCl → MeCCl=S=O → MeCCl$_2$SOCl).[702]

Sulphinate Esters.—Methods of preparation from sulphinyl chlorides[628,703-705] by reaction with alcohols,[703,704] with silyl ethers,[628,703] or with hydroxylamines,[705] from sulphinic acids,[706,707] from sulphoxylates [(RO)$_2$S + H$_2$C=CHCH$_2$Br (at 105–

[689] W. L. Mock and R. M. Nugent, *J. Org. Chem.*, 1978, **43**, 3433.
[690] Y. Tamaru, M. Kagotani, R. Suzuki, and Z. Yoshida, *Chem. Lett.*, 1978, 1329.
[691] Y. Tamaru and Z. Yoshida, *Tetrahedron Lett.*, 1978, 4527.
[692] G. Ricci, S. Dupre, G. Federici, G. Spoto, R. M. Matarese, and D. Cavallini, *Physiol. Chem., Phys.*, 1978, **10**, 435.
[693] S. Oae, K. Shinhama, and Y. H. Kim, *Tetrahedron Lett.*, 1979, 3307.
[694] H. J. Reich, F. Chow, and S. L. Peake, *Synthesis*, 1978, 299.
[695] J. L. Kice and T. W. S. Lee, *J. Am. Chem. Soc.*, 1978, **100**, 5094.
[696] D. H. R. Barton, A. G. Brewster, R. A. H. F. Hui, D. J. Lester, S. V. Ley, and T. G. Back, *J. Chem. Soc., Chem. Commun.*, 1978, 952.
[697] D. H. R. Barton, R. A. H. F. Hui, D. J. Lester, and S. V. Ley, *Tetrahedron Lett.*, 1979, 3331.
[698] D. H. R. Barton, N. J. Cussans, and S. V. Ley, *J. Chem. Soc., Chem. Commun.*, 1978, 393.
[699] M. Shimizu, R. Takeda, and I. Kuwajima, *Tetrahedron Lett.*, 1979, 419.
[700] M. Uchino, K. Suzuki, and M. Sekiya, *Chem. Pharm. Bull.*, 1979, **27**, 1199.
[701] K. Oka and S. Hara, *J. Org. Chem.*, 1978, **43**, 4533.
[702] L. A. Carpino and J. R. Williams, *J. Org. Chem.*, 1979, **44**, 1177.
[703] E. Wenschuh, C. Steyer, and G. Baer, *Z. Chem.*, 1979, **19**, 211.
[704] A. Hessing, M. Jaspers, and I. Schwermann, *Chem. Ber.*, 1979, **112**, 2903.
[705] I. P. Bleeker and J. B. F. N. Engberts, *Recl. Trav. Chim. Pays-Bas*, 1979, **98**, 120.
[706] M. Furukawa, T. Okawara, Y. Noguchi, and M. Nishikawa, *Synthesis*, 1978, 441.
[707] G. Lonchambon, A. Delacroix, R. Garreau, J. N. Veltz, and J. Petit, *Bull. Soc. Chim. Fr., Part 2*, 1979, 541.

115 °C) → $H_2C=CHCH_2S(O)OR$],[708] and by electrolysis of arenethiols or disulphides in AcOH with an alcohol[709] include specific points of interest. Thus the methanolysis of $R^1R^2CHS(O)Cl$ proceeds by nucleophilic substitution at S, not *via* the sulphine (see ref. 702 for contrast),[704] and $Bu^tS(O)ONHCONR_2$, prepared from the sulphinyl chloride and $R_2NCONHOH$, rearranges to $Bu^tSO_2NHCONR_2$ *via* a radical cage mechanism,[705] as demonstrated by [1]H c.i.d.n.p. Three different methods of effecting the overall condensation of a sulphinic acid with an alcohol have been compared.[706]

Alkyl ethenesulphinates, $H_2C=CHS(O)OR$, are obtained by elimination reactions of the β-halogenoethyl or β-tosyloxyethyl analogues.[707] A redox–disproportionation mechanism[710] accounts for the formation of benzils and diaryl disulphides, together with arenesulphonates and thiolsulphonates, through base-catalysed or thermal decomposition of desyl arenesulphinates. Other mechanistic studies deal with solvolysis[711] and stereochemical aspects of displacements at sulphinyl sulphur (see also the 'Sulphoxides' section), which occurs with inversion with higher alkyl esters, but with racemization in the case of methyl esters.[712] Evidence for the general inversion mechanism is provided by the slower $^{18}O-^{16}O$ exchange compared with the rate of racemization of alkyl arenesulphinates by $(Cl_3CCO)_2O$,[713] and this mechanism is relied on in the synthesis of chiral sulphoxides from sulphinates.[490]

Further examples of the use of propargyl sulphinates (see Vol. 5, p. 55) in syntheses of allenes by reaction with organocopper reagents $(RCuBr)MgX \cdot LiBr$[714-717] include a synthesis of (R)-allenes[714] and one of vinyl-allenes.[716]

The role played by benzeneseleninic acid or by phenylseleninyl chloride in promoting the rearrangement of N-aryl-benzohydroxamic acids $PhCONArOH$ to p-hydroxy-benzanilides relies on the formation of seleninate or selenenate esters, respectively, followed by Claisen-type shifts to *ortho*- and then *para*-positions.[718]

Sulphinamides.—A simple synthesis, based on the oxidation of sulphenamides[719] using N-chlorosuccinimide and aqueous $KHCO_3$, involves S-chlorosulphonium ion intermediates. Oxidation of N-acetyl-sulphenamides with m-chloroperoxybenzoic acid gives the corresponding alkyl- or aryl-sulphinyl-acetamides.[720]

16 Sulphonic and Selenonic Acids, and their Derivatives

Preparation of Sulphonic Acids.—The main topics under this heading in the recent literature are very much the same as for the past ten years or more, involving

[708] E. Wenschuh, R. Fahsl, and P. Weingate, *Z. Chem.*, 1978, **18**, 379.
[709] J. Nokami, Y. Fujita, and R. Okawara, *Tetrahedron Lett.*, 1979, 3659.
[710] W. Mueller and K. Schank, *Chem. Ber.*, 1978, **111**, 2870.
[711] S. Braverman and H. Manor, *Phosphorus Sulfur*, 1976, **2**, 213.
[712] M. Mikolajczyk, J. Drabowicz, and H. Slebocka-Tilk, *J. Am. Chem. Soc.*, 1979, **101**, 1302.
[713] J. Drabowicz and S. Oae, *Tetrahedron*, 1978, **34**, 63.
[714] H. Westmijze and P. Vermeer, *Tetrahedron Lett.*, 1979, 4101.
[715] H. Kleijn, C. J. Elsevier, H. Westmijze, J. Meijer, and P. Vermeer, *Tetrahedron Lett.*, 1979, 3101.
[716] H. Kleijn, H. Westmijze, K. Kruithof, and P. Vermeer, *Recl. Trav. Chim. Pays-Bas*, 1979, **98**, 27.
[717] P. Vermeer, H. Westmijze, H. Kleijn, and L. A. Van Dijck, *Recl. Trav. Chim. Pays-Bas*, 1978, **97**, 56.
[718] T. Frejd and K. B. Sharpless, *Tetrahedron Lett.*, 1978, 2239.
[719] M. Haake, H. Gebbing, and H. Benack, *Synthesis*, 1979, 97.
[720] D. N. Harpp, D. F. Mullins, K. Stelioth, and I. Triassi, *J. Org. Chem.*, 1979, **44**, 4196.

aromatic substitution through various strategies, including irradiation of mixtures of hydrocarbon, SO_2, and O_2.[721] Sulphonic acids have also been obtained by the oxidation of 2-thiouracils with MnO_4^-;[722] the electrofluorination of sulpholene to give $C_4F_9SO_2F$ and perfluorosulpholene, which gives $H(CF_2)_4SO_3H$ on alkaline hydrolysis;[723] and by bisulphite-addition routes [formation of an anionic σ-complex between $2,4,6\text{-}(NO_2)_3C_6H_2NMeCH_2CH_2OH$ and HSO_3^-,[724] and the formation of mixtures of $ArCH(\overset{+}{N}H_2R)SO_3^-$ and $ArCH(OH)SO_3^-$ NH_3R^+ from Schiff bases with H_2SO_3[725]]. Long-chain alkanesulphonic acids $Me(CH_2)_{11}{}^{14}CH_2{}^{35}SO_3H$ were obtained from the bromide and $H_2NC(^{35}S)NH_2$, followed by oxidation.[726]

In contradiction to other reports claiming a cycloaddition mechanism, a sulphinylamine gives an α-amino-sulphonic acid $R^1NHCHR^2SO_3H$ with aldehydes through successive hydrolysis and then condensation and bisulphite-addition reactions.[727]

Additional Bibliography: Y. Muramoto, H. Asakura, and H. Suzuki, *Nippon Kagaku Kaishi*, 1979, 297 [*o*- and *p*-IC$_6$H$_4$OEt disproportionate during sulphonation with >80% H$_2$SO$_4$, yielding 2,4-di-iodophenetole together with phenetolesulphonic acids and their iodo-homologues]; F. P. DeHaan, W. D. Covey, G. L. Delker, N. J. Baker, J. F. Feigon, K. D. Miller, and E. D. Stelter, *J. Am. Chem. Soc.*, 1979, **101**, 336 [studies of Olah's π-complex mechanism]; E. N. Krylov, *Zh. Org. Khim.*, 1978, **48**, 2475 [steric interactions determining *ortho*/*para* ratios in sulphonation]; A. Braendstroem, G. Strandlund, and P. O. Lagerstroem, *Acta Chem. Scand., Sect. B*, 1979, **33**, 567 [kinetics of sulphonation of *o*-alkyl-phenols by ClSO$_3$H]; P. Sartori and G. Bauer, *J. Fluorine Chem.*, 1978, **12**, 203 and 1979, **14**, 201 [sulphonation, in liquid SO$_3$, of various polyfluorobenzenes to mono-, di-, and tri-sulphonic acids]; T. Shimura and E. Manda, *Nippon Kagaku Kaishi*, 1978, 1588 (*Chem. Abstr.*, 1979, **90**, 121 275) [contrary to an early report that anthracene with SO$_3$ and pyridine gives only anthracene-1-sulphonic acid, the products are 2- (4%), 1- (14%), 1,5-di- (10%), 1,8-di (60%), and 2,6- and 2,7-di-sulphonic acids (12%)]; P. K. Maarsen, R. Bregman, and H. Cerfontain, *J. Chem. Soc., Perkin Trans. 2*, 1977, 1863 [*m*-NH$_2$C$_6$H$_4$SO$_3$H gives 2,5-di-, 2,4,5-tri-, and 2,3,4,6-tetra-sulphonic acids]; ref. 728 [CF$_3$C(OEt)=CF$_2$ + SO$_3$ → CF$_3$COCF$_2$SO$_3$H].

Reactions of Sulphonic and Selenonic Acids.—$CF_3COCF_2SO_3H$ is a very strong acid, adding readily to $F_2C=CF_2$ below 0 °C and yielding esters with trifluorovinyl ethers.[728] Desulphonation of aminobenzenesulphonic acids can be brought about in CF_3SO_3H if electron-donating groups are located *ortho* or *para* to the sulphonic acid group.[729] Photolysis of sodium anthracene-9-sulphonate produces both desulphonylation (9-anthranol anion + SO_2) and desulphonation (anthracene + SO_3).[730]

A useful practical study[731] has shown that arenesulphonic acids can be extracted from aqueous solutions into immiscible organic solvents as ion-pairs with amines.

The propensity for selenonic acids to react as oxidizing agents has been known for a long time, but this has been regarded more as an interesting contrast between

[721] U. Szalaiko and K. Lewandowski, *Przem. Chem.*, 1979, **58**, 348 (*Chem. Abstr.*, 1979, **91**, 157 202).
[722] F. Freeman, D. L. Bond, S. M. Chernow, P. A. Davidson, and E. M. Karchefski, *Int. J. Chem. Kinet.*, 1978, **10**, 911.
[723] H. Buerger, F. Heyder, G. Pawelke, and H. Niederprum, *J. Fluorine Chem.*, 1979, **13**, 251.
[724] G. F. Bernasconi and H.-C. Wang, *Int. J. Chem. Kinet.*, 1979, **11**, 375.
[725] D. R. Clark and P. Miles, *J. Chem. Res. (S)*, 1978, 124.
[726] A. J. Taylor, A. H. Olavesen, and C. James, *J. Labelled Compd. Radiopharm.*, 1978, **14**, 249.
[727] J. Mirek and S. Rachwal, *Phosphorus Sulfur*, 1977, **3**, 333.
[728] C. G. Krespan, *J. Org. Chem.*, 1979, **44**, 4924.
[729] E. E. Gilbert, *Phosphorus Sulfur*, 1977, **3**, 7.
[730] Y. Izawa, N. Suzuki, A. Inoue, K. Ito, and T. Ito, *J. Org. Chem.*, 1979, **44**, 4581.
[731] A. Brandstrom and G. Strandlund, *Acta Chem. Scand., Ser. B*, 1978, **32**, 489.

the sulphonic and selenonic acids rather than a property to be exploited in synthesis. The selective reactivity of p-ClC$_6$H$_4$SeO$_3$H is shown in its reactions with alkyl sulphides, yielding sulphoxides and the corresponding seleninic acid, which oxidizes further sulphide to sulphoxide even more readily if acid is present.[732]

Sulphonyl Halides.—A novel synthesis exploiting C—O bond cleavage of a sulphonate [ROH → ROSO$_2$CH$_2$CH$_2$NMe$_3$$^+$ X$^-$ → RSC(NH$_2$)=NH$_2$$^+$ X$^-$ → RSO$_2$Cl][733] contrasts with an example illustrating standard methods (RCHFCl → RCHFSCH$_2$Ph → RCHFSO$_2$Cl),[734] Cl$_2$ plus H$_2$O being used in the last step in both cases. 'Sulphochlorination' of alkanes, using SO$_2$ and Cl$_2$ at $-20\,^{\circ}$C, gives mixtures of alkanesulphonyl chlorides in ratios that are characteristic of free-radical attack.[735] ClSO$_3$H is also used in the synthesis of sulphonyl chlorides, a typical example being the substitution of dichlorophenols.[736] Sulphonyl fluorides have been prepared from sulphonyl chlorides, using ZnF$_2$ or AgF at 100 $^{\circ}$C,[737] and the reverse process has been accomplished, using HCl and AlCl$_3$.[738] Electro-fluorinolysis of sulphones has also been established (*e.g.*[723] sulpholene → C$_4$F$_9$SO$_2$F).

Some unusual reactions of aliphatic sulphonyl chlorides can be traced to sulphene formation, *e.g.* MeSO$_2$Cl + Et$_3$N in MeCN → H$_2$C=SO$_2$ → tetramer → (2-py)O$_3$SCH(SO$_2$Me)SO$_2$CH$_2$SO$_2$Me (2-py = 2-pyridyl),[739a] and R^{1}R^{2}CHSO$_2$Cl + CH$_2$(NMe$_2$)$_2$ → Me$_2$NCH$_2$CR^{1}R^{2}SO$_2$NMe$_2$ and other products.[739b] Unexpected transformations of functional groups have been observed in reactions with sulphonyl chlorides, such as a synthesis of di-t-butyl ether (a compound which cannot be synthesized by the classical Williamson synthesis) from Bu^{t}O$^-$ Na$^+$ and RSO$_2$Cl, together with 2-methylpropene, which proceeds *via* the sulphonate ester.[740] CF$_3$SO$_2$Cl acts as a mild chlorinating agent for carbon acids [RCH(CO$_2$Et)$_2$ → RCCl(CO$_2$Et)$_2$], as well as a powerful sulphonylation agent.[741] A novel displacement of the chlorosulphonyl group during an attempted reduction with lithium aluminium hydride has been described (RSO$_2$CH$_2$CH$_2$SO$_2$Cl → RSO$_2$Et rather than the expected RSO$_2$CH$_2$CH$_2$SH).[742]

Sulphonyl fluorides are stable compounds and are reluctant to undergo displacement of a halogen atom; H$_2$C=CHSO$_2$F accordingly gives derivatives of ethanesulphonyl fluoride with active-methylene compounds, amines, and thiols.[743] Photolysis of XCOCF$_2$SO$_2$F (X = Cl, Br, or I) gives XCF$_2$SO$_2$F.[744]

[732] L. G. Faehl and J. L. Kice, *J. Org. Chem.*, 1979, **44**, 2357.

[733] J. F. King and M. Aslam, *Synthesis*, 1980, 285.

[734] G. G. I. Moore, *J. Org. Chem.*, 1979, **44**, 1708.

[735] H. Berthold, H. H. Huenecke, D. Burghardt, M. Hampel, D. Helwig, F. D. Kopinke, S. Krebes, H. Niegel, and W. Pritzkow, *J. Prakt. Chem.*, 1979, **321**, 279; F. D. Kopinke, W. Kroeckel, W. Pritzkow, K. Mathew, and R. Radeglia, *ibid.*, p. 107.

[736] R. J. Cremlyn and T. Cronje, *Phosphorus Sulfur*, 1979, **6**, 413.

[737] I. I. Maletina, A. A. Mironova, T. I. Savina, and Yu. L. Yagupolskii, *Zh. Org. Khim.*, 1979, **15**, 2416.

[738] A. F. Eleev, G. A. Sokolskii, and I. L. Knunyants, *Izv. Akad. Nauk SSSR, Ser. Khim.*, 1978, 2084.

[739] (a) J. S. Grossert, M. M. Bharadwaj, R. F. Langler, T. S. Cameron, and R. E. Cordes, *Can. J. Chem.*, 1978, **56**, 1183; (b) H. Boehme and G. Pindur, *Chem. Ber.*, 1978, **111**, 3294.

[740] H. Masada, T. Yonemitsu, and K. Hirota, *Tetrahedron Lett.*, 1979, 1315.

[741] G. H. Hakimelahi and G. Just, *Tetrahedron Lett.*, 1979, 3643, 3645.

[742] H. O. Fong, W. R. Hardstaff, D. G. Kay, R. F. Langler, R. H. Morse, and D. N. Sandoval, *Can. J. Chem.*, 1979, **57**, 1206.

[743] J. J. Krutak, R. D. Burpitt, W. H. Moore, and J. A. Hyatt, *J. Org. Chem.*, 1979, **44**, 3847.

[744] N. D. Volkov, V. P. Nazaretyan, and L. M. Yagupolskii, *Synthesis*, 1979, 972.

Routine mechanistic studies (solvolysis of $PhCH_2SO_2Cl$[745] and of $ArSO_2Cl$ or $ArSO_2Br$;[746] aminolysis of arenesulphonyl chlorides[747] and of toluene-*p*-sulphonyl bromide, catalysed by imidazole;[748] and the reaction of $ArSO_2Cl$ with phenols[749]) have been reported.

Sulphonate Esters.—Most methods of preparation are straightforward, employing sulphonyl halides (phase-transfer conditions are advantageous[750] for $ArSO_2Cl$ + RCH_2OH), and less familiar sulphonyl peroxides continue to be promoted ($ArCH_2NH_2$ → *O*-sulphonyl-hydroxylamines → imines).[751] More attention is paid here to the recent literature on trifluoromethanesulphonates, whose preparations, though using conventional types of reagent, *e.g.* $(CF_3SO_2)_2O$,[752–754] RSO_2OCOCF_3,[755] or $CF_3SO_2OCOCF_3$,[756] are frequently accompanied by unusual features. Vinyl triflates can be prepared in one step from ketones, $(CF_3SO_2)_2O$, and 2,6-di-t-butyl-4-methylpyridine,[752] and 1-(ethynyl)vinyl triflates from protected ethynyl ketones $R_2CHCOC≡CSiMe_3$.[753] Non-enolizable cycloalkanones give 1,1-bis-triflated cycloalkanes under these conditions,[754] but 2-norcaranone yields 2-(2-trifyloxyethyl)cyclohex-1-enyl triflate through sulphonylation of an enol accompanied by opening of a cyclopropane ring.[754] Forcing conditions (200 °C, for 24 h, in benzene) are required for the preparation of $CF_3SO_3CF_3$ from CF_3SO_3Ag and CF_3I.[758]

While reductive cleavage of a protected 3-*O*-triflyl-hexofuranose by irradiation in the presence of a trace of Na in liquid NH_3 involves C—O bond homolysis,[757] the first example of SO_2—O bond fission of an alkyl triflate has been provided in a use of $CF_3SO_3CF_3$ for the synthesis of trifluoromethyl sulphones, using an enamine as the alkylation agent.[758] Conversion of a methyl alkanesulphonate into the corresponding silyl ester, thence into the mixed anhydride RSO_2OCOCF_3, has been reported,[755] as an alternative to the direct preparation from CF_3SO_3H and TFA with a three-fold excess of P_4O_{10}.[756] Cleavage of a C—O bond is illustrated in the first use of a sulphonyloxy moiety as the leaving group in the nucleophilic substitution of pyrimidines,[759] in reductive displacement by $NaBH_4$ in aprotic solvents (*e.g.* DMSO, HMPA, and sulpholane),[760] and in the conversion of toluene-*p*-sulphonates into peroxides by KO_2 (ROTs + O_2^{-} → ROO· → ROO⁻

[745] I. Lee and W. K. Kim, *Taehan Hwahak Hoechi*, 1978, **22**, 111.

[746] B. G. Gnedin and S. N. Ivanov, *Zh. Org. Khim.*, 1978, **14**, 1497.

[747] B. Giese and K. Heuck, *Chem. Ber.*, 1978, **111**, 1384.

[748] V. G. Zaslavskii, L. M. Litvinenko, T. N. Solomoichenko and V. A. Savelova, *Ukr. Khim. Zh.*, 1980, **46**, 169.

[749] R. V. Vizgert and N. N. Maksimenko, *Zh. Obshch. Khim.*, 1978, **48**, 2776.

[750] W. Szeja, *Synthesis*, 1979, 822.

[751] R. V. Hofmann and E. L. Belfoure, *J. Am. Chem. Soc.*, 1979, **101**, 5687.

[752] P. J. Stang and W. Treptow, *Synthesis*, 1980, 283.

[753] P. J. Stang and T. E. Fisk, *Synthesis*, 1979, 438.

[754] A. Garcia Fraile, A. Garcia Martinez, A. Herrera Fernandez, and J. M. Sanchez Garcia, *An. Quim.*, 1979, **75**, 723; A. Garcia Martinez, I. Espada Rios, and E. Teso Vilar, *Synthesis*, 1979, 382.

[755] A. G. Shipov and Yu. I. Baukov, *Zh. Obshch. Khim.*, 1979, **49**, 1915.

[756] T. R. Forbus and J. C. Martin, *J. Org. Chem.*, 1979, **44**, 313.

[757] T. Tsuchiya, F. Nakamura, and S. Umezawa, *Tetrahedron Lett.*, 1979, 2805.

[758] Y. Kobayashi, T. Yoshida, and I. Kumadaki, *Tetrahedron Lett.*, 1979, 3865.

[759] S. R. James, *J. Chem. Res. (S)*, 1979, 125.

[760] R. O. Hutchins, D. Kandasamy, F. Dux, C. A. Maryanoff, D. Rotstein, B. Goldsmith, W. Burgoyne, F. Cistone, J. Dalessandro, and J. Puglis, *J. Org. Chem.*, 1978, **43**, 2259.

(30)   (31)

$\rightarrow$ ROOR).[761] Cleavage of a C—S bond $\{Ar^1SO_2OAr^2 \rightarrow Ar^1$—$Ar^1$ through the use of PhMgBr + $[(Ph_3P)NiCl_2]^{762}\}$ has been described, and an alternative form of cleavage [(30) $\rightarrow$ (31)] is brought about with Bu^{n}Li at $-72\ °C$.[763] The displacement reaction of PhN(O)=NOTs with PhCH$_2$MgCl did not occur quite as expected, giving PhN(O)=NCHPhCHPhN=N(O)Ph, and is presumed to involve radical intermediates.[764]

Routine mechanistic studies (solvolysis of 1-cyclobutenyl nonafluorobutanoate *via* vinyl cations;[765] hydrolysis of phenyl methanesulphonates and phenyl-sulphonylacetates,[766] alkyl alkanesulphonates,[767] and toluene-*p*-sulphonylmethyl benzenesulphonate;[768] and arenethiolysis of methyl sulphonates[769]) have been reported.

Sulphonyl Hypochlorites.—Further work (see Vol. 5, p. 61) has revealed that CF_3SO_2OCl adds readily to alkenes and displaces Cl or Br from alkyl halides to give triflates, but the potential of this class of compound is severely limited by the danger of explosions when in contact with readily oxidized compounds.[770]

Sulphonamides.—Oxidative amidation of sulphinic acids gives primary sulphon-amides, though the severe reaction conditions (18% oleum with NaN$_3$)[771] restrict the potential of this route. Selective aminosulphonation of anilines, using ClSO$_2$NCO and AlCl$_3$, depends on the formation of cyclic *N*-carbamoyl-sulphonamides.[772] More conventional preparative methods are covered in the conversion of naphthalene-1-thiol into 4-(2-hydroxyethylsulphonyl)naphthalene-1-sulphonamide (an assessment of standard routes has been made),[773] and in the preparation of alkanesulphonyl-[774] and trifluoromethanesulphonyl-hydroxyl-amines[775] from hydroxylamines and the sulphonyl chlorides. In the latter study,[774, 775] conditions favouring *N*- rather than *O*-sulphonylation were established.

[761] R. A. Johnson, E. G. Nidy, and M. V. Merritt, *J. Am. Chem. Soc.*, 1978, **100**, 7960.
[762] E. Wenkert, E. L. Michelotti, and C. S. Swindell, *J. Am. Chem. Soc.*, 1979, **101**, 2246.
[763] W. E. Truce and B. Van Gemert, *J. Am. Chem. Soc.*, 1978, **100**, 5525.
[764] V. N. Yandovskii, U. Traore, I. V. Tselinskii, and E. Yu. Dobrodumova, *Zh. Org. Khim.*, 1978, **14**, 2014.
[765] M. Hanack, E. J. Carnahan, A. Krowczynski, W. Schoberth, L. R. S. Subramanian, and K. Subramanian, *J. Am. Chem. Soc.*, 1979, **101**, 100.
[766] P. Krishnan, S. Sundaram, and N. Venkatasubramanian, *Phosphorus Sulfur*, 1979, **7**, 23.
[767] Yu. G. Skrypnik, *Kinet. Katal.*, 1978, **19**, 1613.
[768] H. A. J. Holterman and J. B. F. N. Engberts, *J. Phys. Chem.*, 1979, **83**, 443.
[769] E. S. Lewis and S. H. Vanderpool, *J. Am. Chem. Soc.*, 1978, **100**, 6421.
[770] Y. Katsuhara and D. D. DesMarteau, *J. Am. Chem. Soc.*, 1979, **101**, 1039.
[771] O. A. Radchenko, *Zh. Org. Khim.*, 1979, **15**, 2420.
[772] Y. Girard, J. G. Atkinson, and J. Rokach, *J. Chem. Soc., Perkin Trans. 1*, 1979, 1043.
[773] G. H. Daub and T. W. Whaley, *J. Org. Chem.*, 1978, **43**, 4659.
[774] U. Hermann, M. Yaktapour, and C. Bleifert, *Z. Naturforsch., Teil B*, 1978, **33**, 574.
[775] V. Kellner and C. Bleifert, *J. Fluorine Chem.*, 1978, **12**, 249.

Bis(trifluoromethanesulphonyl)amides, though effective in C—C bond-forming reactions [PhCH$_2$N(SO$_2$CF$_3$)$_2$ + Me$_2$CuLi → PhCH$_2$Me; and the alkylation of malonates][776] and other nucleophilic substitution reactions (I$^-$ in HMPA → iodides),[777] are susceptible to S—N bond cleavage by bases.[777,778] A novel anionic rearrangement, brought about by NaCN in HMPA at 140 °C, originates in this tendency to cleavage [PhCH$_2$N(SO$_2$Ar)$_2$ → PhCH$_2$$\bar{\text{N}}SO_2$Ar Na$^+$ → PhCH$_2$NHAr + SO$_2$].[778] The stabilization of an amide anion by the sulphonyl group is responsible for the range of reactions of *N*-sodio-*N*-chloro-sulphonamides, Chloramine-T giving *N*-alkylation products (including *N*-sulphonyl-aziridines) with vinyl and allyl chlorides;[779] a remarkable double *ortho*-substitution reaction of *p*-(PhSO$_2$NH)-C$_6$H$_4$NHSO$_2$Ph by Chloramine-B gives 2,5-bis-(phenylsulphonylamino)-*NN'*-bis(phenylsulphonyl)-*p*-benzoquinoneimine.[780]

Cleavage of a sulphonamide to give amines is an essential element in the use of sulphonylation for protection of amines in synthesis. Cathodic reduction[781] and photolysis[782,783] continue to be explored, and though generally serviceable, side-products are often encountered. Photolysis of *N*-alkyl arenesulphonamides at 254 nm in ether gives products of both S—C and S—N bond cleavage (ArH and ArSO$_3$$^-$ R$_2$NH$_2$$^+$, ArSO$_2$SAr, and ArSSAr).[783] *N*-Phenyl trifluoromethanesulphon-amide undergoes cleavage into the imine and CF$_3$SO$_2$H by *N*-alkylation with an α-bromo-ketone.[784] The rearrangement of sulphonanilides to *o*-methylamino-phenyl sulphones that is induced by sulphuric acid has been studied.[785]

Sulphonyl Azides, Azosulphones, and Sulphonyl Isocyanates.—The particular interest in sulphonyl azides lies in their thermolysis into singlet sulphonylnitrenes, and the products of their reactions with the solvent; thus alkylbenzenes and arenesulphonyl azides give azepines and aryl sulphonamides,[786] anthracene gives sulphonamides *via* aziridines,[787] and 3,5-Me$_2$C$_6$H$_3$CH$_2$CH$_2$SO$_2$N$_3$ gives a variety of intramolecular cyclization and intermolecular substitution products.[788]

Thermolysis of phenylazo *p*-tolyl sulphone in the presence of toluene-*p*-sulphonyl iodide is a convenient method for generating sulphonyl radicals[789] (PhN=N-SO$_2$C$_6$H$_4$Me-*p* + *p*-MeC$_6$H$_4$SO$_2$I → PhI + N$_2$ + *p*-MeC$_6$H$_4$SO$_2$·).

The stable 1,4-dipoles R^{1}SO$_2$$\bar{\text{N}}$CONMe$\overset{+}{\text{C}}$(SMe)NMe$_2$ are formed by addition of *S*-methyl-isothioureas to sulphonyl isocyanides.[790]

[776] P. Müller and N. T. M. Phuong, *Tetrahedron Lett.*, 1978, 4727.
[777] R. S. Glass and R. J. Swedo, *J. Org. Chem.*, 1978, **43**, 2291.
[778] P. Müller and N. T. M. Phuong, *Helv. Chim. Acta*, 1979, **62**, 494.
[779] N. A. Rybakova, L. N. Kiseleva, and V. I. Dostovalova, *Izv. Akad. Nauk SSSR, Ser. Khim.*, 1978, 2571.
[780] N. P. Bezverkhii and M. M. Kremlev, *Zh. Org. Khim.*, 1978, **14**, 2596.
[781] R. Kossai, G. Jeminet, and J. Simonet, *Electrochim. Acta*, 1977, **22**, 1395.
[782] J. A. Pincock and A. Jurgens, *Tetrahedron Lett.*, 1979, 1029.
[783] J.-P. Pete and C. Portella, *J. Chem. Res. (S)*, 1979, 20.
[784] R. J. Bergeron and P. G. Hoffman, *J. Org. Chem.*, 1979, **44**, 1835.
[785] S. S. Shrimali, A. Iyer, R. K. Singhal, D. Kishore, and B. C. Joshi, *Rev. Roum. Chim.*, 1979, **24**, 597.
[786] N. R. Ayyangar, M. V. Phatak, A. K. Purchitt, and B. D. Tilak, *Chem. Ind. (London)*, 1979, 853.
[787] J. G. Krause, *Chem. Ind. (London)*, 1978, 271.
[788] R. A. Abramovitch and S. Wake, *Heterocycles*, 1978, **11**, 377.
[789] C. M. M. da Silva Correa, *J. Chem. Soc., Perkin Trans. 1*, 1979, 1519.
[790] E. Schaumann, E. Kausch, J. P. Imbert, and G. Adiwidjaja, *Chem. Ber.*, 1978, **111**, 1475.

17 Disulphides, Diselenides, and Ditellurides

Preparation.—Many examples of the incidental formation of disulphides have been discussed in preceding sections, but few of these have undergone development as potential preparative methods, since there is no shortage of routes to these compounds. However, routes to unsymmetrical disulphides in particular, and from a wider range of starting materials, continue to be developed. Recent papers are collected at the end of this section, with brief details of their contents, where continuing development of known methods is concerned.

Aralkyl trimethylsilyl sulphides $RSSiMe_3$ give disulphides with sulphenyl chlorides,[628] and symmetrical diselenides are formed from the selenium analogues with I_2 ($PhSeSiMe_3 \rightarrow PhSeSePh + Me_3SiI$).[791] The formation of unsymmetrical disulphides from steroidal $\Delta^{1,4}$-diene-3-thiones and 2-nitrobenzenesulphenyl chloride is in accordance with the concept of hard and soft acids and bases, the 'soft' thione nucleophile reacting with the 'soft' sulphenyl halide, but not with the 'hard' sulphonyl chloride.[792] The resulting $\Delta^{1,3,5}$-trienyl 2-nitrophenyl disulphides participate in thiol–disulphide exchange (see next section).

An interesting observation, that $(Bu^n_4N)_2[Fe_4S_4(SR)_4]$ catalyses the oxidation of thiols to disulphides,[793] emerges from a study of this cluster as an analogue of the active site of non-haem iron–sulphur proteins. The adduct formed between thioacetamide and *p*-benzoquinone is capable of 'non-oxidative' formation of disulphides[794] from thiols; this is a curious reaction, possibly with some relevance to processes *in vivo* that involve flavins, but further work is required to determine the correct structure of the effective oxidant. Another unusual conversion of a thiol into the corresponding disulphide involves an arenesulphonyl chloride as reagent, with a three-fold excess of a thiol, in EtOH.[795] Benzenesulphinate anions are converted into disulphides with $EtOP(O)H_2$.[796]

Full details of the preparation of symmetrical diselenides from aldehydes, with $NaBH_4$ and H_2Se in the presence of an amine (see Vol. 5, p. 65), have now been published.[797] Symmetrical *o*-aminophenyl ditellurides are formed from *ortho*-mercuriated azobenzenes with tellurium-(II) or -(IV) salts.[798]

Additional Bibliography: G. A. Olah, M. Arvanaghi, and Y. D. Vankar, *Synthesis*, 1979, 721 [thiols + $Me_2\overset{+}{S}Br\ Br^-$]; Y. H. Kim, K. Shinhama, and S. Oae, *Bull. Chem. Soc. Jpn.*, 1979, **52**, 3117 [di- and poly-sulphides from thiols, disulphides, and thionitrites]; J. Drabowicz and M. Mikolajczyk, *Synthesis*, 1980, 32 [thiols + Br_2 in aqueous $KHCO_3$]; T. A. Chaudri, *Pak. J. Sci. Ind. Res.*, 1976, **19**, 1 [thiols + IN_3 or sulphenyl halides or thiocyanates with NaN_3]; K.-T. Liu and Y.-C. Tong, *Synthesis*, 1978, 669 [thiols + basic alumina exposed to air in benzene]; G. Crank and M. I. H. Makin, *Tetrahedron Lett.*, 1979, 2169 [thiols + KO_2]; T. Yasumura and R. J. Lagow, *Inorg. Chem.*, 1978, **17**, 3108 [pyrolysis of $C_2F_6 + S \rightarrow CF_3S_nCF_3$ ($n = 2$—4)]; D. O. Lambeth, *J. Am. Chem. Soc.*, 1978, **100**, 4808 ['dehydromethionine' (alias *S*-methylisothiazolidine-3-carboxylic acid) + thiols];

[791] M. R. Detty, *Tetrahedron Lett.*, 1979, 4189.
[792] D. H. R. Barton, L. S. L. Choi, R. H. Hesse, M. M. Pechet, and C. Wilshire, *J. Chem. Soc., Perkin Trans. 1*, 1979, 1166.
[793] T. Nagano, K. Yoshikawa, and M. Hirobe, *Tetrahedron Lett.*, 1980, **21**, 297.
[794] V. Horak and W. B. Manning, *J. Org. Chem.*, 1979, **44**, 120.
[795] D. Cipris and D. Pouli, *Synth. Commun.*, 1979, **9**, 207.
[796] H. W. Pinnick, M. A. Reynolds, R. T. MacDonald, and W. D. Brewster, *J. Org. Chem.*, 1980, **45**, 930.
[797] J. W. Lewicki, W. H. H. Gunther, and J. Y. C. Chu, *J. Org. Chem.*, 1978, **43**, 2672.
[798] R. E. Cobbledick, F. W. B. Einstein, W. R. McWhinnie, and F. H. Musa, *J. Chem. Res. (S)*, 1979, 145.

E. S. Huyser and H.-N. Tang, *ACS Symp. Ser., No. 69, Organic Free Radicals*, 1978, 25 [photo-oxidation of thiols sensitized by methylene blue]; F. Bottino, S. Foti, and S. Pappalardo, *J. Chem. Soc., Perkin Trans. 1*, 1979, 1712 [*m*-dimethoxybenzene + SCl₂ or S₂Cl₂ in the presence of Fe gives linear polysulphides ArSₙAr and Ar¹SAr²SAr¹]; W. Buder, *Z. Naturforsch., Teil B*, 1979, **34**, 790 {[(MeO)₃Si(CH₂)₃]₂Sₙ (*n* = 1—4)}; L. Field and R. Ravichandran, *J. Org. Chem.*, 1979, **44**, 2624 [unsymmetrical t-alkyl disulphides through standard routes from thiols with RSSO₂R or RSNO]; J. Lazar and E. Vinkler, *Acta Chim. Acad. Sci. Hung.*, 1979, **101**, 175 [disproportionation of sodium salts of thiolsulphonates in DMF]; Y. Yano, A. Kurushima, and W. Tagaki, *Bull. Chem. Soc. Jpn.*, 1979, **52**, 2739 [PhCH₂SCN → PhSSPh by OH⁻ in phase-transfer conditions]; ref. 805 [disulphides from trisulphides and (RO)₃P].

Reactions of Disulphides, Diselenides, and Ditellurides.—Cleavage of S—S bonds is the predominant reaction of disulphides, and earlier sections have covered the use of these compounds as sulphenylation agents. Thiol–disulphide interchange is a familiar example of this process, and the relative rates of the disulphide-forming step involving Ellman's reagent and 3-hydroxypropanethiol to give the mixed disulphide, and the slower ensuing equilibrium $R^1SSR^2 + R^2S^- \rightleftharpoons R^2SSR^2 + R^1S^-$, have been related to the electron-donating properties of the group R^2.[799] Thiolysis of bis-(*N*-phthalimido) disulphide is catalysed by amines.[800]

Pulse radiolysis of cystamine in aqueous solutions gives the anion radical $RSSR^{\bar{\cdot}}$ as the first product, which undergoes cleavage into $RS \cdot$ and RS^-, from which a variety of products is formed, in proportions that are determined by the various additives.[801]

Extrusion of sulphur from disulphides[802–804] and trisulphides[805] gives sulphides and disulphides, respectively, by treatment with phosphines[802,803] (a polymeric aminophosphine has been advocated[802]), phosphites,[804,805] and MeHgOAc.[804] Ionic pathways continue to be preferred for these reactions,[805] and concurrent electrophilic and nucleophilic mechanisms are in operation, according to interpretations of kinetic data for the reaction with dimethyl disulphide.[804] A new view on nucleophilic attack is advanced in this work,[804] where the encounter involves the σ^* orbital of the S—S bond. Further support is given[806] by c.i.d.n.p. to a radical mechanism that involves $PhCH_2Se \cdot$ for the reaction of dibenzyl selenide with Ph_3P.[806] The process $PhTeTePh + RN_2^+ \ Cl^- \rightarrow RTeCl_2Ph$[807] in the presence of [Cu(OAc)₂] may be more easily understood on this basis.

M.O. predictions, that dibenzyl disulphide is deprotonated faster than the corresponding sulphide, are supported by experiment. A rate factor of *ca.* 7.5 has been determined.[808]

18 Thiolsulphinates

Mild oxidation procedures that are used for the conversion of sulphides into sulphoxides (*m*-chloroperoxybenzoic acid[809] and 2-arylsulphonyl-3-aryl-

[799] R. Freter, E. R. Pohl, J. M. Wilson, and D. J. Hupe, *J. Org. Chem.*, 1979, **44**, 1771.
[800] Y. Abe, T. Horii, S. Kawamura, and T. Nakabayashi, *Bull. Chem. Soc. Jpn.*, 1979, **52**, 3461.
[801] O. I. Micic, M. T. Nenadovic, and P. A. Carapellucci, *J. Am. Chem. Soc.*, 1978, **100**, 2209.
[802] D. N. Harpp, J. Adams, J. G. Gleason, D. Mullins, and K. Steliou, *Tetrahedron Lett.*, 1978, 3989.
[803] D. N. Harpp and A. Granata, *J. Org. Chem.*, 1980, **45**, 271.
[804] R. D. Bach and S. J. Rajan, *J. Am. Chem. Soc.*, 1979, **101**, 3112.
[805] D. N. Harpp and R. A. Smith, *J. Org. Chem.*, 1979, **44**, 4140.
[806] G. Vermeersch, N. Febvay-Garot, S. Caplain, A. Couture, and A. Lablache-Combier, *Tetrahedron Lett.*, 1979, 609.
[807] I. D. Sadekov and A. A. Maksimenko, *Zh. Org. Khim.*, 1978, **14**, 2620.
[808] S. Wolfe and D. Schirlin, *Tetrahedron Lett.*, 1980, **21**, 827.
[809] D. N. Harpp and A. Granata, *Synthesis*, 1978, 782.

oxaziridines[484]) have also been applied to the oxidation of disulphides to thiolsulphinates. Further oxidation to the thiolsulphonate has been noticed.[809,818]

Reactions of thiolsulphinates show a different profile from those for sulphoxides because of the tendency for nucleophilic S—S bond cleavage and protonation at sulphenyl sulphur to occur.[810,811] Acid-catalysed and dialkyl-sulphide-catalysed decomposition of PhS(O)SBu^t in AcOH has as the rate-determining step the attack of the sulphide on the protonated sulphenyl sulphur atom,[811] giving PhSOH and R$_2\overset{+}{S}$SBu^t. Other reactions (PhSOH → Ph$\overset{+}{S}$SR$_2$ → further intermediates) account for the eventual products, *i.e.* PhSSBu^t and PhSO$_2$SPh. The thiolsulphonate is invariably formed through a similar pathway in other reactions [R^{1}S(O)SR^2 + Cl$_3$CSCl → R^{1}SO$_2$SR^1 + R^{2}SSCCl$_3$ + R^{2}SCl[812]].

Oxidation of disulphides and thiolsulphinates with excess N$_2$O$_4$ gives mixtures of thiolsulphonates, indicating S—S bond cleavage into sulphenyl and sulphinyl nitrites and nitrates, followed by recombination.[813]

Pummerer reactions with β-keto-sulphoxides have been shown to involve thiolsulphinates (see Vol. 5, p. 37). Further studies of the Pummerer reaction[814] of ^{13}C- and ^{18}O-labelled benzyl phenylmethanethiolsulphonate [PhCH$_2$S(O)SCH$_2$Ph + Ac$_2$O → PhCH$_2$S(O)CH(SAc)Ph as a 1:1 *erythro*:*threo* mixture] have led to a revised mechanism in which cleavage into the sulphenic acid and thiobenzaldehyde occurs. This initial cleavage is reversible, but the alternative recombination, PhCH$_2$SOH + PhCHS → PhCH$_2$S(O)CH(SH)Ph, accounts well for the scrambling of the label and the stereochemistry. In contrast, thiolsulphinates give transient mixed sulphinic anhydrides RS(O)OCOCF$_3$ with TFAA, as suggested by trapping by alkenes to give β-trifluoroacetoxy-alkyl sulphides.[815]

19 α-Disulphoxides

A new item of indirect evidence for the existence of these compounds as reaction intermediates[816] has been reported in radical-scavenging and spin-trapping experiments for the reaction of an aryl arenethiolsulphinate with AcOOH. The generation in this system of arenesulphinyl radicals from cleavage of an α-disulphoxide provides the best interpretation of the mode of formation of products derived from the arenesulphonyl radicals, and is in support of earlier conclusions by Kice and Oae. A review[817] of this earlier work, together with other reactions (hydrolysis of thiolsulphinates, and oxidation of sulphenylnitrites to thiolsulphonates with N$_2$O$_4$) has appeared.

20 Thiolsulphonates

These are stable oxidation products of disulphides, and can be reached by oxidation of disulphides with *m*-chloroperoxybenzoic acid[809,818] (RCH$_2$SSPh gives

[810] J. L. Kice and C.-C. A. Liu, *J. Org. Chem.*, 1979, **44**, 1918.
[811] T.-L. Ju, J. L. Kice, and C. G. Venier, *J. Org. Chem.*, 1979, **44**, 610.
[812] J. Pintye, G. Stajer, and E. Vinkler, *Acta Chim. Acad. Sci. Hung.*, 1977, **95**, 307.
[813] S. Oae, D. Fukushima, and Y. H. Kim, *Chem. Lett.*, 1978, 279.
[814] N. Furukawa, T. Morishita, T. Akasaka, S. Oae, and K. Uneyama, *Tetrahedron Lett.*, 1978, 1567.
[815] N. Furukawa, T. Morishita, T. Akasaka, and S. Oae, *Tetrahedron Lett.*, 1979, 3973.
[816] B. C. Gilbert, B. Gill, and M. J. Ramsden, *Chem. Ind.* (*London*), 1979, 283.
[817] S. Oae, *Kagaku* (*Kyoto*), 1978, **33**, 240 (*Chem. Abstr.*, 1978, **89**, 23 853).
[818] A. K. Bhattacharya and A. G. Hortmann, *J. Org. Chem.*, 1978, **43**, 2728.

$RCH_2SO_2SPh^{818}$) or from thiolsulphinates with $NaIO_4^{819}$ (see also preceding sections of this Chapter). Hydrolysis and other reactions that generate sulphenic acids $[RS(O)CO_2R + HCl \rightarrow ClCO_2R + RSSR + RSO_2SR,^{820}$ and $ArSBr + OH^- \rightarrow ArSSAr, ArSR,$ and $RSSO_3^-,$ *via* a thiolsulphonate;[821] see also the 'Sulphenic Acids' section] commonly lead to thiolsulphonates. Parent salts $ArSO_2S^- M^+$ are readily prepared.[822]

Thiolsulphonates react as typical sulphenylation reagents towards nucleophiles,[810] and their use in the synthesis of sulphides has been reviewed earlier in this Chapter. Sulphenylation of dithiocarboxylate anions $[RC(S)S^- \rightarrow RC(S)SSMe]$ has been demonstrated.[823] A novel related group of compounds $R^1SSSO_2R^2$ has been prepared by similar S–S bond-forming methods, and undergoes spontaneous conversion into thiolsulphonates in polar solvents, presumably through $R^1S^+ R^2SO_2S^-.[824]$

[819] Y. H. Kim, T. Takata, and S. Oae, *Tetrahedron Lett.*, 1978, 2305.
[820] T. Wakui, Y. Nakamura, and S. Motoki, *Bull. Chem. Soc. Jpn.*, 1978, **51**, 3081.
[821] A. Chaudhuri and S. K. Bhattacharjee, *Curr. Sci.*, 1978, **47**, 731.
[822] H. G. Hansen and A. Senning, *J. Chem. Soc., Chem. Commun.*, 1979, 1135.
[823] R. S. Sukhai and L. Brandsma, *Synthesis*, 1979, 971.
[824] D. N. Harpp, D. K. Ash, and R. A. Smith, *J. Org. Chem.*, 1979, **44**, 4135.

2

Ylides of Sulphur, Selenium, and Tellurium, and Related Structures

BY E. BLOCK, D. L. J. CLIVE, N. FURUKAWA & S. OAE

PART I: Ylides and Carbanionic Compounds of Sulphur *by E. Block*

1 Introduction

A substantial majority of the 375 papers referred to in this section deal with the application of anionic and ylidic sulphur reagents in organic synthesis, particularly *vis-à-vis* C—C bond formation in ring-closure, -contraction, or -expansion reactions, and C=O, C=C, and C≡C bond formation under gentle conditions. Sulphur has been teamed up with silicon, tin, halogen, boron, and aluminium functionalities, and the conditions of carbanion and ylide formation have been refined, to permit precise control over product stereochemistry and regiochemistry (*e.g.* 1,2- *versus* 1,4-addition to enones, *α*- *versus* *γ*-substitution in allylic systems, *α*-proton abstraction *versus* *γ*-proton abstraction, and Michael addition in vinyl-sulphur systems). A number of general reviews have been published on 'sulphur in synthesis' and related subjects,[1] while sections of other reviews deal with lithio-sulphur compounds.[2] Coverage of sulphur anions and ylides is included in a recent book[3] and in chapters in Methodicum Chimicum[4] and in Comprehensive Organic Chemistry.[5] In Volume 5 of these Reports it was observed that 'the use of sulphur-containing carbanions and ylides in synthesis is now so commonplace that complete coverage of the literature within the restraints of space is impractical'. While this statement rings even truer now, it is hoped that most of the notable and representative applications of anionic and ylidic sulphur compounds in organic synthesis published in the past 24 months will be found in these pages.

Two general studies of sulphur-stabilized carbanions, employing n.m.r. and other

[1] E. Block, *Aldrichimica Acta*, 1978, **11**, 51; L. Field, *Synthesis*, 1978, 713; S. F. Martin, *ibid.*, 1979, 633; B. M. Trost, *Acc. Chem. Res.*, 1978, **11**, 453; B. M. Trost, *Chem. Rev.*, 1978, **78**, 363; S. Warren, *Acc. Chem. Res.*, 1978, **11**, 401.
[2] H. W. Gschwend and H. R. Rodriguez, *Org. React.*, 1979, **26**, 1; E. M. Kaiser, *J. Organomet. Chem.*, 1979, **183**, 1.
[3] J. C. Stowell, 'Carbanions in Organic Synthesis,' Wiley–Interscience, New York, 1979.
[4] 'Methodicum Chimicum', Vol. 7B, ed. H. Zimmer and K. Niedenzu, Academic Press, New York, 1978 (Chapters by R. M. Wilson and D. N. Buchanan, and by D. C. Lankin and H. Zimmer).
[5] G. C. Barrett, T. Durst, and C. R. Johnson, in 'Comprehensive Organic Chemistry', Vol. 3, ed. D. N. Jones, Pergamon Press, Oxford, 1979 (three separate chapters).

spectroscopic methods, have been published.[6,7] The agreement between experimental conclusions and *ab initio* calculations, while adequate for the α-sulphenyl carbanion, is not particularly good for the α-sulphinyl and α-sulphonyl carbanions due to interactions with the cation and solvent. One interesting observation is that the order of oxidation potentials for the process $R^- \rightarrow R \cdot + e^-$ is α-sulphenyl carbanion > α-sulphinyl carbanion $\gg$ α-sulphonyl carbanion,[7] in accord with data on free-radical stability.[8] In connection with the metallation of organosulphur compounds it might be noted that *N*-benzylidenebenzylamine is reported to be a useful coloured (red-purple) indicator in metallations involving organolithium compounds.[9]

In contrast with the spate of activity in the area of theoretical calculations on sulphur carbanions and ylides reported in Volume 5, only a single, brief paper has been published in this area in the past two years. This interesting study predicts (and experimentally substantiates) the greater stability of α-disulphide carbanions compared to α-sulphenyl carbanions.[10]

2 Sulphonium Ylides

Synthesis and Properties.—Perhaps the most interesting new compound that might be included in this section is the 'non-ylidic' $F_4S=CH_2$ (1),[11] a colourless, stable gas that boils at $-19\ ^\circ C$ and has been synthesized as shown. The carbon chemical shift of 43.9 p.p.m. is very different from that for phosphonium ylides, suggesting that there is little negative charge on the carbon atom in (1). Carbon is sp^2-hybridized and the barrier to rotation about the C=S bond (estimated bond order is 1.9) is greater than 25 kcal mol^{-1}; this precludes free rotation at room temperature.

$$SF_5CH_2Br \xrightarrow[\text{at} -110\ ^\circ C]{\text{BuLi}} SF_5CH_2Li \longrightarrow F_4S=CH_2 + LiF$$

$$(1)$$

Irradiation of (1) yields ethylene and sulphur tetrafluoride. Further information on this fascinating new molecule is eagerly awaited. Fluoride-induced cleavage of the carbon–silicon bond in (2) exemplifies a new method of preparing sulphur ylides which might prove valuable in cases where the use of strong bases is precluded.[12]

Additional examples of polymer-supported sulphonium ylides have appeared,[13] including a nice example where the ylide is generated under phase-transfer

[6] G. Chassaing and A. Marquet, *Tetrahedron*, 1978, **34**, 1399.
[7] S.-I. Nishi and M. Matsuda, *J. Am. Chem. Soc.*, 1979, **101**, 4632.
[8] E. Block, 'Reactions of Organosulfur Compounds', Academic Press, New York, 1978, Ch. 5.
[9] L. Duhamel and J. C. Plaquevent, *J. Org. Chem.*, 1979, **44**, 3404.
[10] S. Wolfe and D. Schirlin, *Tetrahedron Lett.*, 1980, **21**, 827.
[11] G. Kleemann and K. Seppelt, *Angew. Chem., Int. Ed. Engl.*, 1978, **17**, 516; 1979, **18**, 944.
[12] E. Vedejs and G. R. Martinez, *J. Am. Chem. Soc.*, 1979, **101**, 6452.
[13] S. Tanimoto, Y. Imazu, T. Sugimoto, and M. Okano, *Bull. Inst. Chem. Res. Kyoto Univ.*, 1978, **56**, 345 (*Chem. Abstr.*, 1979, **91**, 21 365).

conditions and the recycling of the polymeric system is possible without loss of reactivity[14]. New syntheses have been reported for trimethylsilyl-[15] and trimethyl-stannyl-sulphonium[16] ylides, mono-[17] and tris-(sulphonio)-cyclopentadienylides,[18] ylide salts,[19] and α-nitrosulphonium ylides.[20] Sulphonium ylides are formed on reaction of sulphides with 1,1-dicyano-epoxides[21] and with 2-diazo-1,3-diketones on irradiation.[22] Other methods include the condensation of DMSO with active-methylene compounds in the presence of 3-sulphopropanoic anhydride[23] and the reaction of stabilized anions with enaminosulphonium salts.[24]

The first X-ray structure of a thiabenzene derivative, *i.e.* 1-benzoyl-2-methyl-2-thianaphthalene, has been reported.[25] The geometry is similar to that of open-chain benzoyl-stabilized ylides with pyramidal sulphur. Other studies point to the ylidic character of '9,10-diphenyl-10-thia-anthracene'.[26] N.m.r. data, including ^{13}C n.m.r. chemical shifts, have been presented for indole-3-sulphonium ylides[27] and carbonyl-stabilized sulphonium ylides and their Pd^{II} complexes.[28] Infrared spectra of α-nitrosulphonium ylides have been analysed.[29] Further details have been provided of a study of the pyramidal inversion of unstabilized sulphonium ylides[30] (see Chap. 5, pp. 243 and 251).

Reactions.—In Volume 5 (p. 74), an example of the formation and alkylation of a sulphonium ylide anion was given. Another example (4) has now been described,[30]

[14] M. J. Farrall, T. Durst, and J. M. J. Frechet, *Tetrahedron Lett.*, 1979, 203.
[15] F. Cooke, P. Magnus, and G. L. Bundy, *J. Chem. Soc., Chem. Commun.*, 1978, 714.
[16] W. Ando, M. Takata, and A. Sekiguchi, *J. Chem. Soc., Chem. Commun.*, 1979, 1121.
[17] K. Friedrich, W. Amann, and H. Fritz, *Chem. Ber.*, 1979, **112**, 1269.
[18] K. H. Schlingensief and K. Hartke, *Liebigs Ann. Chem.*, 1978, 1754.
[19] N. N. Magdesieva and V. A. Danilenko, *Zh. Org. Khim.*, 1979, **15**, 379.
[20] V. V. Semenov and S. A. Shevelev, *Izv. Akad. Nauk SSSR, Ser. Khim.*, 1978, 2355.
[21] A. Robert, M. T. Thomas, and A. Foucaud, *J. Chem. Soc., Chem. Commun.*, 1979, 1048.
[22] V. A. Nikolaev, Y. Frenkh, and I. K. Korobitsyna, *Zh. Org. Khim.*, 1978, **14**, 1433
[23] K. Schank and C. Schunknecht, *Synthesis*, 1978, 678.
[24] T. Severin and I. Brautigam, *Chem. Ber.*, 1979, **112**, 3007.
[25] M. Hori, T. Kataoka, H. Shimizu, S. Ohno, K. Narita, H. Takayanagi, H. Ogura, and Y. Iitaka, *Tetrahedron Lett.*, 1979, 4315.
[26] M. Hori, T. Kataoka, H. Shimizu, Y. Itagaki, and T. Higuchi, *Tetrahedron Lett.*, 1979, 1603.
[27] K. H. Park, G. A. Gray, and G. D. Daves, Jr., *J. Am. Chem. Soc.*, 1978, **100**, 7475; K.-W. Park and G. D. Daves, Jr., *J. Org. Chem.*, 1980, **45**, 780.
[28] G. Fronza, P. Bravo, and C. Ticozzi, *J. Organomet. Chem.*, 1978, **157**, 299.
[29] K. I. Rezchikova, O. P. Shitov, A. P. Seleznev, V. A. Tartakovskii, and V. A. Shlyapochnikov, *Izv. Akad. Nauk SSSR, Ser. Khim.*, 1979, 1129.
[30] D. M. Roush, E. M. Price, L. K. Templeton, D. H. Templeton, and C. H. Heathcock, *J. Am. Chem. Soc.*, 1979, **101**, 2971.

along with other reactions of the precursor sulphonium ylide (3).[30,31] Among the diverse reactions of sulphonium ylides examined in the past two years are photosensitized oxidation;[32] reaction with sulphur monoxide, giving a sulphine [*e.g.* (5)→(6)];[33] Stevens rearrangement in cyclophane syntheses;[34] reaction with Group VA and VIA elements, leading to demethylation and trans-ylidation among other processes;[35] solvolysis, leading to α-substituted sulphides [*e.g.* (7)];[36] fragmentation [*e.g.* (8)[37]]; alkylation of cyclic ylides (see Chap. 5, p. 247);[38] *O*-methylation of oximes;[39] and epoxide formation with diphenylsulphonium cyclopropylide,[40] allylic ylides [*e.g.* (9)],[41] or dimethylsulphonium methylide.[42] While most sulphur ylides react with simple ketones to give epoxides, a few instances are known where olefin

(5)   (6)

(7)

(8)

[31] K. Tokuno, F. Miyoshi, Y. Arata, Y. Itatani, Y. Arakawa, and T. Ohashi, *Yakugaku Zasshi*, 1978, **98**, 1005 (*Chem. Abstr.*, 1979, **90**, 6195).

[32] W. Ando, S. Kohmoto, and K. Nishizawa, *J. Chem. Soc., Chem. Commun.*, 1978, 894; W. Ando, S. Kohmoto, H. Miyazaki, K. Nishizawa, and H. Tsumaki, *Photochem. Photobiol.*, 1979, **30**, 81.

[33] B. F. Bonini, G. Maccagnani, G. Mazzanti, P. Pedrini, and P. Piccinelli, *J. Chem. Soc., Perkin Trans. 1*, 1979, 1720.

[34] (*a*) J. R. Davy, M. N. Iskander, and J. A. Reiss, *Aust. J. Chem.*, 1979, **32**, 1067; (*b*) M. W. Haenel and A. Flatow, *Chem. Ber.*, 1979, **112**, 249; (*c*) D. Kamp and V. Boekelheide, *J. Org. Chem.*, 1978, **43**, 3475; (*d*) D. N. Leach and J. A. Reiss, *ibid.*, p. 2484; (*e*) D. N. Leach and J. A. Reiss, *Tetrahedron Lett.*, 1979, 4501; (*f*) D. N. Leach and J. A. Reiss, *Aust. J. Chem.*, 1979, **32**, 361; (*g*) T. Otsubo, D. Stusche, and V. Boekelheide, *J. Org. Chem.*, 1978, **43**, 3466; (*h*) I. D. Reingold, W. Schmidt, and V. Boekelheide, *J. Am. Chem. Soc.*, 1979, **101**, 2121.

[35] B. A. Arbuzov, Yu. V. Belkin, N. A. Polezhaeva, and G. E. Buslaeva, *Izv. Akad. Nauk SSSR, Ser. Khim.*, 1978, 1643; B. A. Arbuzov, Yu. V. Belkin, and N. A. Polezhaeva, *ibid.*, 1979, 1893; N. N. Magdesieva and V. A. Danilenko, *J. Gen. Chem. USSR (Engl. Transl.)*, 1979, **49**, 1740; H. Matsuyama, M. Matsumoto, M. Kobayashi, and H. Minato, *Bull. Chem. Soc. Jpn.*, 1979, **52**, 1139.

[36] R. Pellicciari, M. Curini, P. Ceccherelli, and B. Natalini, *Gazz. Chim. Ital.*, 1978, **108**, 671

[37] U. J. Kempe, T. Kempe, and T. Norin, *J. Chem. Soc., Perkin Trans. 1*, 1978, 1547.

[38] A. Garbesi, *Tetrahedron Lett.*, 1980, **21**, 547.

[39] S. Tanimoto, T. Yamadera, T. Sugimoto, and M. Okano, *Bull. Chem. Soc. Jpn.*, 1979, **52**, 627.

[40] M. Hanack, E. J. Carnahan, A. Krowczynski, W. Schoberth, L. R. Subramanian, and K. Subramanian, *J. Am. Chem. Soc.*, 1979, **101**, 100; B. M. Trost, Y. Nishimura, K. Yamamoto, and S. S. McElvain, *ibid.*, p. 1328.

[41] E. J. Corey, Y. Arai, and C. Mioskowski, *J. Am. Chem. Soc.*, 1979, **101**, 6748.

[42] (*a*) Y.-I. Lin, S. A. Lang, Jr., C. M. Seifert, R. G. Child, G. O. Morton, and P. F. Fabio, *J. Org. Chem.*, 1979, **44**, 4701; (*b*) T. J. McCarthy, W. F. Conner, and S. M. Rosenfeld, *Synth. Commun.* 1978, **8**, 379; (*c*) D. R. Morton, Jr. and F. C. Brokaw, *J. Org. Chem.*, 1979, **44**, 2880.

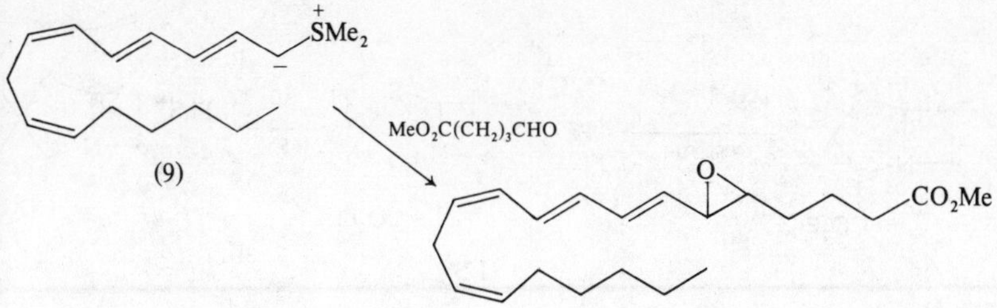

(9)

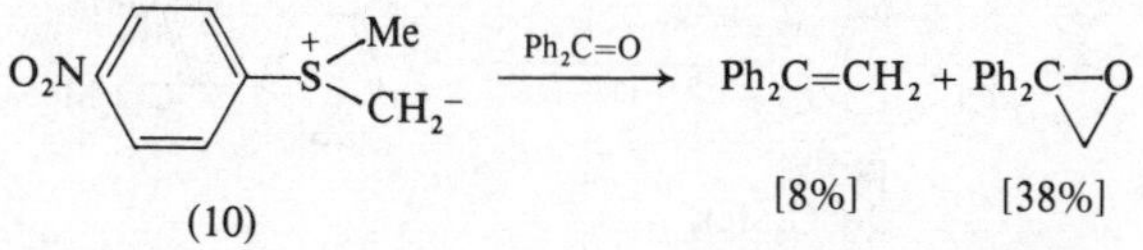

$MeO_2C(CH_2)_3CHO$

(10)

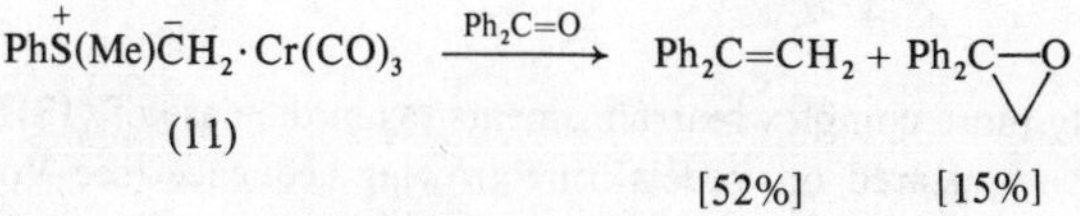

$Ph_2C=O$

$Ph_2C=CH_2$ + $Ph_2C\!-\!O$

[8%]   [38%]

$Ph\overset{+}{S}(Me)\overset{-}{C}H_2 \cdot Cr(CO)_3$ $\xrightarrow{Ph_2C=O}$ $Ph_2C=CH_2$ + $Ph_2C\!-\!O$

(11)

[52%]   [15%]

formation also occurs, apparently by 'Wittig-type' transition states. Thus *p*-nitrophenyl ylide (10) and the chromium–ylide complex (11) both give some 1,1-diphenylethylene on reaction with benzophenone.[43] It is argued that strong electron-withdrawing effects by the *p*-nitrophenyl and tricarbonylchromium groups make attack at the sulphur atom by the negative oxygen competitive with attack by oxygen at carbon. The formation of a highly stabilized olefin is also important since only very low yields of olefins are found upon reaction of (11) with cyclohexanone.[43]

Other reactions of sulphonium ylides include α,β'-elimination,[12,16,44] metal-mediated carbene-transfer to olefins,[45,46] insertion into aromatic C–H bonds[47] or other carbenoid-type processes,[48] formation of Pd^{II} complexes,[49] addition to enones (forming cyclopropyl ketones[50] or heterocycles[51]), reaction with isoquinoline 2-oxide,[52] and [2,3]-sigmatropic rearrangements[53–57] [as in the case of (12)[58]] or

[43] A. Ceccon, F. Miconi, and A. Venzo, *J. Organomet. Chem.*, 1979, **181**, C4.

[44] V. Cere, C. Paolucci, S. Pollicino, E. Sandri, and A. Fava, *J. Org. Chem.* 1979, **44**, 4128.

[45] J. Cuffe, R. J. Gillespie, and A. E. A. Porter, *J. Chem. Soc., Chem. Commun.*, 1978, 641.

[46] R. J. Gillespie and A. E. A. Porter, *J. Chem. Soc., Perkin Trans. 1*, 1979, 2624.

[47] R. J. Gillespie and A. E. A. Porter, *J. Chem. Soc., Chem. Commun.*, 1979, 50.

[48] G. Ege and K. Gilbert, *Tetrahedron Lett.*, 1979, 1567.

[49] H. Nishiyama, *J. Organomet. Chem.*, 1979, **165**, 407.

[50] M. Chiericato, P. Dalla Croce, and E. Licandro, *J. Chem. Soc., Perkin Trans. 1*, 1979, 211.

[51] T. Nishio, M. Sugawara, and Y. Omote, *J. Heterocycl. Chem.*, 1979, **16**, 815.

[52] M. Watanabe, M. Aono, T. Kinoshita, and S. Furukawa, *Yakugaku Zasshi*, 1978, **98**, 198 (*Chem. Abstr.*, 1978, **89**, 6199).

[53] G. Morel, S. Khamsitthideth, and A. Foucaud, *J. Chem. Soc., Chem. Commun.*, 1978, 274.

[54] G. Morel, E. Marchand, and A. Foucaud, *Tetrahedron Lett.*, 1978, 3719.

[55] M. Franck-Neumann and J. J. Lohmann, *Tetrahedron Lett.*, 1978, 3729.

[56] K. Ogura, S. Furukawa, and G.-I. Tsuchihashi, *J. Am. Chem. Soc.*, 1980, **102**, 2125.

[57] H. J. Reich and M. L. Cohen, *J. Am. Chem. Soc.*, 1979, **101**, 1307.

[58] P. J. Giddings, D. I. John, and E. J. Thomas, *Tetrahedron Lett.*, 1980, **21**, 395.

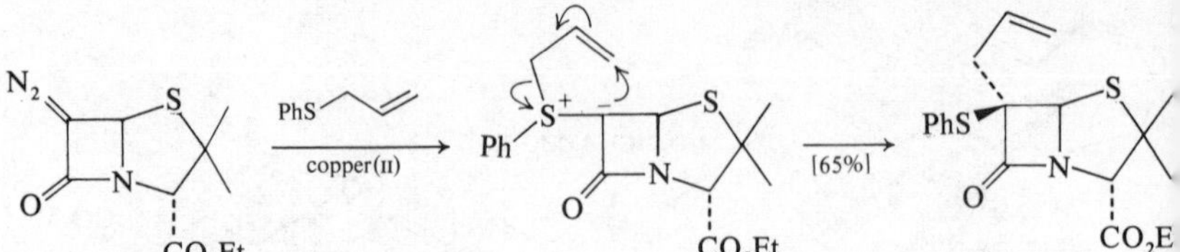

(12)

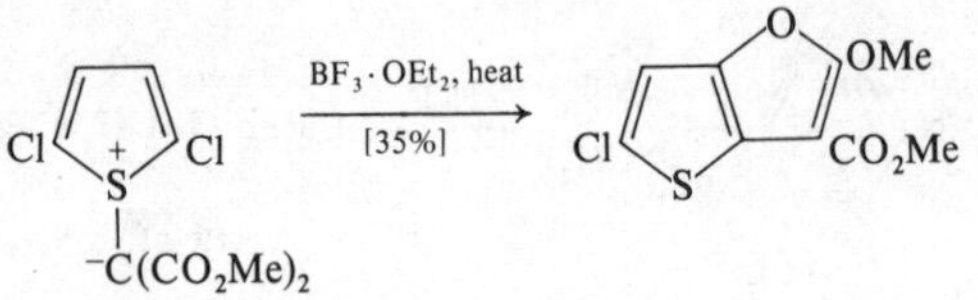

(13)

mechanistically more complex rearrangements [as in the case of (13)[59]]. A number of papers have appeared on Vedejs' ring-growing sequence (see Vol. 4, p. 84),[60] including reactions involving unstabilized sulphonium ylides[61] and stereochemical and conformational studies of the reaction.[44,62,63] (see also Chap. 5, p. 240).

3 Oxosulphonium and Amino-oxosulphonium Ylides

Synthesis and Properties.—Additional details have been published on the preparation of oxosulphonium ylides from sulphoximides[64] (see Chap. 2, Pt III, p. 144) and of phosphorylated oxosulphonium ylides.[65]

Reactions.—1-Methyl-3,5-diphenylthiabenzene 1-oxide is a novel sulphur ylide ligand in tricarbonyl-chromium, -molybdenum, and -tungsten complexes (14),

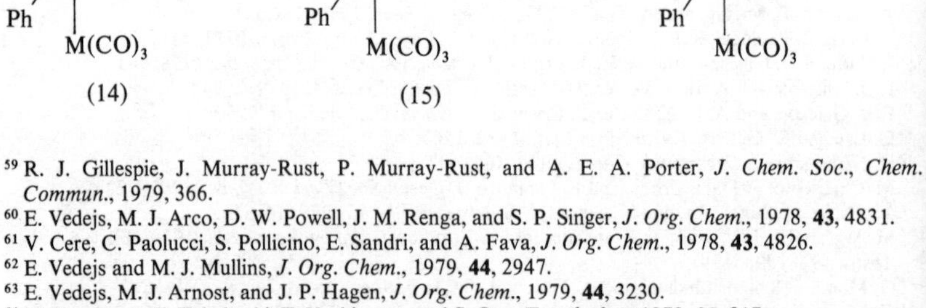

(14)   (15)

[59] R. J. Gillespie, J. Murray-Rust, P. Murray-Rust, and A. E. A. Porter, *J. Chem. Soc., Chem. Commun.*, 1979, 366.
[60] E. Vedejs, M. J. Arco, D. W. Powell, J. M. Renga, and S. P. Singer, *J. Org. Chem.*, 1978, **43**, 4831.
[61] V. Cere, C. Paolucci, S. Pollicino, E. Sandri, and A. Fava, *J. Org. Chem.*, 1978, **43**, 4826.
[62] E. Vedejs and M. J. Mullins, *J. Org. Chem.*, 1979, **44**, 2947.
[63] E. Vedejs, M. J. Arnost, and J. P. Hagen, *J. Org. Chem.*, 1979, **44**, 3230.
[64] N. Furukawa, F. Takahashi, T. Yoshimura, and S. Oae, *Tetrahedron* 1979, **35**, 317.
[65] V. P. Lysenko, I. E. Boldeskul, R. A. Loktionova, and Yu. G. Gololobov, *Zh. Obshch. Khim.*, 1979, **49**, 1230; 1978, **48**, 978; also see L. F. Kasukhin, M. N. Ponomarchuk, V. P. Kysenko, and Yu. G. Gololobov, *Ukr. Khim. Zh.*, 1978, **44**, 519.

which have been found to have a non-planar ring and to undergo lithiation at the
S-methyl group to produce (15).[66] An asymmetric synthesis of cyclopropanes is
based on the addition of dimethyloxosulphonium methylide (17) to the heterocycle
(16).[67] Ylide (17) has also been employed in the synthesis of the spiro-
cyclopentadiene (18),[68] benzoxepin and benzopyran-5-one derivatives,[69]
tropolones,[70] cyclopropanes and vinylcyclopropanes,[71] thiabenzene *S*-oxides,[72]
(±)-laurencin,[73] isoxazoline *N*-oxides,[74] nitro-[75] and cyano-cyclopropanes,[76] and
hydro-azulenones,[77] amongst other compounds.[78]

Michael addition to (dimethylamino)phenylvinyloxosulphonium salts (19) is
thought to involve amino-oxosulphonium ylide intermediates such as (20).[79]

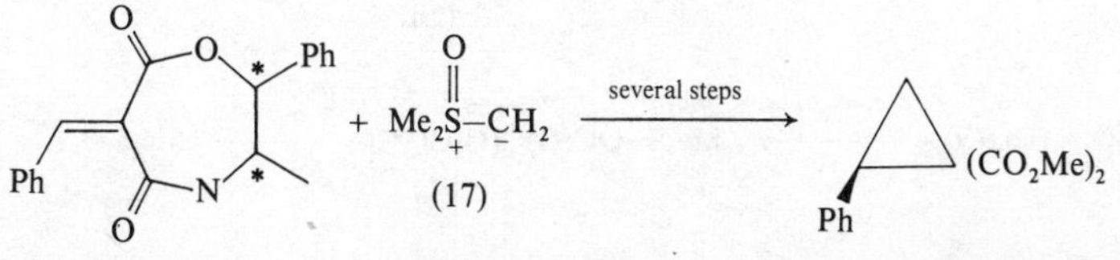

(16)   (17)   [63%; > 90% e.e.]

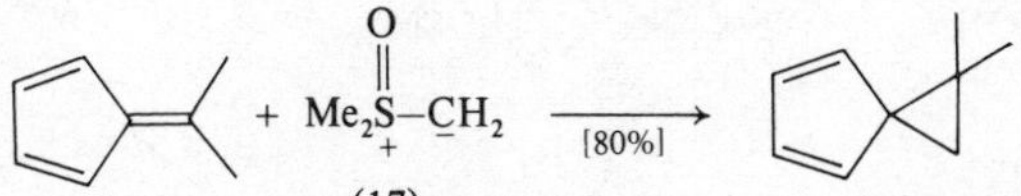

(17)   [80%]   (18)

4 Oxy- and Aza-sulphonium Ylides

Oxysulphonium Ylides.—Reagents that are related to DMSO are now in routine
use for the selective oxidation of alcohols to aldehydes and ketones. The particularly
effective combination of DMSO and oxalyl chloride followed by triethylamine[80] has
been shown to involve the intermediacy of a chlorodimethylsulphonium salt (21).[81]

[66] L. Weber, C. Kruger, and Y.-H. Tsay, *Chem. Ber.*, 1978, **111**, 1709; L. Weber, *ibid.*, 1979, **112**, 99;
L. Weber, *ibid.*, p. 3828.
[67] T. Mukaiyama, K. Fujimoto, and T. Takeda, *Chem. Lett.*, 1979, 1207.
[68] W.-D. Schroer and W. Friedrichsen, *Liebigs Ann. Chem.*, 1978, 1648.
[69] P. Bravo, C. Ticozzi, G. Fronza, R. Bernardi, and D. Maggi, *Gazz. Chim. Ital.*, 1979, **109**, 137; P.
Bravo and C. Ticozzi, *ibid.*, p. 169.
[70] D. A. Evans, D. J. Hart, and P. M. Koelsch, *J. Am. Chem. Soc.*, 1978, **100**, 4593.
[71] D. Brule, J. C. Chalchat, and R. Vessiere, *Bull. Soc. Chim. Fr.*, *Part 2*, 1978, 385.
[72] P. Bennett, J. A. Donnelly, and M. J. Fox, *J. Chem. Soc.*, *Perkin Trans. 1*, 1979, 2990; J. A.
Donnelly, M. J. Fox, and J. G. Hoey, *ibid.*, p. 2629.
[73] T. Masamune, H. Murase, H. Matsue, and A. Murai, *Bull Chem. Soc. Jpn.*, 1979, **52**, 135.
[74] T. Sakakibara and R. Sudoh, *Bull. Chem. Soc. Jpn.*, 1978, **51**, 1193.
[75] T. Sakakibara, R. Sudoh, and T. Nakagawa, *Bull. Chem. Soc. Jpn.*, 1978, **51**, 1189.
[76] J. J. McCullough and C. Manning, *J. Org. Chem.*, 1978, **43**, 2839.
[77] P. Geetha, K. Narasimhan, and S. Swaminathan, *Tetrahedron Lett.*, 1979, 565.
[78] R. Faragher, T. L. Gilchrist, and I. W. Southon, *Tetrahedron Lett.*, 1979, 4113; S. Hagen, T.
Anthonsen, and L. Kilaas, *Tetrahedron*, 1979, **35**, 2583; L. Kohout and J. Fajkos, *Collect. Czech.
Chem. Commun.*, 1978, **43**, 1134.
[79] C. R. Johnson, J. P. Lockard, and E. R. Kennedy, *J. Org. Chem.*, 1980, **45**, 264.
[80] K. Omura and S. Swern, *Tetrahedron*, 1978, **34**, 1651.
[81] A. J. Mancuso, D. S. Brownfain, and D. Swern, *J. Org. Chem.*, 1979, **44**, 4148.

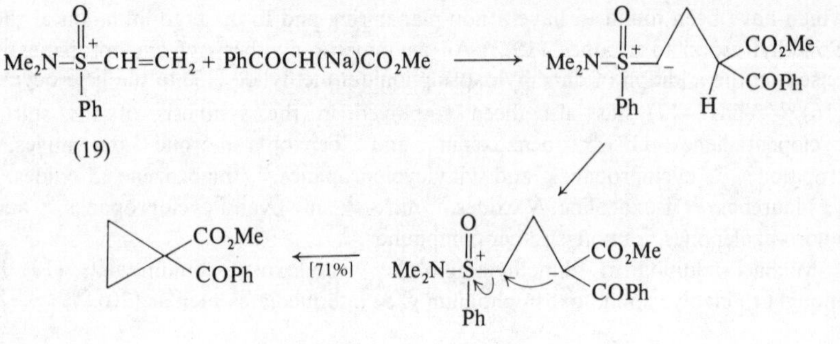

(19)

(20)

[71%]

$$Me_2SO + (COCl)_2 \xrightarrow[\text{at} -60\ °C]{CH_2Cl_2} Me_2\overset{+}{S}-OC(O)C(O)Cl\ Cl^- \longrightarrow Me_2\overset{+}{S}-Cl$$

(21)

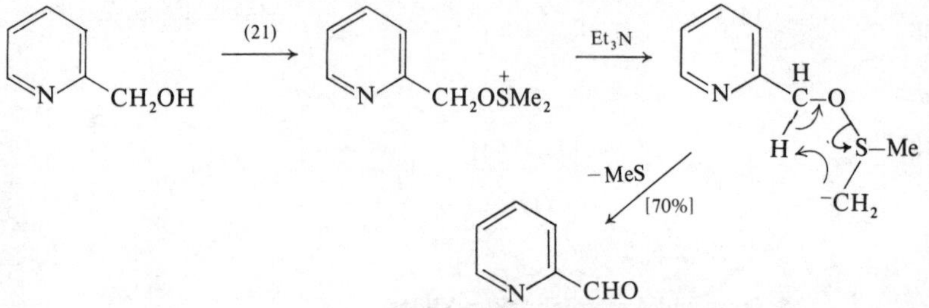

−MeS

[70%]

The DMSO–oxalyl chloride pair gives excellent results in the oxidation of long-chain alcohols to aldehydes,[82] amongst other recent examples[83–86] (see also Chap. 2, Pt III, p. 134). The DMSO–trifluoroacetic anhydride combination has been found to give best results with di-isopropylethylamine as base,[87] and it has been applied to the syntheses of carbohydrate derivatives,[88] ascorbic acid,[89] 7,12-dimethylbenz[a]anthracene metabolites and related systems,[90] and the tricyclic ketone (22).[91] In the latter case, other related oxidation procedures [DMSO–(COCl)₂, Me₂S–N-chlorosuccinimide (NCS), and DMSO–DCCI–py–TFA] were less satisfactory. The pair DMSO–(CCl₃CO)₂O has been used in a synthesis

[82] A. J. Mancuso, S.-L. Huang, and D. Swern, *J. Org. Chem.*, 1978, **43**, 2480.
[83] (a) E. Dimitriadis and R. A. Massy-Westropp, *Aust. J. Chem.*, 1979, **32**, 2003; (b) R. M. Jacobson and R. A. Raths, *J. Org. Chem.*, 1979, **44**, 4013; (c) W. R. Roush, *J. Am. Chem. Soc.*, 1980, **102**, 1390.
[84] E. L. Eliel and W. J. Frazee, *J. Org. Chem.*, 1979, **44**, 3598.
[85] N. Sueda, H. Ohrui, and H. Kuzuhara, *Tetrahedron Lett.*, 1979, 2039.
[86] F. M. Hauser and S. Prasanna, *J. Org. Chem.*, 1979, **44**, 2596.
[87] S. L. Huang, K. Omura, and D. Swern, *Synthesis*, 1978, 297.
[88] K. Sato and J. Yoshimura, *Carbohydr. Res.*, 1979, **73**, 75; J. Yoshimura, K. Sato, and H. Hashimoto, *Chem. Lett.*, 1977, 1327.
[89] T. C. Crawford and R. Breitenbach, *J. Chem. Soc., Chem. Commun.*, 1979, 388.
[90] K. B. Sukumaran and R. G. Harvey, *J. Am. Chem. Soc.*, 1979, **101**, 1353; P. P. Fu, C. Cortez, K. B. Sukumaran, and R. G. Harvey, *J. Org. Chem.*, 1979, **44**, 4265.
[91] O. D. Dailey, Jr. and P. L. Fuchs, *J. Org. Chem.*, 1980, **45**, 216.

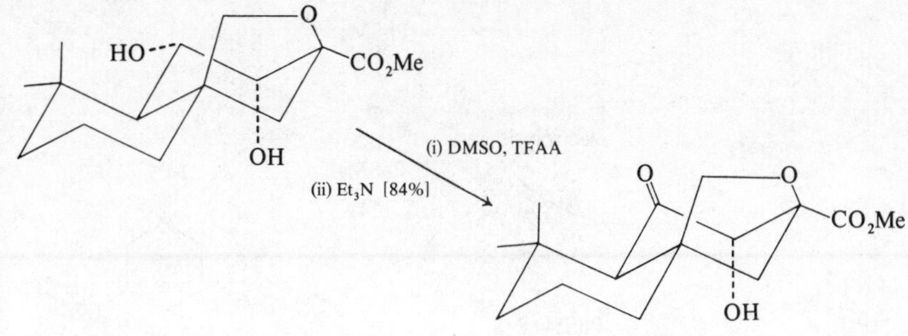

(22)

of gibberellic acid,[92] the Moffatt reagent (DMSO–carbodi-imide–py–TFA) is favoured over other procedures in syntheses of N-methylmaysenine[93] and (±)-warburganal,[94] DMSO–SO$_3$–py has been employed in the synthesis of juncusol,[95] and the pair DMSO–chlorosulphonyl isocyanate has been touted as a particularly mild alternative to existing reagents.[96] A report has been published on the use of Me$_2$S plus NCS (a combination which gives Me$_2$$\overset{+}{S}$Cl),[97] which is effective in converting β-hydroxy-selenides into β-keto-selenides; this is a situation in which DMSO–TFAA and the Moffatt procedure fail.[98] Oxysulphonium ylides are probable intermediates in the conversion of oxirans into α-chloro-ketones using chlorosulphonium salts,[99] in the *ortho*-substitution of phenols *via* aryloxysulphonium salts,[100] and in the novel electrochemical oxidation of alcohols in the presence of catalytic quantities of thioanisole (Scheme 1); a procedure which is thought to involve the cation radical (23) or the dication (24).[101]

Azasulphonium Ylides.—Azasulphonium ylides, which are intermediates in the rearrangements of N-aryl-azasulphonium salts in the *ortho*-functionalization of aromatic amines,[102] can be isolated as stable compounds if suitable substituents are incorporated.[103]

5 Thiocarbonyl Ylides

The Δ^3-1,3,4-thiadiazoline–thiocarbonyl ylide–thiiran–olefin transformation (see Vol. 4, p. 89) has been employed in the synthesis of some unsaturated β-lactams,[104]

[92] E. J. Corey, R. L. Danheiser, S. Chandrasekaran, P. Siret, G. E. Keck, and J. L. Gras, *J. Am. Chem. Soc.*, 1978, **100**, 8031.

[93] E. J. Corey, L. O. Weigel, D. Floyd, and M. G. Bock, *J. Am. Chem. Soc.*, 1978, **100**, 2916.

[94] T. Nakata, H. Akita, T. Naito and T. Oishi, *J. Am. Chem. Soc.*, 1979, **101**, 4400.

[95] A. S. Kende and D. P. Curran, *J. Am. Chem. Soc.*, 1979, **101**, 1857.

[96] G. A. Olah, Y. D. Vankar, and M. Arvanaghi, *Synthesis*, 1980, 141.

[97] E. J. Corey, C. U. Kim, and P. F. Misco, *Org. Synth.*, 1978, **58**, 122.

[98] R. Baudat and M. Petrzilka, *Helv. Chim. Acta*, 1979, **62**, 1406; also see S. Raucher, *Tetrahedron Lett.*, 1978, 2261.

[99] G. A. Olah, Y. D. Vankar, and M. Arvanaghi, *Tetrahedron Lett.*, 1979, 3653.

[100] P. G. Gassman and D. R. Amick, *J. Am. Chem. Soc.*, 1978, **100**, 7611.

[101] T. Shono, Y. Matsumura, M. Mizoguchi, and J. Hayashi, *Tetrahedron Lett.*, 1979, 3861.

[102] P. G. Gassman and H. R. Drewes, *J. Am. Chem. Soc.*, 1978, **100**, 7600.

[103] L. N. Markovskii, Yu. G. Shermolovich, V. V. Vasil'ev, I. E. Boldeskul, S. S. Trach, and N. S. Zefirov, *Zh. Org. Khim.*, 1978, **14**, 1659.

[104] M. D. Bachi, O. Goldberg, and A. Gross, *Tetrahedron Lett.*, 1978, 4167.

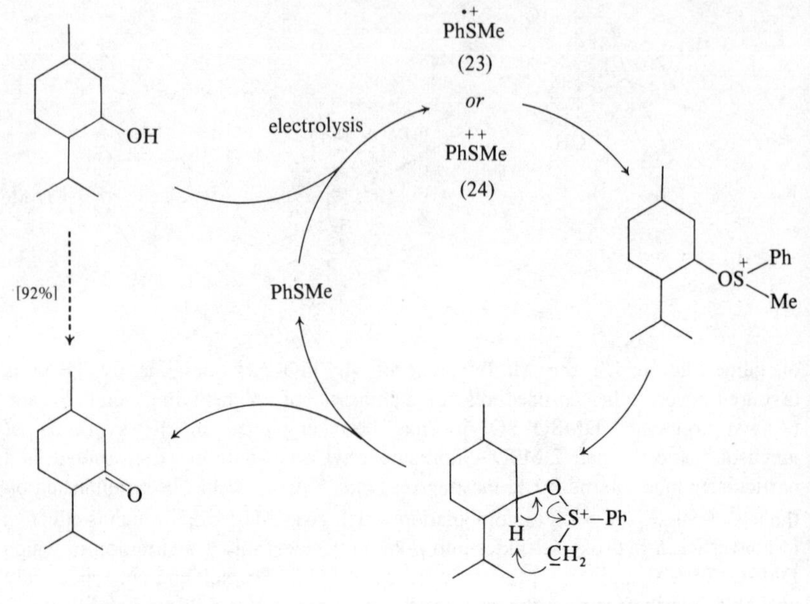

Scheme 1

some hindered olefins such as bis-(2,2,5,5-tetramethylcyclopentylidene),[105] octa-methyl-4,4'-bis-($\Delta^{1,2}$-pyrazolinylidene),[106] an adamantylidene,[107] and thiiran-con-taining cyclophanes (where the reaction stopped at the thiiran stage).[108] Flash vacuum pyrolysis of 1,3-oxathiolan-5-ones (25) has been found to give thiirans *via* thiocarbonyl ylides[109] (see Chap. 5, p. 268). Burgess has reported the interesting discovery that thiocarbonyl ylides such as (26) can undergo a Wittig-type reaction

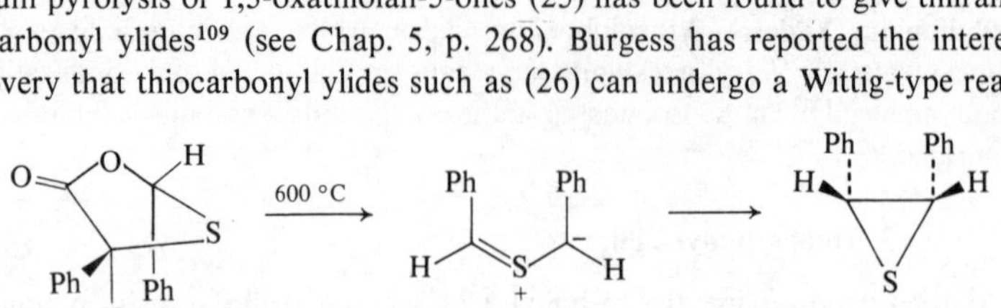

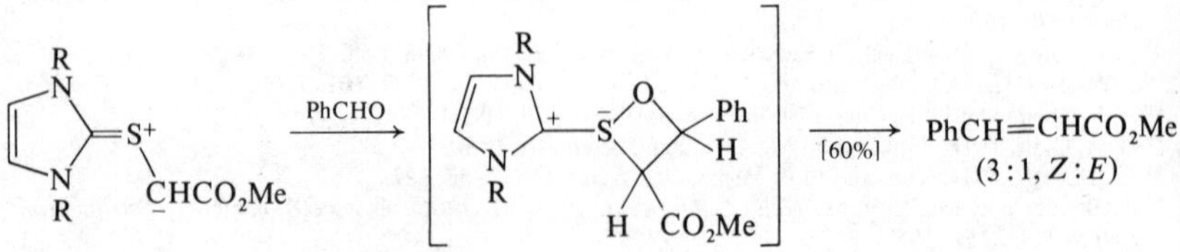

[105] A. Krebs and W. Ruger, *Tetrahedron Lett.*, 1979, 1305.
[106] R. J. Bushby and M. D. Pollard, *J. Chem. Soc., Perkin Trans. 1*, 1979, 2401.
[107] F. Cordt, R. M. Frank, and D. Lenoir, *Tetrahedron Lett.*, 1979, 505.
[108] M. Atzmuller and F. Vögtle, *Chem. Ber.*, 1979, **112**, 138.
[109] T. B. Cameron and H. W. Pinnick, *J. Am. Chem. Soc.*, 1979, **101**, 4755; 1980, **102**, 744.

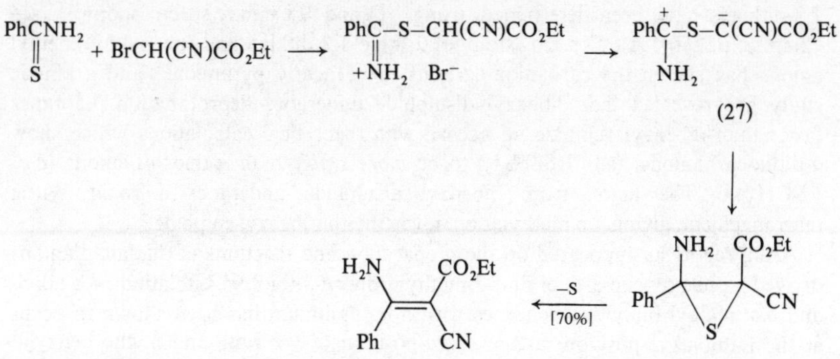

PhCNH$_2$ + BrCH(CN)CO$_2$Et $\longrightarrow$ PhC–S–CH(CN)CO$_2$Et $\longrightarrow$ PhC̈–S–C̄(CN)CO$_2$Et

(with structures for reaction 27)

(27)

H$_2$N   CO$_2$Et

Ph   CN $\xleftarrow[\text{[70%]}]{-S}$ (three-membered ring structure with NH$_2$, CO$_2$Et, Ph, CN, S)

with benzaldehyde.[110] Another olefin-forming reaction that is thought to involve the thiocarbonyl ylide (27) has been described.[111]

6 Sulphenyl Carbanions

Synthesis and Properties.—The surprising result of a crystal-structure determination for the TMEDA complex of 2-lithio-2-methyl-1,3-dithian is that the complex (29) is dimeric.[112] The lithium atoms, which are equatorial, interact about equally strongly with the carbanionoid-carbon of one ring (C–Li = 2.186 Å) as with the sulphur of the other (S–Li = 2.516 Å), so that it is impossible to decide between the alternative descriptions of the complex as a dimer of the C–Li monomer (28) or of the S–Li monomer (30). Steric factors are thought to be at least as important as the anomeric effect in favouring the equatorial position of lithium. Dimeric structures analogous to (29) may well be preserved in solutions of 2-lithio-1,3-dithians in THF, where the oxygen of the solvent molecules can take over the role of the nitrogen atoms of TMEDA.[112]

In other work, the barriers to rotation and to inversion in 2-lithio-2-phenyl-

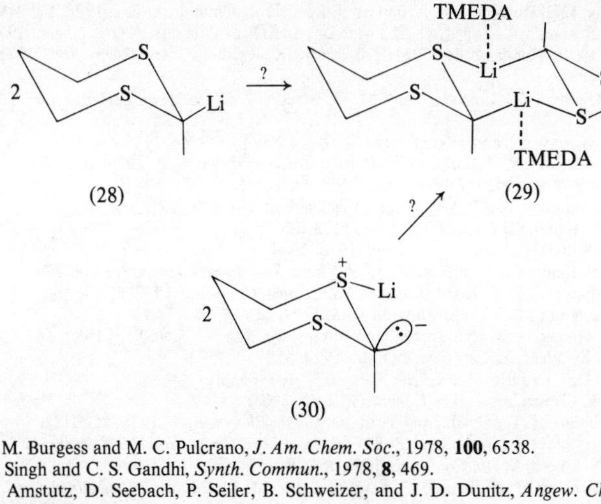

(28)   (29)

(30)

[110] E. M. Burgess and M. C. Pulcrano, *J. Am. Chem. Soc.*, 1978, **100**, 6538.

[111] H. Singh and C. S. Gandhi, *Synth. Commun.*, 1978, **8**, 469.

[112] R. Amstutz, D. Seebach, P. Seiler, B. Schweizer, and J. D. Dunitz, *Angew. Chem., Int. Ed. Engl.*, 1980, **19**, 53.

1,3-dithians have been determined, using ^{13}C and [7]Li n.m.r. spectroscopy[113] (see Chap. 5, p. 260). A [13]C n.m.r. study of 2-lithio-1,3-dithian and (phenylthio)methyl anions has shown the carbanion-carbons to be nearly pyramidal,[6] and a kinetic study has revealed that dibenzyl disulphide undergoes deprotonation 7.5 times faster than dibenzyl sulphide, in accord with theoretical calculations which show α-dithiocarbanions, (*e.g.* $RSSCH_2^-$) to be *more stable* than α-thiocarbanions (*e.g.* $RSCH_2^-$).[10] The anion from dibenzyl disulphide undergoes a rapid Wittig rearrangement, giving, on methylation, α-(methylthio)benzyl sulphide.[10]

A full report has appeared on the preparation and reactions of thiolate dianions derived from propene-3-thiol and 2-methylpropene-3-thiol.[114] Lithiation of *o*-alkyl- and *o,o'*-dialkyl-(alkylthio)benzenes with n-butyl-lithium has been shown to occur at the α-thioalkyl position, at the *ortho*-position of the ring, and at the benzylic position(s), the relative amounts being dependent on the nature of the alkyl groups.[115]

Reactions.—*1,3-Dithianyl Anions.* 1,3-Dithianyl anions continue to find wide-ranging utility in the synthesis of diverse structures, including 2-deoxy-L-lyxose derivatives,[116] eburnamine precursors,[117] 1,2-diformyl-6,6-dimethylcyclohex-2-en-1-ol,[118] cannabinoids,[119] tertiary acyloins,[120] cyclitols,[121] unsaturated carboxylic acids,[122] the terpenes egomaketone[123] and α- and β-curcumene,[124] (20*S*)-20-hydroxycholesterol,[125] norpyrenophorin, pyrenophorin, and vermiculine,[126] compounds with chiral methyl groups,[127] (+)-*trans*-burseran,[128] enones,[129] the macrocyclic lactam *N*-methylmaysenine,[93,130] olivin,[131] macrocyclic acetylacetone crown ethers,[132] [5.1]metacyclophane,[133] (±)-laurencin,[73] monosaccharides,[134] γ-keto-esters and their precursors,[135] muscarine analogues,[136] quinols,[137] *epi-*

[113] L. F. Kuyper and E. L. Eliel, *J. Organomet. Chem.*, 1978, **156**, 245.

[114] M. Pohmakotr, K.-H. Geiss, and D. Seebach, *Chem. Ber.*, 1979, **112**, 1420.

[115] S. Cabiddu, S. Melis, P. P. Piras, and F. Sotgiu, *J. Organomet. Chem.*, 1979, **178**, 291.

[116] E. Hungerbuhler, D. Seebach, and D. Wasmuth, *Angew. Chem., Int. Ed. Engl.*, 1979, **18**, 958.

[117] S. Takano, S. Hatakeyama, and K. Ogasawara, *J. Chem. Soc., Perkin Trans. 1*, 1980, 457.

[118] A. J. G. M. Peterse, J. H. Roskam, and A. de Groot, *Recl. Trav. Chim. Pays-Bas*, 1978, **97**, 249.

[119] C. G. Pitt, H. H. Seltzman, Y. Sayed, C. E. Twine, Jr., and D. L. Williams, *J. Org. Chem.*, 1979, **44**, 677; D. B. Uliss, G. R. Handrick, H. C. Dalzell, and R. K. Razdan, *J. Am. Chem. Soc.*, 1978, **100**, 2929.

[120] L. Colombo, C. Gennari, C. Scolastico, and M. G. Beretta, *J. Chem. Soc., Perkin Trans. 1*, 1978, 1036.

[121] M. Funabashi, K. Kobayashi, and J. Yoshimura, *J. Org. Chem.*, 1979, **44**, 1618.

[122] K. C. Nicolaou, S. P. Seitz, W. J. Sipio, and J. F. Blount, *J. Am. Chem. Soc.*, 1979, **101**, 3884.

[123] A. Hoppmann and P. Weyerstahl, *Tetrahedron*, 1978, **34**, 1723.

[124] H. H. Bokel, A. Hoppmann, and P. Weyerstahl, *Tetrahedron*, 1980, **36**, 651.

[125] M. Koreeda and N. Koezumi, *Tetrahedron Lett.*, 1978, 1641.

[126] B. Seuring and D. Seebach, *Liebigs Ann. Chem.*, 1978, 2044.

[127] B. T. Golding, P. V. Ioannou, and I. F. Eckhard, *J. Chem. Soc., Perkin Trans. 1*, 1978, 774.

[128] K. Tomioka, T. Ishiguro, and K. Koga, *J. Chem. Soc., Chem. Commun.*, 1979, 652.

[129] Y. Nagao, K. Seno, and E. Fujita, *Tetrahedron Lett.*, 1979, 3167.

[130] E. J. Corey, L. O. Weigel, A. R. Chamberlin, and B. Lipshutz, *J. Am. Chem. Soc.*, 1980, **102**, 1439.

[131] J. H. Dodd and S. M. Weinreb, *Tetrahedron Lett.*, 1979, 3593.

[132] A. H. Alberts and D. J. Cram, *J. Am. Chem. Soc.*, 1979, **101**, 3545.

[133] N. Finch and C. W. Gemenden, *J. Org. Chem.*, 1979, **44**, 2804.

[134] H. Ogura, K. Furuhata, H. Takahashi, and Y. Iitaka, *Chem. Pharm. Bull.*, 1978, **26**, 2782.

[135] Y. Tamaru, T. Harada, H. Iwamoto, and Z.-I. Yoshida, *J. Am. Chem. Soc.*, 1978, **100**, 5221; W. D. Woessner and P. S. Solera, *Synth. Commun.*, 1978, **8**, 279.

[136] Z. Lysenko, F. Ricciardi, J. E. Semple, P. C. Wang, and M. M. Joullié, *Tetrahedron Lett.*, 1978, 2679.

[137] K. A. Parker and J. R. Andrade, *J. Org. Chem.*, 1979, **44**, 3964.

7-hydroxymyoporone,[138] butyrolactones,[139] and β-diketone enol ethers.[140] A novel synthesis of (*d,l*)-muscone (31) involves a two-carbon ring-expansion process.[141] 2-Lithio-2-trimethylsilyl-1,3-dithian reacts with sulphur dioxide in a Wittig-like reaction to give the appropriate sulphine (see also Chap. 3, Pt II, p. 160, and Vol. 5, p. 133).[142]

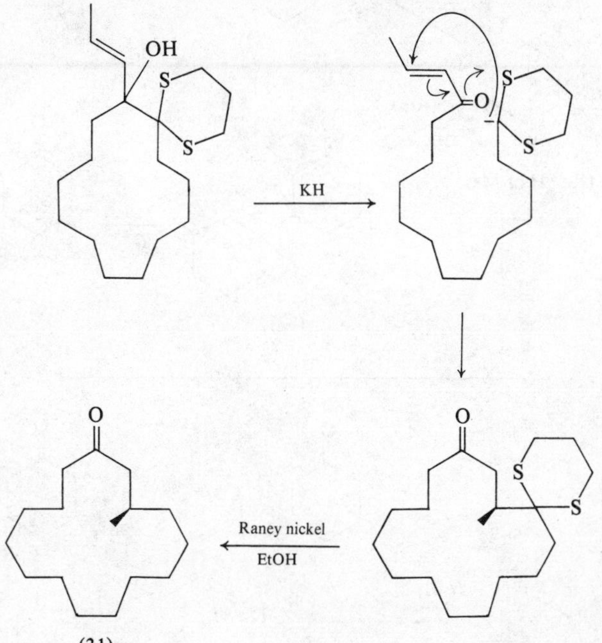

(31)

The 1,2-adducts (32) of 2-litho-1,3-dithians and cyclohex-2-enone can be isomerized to the more stable 1,4-adducts (33) by conversion into the potassium alkoxides (but not the less dissociated sodium or lithium salts).[141] Direct formation of enone 1,4-adducts occurs when HMPA–THF is used as the solvent,[143,144] and enolate trapping can be effected when the initial enone 1,4-adduct is treated with methyl iodide.[143] The allylic anion derived from (34) undergoes α-1,4-addition to cyclohexenone in the presence of CuI (Scheme 2).[145] In the absence of the copper salt, both γ-1,4- and α-1,4-addition occur, with the former predominating. In contrast, allylation of (34) in the presence of CuI occurs primarily at the γ-position, while α-allylation is observed otherwise.[146] The lithio-anion of 2-(β-styryl)-

[138] H. J. Reich, P. M. Gold, and F. Chow, *Tetrahedron Lett.*, 1979, 4433.
[139] G. A. Kraus and B. Roth, *J. Org. Chem.*, 1978, **43**, 2072.
[140] P. S. Tobin, S. K. Basu, R. S. Grosserode, and D. M. S. Wheeler, *J. Org. Chem.*, 1980, **45**, 1250.
[141] S. R. Wilson and R. N. Risra, *J. Org. Chem.*, 1978, **43**, 4903.
[142] M. van der Leij, P. A. T. W. Porskamp, B. H. M. Lammerink, and B. Zwanenburg, *Tetrahedron Lett.*, 1978, 811.
[143] J. Lucchetti, W. Dumont, and A. Krief, *Tetrahedron Lett.*, 1979, 2695.
[144] C. A. Brown and A. Yamaichi, *J. Chem. Soc., Chem. Commun.*, 1979, 100.
[145] F. E. Ziegler and C. C. Tam, *Tetrahedron Lett.*, 1979, 4717.
[146] F. E. Ziegler and C. C. Tam, *J. Org. Chem.*, 1979, **44**, 3428.

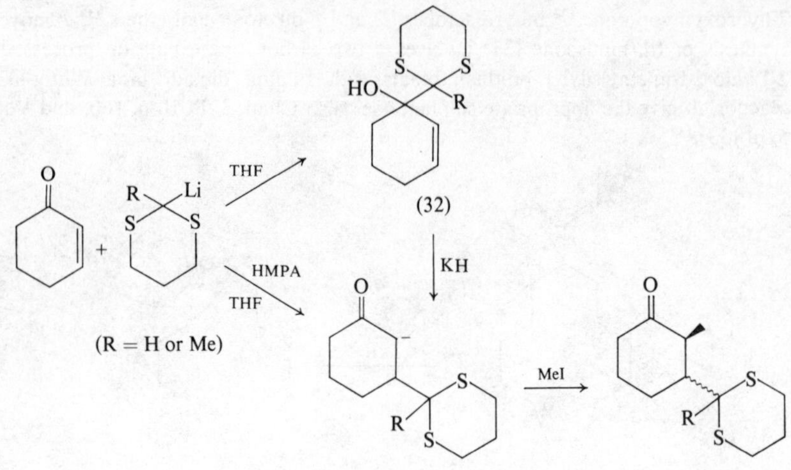

Reagents: i, O; ii, Br; iii, CuI

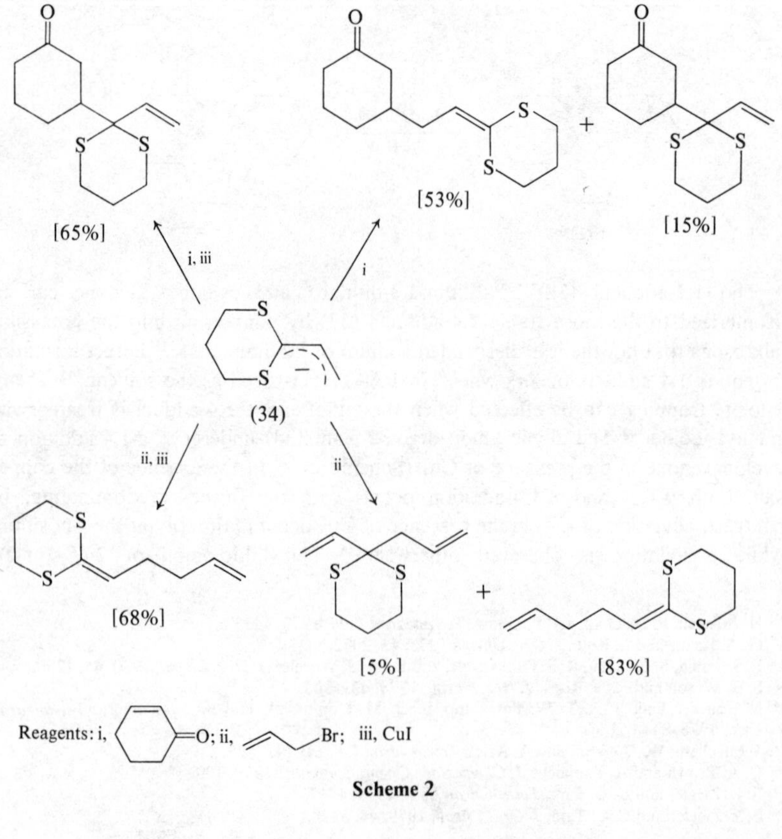

Scheme 2

1,3-dithian, which is the γ-phenyl analogue of (34), has been found to undergo substitution at the α-position with hard electrophiles [*e.g.* D$^+$, Me$_3$SiCl, and (MeO)$_2$SO$_2$] and at the γ-position with softer electrophiles (*e.g.* RI and RBr).[147] 2-Alkylidene-1,3-dithians can be prepared through the reaction of 2-lithio-1,3-dithians with 2,2'-dipyridyl disulphide.[148] An example of an intramolecular Michael addition to a 2-alkylidene-1,3-dithian (35) has been described.[149]

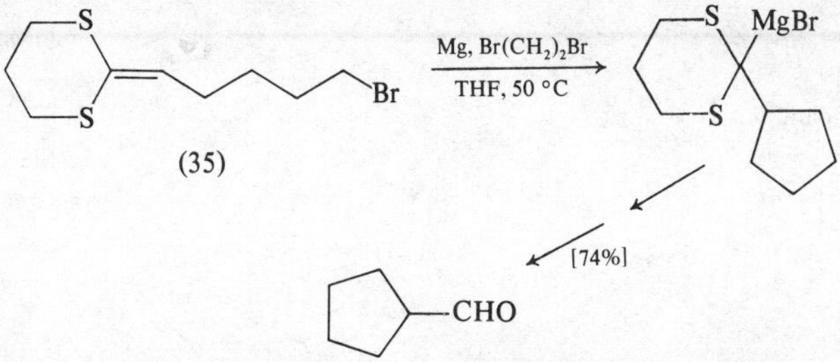

Additional reactions involving 1,3-dithianyl anions include nucleophilic substitution on *N*-methylindoletricarbonylchromium(o)[150a] and on toluenetricarbonylchromium(o);[150b] one-electron transfer, proton abstraction, and alkylation by 2-chlorothiolan and 2-chloro-oxolan[151] and by 2-halogeno-2-nitropropane (one-electron transfer only);[152] cleavage of α-keto-1,3-dithians;[153] intramolecular alkylation of an epoxy-2-lithio-1,3-dithian that is generated by low-temperature transmetallation of a 2-(tributylstannyl)-1,3-dithian;[154] and opening of the furan ring in 2-(2'-furyl)-2-lithio-1,3-dithian (36).[155]

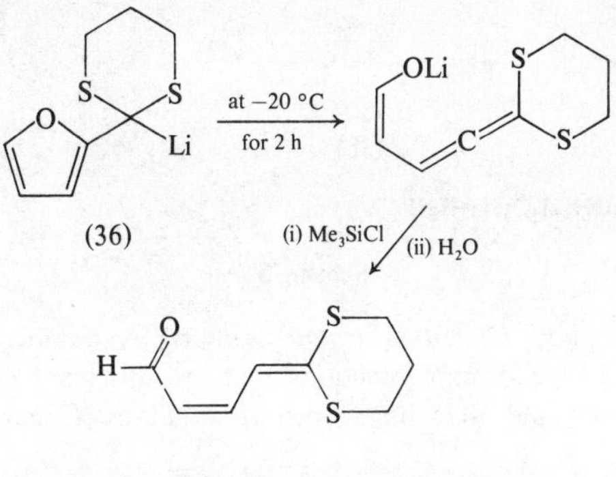

[147] W. S. Murphy and S. Wattanasin, *Tetrahedron Lett.*, 1979, 1827.

[148] Y. Nagao, K. Seno, and E. Fujita, *Tetrahedron Lett.*, 1979, 4403.

[149] N. H. Andersen, P. F. Duffy, A. D. Denniston, and D. B. Grotjahn, *Tetrahedron Lett.*, 1978, 4315.

[150] (*a*) A. P. Kozikowski and K. Isobe, *J. Chem. Soc., Chem. Commun.*, 1978, 1076; (*b*) M. F. Semmelhack, H. T. Hall, Jr., R. Farina, M. Yoshifuji, G. Clark, T. Bargar, K. Hirotsu, and J. Clardy, *J. Am. Chem. Soc.*, 1979, **101**, 3535.

[151] C. G. Kruse, E. K. Poels, and A. van der Gen, *J. Org. Chem.*, 1979, **44**, 2911.

[152] G. A. Russell, M. Jawdasiuk, and M. Makosza, *J. Am. Chem. Soc.*, 1979, **101**, 2355.

[153] S. Takano, M. Takahashi, S. Hatakeyama, and K. Ogasawara, *J. Chem. Soc., Chem. Commun.*, 1979, 556.

[154] D. Seebach, I. Willert, A. K. Beck, and B. T. Grobel, *Helv. Chim. Acta*, 1978, **61**, 2510.

[155] M. J. Taschner and G. A. Kraus, *J. Org. Chem.*, 1978, **43**, 4235.

A number of anionic systems that are related to 1,3-dithian have been studied, including the anion from the unusual 1,3-oxathian (37) [which has been employed with excellent results in an asymmetric synthesis],[84] the 1,3-dithiolan anion (38)[156] [which reacts with trialkylboranes (as well as with other electrophiles)[157] to give, after oxidative work-up, ketones or tertiary alcohols (39), as shown in Scheme 3], and 2-substituted 1,3-dithiolan anions [which fragment to give aromatic dithioesters][158] (see Chap. 3, Pt III, p. 191).

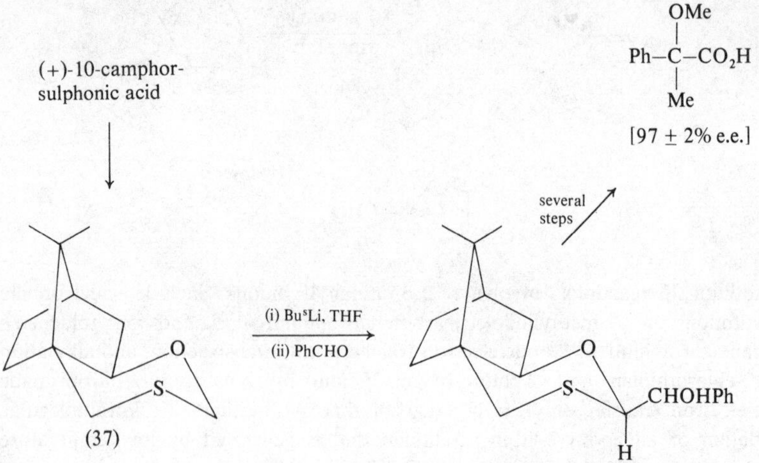

$$Ph-\underset{\underset{Me}{|}}{\overset{\overset{OMe}{|}}{C}}-CO_2H$$

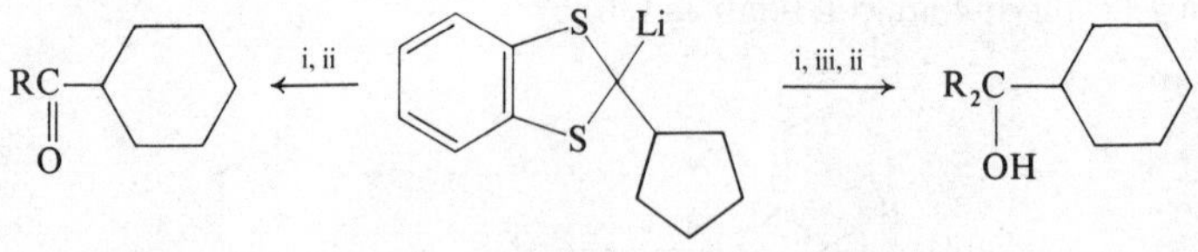

$[97 \pm 2\%\ \text{e.e.}]$

Reagents: i, R_3B; ii, NaOH, H_2O_2; iii, $HgCl_2$

Scheme 3

Acyclic Thioacetal and Orthothioformate Anions. A number of unusual and synthetically useful ring-forming reactions that are thought to involve thioacetal anions such as (40) and (41) have been reported by Cohen.[159] A procedure

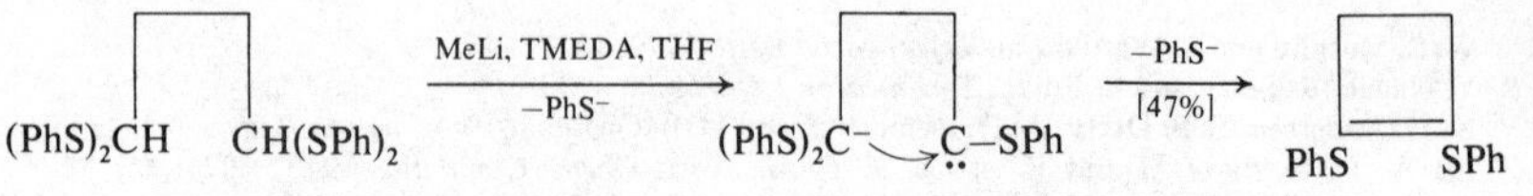

[156] S. Ncube, A. Pelter, and K. Smith, *Tetrahedron Lett.*, 1979, 1893, 1895.
[157] S. Ncube, A. Pelter, K. Smith, P. Blatcher, and S. Warren, *Tetrahedron Lett.*, 1978, 2345.
[158] N. C. Gonnella, M. V. Lakshmikantham, and M. P. Cava, *Synth. Commun.*, 1979, **9**, 17.
[159] (a) T. Cohen and W. M. Daniewski, *Tetrahedron Lett.*, 1978, 2991; (b) T. Cohen, D. Ouellete, and W. M. Daniewski, *ibid.*, p. 5063; (c) T. Cohen, R. B. Weisenfeld, and R. E. Gapinski, *J. Org. Chem.*, 1979, **44**, 4744; T. Cohen and J. R. Matz, *ibid.*, p. 4816.

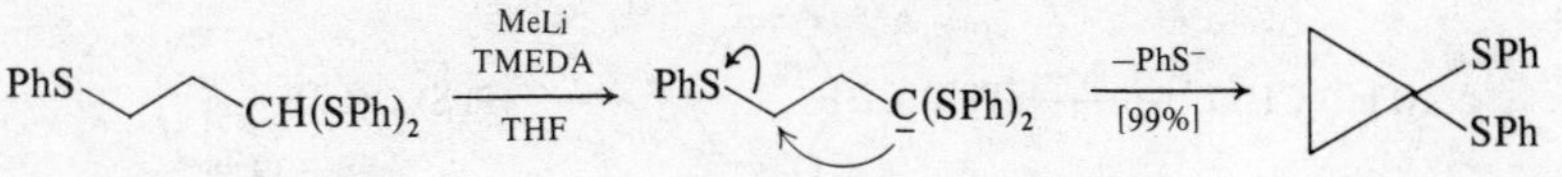

(41)

described in *Organic Syntheses* and another report have appeared on the preparation of 1-phenylthio-1,3-butadienes *via* thioacetal anions,[160] while a related paper provides a synthesis of acyclic dienones.[161] In THF–HMPA, bis-(phenylthio)methyl-lithium adds predominantly (92%) 1,4 to cyclohexenone, in contrast to addition in the absence of HMPA (ratio 55:45 of 1,4-:1,2-addition).[143]

Thiophilic addition to dithioesters followed by capture of the thioacetal anion with an electrophile provides a useful approach to the formation of carbonyl compounds such as (42)[162] and (43),[163] and has been elegantly employed in the total synthesis of the macrocycles (±)-maysine and (±)-4,5-deoxymaysine.[164] Trimethyl borate has been used to convert bis(phenylthio)methyl-lithium into the keten thioacetal (45)[165] *via* the boron-containing carbanion (44), as shown in Scheme 4.[166] Bis(phenylthio)methyl-lithium and related compounds are also the key reagents in the preparation of 2-(phenylthio)-enones, 2-(phenylthio)-butadienes,[167] α-(phenyl-

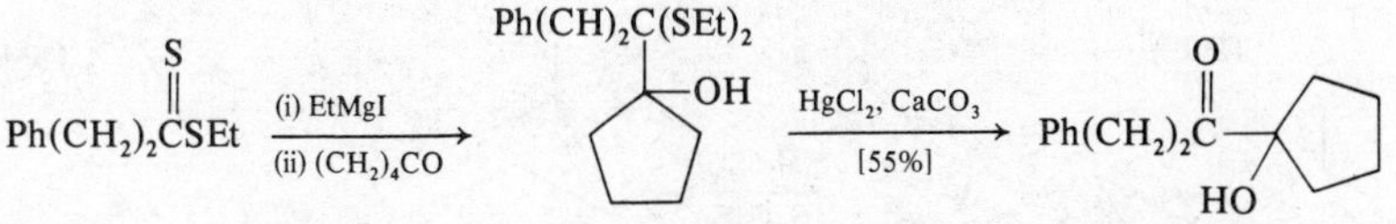

(42)

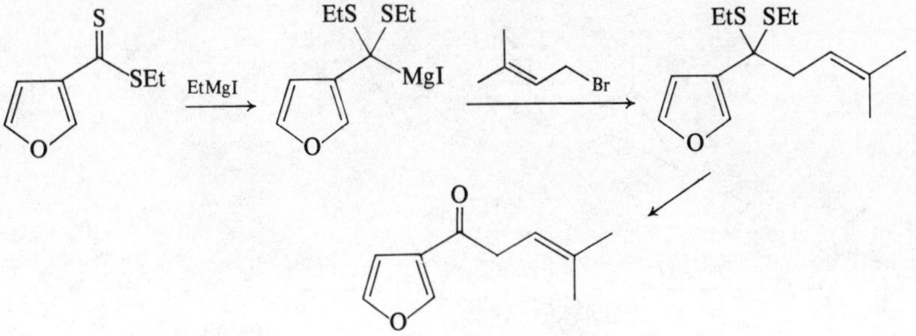

(43)

[160] T. Cohen and R. J. Ruffner, *Org. Synth.*, 1979, **59**, 202; T. Cohen, R. E. Gapinski, and R. B. Hutchins, *J. Org. Chem.*, 1979, **44**, 3599.

[161] M. Pohmakotr and D. Seebach, *Tetrahedron Lett.*, 1979, 2271.

[162] A. I. Meyers, T. A. Tait, and D. L. Comins, *Tetrahedron Lett.*, 1978, 4657.

[163] P. Gosselin, S. Masson, and A. Thuillier, *J. Org. Chem.*, 1979, **44**, 2807.

[164] A. I. Meyers, D. M. Roland, D. L. Comins, R. Henning, M. P. Fleming, and K. Shimizu, *J. Am. Chem. Soc.*, 1979, **101**, 4732; A. I. Meyers, D. L. Comins, D. M. Roland, R. Henning, and K. Shimizu, *ibid.*, p. 7104.

[165] A. Mendoza and D. S. Matteson, *J. Org. Chem.*, 1979, **44**, 1352.

[166] D. S. Matteson and K. Arne, *J. Am. Chem. Soc.*, 1978, **100**, 1325.

[167] P. Blatcher, J. I. Gryson, and S. Warren, *J. Chem. Soc., Chem. Commun.*, 1978, 657; P. Blatcher and S. Warren, *Tetrahedron Lett.*, 1979, 1247.

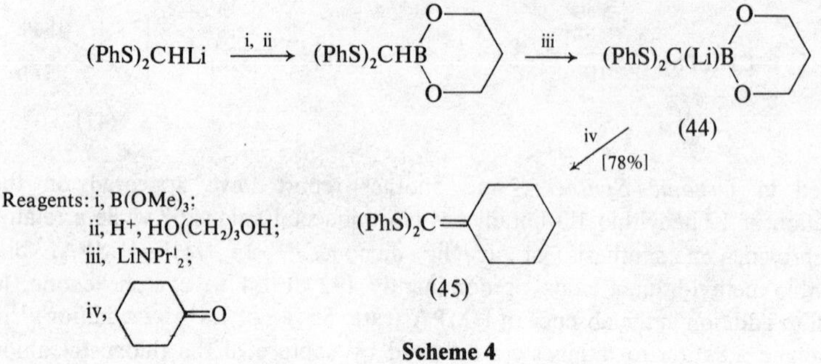

$$(PhS)_2CHLi \xrightarrow{i,\ ii} (PhS)_2CHB \xrightarrow{iii} (PhS)_2C(Li)B$$

(44)

$$\xleftarrow[iv]{[78\%]}$$

Reagents: i, B(OMe)$_3$;
        ii, H$^+$, HO(CH$_2$)$_3$OH;
        iii, LiNPr$^i_2$;
        iv, (cyclohexanone)

$$(PhS)_2C=\text{(cyclohexane ring)}$$

(45)

Scheme 4

thio)-ketones,[168] and (±)-hinokinin.[169] The dianion from bis(ethylthio)acetic acid has also been used in organic synthesis.[170]

Orthothioformate anions have been utilized in the generation of the ester (46)[91] and 1,4-dicarbonyl compounds such as (47).[171,172] A full report has appeared on the decomposition of orthothioformate anions to bis(arylthio)-carbenes[173] (see also Chap. 1, p. 47).

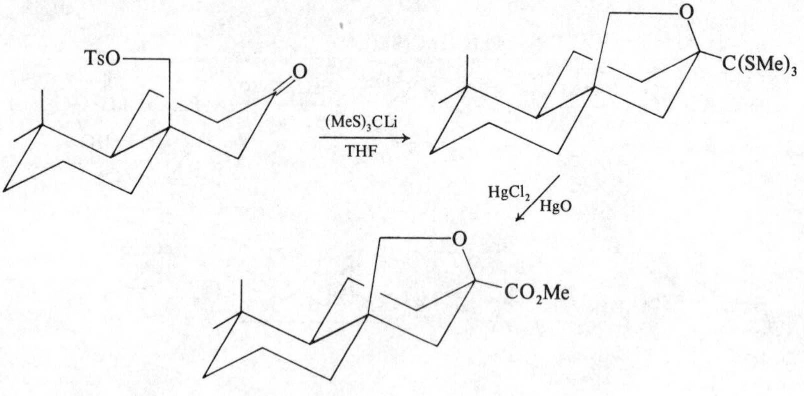

(46) [73%]

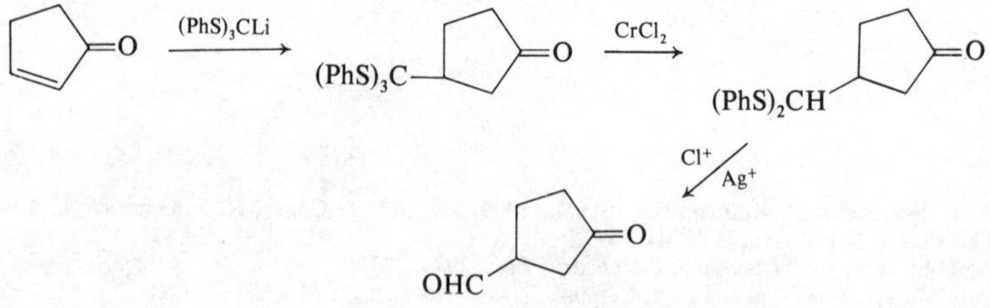

(47) [40%]

[168] P. Blatcher and S. Warren, *J. Chem. Soc., Perkin Trans. 1*, 1979, 1074.
[169] A. Ichihara, N. Nio, Y. Terayama, R. Kimura, and S. Sakamura, *Tetrahedron Lett.*, 1979, 3731.
[170] G. S. Bates, *J. Chem. Soc., Chem. Commun.*, 1979, 161.
[171] T. Cohen and S. M. Nolan, *Tetrahedron Lett.*, 1978, 3533.
[172] R. A. J. Smith and A. R. Lal, *Aust. J. Chem.*, 1979, **32**, 353.
[173] M. Nitsche, D. Seebach, and A. K. Beck, *Chem. Ber.*, 1978, **111**, 3644.

Thioallyl and Related Anions. Regio-control in reactions of thioallylic carbanions has been a subject of considerable interest during the past several years (see also Chap. 1, p. 37). Intramolecular γ-alkylation of the anion (48), prepared from the $\beta\gamma$-unsaturated compound with butyl-lithium in TMEDA, is a key step in an approach to grandisol and its isomers,[174] while exclusive α-alkylation is seen with the anion (49).[63] Both α- and γ-hydroxyalkylation are observed with the propene-3-thiol and 2-methylpropene-3-thiol dianions[114] and with the anion from 3-methyl-3-(phenylthio)propene (50).[175] In the last case, regio-control can be achieved through the addition of $ZnCl_2$. Similar regio-control in the alkylation and hydroxyalkylation of thioallylic carbanions may be realized through the formation of boron[176,177] and aluminium[178] 'ate' complexes (51). In the absence of the 'ate' complex, a mixture of the products of α- and γ-hydroxyalkylation (in the ratio 43:57) is formed.[178] α,γ'-Coupling occurs when the boron 'ate' complex (51) reacts with allylic halides.[176]

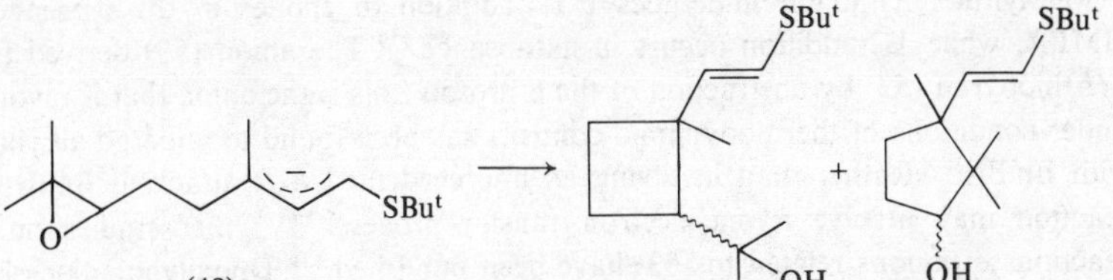

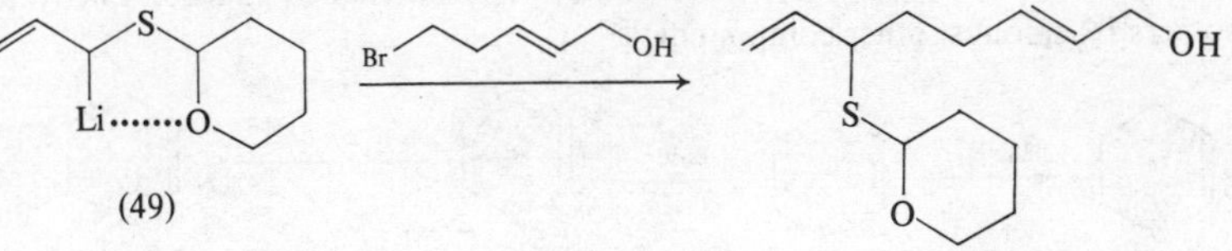

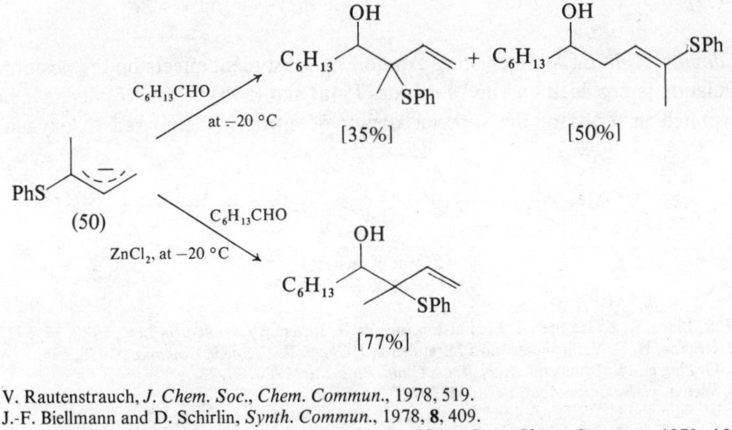

[174] V. Rautenstrauch, *J. Chem. Soc., Chem. Commun.*, 1978, 519.
[175] J.-F. Biellmann and D. Schirlin, *Synth. Commun.*, 1978, **8**, 409.
[176] Y. Yamamoto, H. Yatagai, and K. Maruyama, *J. Chem. Soc., Chem. Commun.*, 1979, 157; Y. Yamamoto, H. Yatagai, and K. Maruyama, *Chem. Lett.*, 1979, 385.
[177] Y. Yamamoto and K. Maruyama, *J. Chem. Soc., Chem. Commun.*, 1980, 239.
[178] Y. Yamamoto, H. Yatagai, and K. Maruyama, *J. Org. Chem.*, 1980, **45**, 195.

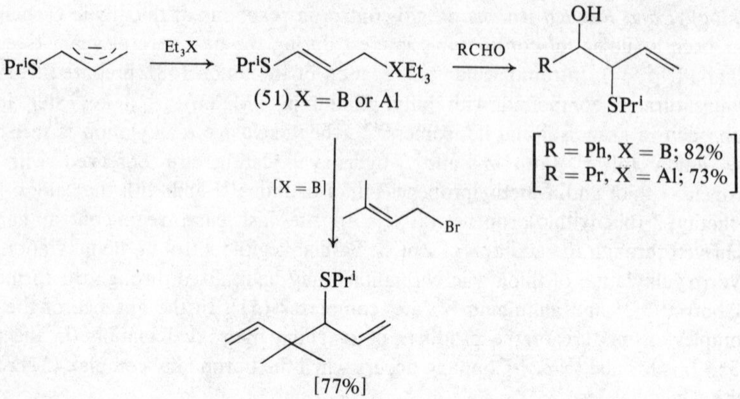

Methylthioallyl anion undergoes α-1,4-addition to enones in the presence of HMPA, while 1,2-addition occurs in its absence.[179] The anion (53) derived from 2*H*-thiopyran (52) by abstraction of the 6-proton (this is the anion that is favoured under conditions of thermodynamic control) has been found to undergo alkylation with Bu^{t}Br.[180] Rather than involving an unprecedented S_N2 attack at Bu^{t}Br, the reaction may involve a one-electron-transfer process.[159d] Other studies on the reactions of anions related to (53) have been published.[181] Thioallylic anions have been employed in syntheses of 2-vinylthian,[61] 1,1-disilylalk-2-enes,[182] and the natural products mokupalide,[183] manicone,[184] ($\pm$)-laurencin,[73] and (6*Z*)-hedycaryol isomers,[185] amongst other compounds.[186]

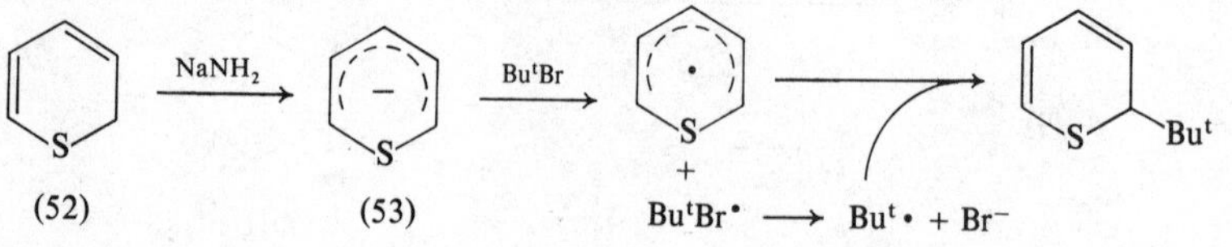

α-Thiovinyl Anions. An interesting example of substituent effects on the position of metallation is provided by vinyl sulphides (54a) and (54b).[187] The results have been interpreted in terms of the superior ability of nitrogen, compared to oxygen, to

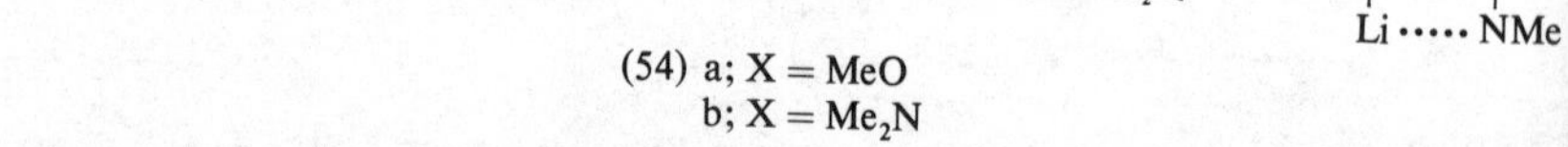

[179] M. R. Binns, R. K. Haynes, T. L. Houston, and W. R. Jackson, *Tetrahedron Lett.*, 1980, **21**, 573.
[180] R. Grafing, H. D. Verkruijsse, and L. Brandsma, *J. Chem. Soc., Chem. Commun.*, 1978, 596.
[181] R. Grafing and L. Brandsma, *Recl. Trav. Chim. Pays-Bas*, 1980, **99**, 23.
[182] H. Wetter, *Helv. Chim. Acta*, 1978, **61**, 3072.
[183] F. W. Sum and L. Weiler, *J. Am. Chem. Soc.*, 1979, **101**, 4401.
[184] T. Mimura, Y. Kimura, and T. Nakai, *Chem. Lett.*, 1979, 1361.
[185] M. Kodama, S.-I. Yokoo, H. Yamada, and S. Ito, *Tetrahedron Lett.*, 1978, 3121.
[186] T. Hayashi, N. Fujitaka, T. Oishi, and T. Takeshima, *Tetrahedron Lett.*, 1980, **21**, 303.
[187] J. J. Fitt and H. W. Gschwend, *J. Org. Chem.*, 1979, **44**, 303.

function as a ligand for the internal chelation of lithium. α-Thiovinyl anions, prepared by the reductive desulphurization of keten thioacetals with lithium naphthalide[188] or by deprotonation of vinyl sulphides, have been used in the synthesis of α-(alkylthio)- or α-(arylthio)-ketones (55) and the enones (56),[189,190] of the annelated cyclopentenone (57),[191] of the heterocycle (58),[192] and of the butenolide (59) prepared from $PhSC(Li)=C(Me)CO_2Me$ and butanal.[193] More work has appeared on reactions of tetrathiafulvalenyl-lithium[194] and an account of the preparation and reactions of α-thioallenyl-lithium compounds has been published.[195]

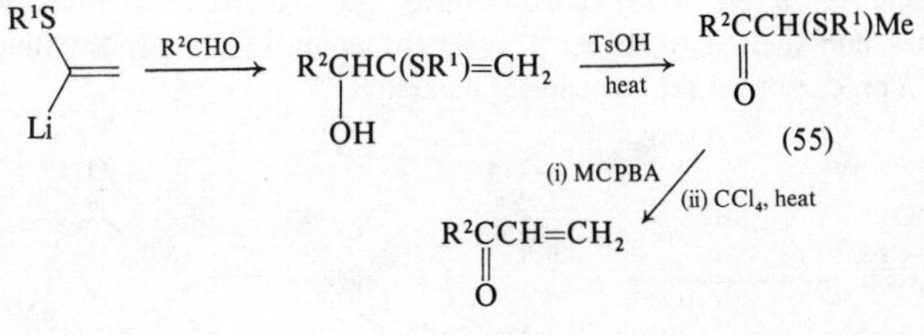

(55)

(56)

[75%]

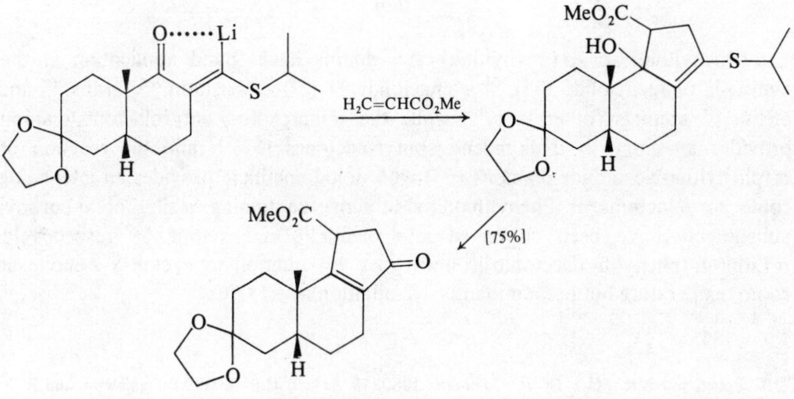

(57)

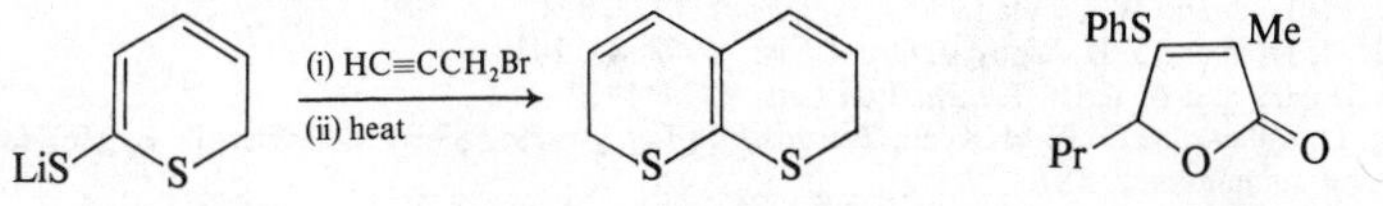

(58)

(59)

[188] T. Cohen and R. B. Weisenfeld, *J. Org. Chem.*, 1979, **44**, 3601.
[189] M. Braun, *Tetrahedron Lett.*, 1978, 3695.
[190] R. C. Cookson and P. J. Parsons, *J. Chem. Soc., Chem. Commun.*, 1978, 821.
[191] J. P. Marino and L. C. Katterman, *J. Chem. Soc., Chem. Commun.*, 1979, 946.
[192] R. Grafing and L. Brandsma, *Recl. Trav. Chim. Pays-Bas*, 1978, **97**, 208; *Synthesis*, 1978, 578.
[193] K. Isobe, M. Fuse, H. Kosugi, H. Hagiwara, and H. Iida, *Chem. Lett.*, 1979, 785.
[194] D. C. Green and R. W. Allen, *J. Chem. Soc., Chem. Commun.*, 1978, 832; *J. Org. Chem.*, 1979, **44**, 1476.
[195] R. C. Cookson and P. J. Parsons, *J. Chem. Soc., Chem. Commun.*, 1978, 822.

Miscellaneous α-Thiocarbanions. 1-Lithio-1-(phenylthio)cyclopropane, which can be prepared by lithium–sulphur exchange or by reductive lithiation of cyclopropanone thioacetals,[196] or by lithiation followed by cyclization of 1,3-bis-(phenylthio)propane,[197] has been used in the generation of cyclobutanone,[198] a tetrahydrochromanone,[199] alkylidenecyclopropanes,[200] and various ketones[201] and alcohols.[197]

α-(Methylthio)acetophenone, in the presence of a base such as K_2CO_3 (which should be too weak to effect deprotonation) still undergoes allylation by what appears to be an S_N2' process. In fact what is probably involved is *S*-allylation followed by ylide formation and [2,3]-sigmatropic rearrangement, as in (60).[56,57] The anions of α-(ethylthio)-ketones have been utilized in preparations of α-diketones and their derivatives,[202] while the anion of an α-(phenylthio)-ketone serves as a precursor for the terpene perilla ketone.[203]

(60)

α-(Phenylthio)- or α-(methylthio)-ester anions have found application in the synthesis of recifeiolide (61),[204] avenaciolide,[205] (*dl*)-aplysistatin,[206] furans,[207] and olefins,[208] amongst other uses,[209] while the α-mercapto-γ-butyrolactone dianion provides an entry to α-alkylidene-γ-butyrolactones (62)[210] and the reaction of α-(phenylthio)acetamide dianions (63) with di-iodomethane provides an interesting route to β-lactams.[211] Phenylthioalkyl-lithiums containing α-silyl or α-boranyl substituents have been converted into olefins[212] and ketones,[166] respectively. α-Lithio-α-(phenylthio)acetonitrile undergoes 1,4-addition to cyclohex-2-enone at room temperature but predominantly 1,2-addition at −78 °C.[213]

[196] T. Cohen and J. R. Matz, *Synth. Commun.*, 1980, **10**, 311; T. Cohen, W. M. Daniewski, and R. B. Weisenfeld, *Tetrahedron Lett.*, 1978, 4665.
[197] K. Tanaka, H. Uneme, S. Matsui, R. Tanikaga, and A. Kaji, *Chem. Lett.* 1980, 287.
[198] B. M. Trost and W. C. Vladuchick, *Synthesis*, 1978, 821.
[199] B. M. Trost and J. H. Rigby, *J. Org. Chem.*, 1978, **43**, 2938.
[200] S. Halazy and A. Krief, *Tetrahedron Lett.*, 1979, 4233.
[201] R. D. Miller and D. R. McKean, *Tetrahedron Lett.*, 1979, 583; R. D. Miller, D. R. McKean, and D. Kaufmann, *ibid.*, p. 587.
[202] Y. Nagao, K. Kaneko, and E. Fujita, *Tetrahedron Lett.*, 1978, 4115.
[203] K. Inomata, M. Sumita, and H. Kotake, *Chem. Lett.*, 1979, 709.
[204] T. Takahashi, S. Hashiguchi, K. Kasuga, and J. Tsuji, *J. Am. Chem. Soc.*, 1978, **100**, 7424.
[205] J. L. Herrmann, M. H. Berger, and R. H. Schlessinger, *J. Am. Chem. Soc.*, 1979, **101**, 1544.
[206] T. R. Hoye and M. J. Kurth, *J. Am. Chem. Soc.*, 1979, **101**, 5065.
[207] K. Inomata, S. Aoyama, and H. Kotake, *Bull. Chem. Soc. Jpn.*, 1978, **51**, 930.
[208] M. Baboulene and G. Sturtz, *J. Organomet. Chem.*, 1979, **177**, 27.
[209] S. Torii, H. Okumoto, and H. Tanaka, *J. Org. Chem.*, 1980, **45**, 1330.
[210] K. Tanaka, H. Uneme, N. Yamagishi, N. Ono, and A. Kaji, *Chem. Lett.*, 1978, 653.
[211] K. Hirai and Y. Iwano, *Tetrahedron Lett.*, 1979. 2031.
[212] T. Agawa, M. Ishikawa, M. Komatsu, and Y. Ohshiro, *Chem. Lett.*, 1980, 335.
[213] N.-Y. Wang, S.-S. Su, and L.-Y. Tsai, *Tetrahedron Lett.*, 1979, 1121.

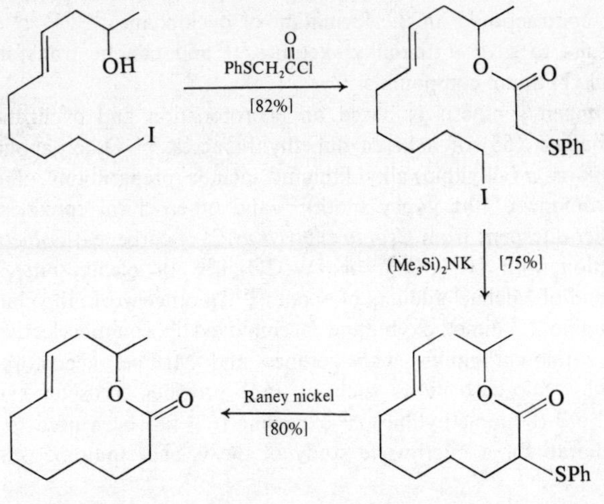

(61)

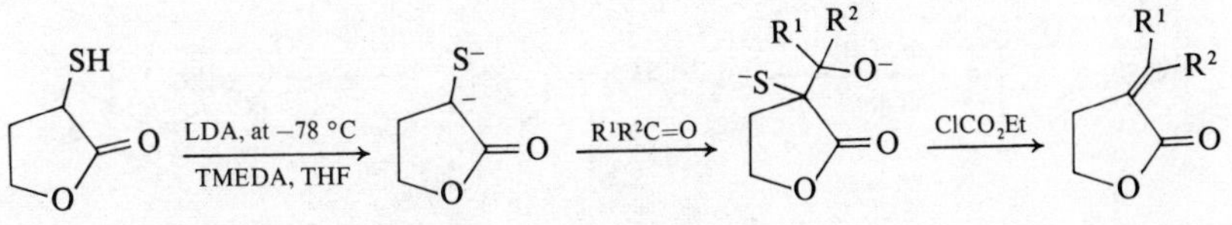

(62)

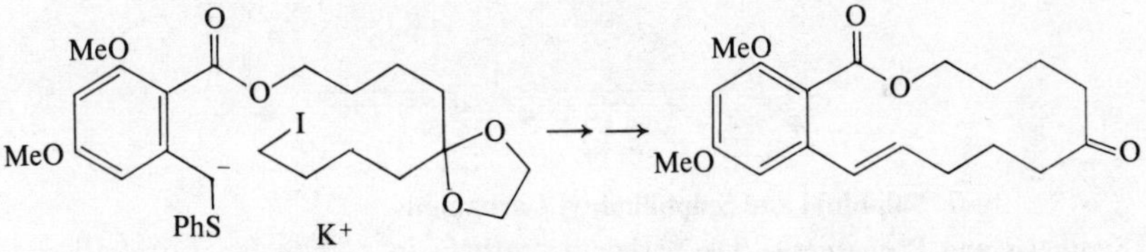

(64)

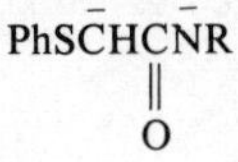

(63)

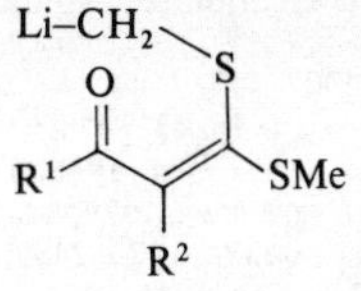

(65)

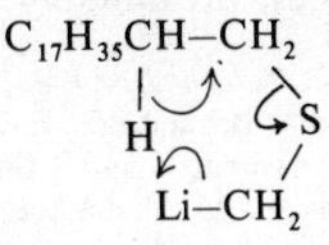

(66)

α-(Arylthio)- or α-(alkylthio)-benzylic anions undergo facile intermolecular[214,215] or intramolecular alkylation in the syntheses of O-methyl-($\pm$)-lasiodiplodin[216] and ($\pm$)-zearalenone dimethyl ether (64),[216] undergo Wittig rearrangements with

[214] J. Ackrell, *J. Org. Chem.*, 1978, **43**, 4892.

[215] B. L. Soltz and J. Y. Corey, *J. Organomet. Chem.*, 1979, **171**, 291.

[216] T. Takahashi, K. Kasuga, M. Takahashi, and J. Tsuji, *J. Am. Chem. Soc.*, 1979, **101**, 5072; T. Takahashi, K. Kasuga, and J. Tsuji, *Tetrahedron Lett.*, 1978, 4917.

resultant ring contraction[124] in the formation of cyclophanes,[34a,217,218] react with carboxylate salts to give α-(thioalkyl)-ketones,[219] and can be transformed into sulphur-free alkyl-lithium compounds.[220]

A new thiophen synthesis is based on deprotonation and cyclization of the resultant carbanion (65) of a keten dimethylthioacetal.[221] Other applications of α-(phenylthio)- or α-(alkylthio)-alkyl-lithiums include preparations of disparlure (the sex pheromone of the gypsy moth)[222] and other chiral epoxides,[223] of an anti-peptic-ulcer diterpene from *Croton sublyratus*,[224] of carbene–thiometal chelates through reaction with [Cr(CO)$_6$] or [W(CO)$_6$],[225] of olefins *via* β-hydroxy-sulphides,[226] and of Michael adducts of enones.[227] In other work, the chiral solvent 1,4-dimethylamino-2,3-dimethoxybutane is employed in enantioselective addition reactions of α-thio-carbanions with ketones and Michael acceptors.[228] α,β'-Elimination of α-thio-carbanions such as (66) provides a useful synthesis of olefins,[229] and a 2-(lithiomethylthio)-Δ^2-oxazoline (67) has been used to prepare a ^{13}C-labelled thiiran for a microwave study of the valence tautomerism of allene episulphide.[230]

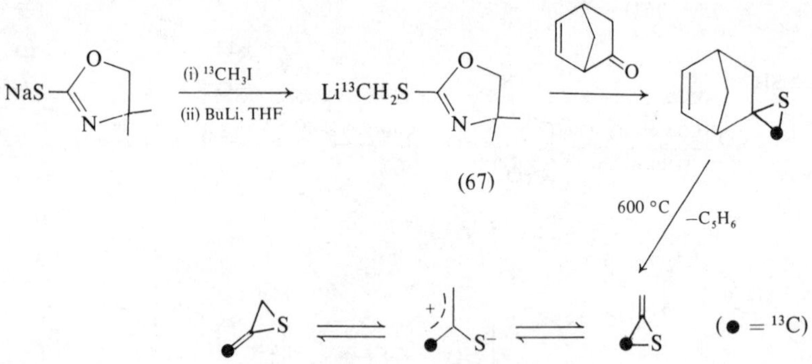

(67)

600 °C / $-C_5H_6$

(● = ^{13}C)

7 Sulphinyl and Sulphiliminyl Carbanions

Synthesis and Properties.—The carbanion carbons in α-lithiothian 1-oxide[231] and PhS(O)CH$_2$Li[6] have been found by ^{13}C n.m.r. spectroscopy to be nearly planar (*sp^2*-hybridized), presumably due to a chelated structure such as (68).[6,7] Interest

[217] M. W. Haenel, *Tetrahedron Lett.*, 1978, 4007.

[218] D. Kamp and V. Boekelheide, *J. Org. Chem.*, 1978, **43**, 3470.

[219] S. Kano, T. Yokomatsu, and S. Shibuya, *J. Org. Chem.*, 1978, **43**, 4366.

[220] C. G. Screttas and M. Micha-Screttas, *J. Org. Chem.*, 1979, **44**, 713.

[221] J. P. Marino and J. L. Kostusyk, *Tetrahedron Lett.*, 1979, 2489, 2493.

[222] W. H. Pirkle and P. L. Rinaldi, *J. Org. Chem.*, 1979, **44**, 1025.

[223] W. H. Pirkle and P. L. Rinaldi, *J. Org. Chem.*, 1978, **43**, 3803.

[224] A. Ogiso, E. Kitazawa, M. Kurabayashi, A. Sato, S. Takahashi, H. Noguchi, H. Kuwano, S. Kobayashi, and H. Mishima, *Chem. Pharm. Bull.*, 1978, **26**, 3117.

[225] H. G. Raubenheimer, S. Lotz, H. W. Viljoen, and A. A. Chalmers, *J. Organomet. Chem.*, 1978, **152**, 73.

[226] J. N. Denis, W. Dumont, and A. Krief, *Tetrahedron Lett.*, 1979, 4111.

[227] L. Wartski, M. El Bouz, J. Seyden-Penne, W. Dumont, and A. Krief, *Tetrahedron Lett.*, 1979, 1543.

[228] D. Seebach and W. Langer, *Helv. Chim. Acta*, 1979, **62**, 1701; W. Langer and D. Seebach, *ibid.*, p. 1710.

[229] J.-F. Biellmann, H. d'Orchymont, and J.-L. Schmitt, *J. Am. Chem. Soc.*, 1979, **101**, 3283.

[230] E. Block, R. E. Penn, M. D. Ennis, T. A. Owens, and S. L. Yu, *J. Am. Chem. Soc.*, 1978, **100**, 7436.

[231] R. Lett and G. Chassaing, *Tetrahedron*, 1978, **34**, 2705.

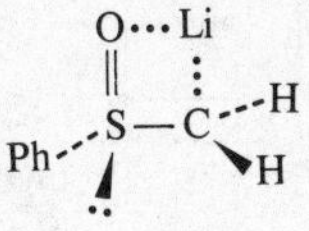

(68)

continues in the stereochemistry of reactions of α-sulphinyl carbanions with 'chelating' and 'non-chelating' reagents (*e.g.* D_2O and MeI, respectively; see Vol. 5, p. 91). Thus it has been found that *solid* lithiobenzyl methyl sulphoxide reacts completely stereospecifically with inversion with MeI vapour at 25 °C whereas the analogous reaction in solution is only stereoselective.[232] The solid lithio-compound does not react stereospecifically with chelating reagents such as D_2O or EtOD. In solution, the normal stereochemistry of the deuteriation of lithiobenzyl methyl sulphoxide (retention) can be reversed by the prior addition of Et_3Al, which is thought to afford an 'ate complex' (69).[178] In contrast to α-lithiothian 1-oxide and acyclic α-sulphinyl carbanions, α-lithiothiepan 1-oxide and α-lithiothiocan 1-oxide both react with MeI with predominant retention of stereochemistry (*cis*-methylation).[61,233]

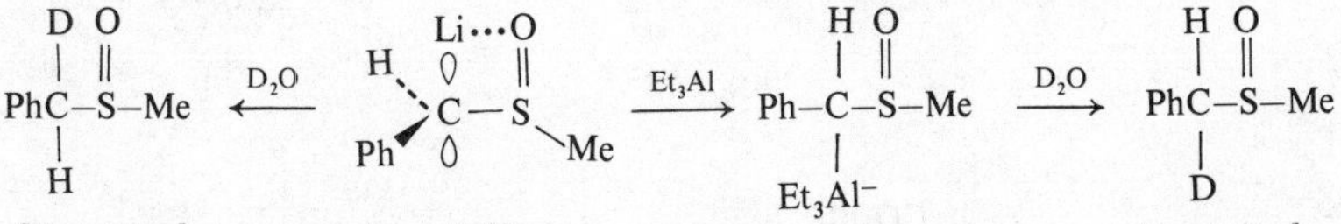

Reactions.—Asymmetric syntheses involving α-sulphinyl carbanions have been described, including the preparation of β-methoxy-β-phenylacetaldehyde *via* a chiral dithioacetal mono-*S*-oxide anion[234] and of lactones,[235] β-hydroxy-lactones, and esters such as (*R*)-mevalonolactone.[236] α-Sulphinyl carbanions are intermediates in the reaction of sulphines with carbanions (thiophilic addition) which can lead ultimately to ketones[237] or olefins[238,239] (see also Chap. 3, Pt II, p. 162).

An $\alpha\beta$-unsaturated sulphoxide (72) is the final product isolated when *cis*-1,2-diphenylthiiran *S*-oxide (70) is treated with butyl-lithium, followed by alkylation.[240] The lithio-thiiran *S*-oxide (71) is a presumed intermediate. A number of papers have dealt with the reactions of alkyl-lithiums or other strong bases with $\alpha\beta$-unsaturated sulphoxides. The reactions observed include (*a*) α-deprotonation, affording α-sulphinylvinyl anions, which turn out to be surprisingly configuration-

[232] J. F. Biellmann, J. F. Blanzat, and J. J. Vicens, *J. Am. Chem. Soc.*, 1980, **102**, 2460.
[233] V. Cere, S. Pollicino, E. Sandri, and A. Fava, *Tetrahedron Lett.*, 1978, 5239.
[234] L. Colombo, C. Gennari, C. Scolastico, G. Guanti, and E. Narisano, *J. Chem. Soc., Chem. Commun.*, 1979, 591.
[235] F. Matloubi and G. Solladié, *Tetrahedron Lett.*, 1979, 2141.
[236] E. Abushanab, D. Reed, F. Suzuki, and C. J. Sih, *Tetrahedron Lett.*, 1978, 3415; C. Mioskowski and G. Solladié, *Tetrahedron*, 1980, **36**, 227.
[237] G. E. Veenstra and B. Zwanenburg, *Tetrahedron*, 1978, **34**, 1585.
[238] M. van der Leij, H. J. M. Strijtveen, and B. Zwanenburg, *Recl. Trav. Chim. Pays-Bas*, 1980, **99**, 45.
[239] J. A. Loontjes, M. van der Leij, and B. Zwanenburg, *Recl. Trav. Chim. Pays-Bas*, 1980, **99**, 39.
[240] B. F. Bonini, G. Maccagnani, G. Mazzanti, and P. Piccinelli, *Tetrahedron Lett.*, 1979, 3987.

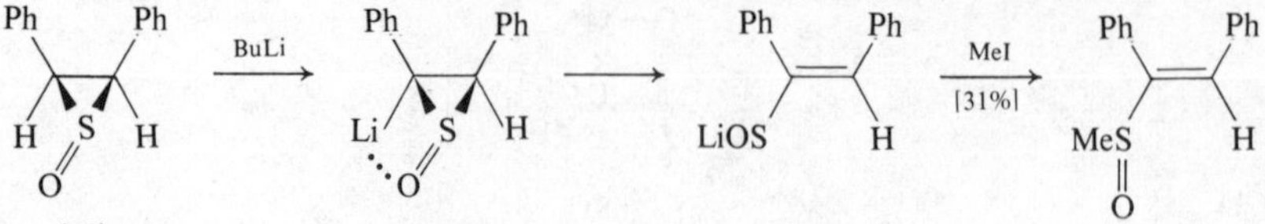

ally unstable,[241-243] (*b*) Michael addition, providing an approach to prostaglandins[244] and medium-sized rings such as (73),[245] and, occasionally, (*c*) loss of sulphur.[246] α-Sulphinylallenyl anions have also been generated.[195]

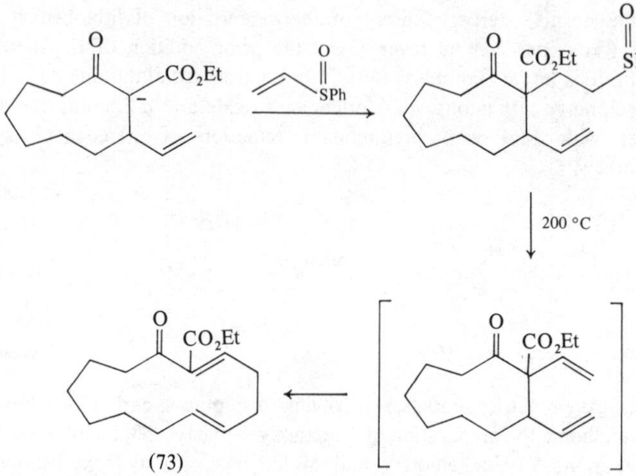

The easily formed α-chloro-α-phenylsulphinyl α-lithio-compounds (74) provide the starting point for syntheses of epoxy-sulphones (76; $R^1 = R^2 = H$),[247] enones (75; $R^1 = R^2 = Me$),[248] vinyl chlorides (77),[249] and thioesters (78), as shown in Scheme 5.[250] Methyl methylthiomethyl sulphoxide anion has been used to make 2-arylpropanoate esters[251] and 1,4-diketone precursors,[252] while other types of substituted α-sulphinyl carbanions have been utilized in the preparation of

[241] G. H. Posner, P.-W. Tang, and J. P. Mallamo, *Tetrahedron Lett.*, 1978, 3995.
[242] R. R. Schmidt, H. Speer, and B. Schmid, *Tetrahedron Lett.*, 1979, 4277.
[243] H. Okamura, Y. Mitsuhira, M. Miura, and H. Takei, *Chem. Lett.*, 1978, 517.
[244] R. David and K. G. Untch, *J. Org. Chem.*, 1979, **44**, 3755.
[245] J. Bruhn, H. Heimgartner, and H. Schmid, *Helv. Chim. Acta*, 1979, **62**, 2630.
[246] G. Neef, U. Eder, and A. Seeger, *Tetrahedron Lett.*, 1980, **21**, 903.
[247] T. Durst, K-C. Tin, F. de Reinach-Hirtzbach, J. M. Decesare, and M. D. Ryan, *Can. J. Chem.*, 1979, **57**, 258.
[248] D. G. Taber and B. P. Gunn, *J. Org. Chem.*, 1979, **44**, 450.
[249] V. Reutrakul and P. Thamnusan, *Tetrahedron Lett.*, 1979, 617.
[250] K. M. More and J. Wemple, *J. Org. Chem.*, 1978, **43**, 2713.
[251] K. Ogura, S. Mitamura, K. Kishi, and G.-I. Tsuchihashi, *Synthesis*, 1979, 880; K. Ogura, Y. Ito, and G.-I. Tsuchihashi, *Bull. Chem. Soc. Jpn.*, 1979, **52**, 2013.
[252] K. Ogura, M. Yamashita, and G.-I. Tsuchihashi, *Tetrahedron Lett.*, 1978, 1303.

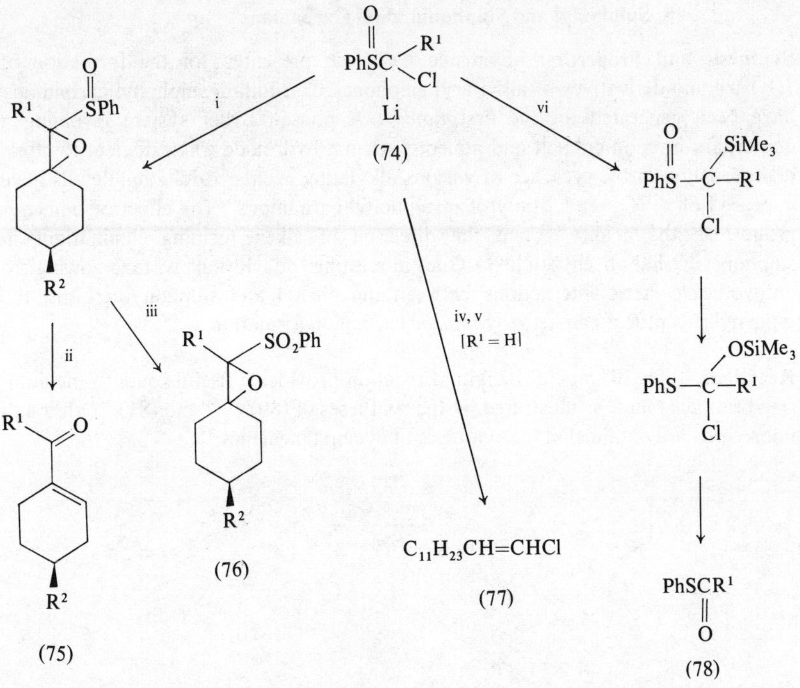

Reagents: i, R^2 ⬤—⬡=O; ii, heat; iii, [O]; iv, $C_{12}H_{25}I$; v, heat, in xylene; vi, Me_3SiCl

Scheme 5

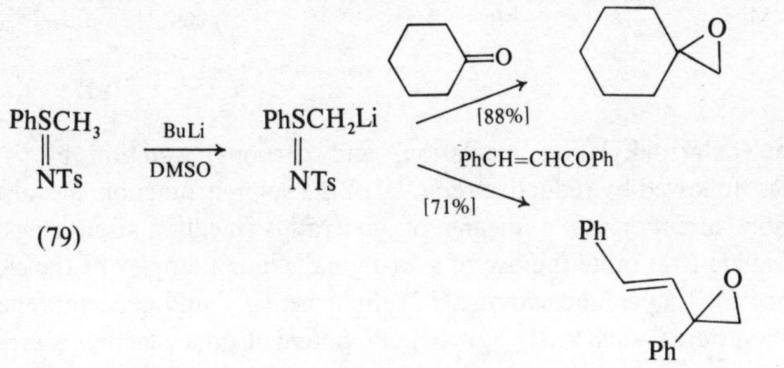

(E)-hotrienol,[253] brefeldin A,[254] cyclobutenes,[255] dienes,[256] $\alpha\beta$-unsaturated sulphides,[257] and allylic alcohols.[258] N-Tosylsulphilimine (79) can be readily converted into its anion, which converts ketones into oxirans.[259]

[253] E. Guittet and S. Julia, *Synth. Commun.*, 1979, **9**, 317.

[254] P. A. Bartlett and F. R. Green, III, *J. Am. Chem. Soc.*, 1978, **100**, 4858.

[255] S. R. Wilson, L. R. Phillips, Y. Pelister, and J. C. Huffman, *J. Am. Chem. Soc.*, 1979, **101**, 7373.

[256] B. M. Trost, L. Weber, P. Strege, T. J. Fullerton, and T. J. Dietsche, *J. Am. Chem. Soc.*, 1978, **100**, 3426.

[257] T. Kojima and T. Fujisawa, *Chem. Lett.*, 1978, 1425.

[258] R. W. Hoffmann, S. Goldmann, N. Maak, R. Gerlach, F. Frickel, and G. Steinbach, *Chem. Ber.*, 1980, **113**, 819.

[259] C. R. Johnson, K. Mori, and A. Nakanishi, *J. Org. Chem.*, 1979, **44**, 2065.

8 Sulphonyl and Sulphonimidoyl Carbanions

Synthesis and Properties.—Evidence has been presented for the formation of 1,1,1-trilithio-derivatives of alkyl aryl sulphones.[260] α-Iodo-α-sulphonyl carbanions have been generated for the first time.[261] A phase-transfer system involving a quaternary ammonium salt and aqueous sodium hydroxide was sufficient to effect deprotonation in the presence of various alkylating agents. Additional details have appeared of a ^{13}C n.m.r. study of α-sulphonyl carbanions.[6] The effect of α-phenyl groups on the acidity[262] and the direction of alkene-forming eliminations in sulphones[263] has been studied. One interesting conclusion is that, owing to unfavourable steric interactions between the phenyl and sulphonyl groups, the α-phenyl substituent can actually *impede* carbanion formation.

Reactions.—The Ramberg–Bäcklund reaction provides a useful route to medium-sized ring polyenes, as illustrated by the syntheses of (80)[63,264] and (81),[265] although poor yields are obtained in the synthesis of cyclophanedienes.[34b]

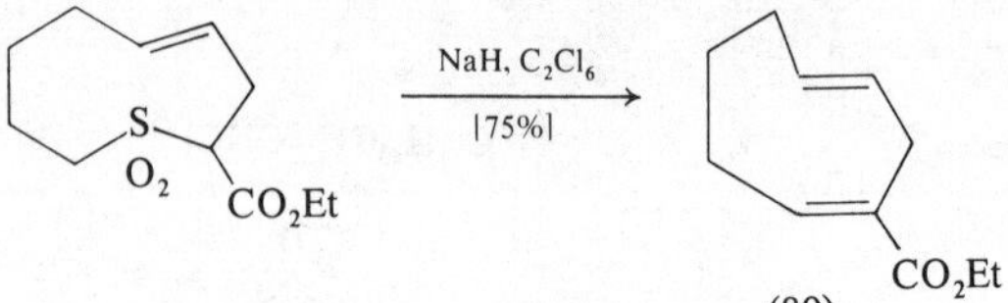

$$\xrightarrow[\;|75\%|\;]{\text{NaH, C}_2\text{Cl}_6}$$

(80)

$$\xrightarrow[\text{Bu}^t\text{OH, CCl}_4]{\text{KOH}}$$

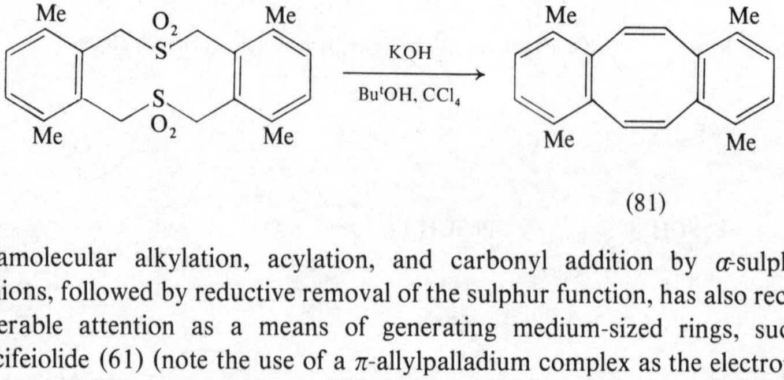

(81)

Intramolecular alkylation, acylation, and carbonyl addition by α-sulphonyl carbanions, followed by reductive removal of the sulphur function, has also received considerable attention as a means of generating medium-sized rings, such as ($\pm$)-recifeiolide (61) (note the use of a π-allylpalladium complex as the electrophile; see Scheme 6),[266] cycloundecanone (82) (Scheme 7),[267] and neocembrene.[268] Other polycyclic systems, such as ($\pm$)-zizaene,[269] potential dodecahedrane precursors,[270]

[260] A. Roggero, T. Salvatori, A. Proni, and A. Mazzei, *J. Organomet. Chem.*, 1979, **177**, 313.
[261] A. Jonczyk and T. Pytlewski, *Synthesis*, 1978, 883.
[262] F. Hibbert, *J. Chem. Soc., Perkin Trans. 2*, 1978, 1171.
[263] P. J. Thomas and C. J. M. Stirling, *J. Chem. Soc., Chem. Commun.*, 1978, 975.
[264] E. Vedejs and S. P. Singer, *J. Org. Chem.*, 1978, **43**, 4884.
[265] M.-K. Au, T.-W. Siu, T. C. W. Mak, and T.-L. Chan, *Tetrahedron Lett.*, 1978, 4269.
[266] B. M. Trost and T. R. Verhoeven, *Tetrahedron Lett.*, 1978, 2275; *J. Am. Chem. Soc.*, 1979, **101**, 1595.
[267] Y. Ohtsuka and T. Oishi, *Tetrahedron Lett.*, 1979, 4487.
[268] H. Takayanagi, T. Uyehara, and T. Kato, *J. Chem. Soc., Chem. Commun.*, 1978, 359.
[269] E. Piers and J. Bannville, *J. Chem. Soc., Chem. Commun.*, 1979, 1138.
[270] N. J. Hales and L. A. Paquette, *J. Org. Chem.*, 1979, **44**, 4603.

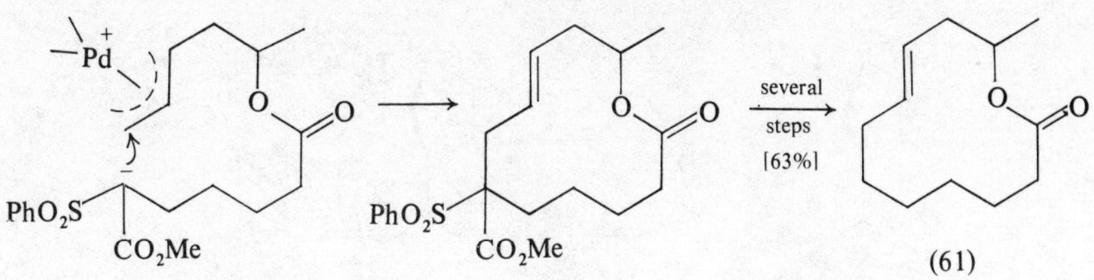

(61)

Scheme 6

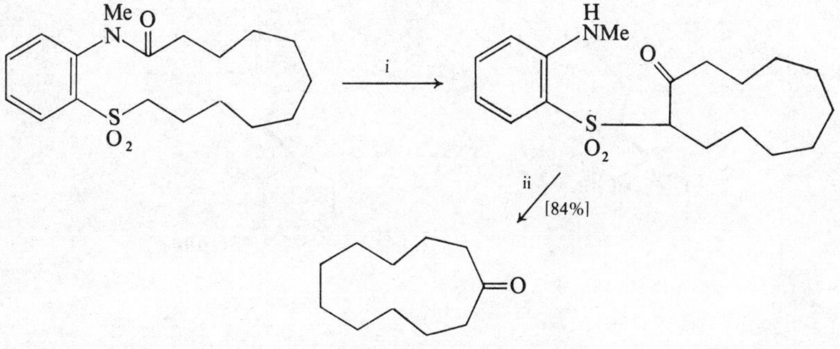

(82)

Reagents: i, KOBu^t, THF, DMSO; ii, Al amalgam

Scheme 7

and hinesol (83),[271] have been synthesized, together with γ-butyrolactones[272] and other cyclic carbonyl compounds.[273,274]

Incorporation of the trimethylsilyl function into sulphur-containing molecules permits elimination reactions under neutral conditions, as in the case of (84),[275] amongst other examples,[276] and facilitates Michael addition to vinyl sulphones such as (85),[277] where steric control is achieved through co-ordination of the lithium to the polyether function. A related example, involving steric repulsion by a trialkylsiloxy-group, has appeared.[278] If the trimethylsilyl group in systems such as (85) is replaced by hydrogen, abstraction of the α-vinyl proton may occur, affording α-sulphonylvinyl carbanions, which can be trapped with electrophiles.[277b,279] Grignard reagents undergo Michael addition to vinyl sulphones such as (86), which then cyclizes.[280,281] A spectacular example, apparently involving abstraction of an

[271] D. A. Chass, D. Buddhasukh, and P. D. Magnus, *J. Org. Chem.*, 1978, **43**, 1750.
[272] V. Reutrakul, P. Tuchinda, and K. Kusamran, *Chem. Lett.*, 1979, 1055.
[273] E. A. Harrison, Jr., and K. C. Rice, *J. Org. Chem.*, 1979, **44**, 2977.
[274] K. Schank and H.-G. Schmitt, *Chem. Ber.*, 1978, **111**, 3497.
[275] P. J. Kocienski, *Tetrahedron Lett.*, 1979, 2649.
[276] H. J. Reich, J. J. Rusek, and R. E. Olson, *J. Am. Chem. Soc.*, 1979, **101**, 2225.
[277] M. Isobe, M. Kitamura, and T. Goto, (a) *Tetrahedron Lett.*, 1979, 3465; (b) *Chem. Lett.*, 1980, 331.
[278] J. C. Saddler, P. C. Conrad, and P. L. Fuchs, *Tetrahedron Lett.*, 1978, 5079.
[279] J. J. Eisch and J. E. Galle, *J. Org. Chem.*, 1979, **44**, 3279.
[280] J. J. Eisch and J. E. Galle, *J. Org. Chem.*, 1979, **44**, 3277.
[281] E. Ghera and Y. Ben-David, *Tetrahedron Lett.*, 1979, 4603.

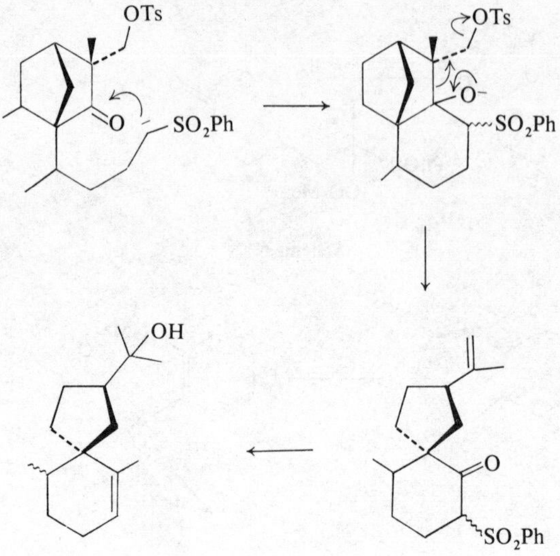

(83)

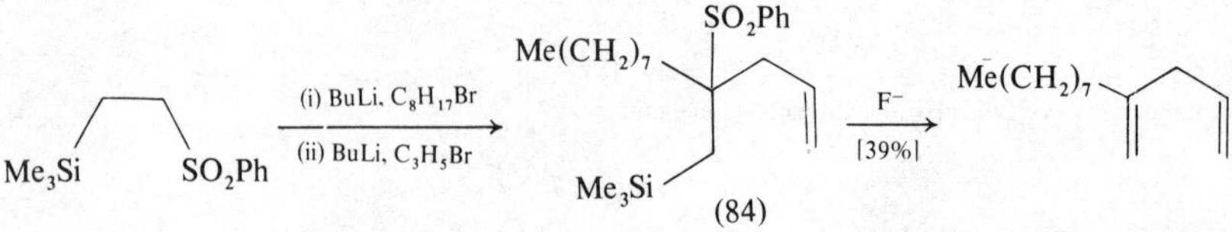

(84)

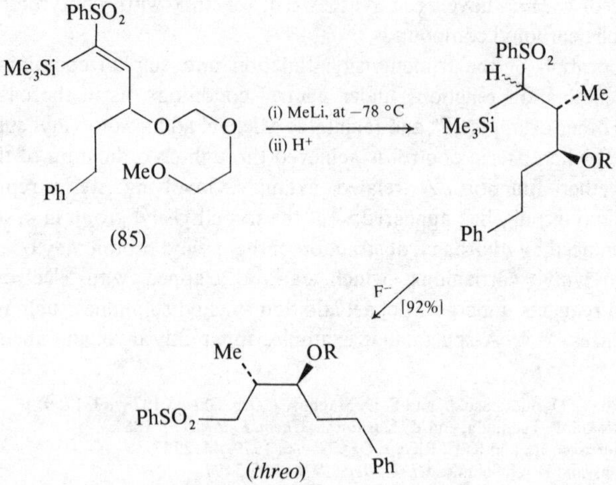

(85)

(*threo*)

α-sulphonylvinyl proton to give an anion, which then undergoes self Michael addition, is the reaction of bis(allenyl) sulphone (87) with butyl-lithium, giving a 2,6-dithia-adamantane derivative.[282]

[282] S. Braverman, D. Reisman, and M. Sprecher, *Tetrahedron Lett.*, 1979, 901.

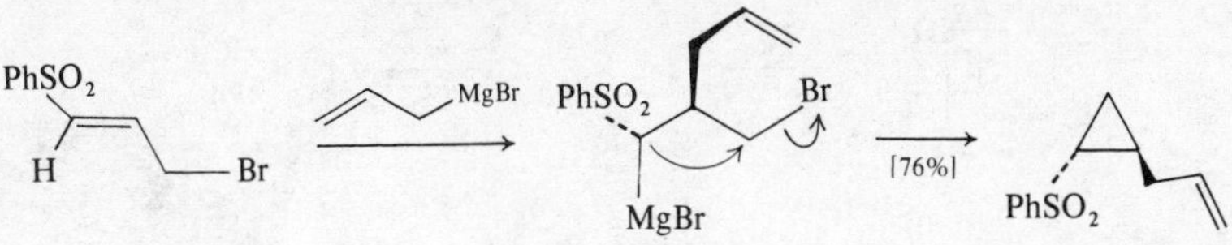

(86)

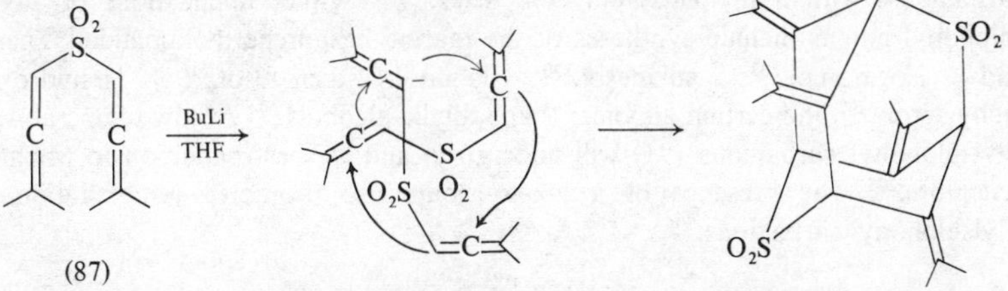

(87)

Alkylation of anions (88)[283] and (89),[284,285] followed by pyrolysis, provides novel methods of synthesis of carbonyl compounds and polycyclic hydrocarbons (including steroids), respectively. A synthesis of polycyclic hydrocarbons related to that based on the anion (89) involves alkylation of a benzocyclobutenyl-sulphonyl carbanion.[286] Alkylation of a thietan 1,1-dioxide carbanion (90), followed by photolysis of the product, affords cyclopropanes.[287]

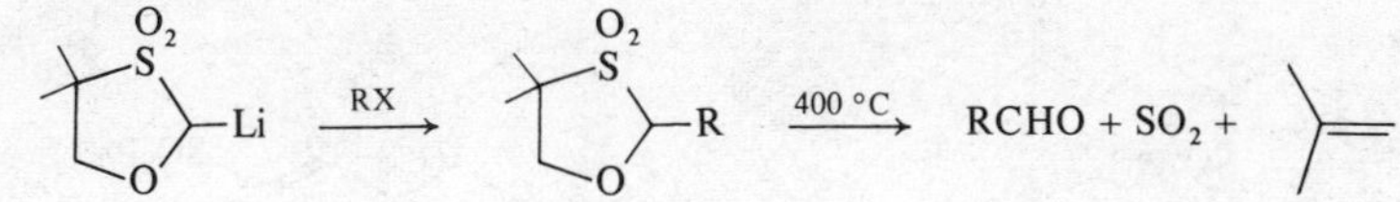

(88)

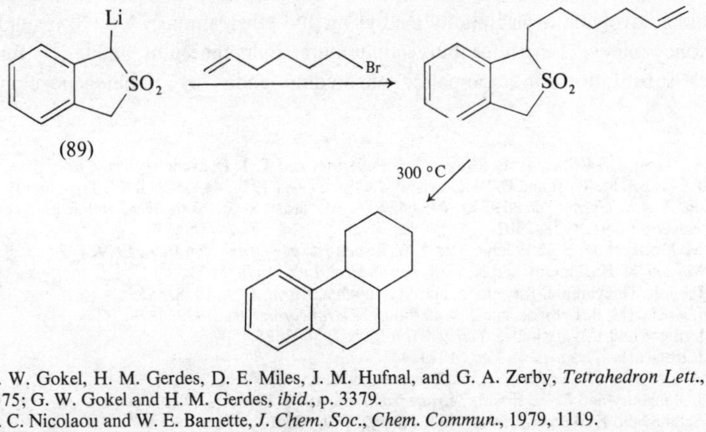

(89)

[283] G. W. Gokel, H. M. Gerdes, D. E. Miles, J. M. Hufnal, and G. A. Zerby, *Tetrahedron Lett.*, 1979, 3375; G. W. Gokel and H. M. Gerdes, *ibid.*, p. 3379.
[284] K. C. Nicolaou and W. E. Barnette, *J. Chem. Soc., Chem. Commun.*, 1979, 1119.
[285] W. Oppolzer, D. A. Roberts, and T. G. C. Bird, *Helv. Chim. Acta*, 1979, **62**, 2017.
[286] B. D. Gowland and T. Durst, *Can. J. Chem.*, 1979, **57**, 1462.
[287] J. D. Finlay, D. J. H. Smith, and T. Durst, *Synthesis*, 1978, 579.

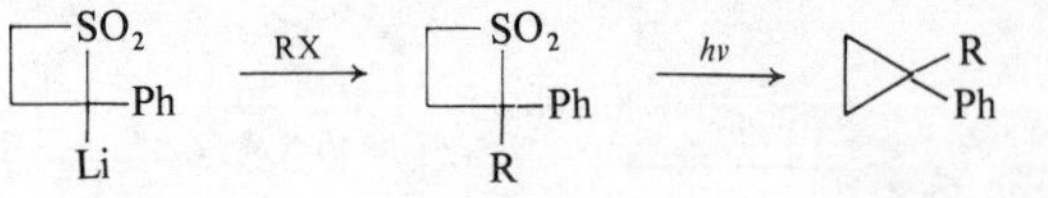

(90)

All-*trans*-geranylgeraniol and other olefins have been synthesized through the reaction of allylic sulphonyl carbanions (or other functionalized α-sulphonyl carbanions) with π-allylpalladium complexes.[256,288] Other applications of allylic sulphonyl anions include syntheses of the marine hexaprene mokupalide,[183] nerol and isogeraniol,[289] solanesol,[290] labda-7,14-dien-13-ol,[291] 1α-hydroxy-tachysterol₃[292] and certain alkynes,[293] and allylic alcohols.[294] At low temperatures, allylsulphonyl carbanions (91) will undergo γ- and well as α-addition to Michael acceptors.[295] The presence of a γ-keto-group also promotes γ-substitution of allylsulphonyl carbanions.[296]

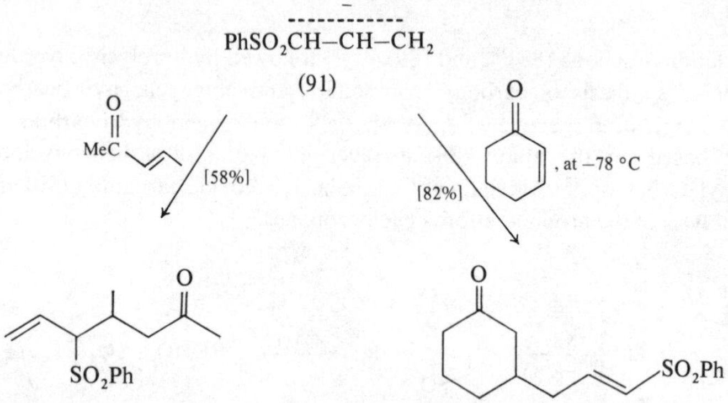

$$\overline{PhSO_2CH-CH-CH_2}$$

(91)

Miscellaneous reactions of α-sulphonyl carbanions with electrophiles include processes involving arylsulphonylmethyl methyl ether anions,[274,297,298] α-sulphinyl sulphone anions,[299] perfluorobutylsulphonylmethylmagnesium halides,[300] nucleophilic substitution in aromatic nitro-compounds by α-halogenosulphonyl

[288] B. M. Trost, L. Weiber, P. E. Strege, T. J. Fullerton, and T. J. Dietsche, *J. Am. Chem. Soc.*, 1978, **100**, 3416; B. M. Trost and F. W. Gowland, *J. Org. Chem.*, 1979, **44**, 3448; B. M. Trost and D. M. T. Chan, *J. Am. Chem. Soc.*, 1979, **101**, 6432; B. M. Trost, T. R. Verhoeven, and J. M. Fortunak, *Tetrahedron Lett.*, 1979, 2301.
[289] A. M. Moiseenkov, E. V. Polunin, and A. V. Semenovsky, *Tetrahedron Lett.*, 1979, 4759.
[290] Y. Masaki, K. Hashimoto, and K. Kaji, *Tetrahedron Lett.*, 1978, 5123.
[291] S. Torii, K. Uneyama, I. Kawahara, and M. Kuyama, *Chem. Lett.*, 1978, 455.
[292] P. J. Kocienski, B. Lythgoe, and I. Waterhouse, *Tetrahedron Lett.*, 1979, 4419.
[293] B. Lythgoe and I. Waterhouse, *Tetrahedron Lett.*, 1978, 2625.
[294] P. J. Kocienski, *Tetrahedron Lett.*, 1979, 441.
[295] G. A. Kraus and K. Frazier, *Synth. Commun.*, 1978, **8**, 483.
[296] P. T. Lansbury and R. W. Erwin, *Tetrahedron Lett.*, 1978, 2675.
[297] K. Schank and F. Werner, *Liebigs Ann. Chem.*, 1979, 1977.
[298] K. Schank, F. Schroeder, and A. Weber, *Liebigs Ann. Chem.*, 1979, 547.
[299] H. Bohme and B. Clement, *Tetrahedron Lett.*, 1979, 1737.
[300] K. Laping and M. Hanack, *Tetrahedron Lett.*, 1979, 1309.

carbanions,[301] Michael additions[86,281,302] and new syntheses of nitriles,[303] α-oxo-sulphones.[297] but-2-enolides,[304] $\alpha\beta$-unsaturated sulphones,[305–307] vitamin D_3 metabolites,[308] cyclopropyl compounds,[309] dienes by 1,2-addition to $\alpha\beta$-unsaturated aldehydes,[310] and alkynes.[311]

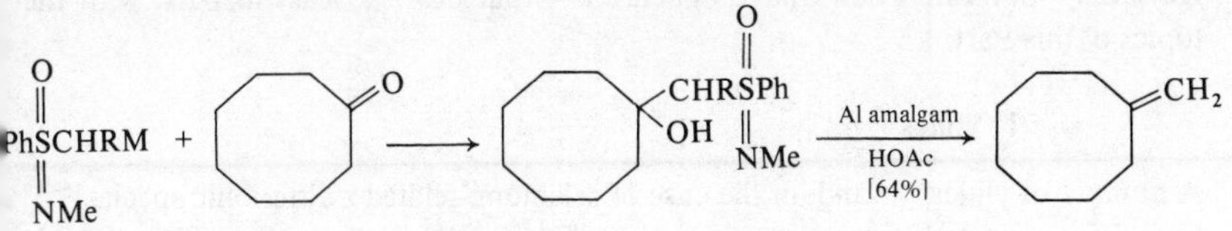

(92)

N-Methylphenylsulphonimidoylmethyl-lithium (92; R = H, M = Li) and its derivatives provide a useful alternative to the Wittig reaction, occasionally succeeding where the Wittig reaction proves troublesome,[312] as in the synthesis of gascardic acid[313] or the prostacycline model compound (93).[42c] α-Halogeno-sulphoximides undergo processes analogous to the Ramberg–Bäcklund rearrangement and the Darzens-type condensation on treatment with base, presumably *via* the appropriate sulphonimidoyl carbanions[314,315] (see also Chap. 2, Pt III, p. 145).

[92; R = $(CH_2)_4OR^1$, M = MgBr]

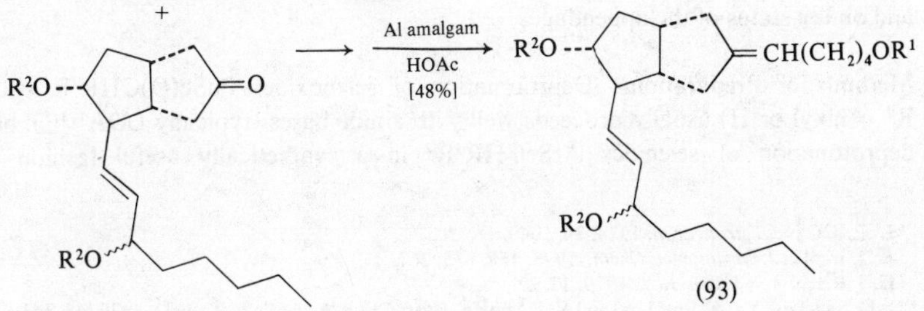

(93)

Acknowledgement. During the period in which this review was prepared, the author received support from NSF and the donors of the Petroleum Research Fund, administered by the American Chemical Society, and this is gratefully acknowledged.

[301] J. Golinski and M. Makosza, *Tetrahedron Lett.*, 1978, 3495.

[302] J. Wildeman, P. C. Borgen, and H. Pluim, *Tetrahedron Lett.*, 1978, 2213.

[303] P. A. Wade, S. D. Morrow, S. A. Hardinger, M. S. Saft, and H. R. Hinney, *J. Chem. Soc., Chem. Commun.*, 1980, 287.

[304] G. K. Cooper and L. J. Dolby, *J. Org. Chem.*, 1979, **44**, 3414.

[305] H. Kotake, K. Inomata, and M. Sumita, *Chem. Lett.*, 1978, 717.

[306] M. Julia, A. Righini, and D. Uguen, *J. Chem. Soc., Perkin Trans. 1*, 1978, 1646.

[307] Y. Ueno, H. Setoi, and M. Okawara, *Chem. Lett.*, 1979, 47.

[308] B. M. Trost, P. R. Bernstein, and P. C. Funfschilling, *J. Am. Chem. Soc.*, 1979, **101**, 4378.

[309] Y.-H. Chang and H. W. Pinnick, *J. Org. Chem.*, 1978, **43**, 373; also see C. L. Bumgardner, J. R. Lever, and S. T. Purrington, *ibid.*, 1980, **45**, 749.

[310] P. J. Kocienski, B. Lythgoe, and S. Ruston, *J. Chem. Soc., Perkin Trans. 1*, 1978, 829.

[311] P. A Bartlett, F. R. Green, III, and E. H. Rose, *J. Am. Chem. Soc.*, 1978, **100**, 4852.

[312] C. R. Johnson and R. A. Kirchhoff, *J. Am. Chem. Soc.*, 1979, **101**, 3602.

[313] R. K. Boeckman, Jr., D. M. Blum, and S. D. Arthur, *J. Am. Chem. Soc.*, 1979, **101**, 5060.

[314] C. R. Johnson and H. G. Corkins, *J. Org. Chem.*, 1978, **43**, 4140.

[315] H. G. Corkins, L. Veenstra, and C. R. Johnson, *J. Org. Chem.*, 1978, **43**, 4233.

PART II: Ylides and Carbanionic Compounds of Selenium and Tellurium
by D. L. J. Clive

Reviews.—Several reviews have appeared[1-5] that deal, at least in part, with the topics of this Part.

1 Ylides

A number of ylides[6-14] and, in the case of selenium, related zwitterionic species[15-19] have been reported. Compounds (1),[6] (2),[10] and (3)[15,16] are representative of this literature, which deals with preparative details and simple chemical properties.

Ylide formation can sometimes limit the synthetic utility of selenium-substituted enolates.[20] For example, alkylation of (4) with prenyl iodide gave a mixture of (5) and (6) in which the latter, formed from the ylide (7), predominates.[20]

2 Selenium-stabilized Carbanions

Sufficient information on selenium-stabilized carbanions is now available both to require and to allow classification of the field. The carbanions may be of several types, depending on the oxidation level (selenide or selenoxide) of the heteroatom and on the status of the appendages.

Methods of Preparation.—Deprotonation of selenoxides $PhSe(O)CHR^1R^2$ (R^1, R^2 = alkyl or H) usually proceeds well with amide bases (typically LDA[1,3,5]), but deprotonation of selenides $ArSeCHR^1R^2$ in a synthetically useful fashion is

[1] D. L. J. Clive, *Tetrahedron*, 1978, **34**, 1049.

[2] K. J. Irgolic, *J. Organomet. Chem.*, 1978, **158**, 235, 267.

[3] H. J. Reich, *Acc. Chem. Res.*, 1979, **12**, 22.

[4] I. D. Sadekov, Ya. A. Bushkov, and V. I. Minkin, *Russ. Chem. Rev. (Engl. Transl.)*, 1979, **48**, 343.

[5] H. J. Reich, in 'Oxidation in Organic Chemistry', Part C, ed. W. S. Trahanovsky, Academic Press, New York, 1978, p. 1.

[6] B. A. Arbuzov, Yu. V. Belkin, N. A. Polezhaeva, and G. E. Buslaeva, *Bull. Acad. Sci. USSR, Div. Chem. Sci.*, 1979, 1072.

[7] N. Ya. Derkach, N. P. Tishchenko, and V. G. Voloshchuk, *J. Org. Chem. USSR (Engl. Transl.)*, 1978, **14**, 896.

[8] N. N. Magdesieva, R. A. Kyandzhetsian, and O. A. Rakitin, *Ref. Dokl. Soobshch.—Mendeleevsk. S'ezd Obshch. Prikl. Khim., 11th*, 1975, **2**, 89–90 (*Chem. Abstr.*, 1978, **88**, 169 715).

[9] N. N. Magdesieva and R. A. Kyandzhetsian, *Otkrytiya, Izobret., Prom. Obraztsy, Tovarnye Znaki*, 1979, 235 (*Chem. Abstr.*, 1979, **90**, 186 792).

[10] I. D. Sadekov, A. I. Usachev, and V. I. Minkin, *J. Gen. Chem. USSR (Engl. Transl.)*, 1978, **48**, 427.

[11] I. D. Sadekov, A. I. Usachev, A. A. Maksimenko, and V. I. Minkin, *J. Gen. Chem. USSR (Engl. Transl.)*, 1978, **48**, 853.

[12] S. Tamagaki, I. Hatanaka, and S. Kozuka, *Mem. Fac. Eng. Osaka City Univ.*, 1977, **18**, 81 (*Chem. Abstr.*, 1979, **90**, 22 481).

[13] K. Friedrich, W. Amann, and H. Fritz, *Chem. Ber.*, 1979, **112**, 1267.

[14] V. V. Semenov, L. G. Mel'nikova, S. A. Shevelev, and A. A. Fainzil'berg, *Izv. Akad. Nauk SSSR, Ser. Khim.*, 1980, 138.

[15] N. N. Magdesieva and V. A. Danilenko, *J. Org. Chem. USSR (Engl. Transl.)*, 1979, **15**, 332.

[16] N. N. Magdesieva and V. A. Danilenko, *J. Gen. Chem. USSR (Engl. Transl.)*, 1979, **49**, 1740.

[17] N. N. Magdesieva and V. A. Danilenko, *J. Org. Chem. USSR (Engl. Transl.)*, 1979, **15**, 1857.

[18] N. N. Magdesieva and R. A. Kyandzhetsian, *J. Org. Chem., USSR (Engl. Transl.)*, 1979, **15**, 2169.

[19] N. N. Magdesieva and V. A. Danilenko, *J. Org. Chem. USSR (Engl. Transl.)*, 1979, **15**, 2428.

[20] H. J. Reich and M. L. Cohen, *J. Am. Chem. Soc.*, 1979, **101**, 1307.

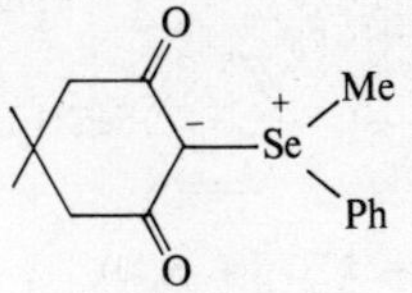

(1) 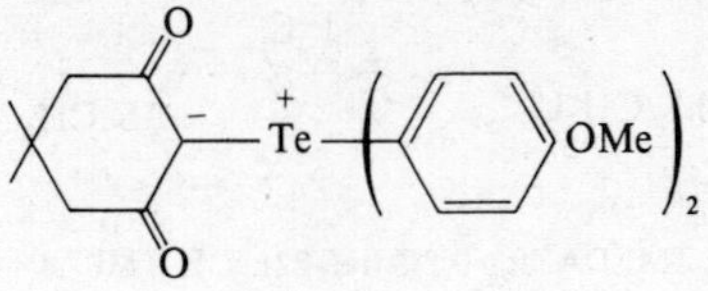   (2) 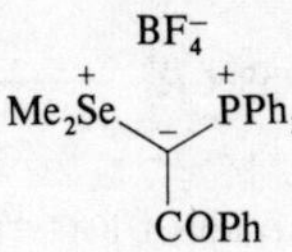   (3)

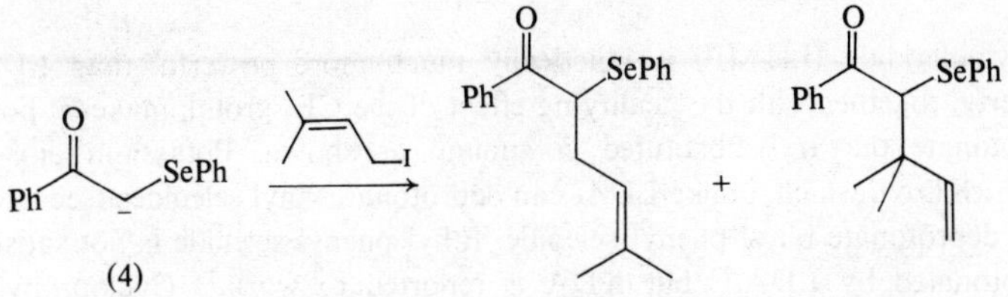

(4)

(5)   (6)

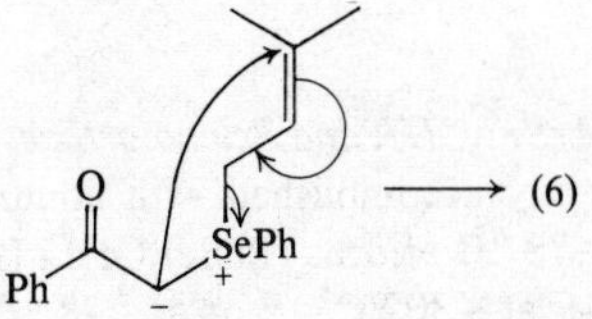

⟶ (6)

(7)

possible only for certain constituents R^1 and R^2. Even then, a careful choice of conditions must be made.[1,3,5]

In the case of $ArSeCHR^1R^2$, two competing reactions are possible when the selenide is treated with a base: deprotonation or nucleophilic attack at selenium. The latter mechanism leads to C—Se bond cleavage and, when deliberately applied to selenoketals, it provides a good route to those carbanions whose substitution pattern renders them unavailable by deprotonation (Scheme 1). The accessibility of carbanions by deprotonation can be summarized in the following way. (The analysis does not include selenophens,[21] and the yields that are quoted below represent minimum values because they refer in each case to a further transformation product of the carbanion.)

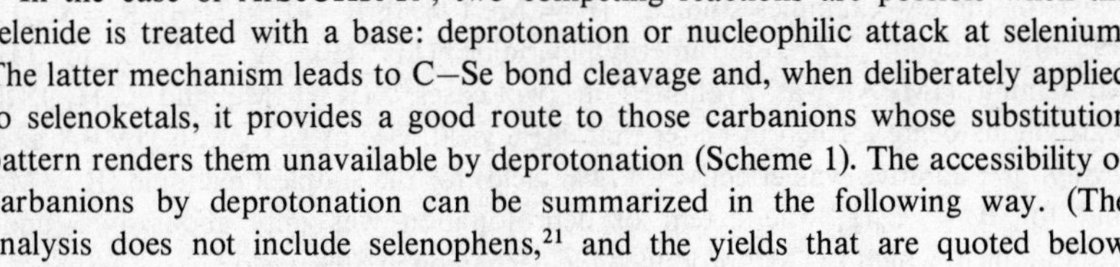

Scheme 1

(*i*) *Deprotonation of Aryl Alkyl Selenides ArSeR.* Scheme 2 shows the transformations in this category that have been reported.[3,22,23] Lithium 2,2,6,6-tetra-

[21] J. Morel, C. Paulmier, D. Semard, and P. Pastour, *C.R. Hebd. Seances Acad. Sci., Ser. C.*, 1970, **270**, 825; L. Christiaens, R. Dufour, and M. Renson, *Bull. Soc. Chim. Belg.*, 1970, **79**, 143.
[22] D. Seebach and N. Peleties, *Chem. Ber.*, 1972, **105**, 511.
[23] H. J. Reich, F. Chow, and S. K. Shah, *J. Am. Chem. Soc.*, 1979, **101**, 6638.

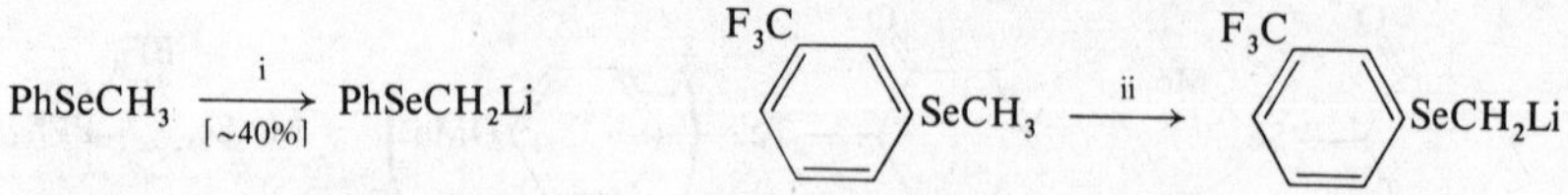

PhSeCH$_3$ $\xrightarrow[\text{[~40\%]}]{\text{i}}$ PhSeCH$_2$Li

Reagents: i, BuLi, THF, TMEDA, at >0 °C (ref. 22); ii, Li TMP, at −40 or −55 °C (refs. 3, 23)

Scheme 2

methylpiperidide (LiTMP) is kinetically much more powerful than LDA. This property, together with the acidifying effect of the CF$_3$ group, makes it possible to deprotonate the aryl-substituted compound as shown. Potassium di-isopropyl-amide (KDA), which, unlike LDA, can deprotonate vinyl selenides (see later), does not[24] deprotonate butyl phenyl selenide. Ethyl phenyl selenide is not satisfactorily deprotonated by LDA,[23] but KDA is reported to work.[25] Cyclopropyl phenyl selenide is not deprotonated (at least in THF) by LDA, LiTMP, or lithium diethylamide.[23]

(ii) Deprotonation of Selenoketals (ArSe)$_2$CHR and Seleno-orthoesters (RSe)$_3$CH.
Deprotonation of (PhSe)$_2$CH$_2$ is readily accomplished with lithium amide bases: lithium di-isobutylamide in THF at −78 °C affords (PhSe)$_2$CHLi (91%).[22] LDA is also very efficient (98% yield) in THF at −30 °C[26, 27] or at −78 °C,[23] and ether is another satisfactory solvent.[23]

Selenoketals of the type (PhSe)$_2$CHR (R = alkyl) are more difficult to metallate than the unsubstituted compound.[27] Butyl-lithium is unsuitable,[27] but appropriate conditions have been found: KDA[24] in THF at −78 °C is an effective reagent, at least for the few examples studied[24] [R = Me (98%), R = Pr^i (71%), R = C$_{10}$H$_{21}$ (98%)]. Lithium 2,2,6,6-tetramethylpiperidide (LiTMP), at −30 °C, in THF containing HMPA, was evaluated in two cases[27] (R = Me and C$_6$H$_{13}$), the carbanions being formed in better than 80% yield. Use of LDA with HMPA as an *essential*[28] additive was effective[27] (74% yield) for the simplest example (R = Me), but for R = C$_6$H$_{13}$ the extent of deprotonation was only about 44% under conditions in which LiTMP produced the carbanion in 86% yield.[27]

Bis(methylseleno)ketals, (MeSe)$_2$CHR (R = H or alkyl), are not metallated by LiTMP or LDA under conditions that were studied[27] with the bis(phenyl-seleno)-analogues. However, the ring-closure shown in reaction (1) occurs efficiently in the presence of LDA.[29]

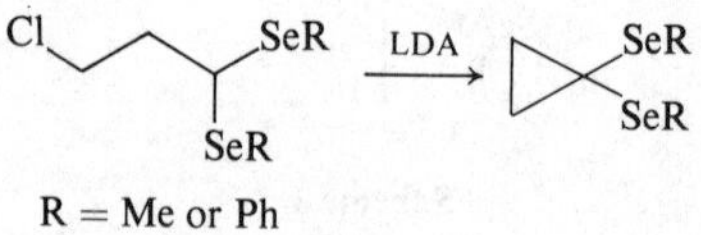

(1)

R = Me or Ph

[24] S. Raucher and G. A. Koolpe, *J. Org. Chem.*, 1978, **43**, 3794.
[25] Ref. 23, footnote 19.
[26] Ref. 24, footnote 16.
[27] D. Van Ende, A. Cravador, and A. Krief, *J. Organomet. Chem.*, 1979, **177**, 1.
[28] *Cf.* Ref. 23.
[29] S. Halazy, J. Lucchetti, and A. Krief, *Tetrahedron Lett.*, 1978, 3971.

Tris(methylseleno)methane is deprotonated by LDA[27,29,30] and tris(phenyl-seleno)methane by lithium di-isobutylamide.[22]

(iii) Deprotonation of Benzyl Selenides PhSeCHRPh (R = alkyl). Deprotonation of benzyl phenyl selenide is readily effected (81%) with LDA at −78 °C,[23,31,32] but the anion is not formed with potassium t-butoxide,[31] and C−Se bond cleavage results with BuLi.[31] The higher homologue PhSeCH(Me)Ph is not appreciably deprotonated by LDA.[23,32] [For benzyl selenides with additional anion-stabilizing groups, see Section (vi).]

(iv) Deprotonation of Vinyl Selenides RCH=CHSePh. Potassium di-isopropyl-amide is suitable[24] for making selenium-stabilized vinyl carbanions [reaction (2)], but not LDA, potassium hexamethyldisilazane, or BuLi. These carbanions should be used promptly, to minimize the ejection of phenylselenolate anion.[24] In the case where R = Me, allylic deprotonation occurs instead.[24] In other studies,[3,33,34] lithium amide bases (including LDA) were used for the deprotonation of vinyl selenides: LDA at −50 °C,[3] or at −78 °C,[33] (in the presence[33] or absence[33] of HMPA[28]) for phenyl vinyl selenide; LDA at −78 °C for *m*-(trifluoromethyl)phenyl vinyl selenide[3] and for compound (8)[3]; and LiTMP at −78 °C for compound (9).[3] [For deprotonation of an allenic selenide, see Section (v)].

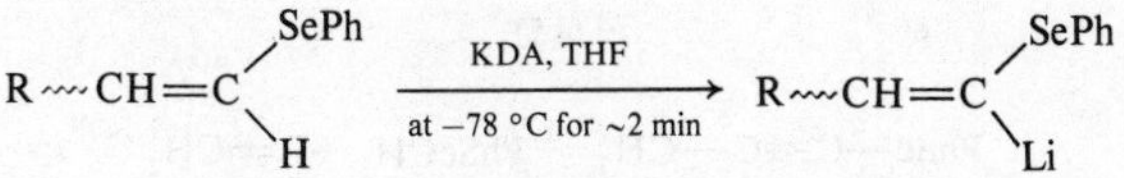

$$R \sim CH = C \overset{SePh}{\underset{H}{\Big\langle}} \quad \xrightarrow[\text{at } -78\,°C \text{ for } \sim 2 \text{ min}]{\text{KDA, THF}} \quad R \sim CH = C \overset{SePh}{\underset{Li}{\Big\langle}} \qquad (2)$$

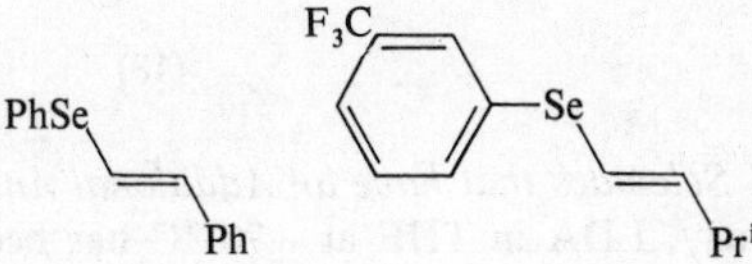

(8)   (9)

(v) Deprotonation of Allylic, Propargylic, and Allenic Selenides. Several examples in this category have been studied,[1,3,35] and LDA appears to be generally suitable except when the α-position carries an alkyl group[1,36] (Scheme 3[35,36]). Propargyl selenides (10) and (11) are deprotonated by LDA at −78 °C to yield the lithium salts (12) and (13), respectively.[3,37] Glyme may also be used as the solvent, at least for generating (12).[37] α-Methyl compounds (14) are not deprotonated by LDA.[23] The allenic carbanion (15) is produced[3] by the action of LDA, at −78 °C, on (16), (17), or (18), two equivalents of the base being needed in the latter case. [For related examples, see Section (vi)].

[30] D. Van Ende, W. Dumont, and A. Krief, *J. Organomet. Chem.*, 1978, **149**, C10.
[31] R. H. Mitchell, *J. Chem. Soc., Chem. Commun.*, 1974, 990.
[32] H. J. Reich and S. K. Shah, *J. Am. Chem. Soc.*, 1975, **97**, 3250.
[33] M. Sevrin, J. N. Denis, and A. Krief, *Angew. Chem., Int. Ed. Engl.*, 1978, **17**, 526.
[34] Ref. 23, footnote 21.
[35] H. J. Reich, R. E. Olson, and M. C. Clark, *J. Am. Chem. Soc.*, 1980, **102**, 1423.
[36] H. J. Reich, *J. Org. Chem.*, 1975, **40**, 2570.
[37] H. J. Reich and S. K. Shah, *J. Am. Chem. Soc.*, 1977, **99**, 263.

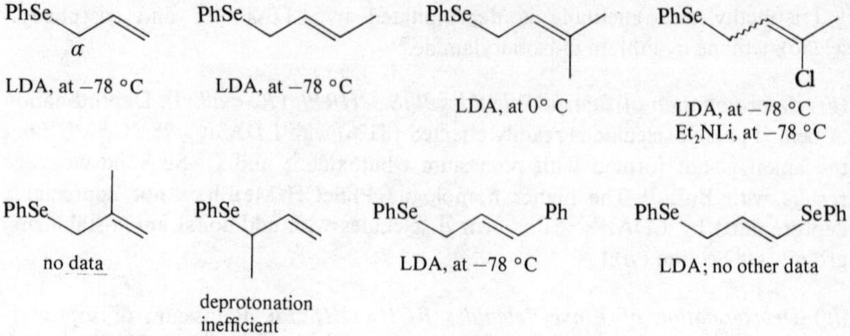

PhSe⌇⌇⌇ (α)   PhSe⌇⌇⌇   PhSe⌇⌇⌇   PhSe⌇⌇⌇
                                                                    Cl

LDA, at −78 °C   LDA, at −78 °C   LDA, at −78 °C
                                    LDA, at 0° C   Et$_2$NLi, at −78 °C

PhSe⌇⌇⌇   PhSe⌇⌇⌇   PhSe⌇⌇⌇Ph   PhSe⌇⌇⌇SePh

no data   LDA, at −78 °C   LDA; no other data

              deprotonation
              inefficient

Selenides and the conditions under which they were deprotonated

Scheme 3

PhSeCH$_2$C≡CH   PhSeCH$_2$C≡CMe   PhSeCH(Li)C≡CLi   PhSeCH(Li)C≡CMe
    (10)   (11)   (12)   (13)

        PhSeCH(Me)C≡CR   PhSeC=C=CH$_2$ PhSeCH=C=CH$_2$
            (14) | (16)
                                 Li
                                (15)

        PhSe—C≡C—CH$_3$ PhSeCH$_2$—C=CH$_2$
            (17) |
                                      Cl
                                     (18)

(*vi*) *Deprotonation of Selenides that have an Additional Anion-stabilizing Group;* *ArSeCH(R)X*. Generally, LDA in THF at −78 °C has been used; Table 1 lists the main examples.[3,20,23,32,35,36,38-47]

(*vii*) *Formation of Selenium-stabilized Carbanions by Michael Addition.* Another approach to selenium-stabilized carbanions is *via* β-addition of a carbanion to a vinyl selenide.[33,48,49] Under the correct conditions (dimethoxymethane or ether as solvent, and a reaction temperature of 0 °C), BuLi, Pr^{i}Li, and Bu^{t}Li react in a Michael fashion as shown in reaction (3).

[38] J. Lucchetti and A. Krief, *Tetrahedron Lett.*, 1978, 2693.
[39] K. B. Sharpless, R. F. Lauer, and A. Y. Teranishi, *J. Am. Chem. Soc.*, 1973, **95**, 6137.
[40] F. G. Bordwell, J. E. Bares, J. E. Bartmess, G. E. Drucker, J. Gerhold, G. J. McCollum, M. Van Der Puy, N. R. Vanier, and W. S. Matthews, *J. Org. Chem.*, 1977, **42**, 326.
[41] N. Petragnani and H. M. C. Ferraz, *Synthesis*, 1978, 476.
[42] H. J. Reich and S. K. Shah, *J. Org. Chem.*, 1977, **42**, 1773.
[43] K. Sachdev and H. S. Sachdev, *Tetrahedron Lett.*, 1977, 814; *ibid.*, 1976, 4223.
[44] B.-T. Gröbel and D. Seebach, *Chem. Ber.*, 1977, **110**, 852.
[45] P. A. Grieco and Y. Yokoyama, *J. Am. Chem. Soc.*, 1977, **99**, 5210.
[46] Y. Masuyama, Y. Ueno, and M. Okawara, *Chem. Lett.*, 1977, 835.
[47] J. V. Comasseto and N. Petragnani, *J. Organomet. Chem.*, 1978, **152**, 295.
[48] S. Raucher and G. A. Koolpe, *J. Org. Chem.*, 1978, **43**, 4252.
[49] T. Kauffmann, H. Ahlers, H.-J. Tilhard, and A. Woltermann, *Angew. Chem., Int. Ed. Engl.*, 1977, **16**, 710.

Table 1 *Conditions under which the compounds ArSeCH(R)X have been deprotonated to produce carbanions*

Selenide	Conditions	Ref.
PhSeCH$_2$CO$_2$R (R = alkyl)	LDA	38
	c-C$_6$H$_{11}$N(Li)Pr^i	39
p-ClC$_6$H$_4$SeCH$_2$CO$_2$Me	LDA	38
MeSeCH$_2$CO$_2$Me	LDA	38
PhSeCH$_2$COPh	LDA	20
	KH, DMSO	40
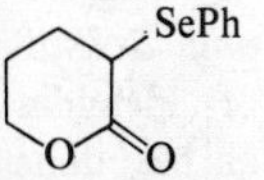 2-(phenylseleno)cyclohexanone (ring bearing SePh and C=O, O in ring)	LDA	38
PhSeCHRCO$_2^-$	LDA, at -78 °C (R = H) or 0—40 °C (R = Me)	23, 41
PhSeCH$_2$SiMe$_3$	Bu^{s}Li, TMEDA, hexane, at 25 °C	32, 42
	LDA, TMEDA, THF	43
m-CF$_3$C$_6$H$_4$SeCH$_2$SiMe$_3$	LiTMP, THF, at -40 °C (or Et$_2$O)	23, 42
	LDA	28, 42
m-CF$_3$C$_6$H$_4$SeCH$_2$OMe	LiTMP, at -78 °C	3, 23
PhSeCH$_2$OMe	LiTMP (Deprotonation incomplete)	23
PhSeCHPh | SiMe$_3$	Et$_2$NLi, at 0 °C LDA	32, 42 44
m-CF$_3$C$_6$H$_4$SeCHPh | SiMe$_3$	Et$_2$NLi, at -78 °C	42
PhSeCHSePh | SiMe$_3$	LDA, at 0 °C	44
PhSeCHCH=CH$_2$ | SiMe$_3$	Et$_2$NLi, at -78 °C	36
PhSe—CH=C(SePh)— SiMe$_3$	Et$_2$NLi	35
PhSeCHC$_6$H$_{13}$ | CN	LDA	45
PhSeCHR (R = H or alkyl) | CN	aq. NaOH, Bu$_4$N$^+$ I$^-$	46
PhSeCHR (R = H or alkyl) | (O)P(OEt)$_2$	BuLi, THF, at -78 °C	47
	NaH, THF–HMPA, at $+80$ °C	47

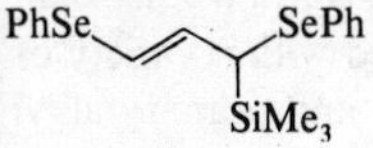

$$\text{BuLi} + \text{PhSeCH}{=}\text{CH}_2 \xrightarrow{\text{DME}} \text{BuCH}_2{-}\text{CHSePh} \qquad (3)$$

Applications of Selenium-stabilized Carbanions. The previous sections show that many selenides are not effectively deprotonated, and so the corresponding α-lithio-selenides must be prepared by other routes, such as cleavage of selenoketals (see Scheme 1) or of seleno-orthoesters. An alternative (but indirect) procedure is the low-temperature (usually at $-78\ °C$) deprotonation (by LDA) of selenoxides that are generated *in situ* at temperatures below $0\ °C$.[1,3,5] A further approach involves Michael addition to vinyl selenides.

Applications Based on C$-$Se *Bond Cleavage in Selenoketals and Seleno-orthoesters.* Primary allylic alcohols are available by homologation of aldehydes (Scheme 4).[50] In this method the selenoxide fragmentation $(19) \rightarrow (20)$ is best done with t-butyl hydroperoxide in the presence of alumina.

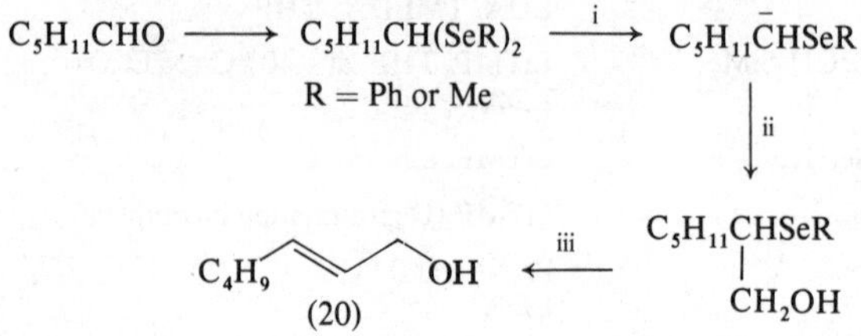

$$\text{C}_5\text{H}_{11}\text{CHO} \longrightarrow \underset{R = \text{Ph or Me}}{\text{C}_5\text{H}_{11}\text{CH(SeR)}_2} \xrightarrow{\ i\ } \text{C}_5\text{H}_{11}\bar{\text{C}}\text{HSeR}$$

Reagents: i, BuLi; ii, HCHO; iii, oxidation

(19)

Scheme 4

Cyclopropanone selenoketals such as $(21;\ R = \text{Ph or Me})$, which are prepared[29,51] by ring-closure processes, behave normally when treated with BuLi in THF[29,51] to give the corresponding α-lithio-selenides (22). These species are highly nucleophilic, but they are more basic than other selenium-stabilized carbanions.[29] Enhanced nucleophilicity towards carbonyl compounds, at least for $(22;\ R = \text{Me})$, is observed[52] by generating the carbanion in ether; in this solvent, Bu^tLi (rather than Bu^nLi) is required.[52] The α-lithio-selenides (22) have been treated with a variety of simple electrophiles, such as CO_2,[52] aldehydes,[29,52] ketones,[29,52] and primary alkyl halides.[53] In the latter case, the presence of HMPA is necessary in the alkylation step if acceptable yields are to be produced. Scheme 5 illustrates two typical reactions. The formation of (23) merits comment because the ketone employed is

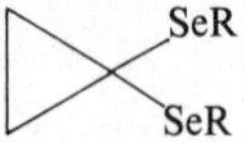

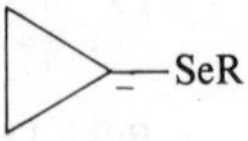

(21) (22)

[50] D. Labar, W. Dumont, L. Hevesi, and A. Krief, *Tetrahedron Lett.*, 1978, 1145.
[51] H. J. Reich, S. K. Shah, and F. Chow, *J. Am. Chem. Soc.*, 1979, **101**, 6648.
[52] S. Halazy and A. Krief, *J. Chem. Soc., Chem. Commun.*, 1979, 1136.
[53] S. Halazy and A. Krief, *Tetrahedron Lett.*, 1979, 4233.

readily enolized. Hydroxy-selenides such as (23) yield olefins under a variety of conditions and, for example, (23) is converted into (24) in the presence of phosphorus tri-iodide and triethylamine.[52] Selenides such as (25) also afford regiochemically pure olefins by the sequence shown.[53] The transformation (25) → (26) can be carried through the phenyl series, starting with (21; R = Ph),[53] and the methodology represented by Scheme 5 provides a simple route to α-olefins, starting from alkyl halides (*e.g.* $C_8H_{17}CH_2Br$) and $MeSeCH_2Li$.[53]

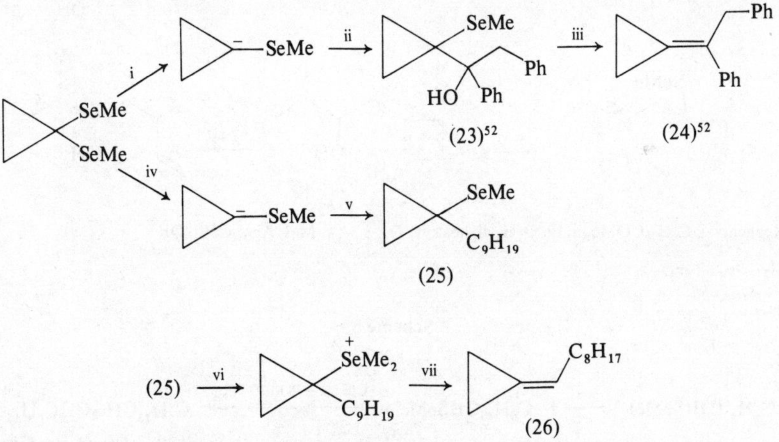

Reagents: i, Bu^{t}Li, Et$_2$O; ii, PhCOCH$_2$Ph; iii, PI$_3$, Et$_3$N; iv, BuLi, THF; v, C$_9$H$_{19}$Br, HMPA; vi, MeSO$_3$F; vii, excess Bu^{t}OK

Scheme 5

Cyclobutanone selenoketals (27; R = Ph or Me) behave in a similar way.[54] Treatment with BuLi in THF generates the lithium salts (28), and these have been quenched with simple electrophiles (*e.g.* undecanal, undecan-2-one, 1-halogeno-nonane, and 1,2-epoxyoctane).[54] The reactions constitute a useful route to dienes (Scheme 6).

Bis(seleno)ketals are converted into olefins by copper(I)-mediated coupling of the derived α-lithio-selenides,[55] as shown in Scheme 7. Bis(phenylseleno)ketals may also be used, and two *different* α-lithio-selenides can be coupled.

[54] S. Halazy and A. Krief, *Tetrahedron Lett.*, 1980, **21**, 1997.
[55] J. Lucchetti, J. Remion, and A. Krief, *C.R. Hebd. Seances Acad. Sci., Ser. C*, 1979, **288**, 553.

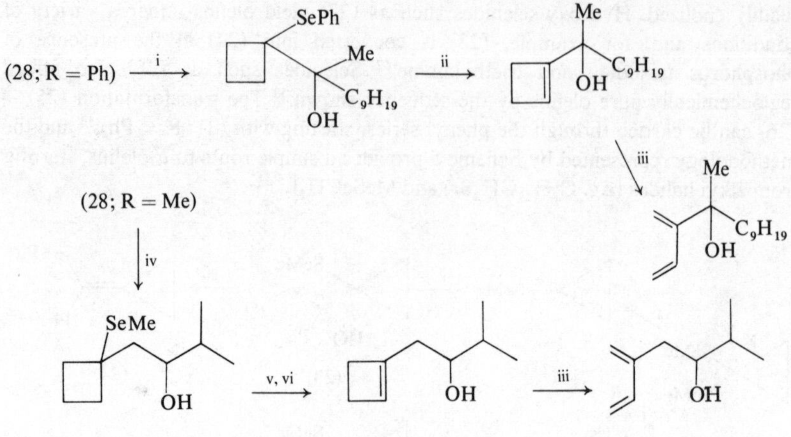

Reagents: i, $C_9H_{19}COMe$; ii, Bu^tO_2H; iii, heat; iv, O ; v, MeI, Ag^+; vi, Bu^tOK

Scheme 6

$$C_6H_{13}CH(SeMe)_2 \longrightarrow C_6H_{13}\bar{C}HSeMe \xrightarrow{\text{CuI at } -78 \text{ to } 20\ ^\circ\text{C}} C_6H_{13}CH{=}CHC_6H_{13}$$

$E + Z$

Scheme 7

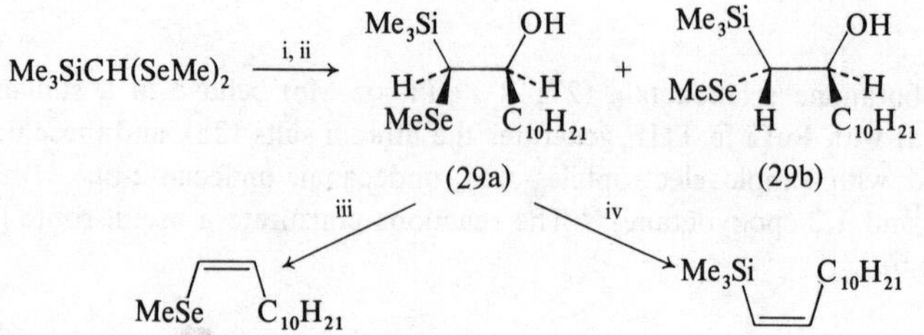

Reagents: i, BuLi; ii, $C_{10}H_{21}CHO$; iii, Bu^tOK; iv, $POCl_3$, Et_3N

Scheme 8

Isomeric α-silyl-β-hydroxy-selenides, such as (29), prepared by the selenoketal–BuLi methodology, are often easily separated,[56] and treatment with Bu^tOK yields olefins (*i.e.* vinyl selenides) while exposure to $POCl_3$ and Et_3N or to $SOCl_2$ and Et_3N affords vinylsilanes (Scheme 8). α-Silyl-β-hydroxy-selenides can also be converted into vinyl bromides and into ketones.[56]

Selenoketals also constitute starting materials for the preparation of oxetans and tetrahydrofurans[57] (Scheme 9).

[56] W. Dumont, D. Van Ende, and A. Krief, *Tetrahedron Lett.*, 1979, 485.
[57] M. Sevrin and A. Krief, *Tetrahedron Lett.*, 1980, **21**, 585.

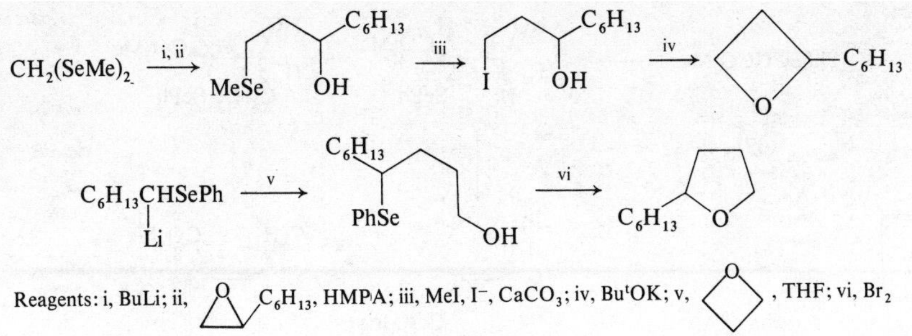
CH2(SeMe)2
i, ii
C6H13
MeSe
OH
iii
C6H13
I
OH
iv
C6H13
O
C6H13CHSePh
Li
v
C6H13
PhSe
OH
vi
C6H13
O
Reagents: i, BuLi; ii, C6H13, HMPA; iii, MeI, I−, CaCO3; iv, ButOK; v, THF; vi, Br2

Scheme 9

Full experimental details are available[23] for the sequence $PhSeCH_2Li$ + ketone or aldehyde → *β*-hydroxy-selenide → olefin. The action of BuLi on (30) generates (31),[23] and the vinyl carbanion (32) is available in the same way from (33).[58]

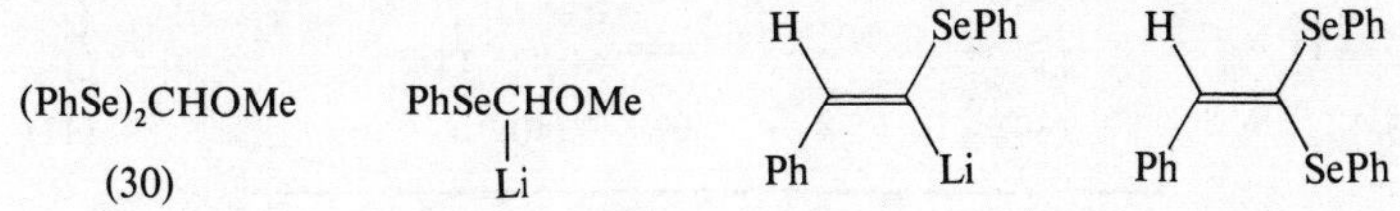
(PhSe)2CHOMe
PhSeCHOMe
Li
H
SePh
Ph
Li
H
SePh
Ph
SePh

(30) (31) (32) (33)

Carbanions produced by the reaction of seleno-orthoesters with BuLi have been treated with common electrophiles (*e.g.* Me_3SiCl, propylene oxide, primary alkyl halides, aldehydes, and ketones).[27] The syntheses of (34) and (35) are representative.[27] The anion (36) undergoes 1,2-addition with enones, but, in the presence of HMPA, the thermodynamically more stable 1,4-adduct is obtained.[59]

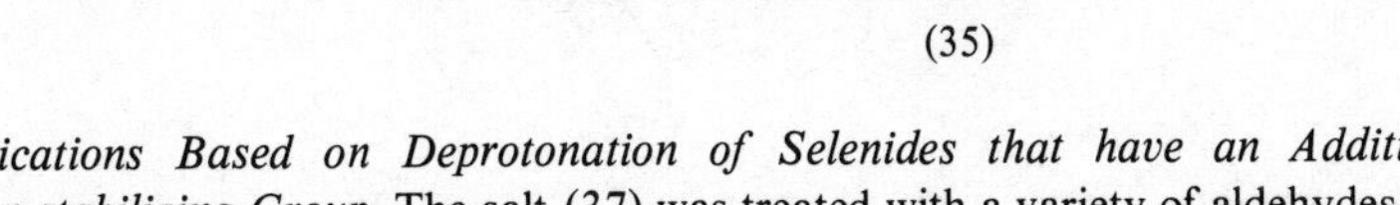
(MeSe)3CMe
(MeSe)2CMe
C6H13CHO
CH(OH)C6H13
(MeSe)2C
Me

(36) (34)

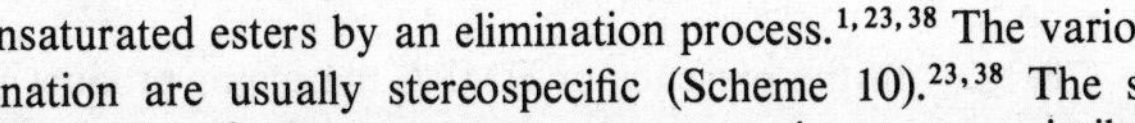
(MeSe)3CH
(MeSe)2CH
Me3SiCl
(MeSe)2CHSiMe3

(35)

Applications Based on Deprotonation of Selenides that have an Additional Anion-stabilizing Group. The salt (37) was treated with a variety of aldehydes and, where the resulting *β*-hydroxy-selenides[38] were separable, they easily afforded *αβ*-unsaturated esters by an elimination process.[1,23,38] The various methods for the elimination are usually stereospecific (Scheme 10).[23,38] The salts (38),[38] (39),[38] (40),[38] and (41)[23] (all generated by deprotonation) behave similarly, and the methyl analogue of the last compound has been used[41] for the construction of *α*-methylene lactones (Scheme 11). Salts (37), (39), and (40) give mainly 1,2-adducts with

[58] B.-T. Gröbel and D. Seebach, *Chem. Ber.*, 1977, **110**, 867.
[59] J. Lucchetti, W. Dumont, and A. Krief, *Tetrahedron Lett.*, 1979, 2695.

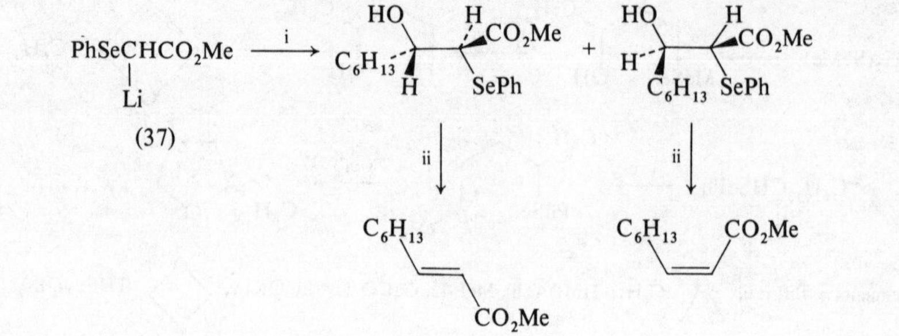

Reagents: i, C₆H₁₃CHO; ii, SOCl₂, Et₃N

Scheme 10

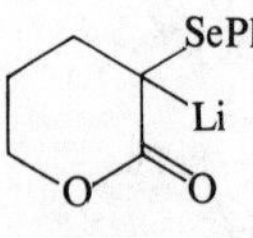

(38) 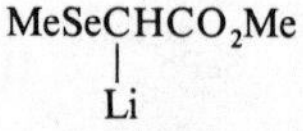   (39) 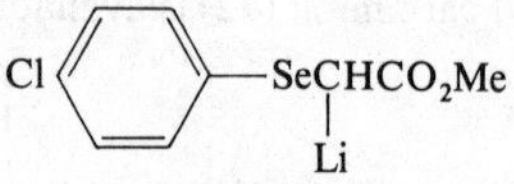   (40) 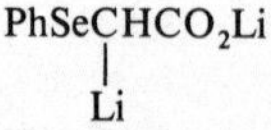   (41)

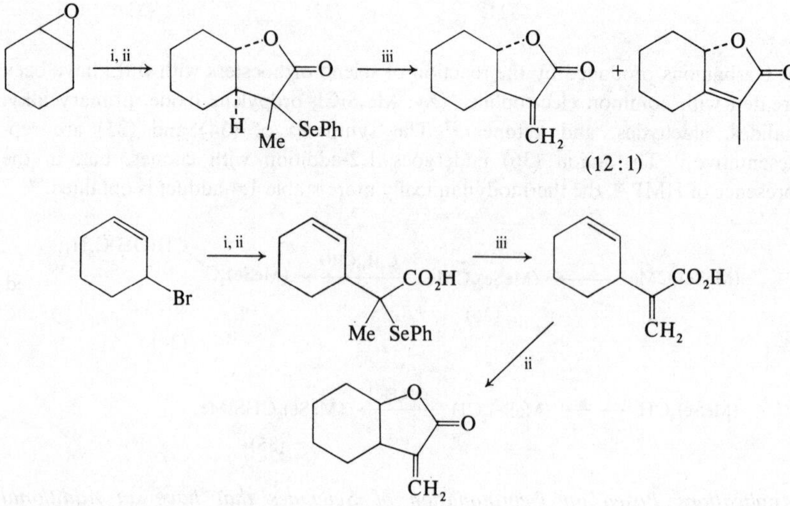

Reagents: i, PhSeC(Me)LiCO₂Li; ii, H⁺; iii, oxidation

Scheme 11

enones in THF at −78 °C.[60] In the presence of HMPA the major product is the 1,4-adduct.[60]

The phosphorus reagent (42) can be alkylated[47] with alkyl halides. The products (43) can be deprotonated (by NaH or BuLi) and then used in Wittig reactions to afford vinyl selenides as *E/Z* mixtures.[47]

[60] J. Lucchetti and A. Krief, *Tetrahedron Lett.*, 1978, 2697.

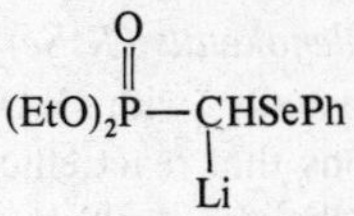

$$(EtO)_2\overset{\displaystyle O}{\overset{\|}{P}}-\underset{\displaystyle Li}{CHSePh}$$

(42)

$$(EtO)_2\overset{\displaystyle O}{\overset{\|}{P}}-\underset{\displaystyle R}{CHSePh}$$

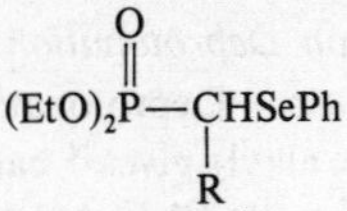

(43)

Prenylation of (4) (see p. 112), generated by the action of LDA,[20] has been mentioned in the section on ylides. The α-lithio-selenide (44),[23] made with LiTMP, has been used to prepare enol ethers (Scheme 12).[23] Deprotonation (by LDA) of the allylic selenide (45)[35] is the first step in the synthesis of the silyl ketone (46).[35] The α,γ-dilithiopropargyl selenide (12) [see p. 116] has also been converted into an

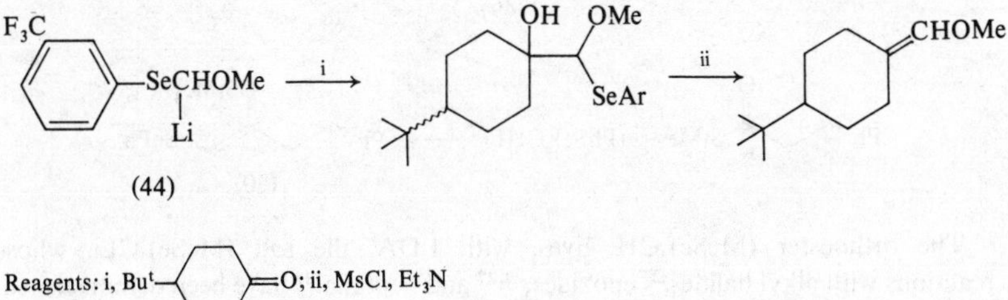

Reagents: i, Bu^t-⟨cyclohexyl⟩=O; ii, MsCl, Et$_3$N

Scheme 12

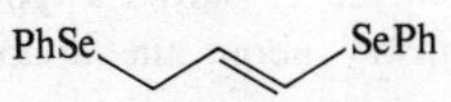

(45)

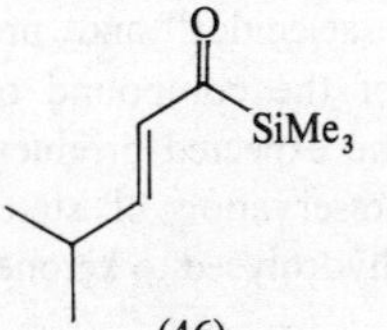

(46)

αβ-unsaturated silyl ketone[61] and has been elaborated into a metabolite of mould-damaged sweet potatoes.[62] The synthesis is based on the finding[37] that propargyl selenoxides rearrange to α-phenylseleno-enones at a low temperature (*ca.* −30 °C), as shown in reaction (4).

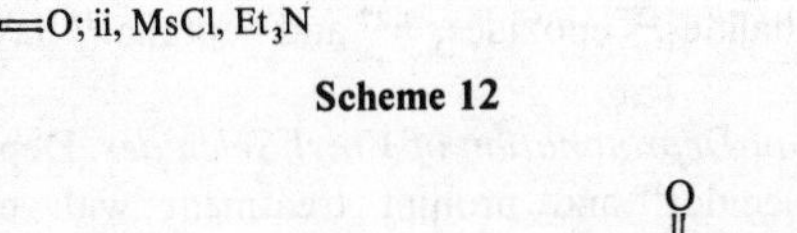

(4)

Deprotonation (by LDA) of (47) followed by reaction with benzaldehyde, formaldehyde, or propanal gives (48; R = Ph, H, or Et).[44]

$(PhSe)_2CHSiMe_3$ $(PhSe)_2C=CHR$

(47)   (48)

[61] H. J. Reich, J. J. Rusek, and R. E. Olson, *J. Am. Chem. Soc.*, 1979, **101**, 2225.
[62] H. J. Reich, P. M. Gold, and F. Chow, *Tetrahedron Lett.*, 1979, 4433.

Applications Based on Deprotonation of Selenoketals $(R^1Se)_2CHR^2$ $(R^2 = alkyl)$ *and Seleno-orthoesters.* Deprotonation, usually with KDA, of selenoketals $(PhSe)_2CHR^2$ $(R^2 = alkyl)$ gives[24] carbanions that react efficiently with common electrophiles [primary alkyl[24,27] and benzyl halides,[24] aldehydes,[24,27] ketones,[24,27] and (for $R^2 = $ alkyl or H) epoxides[24,27,51]]. A few of the resulting selenoketals have been hydrolysed under very mild conditions.[24] Carbanions derived from selenoketals undergo intramolecular reactions [*e.g.* (49)][29] when appropriately substituted, and can be used in the synthesis of silyl enol ethers, *e.g.* (50).[61]

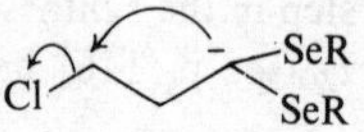

(49)

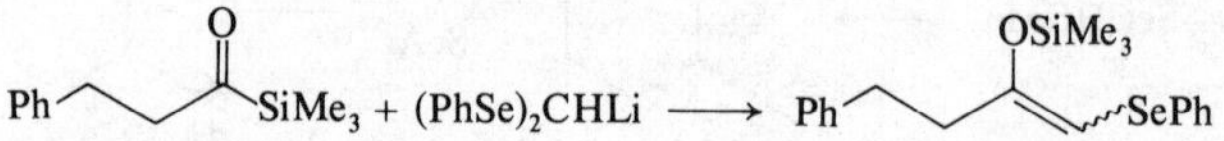

(50)

The orthoester $(MeSe)_3CH$ gives, with LDA, the salt $(MeSe)_3CLi$, whose reactions with alkyl halides,[27] epoxides,[29,27] and heptanal[27] have been demonstrated.

Applications Based on Deprotonation of Vinyl Selenides. Deprotonation (by KDA) of phenyl vinyl selenide[24] and prompt treatment with electrophiles such as representatives of the compound classes alkyl halides, aldehydes, enones, and epoxides gives the expected products.[63] Scheme 13 shows a typical example and illustrates the preservation of stereochemistry about the double bond.[24] Vinyl selenides can be hydrolysed to ketones.[24]

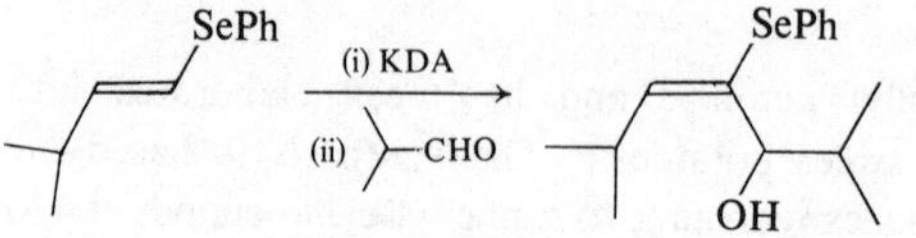

Scheme 13

Applications Based on Deprotonation of Benzyl Selenides PhSeCHRPh ($R = H$ or *alkyl*). Only a low degree of stereoselection is observed[23] in the reaction of PhSeCH(Li)Ph with benzaldehyde and propanal. The derived alcohols of (RS,SR) and (RR,SS) configuration were converted into olefins (using MsCl and Et$_3$N), the former giving the (Z)-olefin and the latter the (E)-isomer.[23] In further experiments the salt PhSeCH(Li)Ph was alkylated with several alkyl halides and with propylene oxide.[23]

Applications Based on Michael Addition to Phenyl Vinyl Selenide. The α-lithio-selenides (51), (52), and (53), which were generated[48] by treating phenyl vinyl selenide with the appropriate organolithium in dimethoxymethane[48] or in ether,[33,48]

[63] See also ref. 8.

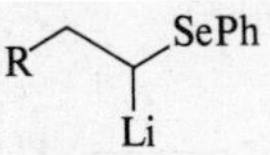

(51) R = Bu^n
(52) R = Pr^i
(53) R = Bu^t

were quenched with common electrophiles chosen from the following types: alkyl halides (in the presence of HMPA), PhSeBr, Me$_3$SiCl, aldehydes, ketones, PhCN, and D$_2$O. The C—C connective route[48] shown in Scheme 14 inserts a two-carbon unit between the organolithium compound and the electrophile.

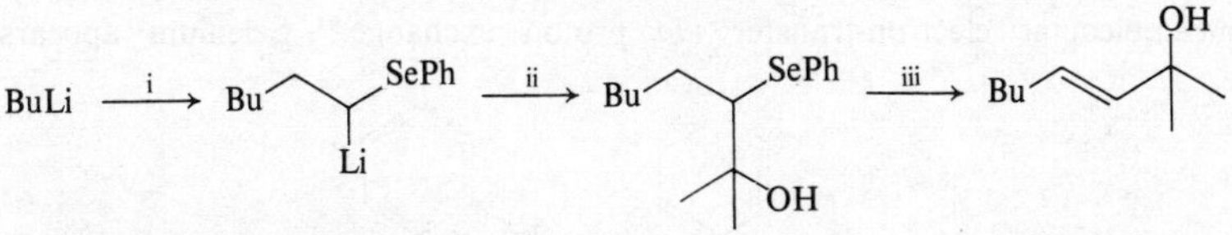

Reagents: i, H$_2$C=CHSePh; ii, Me$_2$CO; iii, oxidation

Scheme 14

Applications Based on α-Lithio-selenoxides. Full details have appeared[51] on the uses of α-lithio-selenoxides, generated by low-temperature deprotonation of selenoxides with LDA. They are probably less nucleophilic than the corresponding α-lithio-selenides. Their reactions leading to olefins, allylic alcohols, and (less well explored) enones are summarized in Scheme 15, and they have also been applied in the synthesis[61] of silyl enol ethers (Scheme 16).

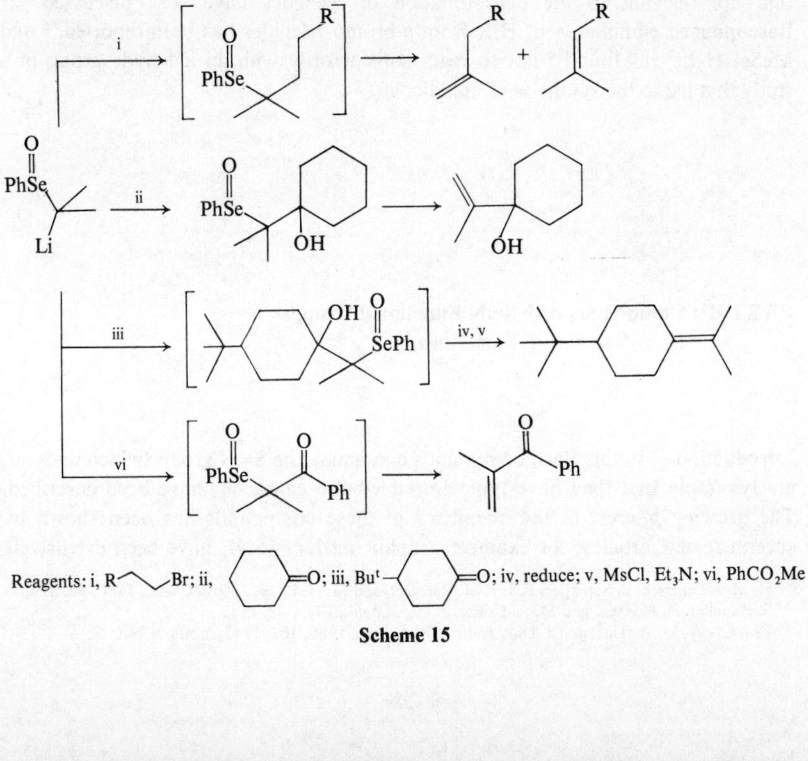

Reagents: i, R ⌢ Br; ii, ⬡=O; iii, Bu^t-⬡=O; iv, reduce; v, MsCl, Et$_3$N; vi, PhCO$_2$Me

Scheme 15

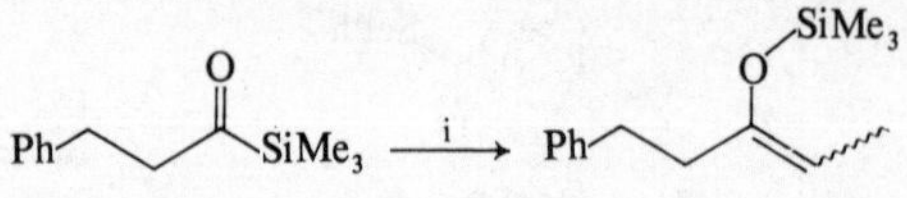

Reagent: i, PhSeCHLiMe
Scheme 16

Miscellaneous Studies on Selenium-stabilized Carbanions.—Generation of the carbanion (54) led to products derived from (55), which is formed by an intramolecular electron-transfer *via* proton exchange.[64] Selenium appears to

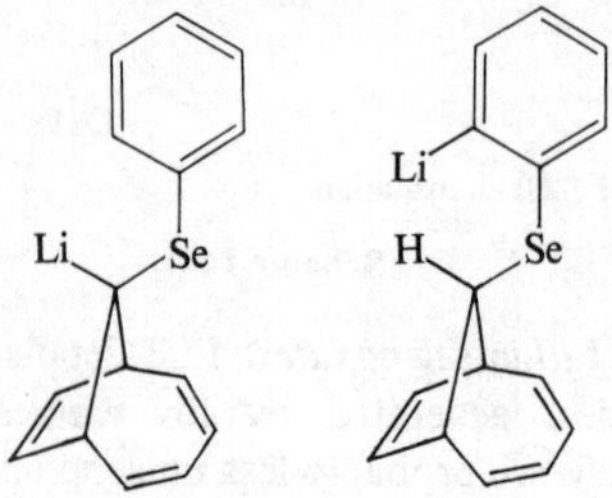

(54)   (55)

stabilize an adjacent carbanion less effectively than sulphur.[3,23] Some pK_a values that are relevant to the deprotonation of selenides have been discussed.[23,51] Base-induced elimination of HBr from β-bromo-selenides has been reported,[65] and MeSeCH$_2$Li was found[66] not to react satisfactorily with an aldehyde group in a study that led to the synthesis of aphidicolin.

PART III: Compounds with S=N Functional Groups
        by S. Oae and N. Furukawa

Introduction.—In this Part, compounds containing the S=N group (which does not always imply that they have typical double-bond character) have been described. The growing interest in the chemistry of these compounds has been shown by several review articles; for example, sulphur nitrides (SN)$_x$ have been extensively

[64] H. M. J. Gillissen, P. Schipper, P. J. J. M. Van Ool, and M. H. Buck, *J. Org. Chem.*, 1980, **45**, 319.
[65] S. Raucher, R. Hansen, and M. A. Colter, *J. Org. Chem.*, 1978, **43**, 4885.
[66] E. J. Corey, M. A. Tius, and J. Das, *J. Am. Chem. Soc.*, 1980, **102**, 1742, footnote 24*d*.

investigated because of their interesting structures[1-4] and their physical properties, such as superconductivity.[5,6] Cyclic sulphur—nitrogen compounds have also been summarized.[7-9] The chemistry of sulphimides,[10] sulphoximides,[11] and sulphodiimides[12] has also been reviewed. The biologically important role of methionine sulphoximides has been described.[13]

1 Di-co-ordinate Sulphur

Sulphinyl-amines and -amides.—*Structure.* X-Ray crystallographic analyses[14,15] and infrared,[16] nitrogen-15 n.m.r.,[17] and dipole-moment measurements[18] of several N-sulphinylanilines and N-sulphinyl-sulphonamides have been performed. These spectral analyses have revealed that the compounds (1) and (2; Ar = Ph) have *syn*-coplanar configurations around the N=S linkage; this is in accordance with the previous observations except in the case of the mesityl analogue (2; Ar = 2,4,6-Me$_3$C$_6$H$_2$), in which the mesityl ring is perpendicular to the N=S=O plane.

$$\text{PhSO}_2\text{N=S=O} \qquad \text{ArN=S=O} \qquad \text{RN=S=O}$$

$$(1) \qquad\qquad (2) \qquad\qquad (3)\ \text{R = Me, Et, MeS, or Me}_2\text{N}$$

Preparation and Reactions. The first preparation and characterization of potassium thiazate (KNSO) has been reported.[19] Perfluoroaromatic N-sulphinyl compounds (3; R = C$_6$F$_5$S or C$_6$F$_5$SO) have been prepared by treating the corresponding perfluoroaromatic sulphur acid amides C$_6$F$_5$S(O)$_n$NH$_2$ (n = 0 or 1) with SOCl$_2$ or the sulphenyl chloride with Me$_3$SiNSO.[20] Treatment of the sulphinyl compounds (3; R = MeS or Me$_2$N) with Me$_3$O$^+$ BF$_4^-$ yields the corresponding N- or S-methylated salts, which undergo [4 + 2]-cycloaddition reactions with dienes to afford the

[1] G. B. Street and W. D. Gill, *NATO Conf. Ser.*, [*Ser*] 6, 1978, **1**, 301 (*Chem. Abstr.* 1979, **91**, 67 639).
[2] T. Nakajima, *Yoyuen*, 1979, **22**, 7 (*Chem. Abstr.* 1979, **91**, 116 460).
[3] C. Bernard and G. Robert, *Bull. Soc. Chim. Fr., Part 2*, 1978, 395.
[4] D. G. Stojakovic, *Glas. Hem. Drus. Beograd*, 1978, **43**, 775 (*Chem. Abstr.* 1979, **91**, 166 870).
[5] M. M. Labes, P. Love, and L. F. Nichols, *Chem. Rev.*, 1979, **79**, 1.
[6] A. J. Banister, *Phosphorus Sulfur*, 1979, **6**, 421.
[7] A. J. Banister, *Phosphorus Sulfur*, 1978, **5**, 147.
[8] I. Haiduc, *Method. Chim.*, 1978, **7**, 789.
[9] H. W. Roesky, *Adv. Inorg. Chem. Radiochem.*, 1979, **22**, 239.
[10] (*a*) S. Oae, 'Organic Chemistry of Sulfur,' Plenum Press, New York, 1977, p. 383; (*b*) A. Kucsman and I. Kapovits, *Phosphorus Sulfur*, 1977, **3**, 9; (*c*) N. Furukawa, *Kagaku No Ryoiki*, 1978, **32**, 31; (*d*) R. Appel and J. Kohnke, *Method. Chim.*, 1978, **7**, 743; (*e*) C. R. Johnson, in 'Comprehensive Organic Chemistry' Vol. 3, ed. D. N. Jones, Pergamon Press, Oxford, 1979, p. 215.
[11] Ref. 10*e*, p. 233.
[12] Ref. 10*e*, p. 237.
[13] A. Meister, *Enzyme-Act. Irreversible Inhibitors, Proc. Int. Symp.*, 1978, 187 (*Chem. Abstr.* 1980, **92**, 71 523).
[14] B. Degerer and A. Gieren, *Naturwissenschaften*, 1979, **66**, 470.
[15] G. Deleris, C. Courseille, J. Kowalski, and J. Dunogues, *J. Chem. Res. (S)*, 1979, 122.
[16] R. Meiji, A. Oskman, and D. J. Stufkens, *J. Mol. Struct.*, 1979, **51**, 37.
[17] I. Yavari, J. S. Staral, and J. D. Roberts, *Org. Magn. Reson.*, 1979, **12**, 340.
[18] W. Waclawek and C. Puchala, *Bull. Acad. Pol. Sci., Ser. Sci. Chim.*, 1978, **26**, 337.
[19] D. A. Armitage and J. C. Brand, *J. Chem. Soc., Chem. Commun.*, 1979, 1078.
[20] I. Glander and A. Golloch, *J. Fluorine Chem.*, 1979, **14**, 403.

corresponding adducts (4)[21] (see Vol. 5, p. 102). Trifluoromethanesulphonyl sulphinylamide, CF_3SO_2NSO, affords the dimer (5) upon treatment with SO_3.[22]

(4) R = $\overset{+}{S}Me_2$ or $\overset{+}{N}Me_3$ (5)

Several other *N*-sulphinyl-amines or -amides can be prepared or generated, without isolation, and added to olefins,[23] ketones,[24] or enamines and ynamines.[25] Similarly, *N*-sulphinyl-sulphonamides (1) and sulphinyl-amines (2; Ar = Ph or *p*-$NO_2C_6H_4$) react with hexafluoroacetone or polyfluoro-olefins in aprotic solvents in the presence of CsF [see reactions (1) and (2)], affording, directly, the imino-compounds; these are undoubtedly formed *via* initial formation of the [2 + 2]-cyclo-adducts.[26]

$$(CF_3)_2CO + ArNSO \xrightarrow{\text{CsF, MeCN}} (CF_3)_2C=NAr \qquad (1)$$
$$(2)$$

$$(CF_3)_2C=CF_2 + PhSO_2NSO \xrightarrow[\text{(ii) HCl}]{\text{(i) CsF, MeCN}} (CF_3)_2CHCF=NSO_2Ph \qquad (2)$$
$$(1)$$

Isocyanides are prepared in high yields by treating various aliphatic, aromatic, and heterocyclic sulphinylamines with dichlorocarbene (generated under phase-transfer conditions), using KOH as a base together with 18-crown-6 or dibenzo-18-crown-6 in benzene solution.[27] Aliphatic *N*-sulphinyl-amines can be converted into the corresponding primary amines *via* a [3,3]sigmatropic rearrangement of their α-allylated derivatives. Allylic sulphinyl-amides are formed initially, as shown in Scheme 1.[28]

New metal–R insertion complexes between $[(\eta^5\text{-}C_5H_5)Fe(CO)_2R]$ (R = alkyl or aryl) and (1) or sulphur di-imides have been reported.[29,30] The photolysis of *N*-phenyl-*N'*-sulphinylhydrazine in aromatic solvents generates phenyl radicals; these undergo addition to the solvents, affording biphenyls.[31]

N-Thiosulphinyl-amines (6)[32] or -anilines (7; $R^2 = N=S=S$)[33] have been prepared

[21] M. Rössert, W. Kraus, and G. Kresze, *Tetrahedron Lett.*, 1978, 4669.
[22] H. W. Roesky, R. Aramaki, and L. Schoenfelder, *Z. Naturforsch., Teil B*, 1978, **33**, 1072.
[23] J. A. Kloek and K. L. Leschinsky, *J. Org. Chem.*, 1980, **45**, 721.
[24] J. E. Semple and M. M. Joullié, *J. Org. Chem.*, 1978, **43**, 3066.
[25] T. Nagai, T. Shingaki, M. Inagaki, and T. Oshima, *Bull. Soc. Chem. Jpn.*, 1979, **52**, 1102.
[26] Yu. V. Zeifman, E. G. Ter-Gabriélyan, D. P. Del'tsova, and N. P. Gambaryan, *Izv. Akad. Nauk SSSR, Ser. Khim.*, 1979, 396.
[27] R. Mayer, R. Beckert, and G. Domschke, Ger. P. 128 531 (1978).
[28] F. M. Schell, J. P. Carter, and C. Wiaux-Zamar, *J. Am. Chem. Soc.*, 1978, **100**, 2894.
[29] R. G. Severson and A. Wojcicki, *J. Organomet. Chem.*, 1978, **149**, C66.
[30] R. G. Severson, T .W. Leung, and A. Wojcicki, *Inorg. Chem.*, 1980, **19**, 915.
[31] G. Deluca, G. Renzi, and A. Pizzabiocca, *Chem. Ind.* (*London*), 1979, 899.
[32] N. Soma, S. Moriyama, T. Yoshioka, and T. Nishimura, Japan. Kokai 78 124 273.
[33] Y. Inagaki, R. Okazaki, and N. Inamoto, *Bull. Chem. Soc. Jpn.*, 1979, **52**, 1998.

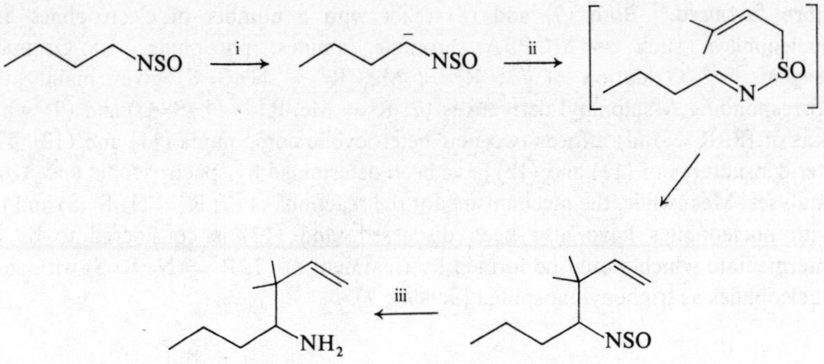

Reagents: i, Bu^{t}OK, DME, at 0 °C; ii, [structure] Br; iii, HCl

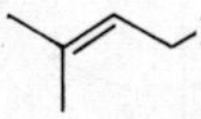

Scheme 1

and both their physical and their chemical properties extensively studied. Among the compounds (7; R^1 = Me, R^2 = N=S=S), (7; R^1 = Me, R^2 = N=S=O), and (7; R^1 = Me, R^2 = Br), the dipole moment of the N=S=O group is the largest, while the charge separation of the N=S=S group is about half that of the N=S=O group.[34,35]

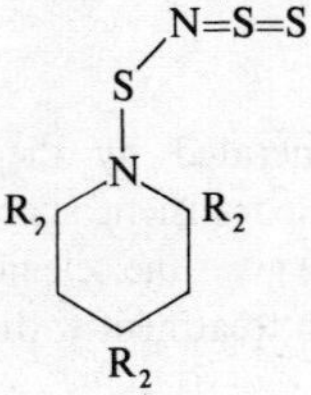

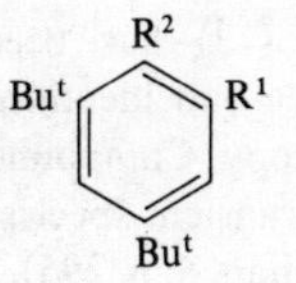

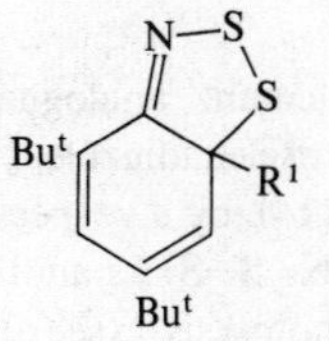

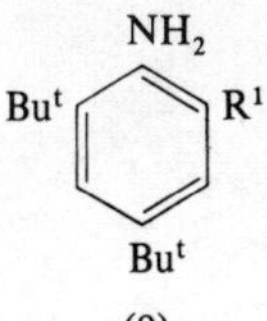

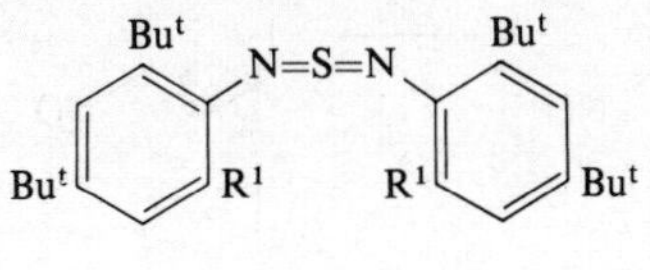

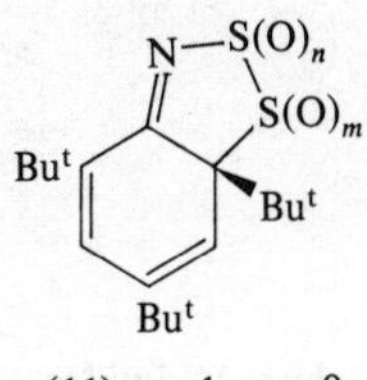

Thermolysis of (7; R^2 = N=S=S) affords a new type of heterocycle (8) together with the parent aniline (9), while photolysis of (7; R^2 = N=S=S) gives (9) and the sulphur di-imide (10).[36] The compound (7; R^1 = Bu^t, R^2 = N=S=S) has been found to be in equilibrium with (8; R^1 = Bu^t), and the thermodynamic parameters of the equilibrium (which were determined by n.m.r. spectroscopy) indicate that (7) is the

[34] Y. Inagaki, R. Okazaki, and N. Inamoto, *Heterocycles.*, 1978, **9**, 1613.
[35] Y. Inagaki, R. Okazaki, N. Inamoto, and T. Shimozawa, *Chem. Lett.*, 1978, 1217.
[36] Y. Inagaki, R. Okazaki, and N. Inamoto, *Bull. Chem. Soc. Jpn.*, 1979, **52**, 2002.

more favoured.[37] Both (7) and (8) react with a number of electrophiles and nucleophiles, such as MCPBA, bromine, amines, phosphines, or Grignard reagents.[38,39] Oxidation of (7; R^1 = Me, R^2 = N=S=S) gives mainly the corresponding N-sulphinyl derivatives (7; R^1 = Me, R^2 = N=S=O) and (9), while that of (8; R = Bu^t) affords two new heterocyclic compounds (11) and (12). The steric structures of (11) and (12) have been determined by spectroscopic and X-ray analyses. Meanwhile, the mechanisms for the reactions of (7; R^2 = N=S=S) and (8) with nucleophiles have also been discussed, and (13) is considered to be an intermediate which would be formed by treatment of (7; R^2 = N=S=S) with such nucleophiles as triphenylphosphine (Scheme 2).

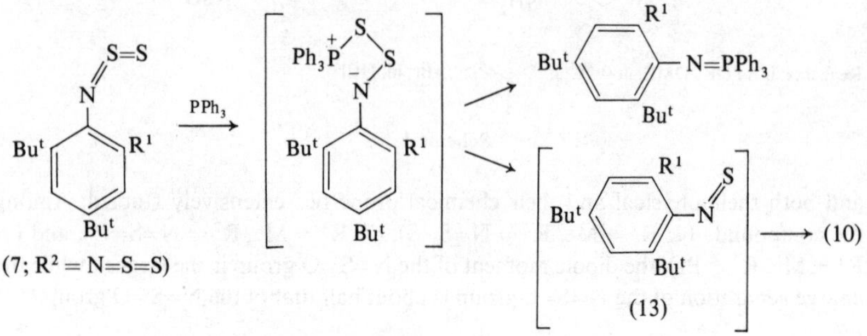

Scheme 2

The selenium analogue of (13) has been generated by the photolysis of 2,1,3-benzoselenadiazole 1-oxide and identified as a transient species, having the structure (14), by u.v. spectroscopy. Compound (14) gives the selenium analogue of (7; R^2 = N=S=S) as another transient species upon treatment with Se$_8$, generated by flash photolysis[40] (see also Chap. 6, p. 295).

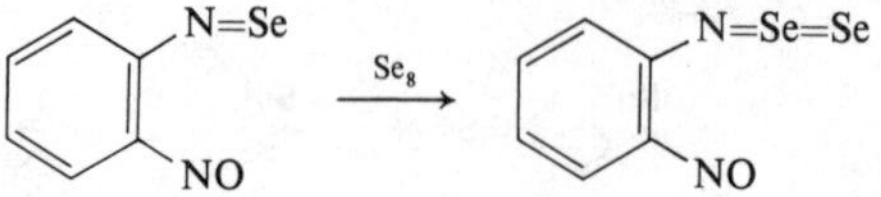

(14)

Sulphur Di-imides.—*Preparation and Properties of* $(SN)_x$. This and related sulphur–nitrogen compounds have created interest both in their preparation and their physical properties (see Vol. 5, p. 103). Pyrolysis of S_4N_4 gives either $(SN)_x$ or S_2N_2, depending on the reaction conditions.[41,42] The physical and spectroscopic

[37] Y. Inagaki, R. Okazaki, N. Inamoto, K. Yamada, and H. Kawazura, *Bull. Chem. Soc. Jpn.*, 1979, **52**, 2008.
[38] Y. Inagaki, R. Okazaki, and N. Inamoto, *Bull. Chem. Soc. Jpn.*, 1979, **52**, 3615.
[39] Y. Inagaki, T. Hosogai, R. Okazaki, and N. Inamoto, *Bull. Chem. Soc. Jpn.*, 1980, **53**, 205.
[40] C. L. Pedersen, *Tetrahedron Lett.*, 1979, 745.
[41] P. Love, G. Myer, H. I. Kao, M. M. Labeo, W. R. Junker, and C. Elbaum, *Ann. N.Y. Acad. Sci.*, 1978, **313**, 745.
[42] R. D. Smith, *J. Chem. Soc., Dalton Trans.*, 1979, 478.

properties of $(SN)_x$, such as thermopower,[43] mass spectroscopy,[44] optical spectroscopy,[45] molecular-beam electric deflection analysis,[46] e.p.r. spectroscopy,[47] electrical conductivity,[48] and superconductivity,[49] have been reported. The structure and properties of halogenated $(SN)_x$ have also been investigated.[50-52]

Tetrasulphur Tetranitride. This compound has been used as a sulphur-transfer agent, reacting, for example, with ketones or acetylenes to afford thiadiazoles[53,54] (see also Chap. 6, p. 292). Furthermore, it reacts with triphenylphosphine, yielding $(Ph_3P=N)_2 \cdot S_4N_4$, which has been characterized by X-ray spectroscopic analysis.[55] The reaction with $[Pt(PPh_3)_4]$ gives some novel platinum complexes, such as $[Pt(N_4S_4)(PPh_3)_2]$.[56] Brominations of S_4N_4 afford either $(SNBr_{0.4})_x$ (in the gas-phase reaction), $S_4N_3Br_3$ (in the liquid-phase reaction), or S_4N_3Br and $CS_3N_2Br_2$ (in solution in CS_2).[57] A new sulphur nitride, S_5N_6, has been prepared and its X-ray, Raman, and u.v.—visible spectra have been presented.[58,59] Some other sulphur nitride derivatives, such as S_4N_5X (X = Cl, F, or $SbCl_6$),[60] S_4N_3X (X = $TeBr_5$, $TeBr_6$, or $SbCl_4$),[61] and $Pb(S_2N_2)$,[62] have been prepared. The structures of $[Bu_4N]^+[S_3N_3]^-$ and $[(Ph_3P)_2N]^+[S_4N]^-$ have been determined by X-ray spectroscopic analysis; both of the anions $[S_3N_3]^-$ and $[S_4N]^-$ were shown to have essentially planar structures.[63,64]

Preparation of Other Sulphur Di-imides. Pyrolysis of N-sulphinylaniline has been found to give sulphur di-imides in the presence of a nickel(o) catalyst.[65] Sulphur di-imides that have the general structure $RN=S=NR$ (R = Me_3Si, Me_3Sn, or C_6F_5)

[43] L. J. Azvedo, P. M. Chikin, W. G. Clark, W. W. Fuller, and J. Hammann, *Phys. Rev. B: Condens. Matter*, 1979, **22**, 4450.
[44] N. Ueno, K. Kugita, O. Koga, S. Suzuki, K. Yoshino, and K. Kaneto, *Jpn. J. Appl. Phys.*, 1979, **18**, 1597.
[45] Z. Iqbal and D. S. Downs, *Chem. Phys.*, 1979, **44**, 137.
[46] R. R. Cavanagh, R. S. Altman, D. R. Herschbach, and W. Klemperer, *J. Am. Chem. Soc.*, 1979, **101**, 4734.
[47] P. Love and M. M. Labes, *J. Chem. Phys.*, 1979, **70**, 5147.
[48] R. Nowak, W. Kutner, H. R. Mark, Jr., and A. G. MacDiarmid, *Ann. N.Y. Acad. Sci.*, 1978, **313**, 767.
[49] P. Barrett, R. G. Chambers, P. Feenan, W. G. Herrenclen-Harker, M. G. Priestley, R. W. Trimder, and S. F. J. Read, *J. Phys. F*, 1980, **10**, L89.
[50] Z. Iqbal, R. H. Baughman, J. Keppinger, and A. G. MacDiarmid, *Ann. N.Y. Acad. Sci.*, 1978, **313**, 775.
[51] C. Bernard, P. Tougain, and G. Robert, *Ann. Chim. (Paris)*, 1979, **4**, 591.
[52] J. W. Macklin, G. B. Street, and W. D. Gill, *J. Phys. Chem.*, 1979, **70**, 2425.
[53] S. Mataka, A. Hosoki, K. Takahashi, and M. Tashiro, *Synthesis*, 1979, 524.
[54] S. Mataka, K. Takahashi, Y. Yamada, and M. Tashiro, *J. Heterocycl. Chem.*, 1979, **16**, 1009.
[55] J. Bojes, T. Chivers, G. Maclean, and R. T. Oakley, *Can. J. Chem.*, 1979, **57**, 3171.
[56] A. A. Bhattacharyya, J. A. Maclean, Jr., and A. G. Turner, *Inorg. Chem.*, 1979, **34**, L199.
[57] G. Wolmerchaeuser, G. B. Street, and R. D. Smith, *Inorg. Chem.*, 1979, **18**, 383.
[58] T. Chivers and J. Proctor, *J. Chem. Soc., Chem. Commun.*, 1978, 642.
[59] T. Chivers and J. Proctor, *Can. J. Chem.*, 1979, **57**, 1286.
[60] T. Chivers, L. Fielding, W. G. Laidlaw, and M. Trsic, *Inorg. Chem.*, 1979, **18**, 3379.
[61] R. C. Paul, R. P. Sharma, R. D. Verma, and J. K. Puri, *Indian J. Chem., Sect. A*, 1979, **18**, 516.
[62] D. T. Haworth and G. Y. Liu, *J. Inorg. Nucl. Chem.*, 1980, **42**, 137.
[63] J. Bojes, T. Chivers, W. G. Laidlaw, and M. Trsic, *J. Am. Chem. Soc.*, 1979, **101**, 4517.
[64] T. Chivers and R. T. Oakley, *J. Chem. Soc., Chem. Commun.*, 1979, 752.
[65] D. Walther, E. D. Dinjus, and H. Wolf, *Z. Chem.*, 1979, **19**, 381.

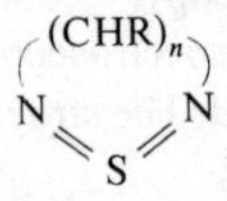

(15) R = H, $n = 2$ (18) 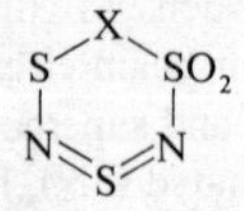   (19) X = NBu^n (21)
(16) R = H, $n = 3$ (20) X = N=S=N
(17) R = CF$_3$, $n = 2$

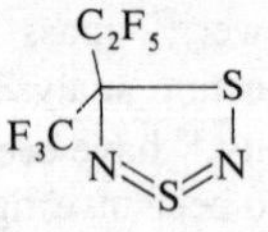

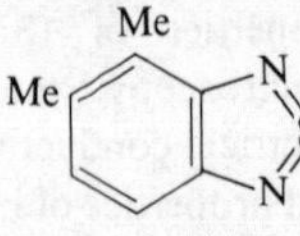

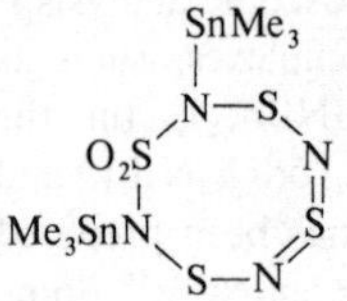

(22) 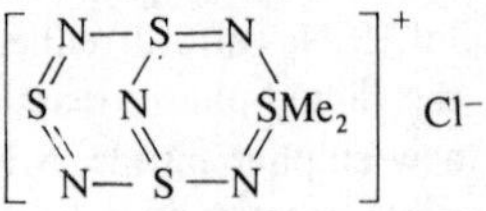   (23) 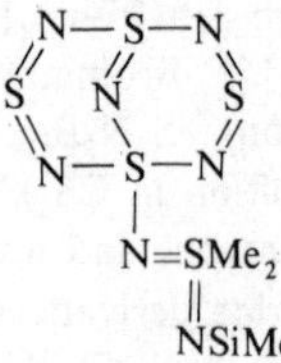   (24)

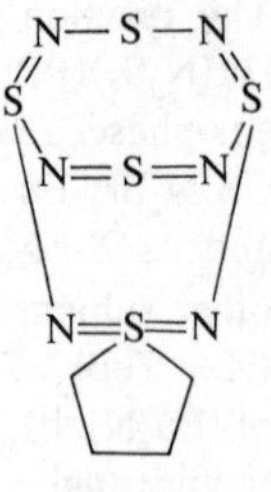

(25)

have been shown to be good starting materials for the preparation of both acyclic and cyclic sulphur di-imides, of various types, *e.g.* (15)—(25).[66–72]

Reactions. A reformed systematic nomenclature of sulphur–nitrogen heterocycles has been proposed.[73] Besides the preparation of these cyclic sulphur–nitrogen compounds, Roesky *et al.* have investigated extensively the preparation and reactions of new types of cyclic and bicyclic sulphur–nitrogen heterocycles and elucidated their structural and bonding characteristics, mainly by X-ray crystallographic analysis; *e.g.*, the structures of $[S_3N_5Me_2CO]^+$ $[AsF_6]^-$, $S_4N_4O_2(SnMe_3)_2$, $S_4N_4SO_3$, and the $S_3N_2^+$ ion.[74–79] Correlations between structures and bonding parameters such as bond angles and bond lengths have been presented.[80,81]

[66] I. Legyl, G. Kreze, M. Berger, W. Kobahn, and H. Schaefer. *Acta Chim. Acad. Sci. Hung.*, 1978, **96**, 275.
[67] U. Kingebiel and D. Bentmann, *Z. Naturforsch., Teil B*, 1979, **34**, 123.
[68] H. W. Roesky, W. Schmieder, and K. Ambrosius, *Z. Naturforsch., Teil B*, 1979, **34**, 197.
[69] W. Bludssus and R. Mews, *J. Chem. Soc., Chem. Commun.*, 1979, 35.
[70] H. W. Roesky, M. Witt, M. Diehl, J. W. Batt, and H. Fuess, *Chem. Ber.*, 1979, **112**, 1372.
[71] H. W. Roesky, M. N. Rao, T. Nakajima, and W. S. Sheldrick, *Chem. Ber.*, 1979, **112**, 3531.
[72] H. W. Roesky, C. Graf, M. N. S. Rao, B. Krebs, and G. Henkel, *Angew. Chem., Int. Ed. Engl.*, 1979, **18**, 780.
[73] H. G. Heal and A. J. Banister, *Phosphorus Sulfur*, 1978, **5**, 95.
[74] H. W. Roesky, T. Muller, and E. Rodek, *J. Chem. Soc., Chem. Commun.*, 1979, 439.
[75] H. W. Roesky, M. Diehl, B. Krebs, and M. Hein, *Z. Naturforsch., Teil B*, 1979, **34**, 814.
[76] H. W. Roesky, M. Witt, B. Krebs, and H. J. Korte, *Angew. Chem.*, 1979, **91**, 444.
[77] H. W. Roesky, M. Witt, J. W. Batt, H. Fuess, F. J. Balta Calleja, and F. Ania, *Z. Anorg. Allg. Chem.*, 1979, **458**, 225.
[78] E. Rodek, N. Amin, and H. W. Roesky, *Z. Anorg. Allg. Chem.*, 1979, **457**, 123.
[79] B. Krebs, G. Henkel, S. Pohl, and H. W. Roesky, *Chem. Ber.*, 1980, **113**, 226.
[80] H. W. Roesky, *Angew. Chem.*, 1979, **91**, 112.
[81] A. J. Banister and J. A. Durrant, *J. Chem. Res. (S)*, 1978, 152.

Structure. Ultraviolet[82] and mass-spectral analyses[83] have been reported. X-Ray analyses of some N=S=N compounds have also been performed.[3] In the compounds (26)[83] and (27),[84] both N=S=S and N=S=N groups form nearly coplanar configurations. The S=N bond lengths have been determined, and are shown. The S—N bond length of (27), *i.e.* 1.592 Å, is half-way between that of a double bond and that of a single bond, which that of S=S displays somewhat more double-bond character.

$$Bu^tN=\overset{+}{\underset{\underset{Me}{|}}{S}}=NBu^t\ BF_4^-$$

(26)

$$\underset{1.607\ \text{Å}\quad 1.592\ \text{Å}}{Ph_3P=N-S-N}\overset{1.587\ \text{Å}\quad 1.908\ \text{Å}}{=S=S}$$

(27)

The crystal structure of naphtho[1,8-*cd*;4,5-*c'd'*]bis(thiadiazine) has been determined. The results suggest a relatively small contribution of the valence-bond structure involving the naphthalene nucleus.[85] A calculation of the reflectivity of $(SN)_x$ has been performed.[86] A valence structure analysis of S_2N_2 by means of INDO calculations has been carried out.[87] The structures of $S_4N_5^-$, $S_4N_5^+$, and S_5N_6 have also been discussed, starting from M.O. theory.[88] The orientations of the cycloadditions of S_4N_4, As_4N_4, and As_4Se_4 with olefins have been discussed. The prediction for S_4N_4 and olefins rationalizes the experimental observations, while no experimental results have been obtained for As_4S_4 and As_4Se_4; rather different results are expected, in view of the dissimilarity of the orbital patterns of the LUMO's of As_4X_4 (X = S or Se) and those of S_4N_4.[89]

Thione S-Imides.—Compounds of this type, which have a —C=S=N— group within the molecule, have been prepared recently (see Vol. 5, p. 107). Thione S-imides (28) have been shown to react readily as 1,2- or 1,3-dipolar compounds with olefins, imines, oximes, aldehydes, or thiones, yielding 1,3-dipolar cyclo-adducts or Diels–Alder-type adducts[90–92] (see also Chap. 3, Pt II, p. 163).

2 Tri-co-ordinate Sulphur

Sulphonylamines and Sulphur Tri-imides.—A gas-phase electron-diffraction study of the molecular structure of imido-disulphonyl chloride, $NH(SO_2Cl)_2$, indicates that this exists as a single conformer, having C_2 symmetry.[93] Further details on the sulphur tri-imides $(Bu^tN=)_2S=NSiMe_3$ and $S(=NBu^t)_3$ have been reported[94] (see

[82] J. Fabian, R. Mayer, and S. Bleisch, *Phosphorus Sulfur*, 1979, **7**, 61.
[83] A. Gieren and B. Dederer, *J. Chem. Res. (S)*, 1979, 41.
[84] T. Chivers, R. T. Oakley, A. W. Cordes, and P. Swepston, *J. Chem. Soc., Chem. Commun.*, 1980, 35.
[85] A. Gieren, V. Lamm, R. C. Haddon, and M. L. Kaplan, *J. Am. Chem. Soc.*, 1979, **101**, 7279.
[86] H. Numata, *Phys. Lett. A*, 1979, **72**, 453.
[87] H. Fujimoto and T. Yokoyama, *Bull. Chem. Soc. Jpn.*, 1980, **53**, 800.
[88] R. Bartzko and R. Gleiter, *Chem. Ber.*, 1980, **113**, 1138.
[89] T. Yamabe, K. Tanaka, A. Tachikawa, K. Fukui, and H. Kato, *J. Phys. Chem.*, 1979, **83**, 767.
[90] T. Saito and S. Motoki, *Chem. Lett.*, 1978, 591.
[91] T. Saito and S. Motoki, *J. Org. Chem.*, 1979, **44**, 2493.
[92] T. Saito, I. Oikawa, and S. Motoki, *Bull. Chem. Soc. Jpn.*, 1980, **53**, 1023.
[93] B. Beagley, R. Moutran, S. P. Narula, and V. Ulbrecht, *J. Mol. Struct.*, 1979, **56**, 207.
[94] F. M. Tesky, R. Mews, and O. Glemser, *Z. Anorg. Allg. Chem.*, 1979, **452**, 103.

Vol. 5, p. 107) and the tri-imides (29) have also been synthesized, from $CF_3CF_2N=SF_2$ and the diamine.[95] Several dithiadiazolines (30; R = p-$NO_2C_6H_4$, p-ClC_6H_4, Pr, $SiMe_3$, or Ts) have been prepared, in 30—75% yields, by treating 4-benzyl-5-tosylimino-1,2,3,4-thiatriazoline with RNSO, while hydrolysis of these dithiadiazolines has been found to give $PhCH_2NHC(=NTs)SNHR$.[96]

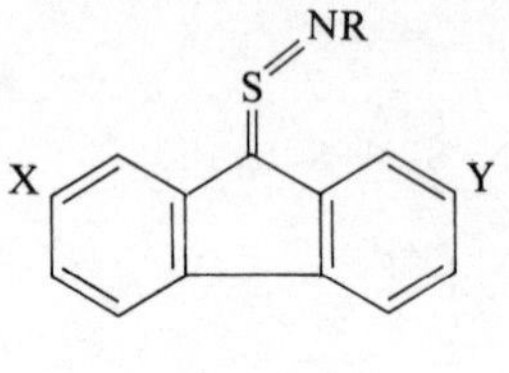

(28) 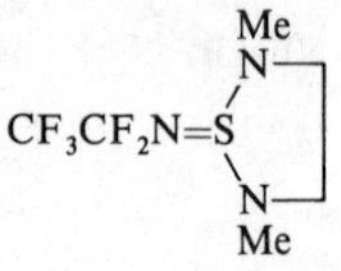 (29) 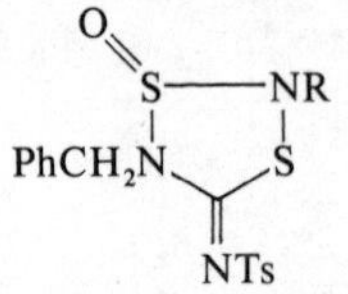 (30)

Sulphimides and Azasulphonium Salts.—*Preparation and Structure of Sulphimides.* Several new procedures for the preparation of sulphimides have been published. Sulphonamides, on treatment with DMSO and oxalyl chloride, can be converted into the corresponding N-substituted sulphimides $ArSO_2N=SMe_2$. This reaction can be applied to compounds containing substituents having strong electron-withdrawing effects, *e.g.* sulphonyl or p-nitrophenyl.[97] Various thietan derivatives (31), such as sulphoxides, sulphimides, sulphonium salts, and sulphoximides, have been prepared with dry Chloramine-T, and their isomers separated.[98]

(31) X = O, Y = lone pair   (32)
    X = NTs, Y = lone pair
    X = NTs, Y = O

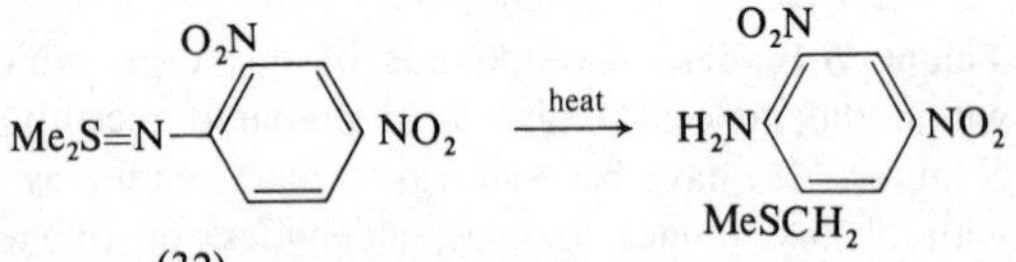

Another improved synthesis of various symmetric and unsymmetric N-(toluene-p-sulphonyl)-sulphimides is by the treatment of the sulphides with solid Chloramine-T under phase-transfer catalytic conditions, using tetrabutylammonium salts as the catalyst, in CH_2Cl_2.[99] S-Vinyl-sulphimides can be prepared by the following three methods; (*a*) treatment of vinyl sulphides with Chloramine-T, (*b*) base-catalysed elimination of HCl from 2-(halogenoethyl)aryl-sulphimides, and (*c*) elimination of acetic acid from β-acetoxy-sulphimides with bases.[100] N-2,4-Dinitrophenyl-SS-dimethylsulphimide (32) can be prepared by the sequential treatment of DMSO with P_2O_5 in DMF, then with 2,4-dinitroaniline, and finally

[95] O. Glemser and J. M. Shreeve, *Inorg. Chem.*, 1979, **18**, 213.
[96] G. L'abbé, A. van Asch, J. P. Declercq, G. Germain, and M. Van Meerssche, *Bull. Soc. Chim. Belg.*, 1978, **87**, 285.
[97] S. L. Huang and D. Swern, *J. Org. Chem.*, 1978, **43**, 4537.
[98] M. Buza, K. K. Andersen, and M. D. Pazdon, *J. Org. Chem.*, 1978, **43**, 3827.
[99] C. R. Johnson, K. Mori, and A. Nakanishi, *J. Org. Chem.*, 1979, **44**, 2065.
[100] T. Yamamoto, M. Kakimoto, and M. Okawara, *Bull. Chem. Soc. Jpn.*, 1979, **52**, 841.

with NEt_3. Heating (32) with Et_3N in DMF at 100 °C slowly afforded the rearranged product, *i.e.* 4,6-dinitro-2-(methylthiomethyl)aniline,[101] while heating the sulphimide (32) with phenols gave the *o*-methylthiomethylated phenols.[102]

The copper-catalysed thermolysis of toluene-*p*-sulphonyl azide in the presence of various sulphides also affords the corresponding sulphimides.[103] A kinetic study on the reactions of Chloramine-T with *ortho*-substituted sulphides indicates a marked steric hindrance by *ortho*-substituents, while the anchimeric effect of *ortho*-groups that have a carbonyl function decreases in the order $CH_2CO_2H \approx CH_2CO_2Me <$ $CH_2CO_2^- < CO_2Me \approx CO_2H < CO_2^-$.[104]

Ethoxycarbonylnitrene (generated photochemically from the azide) also reacts with 4-t-butylthian, affording a nearly 1:1 mixture of both *cis*- and *trans*-sulphimides. This non-stereoselectivity in the reaction of the nitrene is in marked contrast with that of the reaction of carbenes with the thian; the thian reacts with both bis(alkoxycarbonyl)- and bis(diacetyl)-carbenes, giving the *trans*-ylides as the sole products[105] (see Vol. 4, p. 112). Various sulphimides of cyclic sulphides, *e.g.* thians, *cis*- and *trans*-thiadecalins, oxathians, and dithians, can be prepared by treating the sulphides either with anilines and *N*-chlorosuccinimide or with Bu^tOCl. In this reaction, the formation of (33; $R^1 = Bu^t$, $R^2 = p\text{-}ClC_6H_4$) with an equatorial S—N bond is more favoured than that of (33) with an axially orientated S—N linkage. Meanwhile, the sulphimides (33) having axial S—N bonds can be prepared from conformationally rigid sulphoxides with equatorial S—O bonds. Formation of sulphimides of rigid ring systems such as (34) proceeds by an S_N2 reaction of the cyclic sulphide on the *N*-chloro-aniline, while reactions of conformationally mobile cyclic sulphides with *N*-chlorosuccinimide may occur *via* succinimidoyl azasulphonium ions as intermediates. Conformational equilibria of these sulphimides have been determined by low-temperature ^{13}C n.m.r. spectroscopy. The sulphimides (35) have been prepared and their configurations and conformations also determined by ^1H and ^{13}C n.m.r. spectroscopy. The compounds (35; $R^1 = H$, $R^2 = p\text{-}ClC_6H_4$) rearrange stereospecifically to the corresponding 2-(2′-amino-5′-chlorophenyl)-1,3-dithians.[106-109] *X*-Ray photoelectron spectroscopic measurements of various sulphimides have revealed that these thermally stable sulphimides

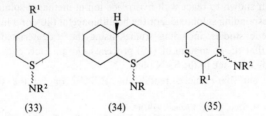

(33)   (34)   (35)

[101] T. Yamamoto, Y. Harigaya, and M. Okawara, *Tetrahedron*, 1978, **34**, 3097.
[102] T. Yamamoto and M. Okawara, *Bull. Chem. Soc. Jpn.*, 1978, **51**, 2443.
[103] P. Svoronos and V. Horak, *Synthesis*, 1979, 596.
[104] F. Ruff, I. Kapovits, J. Rábai, and Á. Kucsman, *Tetrahedron*, 1978, **34**, 2767.
[105] D. C. Appleton, D. C. Bull, J. McKenna, J. M. McKenna, and A. R. Walley, *J. Chem. Soc., Perkin Trans. 2*, 1980, 385.
[106] P. K. Claus, W. Rieder, and F. W. Vierhapper, *Monatsh. Chem.*, 1978, **109**, 609.
[107] J. Bailer, P. K. Claus, and F. W. Vierhapper, *Tetrahedron*, 1979, **35**, 1373.
[108] P. K. Claus, F. W. Vierhapper, and R. L. Willer, *J. Org. Chem.*, 1979, **44**, 2863.
[109] J. Bailer, P. K. Claus, and F. W. Vierhapper, *Tetrahedron*, 1980, **36**, 901.

are best represented by a semi-polar canonical structure, since the contribution of $3d$–$2p$ π-interaction is of little importance in the structure of sulphimides.[110]

Reactions of Sulphimides and Azasulphonium Salts. Pyrolysis of sulphimides that have β-protons has been reviewed in detail.[111]

In contrast to the high stability of *N*-tosyl-*SS*-diaryl-sulphimides, the corresponding *N*-acyl-*SS*-diaryl-sulphimides undergo both thermal and photochemical decompositions to afford the corresponding isocyanates and diaryl sulphides quantitatively. Photochemical decomposition has been confirmed to generate partially the singlet nitrene, while kinetic investigations suggest that the thermolysis proceeds exclusively *via* a concerted migration of the alkyl group from carbonyl-carbon to imino-nitrogen that is similar to that in the Curtius rearrangement of acid azides.[112–114] Photolysis of *SS*-dimethyl-*N*-aryl-benzimidoyl-sulphimide that has two *ortho*-alkyl substituents (36) generates imidoylnitrenes [ArN=C(Ph)N:]; these give carbodi-imides and cyclopenta[*d*]pyrimidines by initial cyclization followed by a [1,5]-imidoyl shift.[115,116] *N*-Chloro- and *N*-bromo-sulphimides (37; X = Cl or Br), prepared from the sulphimide with *N*-chloro- or *N*-bromo-succinimide, have been shown to react with sulphides, phosphines, and tertiary amines, yielding a new type of sulphimide derivative (38; Y = $\overset{+}{S}R_2$, $\overset{+}{P}R_3$, or $\overset{+}{N}R_3$). *X*-Ray analysis of the salt [38; Y = $\overset{+}{S}Ph(Me)$, X = ClO$_4$] indicates that the (Ph$_2$)S—N bond is shorter than the N—S[(Ph)Me] bond, demonstrating that the charge is localized mainly on the sulphur atom of the group Y.[117] The mechanism of formation of (38) has been investigated kinetically and stereochemically.[118]

(36)   (37)   (38)

N-Tosyl-*S*-alkyl-*S*-aryl-sulphimides or the cycloalkyl-sulphimides (39; n = 2, 3, or 4) have been shown to react with hydroxide ion in methanol solution, affording either the corresponding sulphoxide, or the hemithioacetal (40), or a mixture of both products. Kinetic studies, including studies using the α-deuteriated compounds, have revealed that the formation of (40) proceeds *via* an *E*lcB path that involves rate-determining cleavage of the S—N bond.[119]

Sulphimides can be reduced readily by P$_2$S$_5$,[120] or by the thiol–Me$_2$SiCl

[110] S. Tsuchiya and M. Seno, *J. Org. Chem.*, 1979, **44**, 2850.

[111] S. Oae and N. Furukawa, *Tetrahedron*, 1977, **33**, 2359 (Tetrahedron Report No. 37).

[112] N. Furukawa, T. Nishio, M. Fukumura, and S. Oae, *Chem. Lett.*, 1978, 209.

[113] N. Furukawa, M. Fukumura, T. Nishio, and S. Oae, *Phosphorus Sulfur*, 1978, **5**, 231.

[114] N. Furukawa, M. Fukumura, T. Nishio, and S. Oae, *Bull. Chem. Soc. Jpn.*, 1978, **51**, 3599.

[115] T. L. Gilchrist, C. J. Moody, and C. W. Rees, *J. Chem. Soc., Perkin Trans. 1*, 1979, 1871.

[116] C. W. Rees, *Pure Appl. Chem.*, 1979, **51**, 1243.

[117] Y. Nishikawa, Y. Matsuura, M. Kakudo, T. Akasaka, N. Furukawa, and S. Oae, *Chem. Lett.*, 1978, 447.

[118] T. Akasaka, T. Yoshimura, N. Furukawa, and S. Oae, *Phosphorus Sulfur*, 1978, **4**, 211.

[119] T. Masuda, N. Furukawa, T. Aida, and S. Oae, *Phosphorus Sulfur*, 1979, **6**, 429.

[120] I. W. J. Still and K. Turnbull, *Synthesis*, 1978, 540.

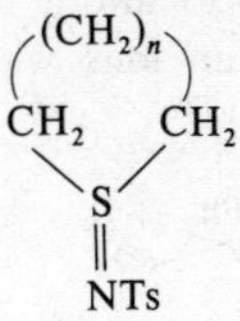

$$\begin{array}{ccc}
\overbrace{(CH_2)_n} & \overbrace{(CH_2)_n} & \\
CH_2 \quad CH_2 & CH_2 \quad CHOMe & \\
\diagdown S \diagup & \diagdown S \diagup & \\
\| & & \\
NTs & (40) & (41) \\
(39) & &
\end{array}$$

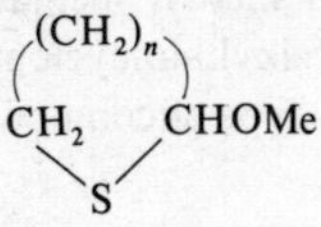

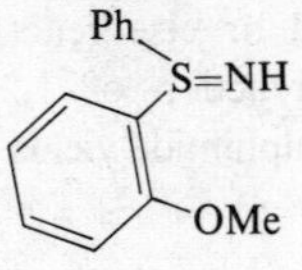

system,[121] at room temperature, to the corresponding sulphides. The Michael-type addition of 'free' diphenylsulphimide to various electrophilic olefins gives mainly the corresponding *trans*-aziridines and enamino-ketones. When optically active 'free' (+)-(R)-S-o-methoxyphenyl-S-phenylsulphimide (41), prepared either by asymmetric induction or by resolution of the racemate by seeding with the optically pure compound (41), is treated with an electrophilic olefin such as benzalacetophenone, optically active 2-benzoyl-3-phenylaziridine of *ca.* 30% optical purity is obtained.[122] Recently, aryl 'free' sulphimides have been found to react with various aryl or alkyl aldehydes, affording aryl or alkyl nitriles in excellent yields [reaction (3)].[123] Transformation of *N*-benzyl-*SS*-difluorosulphimide ($F_2S=N-CH_2Ph$) into the nitrile sulphide ($PhC\equiv N\rightarrow S$) has been shown to be catalysed by fluoride ion.[124]

$$Ph_2S=NH + RCHO \longrightarrow RCN + Ph_2S + H_2O \qquad (3)$$
$$(R = alkyl \text{ or } aryl)$$

Sulphimides have been shown to undergo cycloaddition (see Vol. 5, p. 109). *N*-Aryl-*SS*-dimethylsulphimides react with nitrile oxides *p*-$R^2C_6H_4C\equiv N\rightarrow O$, affording either (42) or (43).[125] Meanwhile, *N*-heterocyclyl-sulphimides (44) can be prepared, and they afford new heterocycles upon treatment with ketens, through a 1,3-dipolar cycloaddition which eventually gives compound (45).[126,127] *N*-Phthalimido-*SS*-dimethylsulphimide has been prepared by oxidation of a mixture of *N*-aminophthalimide and dimethyl sulphide with $Pb(OAc)_4$. Further reaction, in refluxing benzene, generates the *N*-phthalimido-nitrene, which dimerizes to the

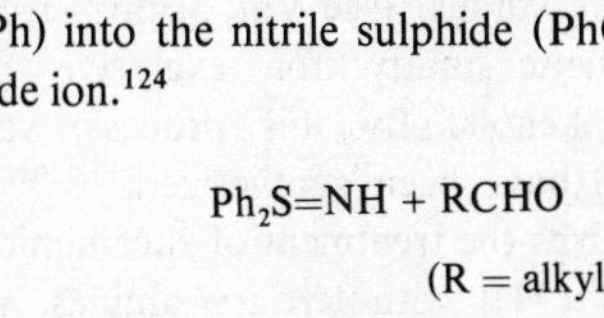

(42) 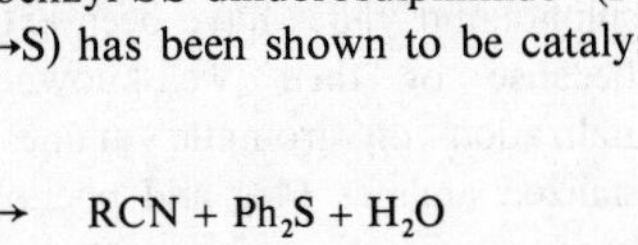 (43)

[121] T. Numata, H. Togo, and S. Oae, *Chem. Lett.*, 1979, 329.
[122] N. Furukawa, T. Yoshimura, M. Ohtsu, T. Akasaka, and S. Oae, *Tetrahedron*, 1980, **36**, 73.
[123] N. Furukawa, M. Fukumura, T. Akasaka, T. Yoshimura, and S. Oae, *Tetrahedron Lett.*, 1980, **21**, 761.
[124] M. J. Sanders, S. L. Dye, A. G. Miller, and J. R. Grunwell, *J. Org. Chem.*, 1979, **44**, 510.
[125] S. Shiraishi, T. Shigemoto, and S. Ogawa, *Bull. Chem. Soc. Jpn.*, 1978, **51**, 563.
[126] M. Sakamoto, K. Miyazawa, K. Kuwabara, and Y. Tomimatsu, *Heterocycles*, 1979, **12**, 231.
[127] M. Sakamoto, K. Miyazawa, K. Kuwabara and Y. Tomimatsu, *Chem. Pharm. Bull.*, 1979, **27**, 2116.

azo-compound or else reacts with methyl methacrylate to afford the aziridine (46).[128] The reaction of 1,2,3-tri(alkylthio)cyclopropenium ions with N-acetyl-SS-dimethylsulphimide yields a new type of compound (47).[129]

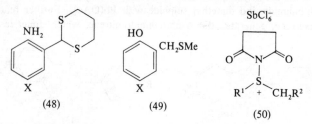

(44) X = O or S   (45) X = O or S

(46)   (47)

Azasulphonium ylides have been extensively studied (see Vol. 5, pp. 81–2 and 111) because of their well-known synthetic utility for exclusive *ortho*-functionalization of aromatic amines or phenols. By this process, various functionalized anilines (48) and phenols (49) have been synthesized.[102, 109, 130, 131] The other main direction of investigation involves the treatment of succinimido- or hydantoinyl-sulphonium salts [*e.g.* (50) and (51)] with tertiary amines, which affords interesting [2,3]- or [1,3]-sigmatropically rearranged products [*e.g.* (52)].[132, 133] The amino-sulphonium salt (51), upon treatment with a carboxylic acid, forms the corresponding carboxy-sulphonium salt as an intermediate, which then reacts with amines to form the acid amides.[134] The synthesis of 9-alkyl-[135] or of 9-phenyl-thioxanthene-N-(toluene-*p*-sulphonyl)sulphimides (53) has been reported.[136] The configuration of (53) has been determined by [1]H n.m.r.

(48)   (49)   (50)

[128] M. Edward, T. G. Gilchrist, C. J. Harris, and C. W. Rees, *J. Chem. Res. (S)*, 1979, 114.
[129] S. Inoue, G. Yasuda, and T. Hori, *Bull. Chem. Soc. Jpn.*, 1978, **51**, 3653.
[130] P. G. Gassman and H. R. Drewes, *J. Am. Chem. Soc.*, 1978, **100**, 7600.
[131] P. G. Gassman and D. R. Amick, *J. Am. Chem. Soc.*, 1978, **100**, 7611.
[132] E. Vilsmaier, J. Schults, and S. Zimmerer, *Chem. Ber.*, 1979, **112**, 2231.
[133] E. Vilsmaier, W. Treger, W. Sprugel, and K. Gagel, *Chem. Ber.*, 1979, **112**, 2997.
[134] K. Takeda, T. Kobayashi, and H. Ogura, *Chem. Pharm. Bull.*, 1979, **27**, 536.
[135] Y. Tamura, Y. Nishikawa, C. Mukai, K. Sakamoto, M. Ikeda, and M. Kise, *J. Org. Chem.*, 1979, **44**, 1684.
[136] Y. Tamura, C. Mukai, Y. Nishikawa, and M. Ikeda, *J. Org. Chem.*, 1979, **44**, 3296.

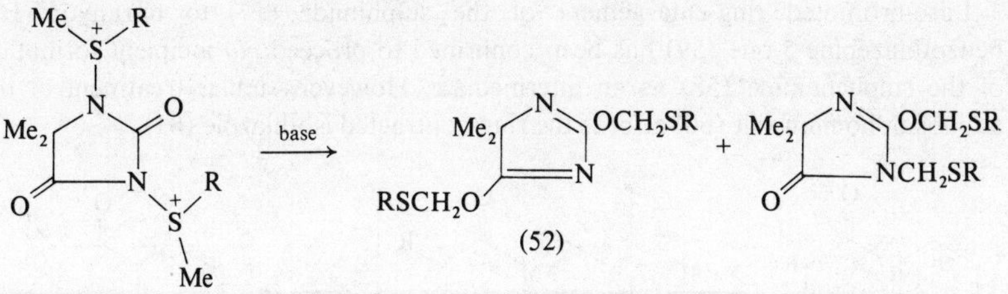

(52)

(51)

spectroscopy and compared with that of the corresponding sulphoxide. Compound (53) has been shown to rearrange to the 9-*N*-toluene-*p*-sulphonamido-9-substituted-thioxanthene (54) upon treatment with bases. The rates of this base-catalysed rearrangement fall in the order *trans*-Me > *cis*-Me > *cis*-Et > *trans*-Pr^i.

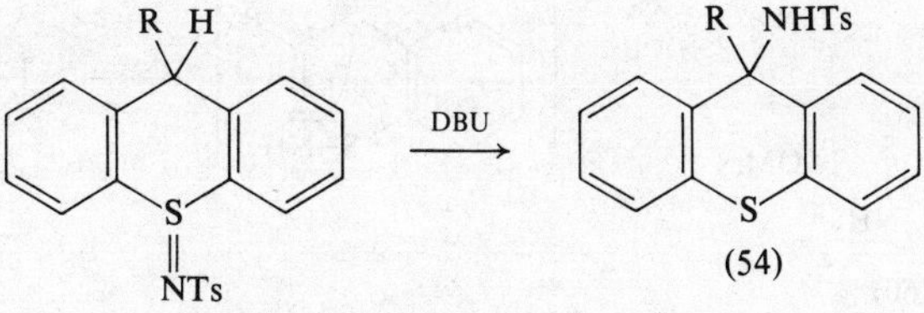

(53) R = Me (*cis*, *trans*), Et (*cis*, *trans*),
Pr^i (*trans*), or Ph

The cyclic sulphimides (55)[137] and (56)[138] have been synthesized, and both thermal and base-catalysed rearrangements of these sulphimides have been shown to result in ring-enlargement to afford heterocyclic compounds.

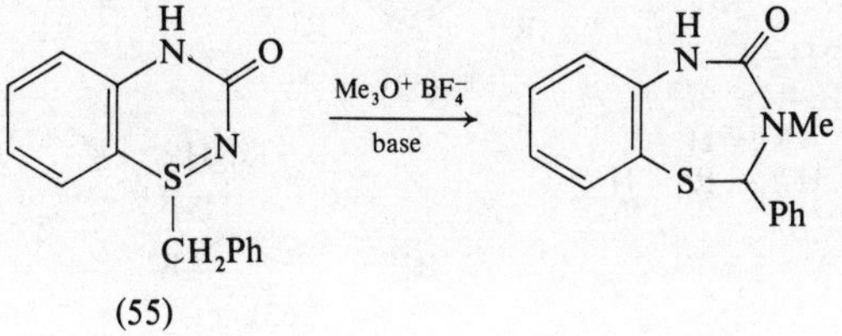

(55)

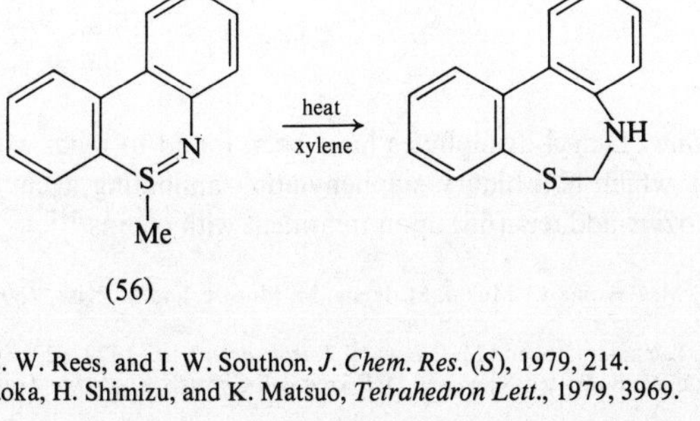

(56)

[137] T. L. Gilchrist, C. W. Rees, and I. W. Southon, *J. Chem. Res. (S)*, 1979, 214.
[138] M. Hori, T. Kataoka, H. Shimizu, and K. Matsuo, *Tetrahedron Lett.*, 1979, 3969.

Base-promoted ring-enlargement of the sulphimide (57) to tetrahydro-1,2-benzothiazepine-5-one (59) has been confirmed to proceed *via* incipient formation of the sulphenamide (58) as an intermediate. However, similar treatment of the aminosulphonium salt (60) affords the ring-contracted isothiazole (61).[139]

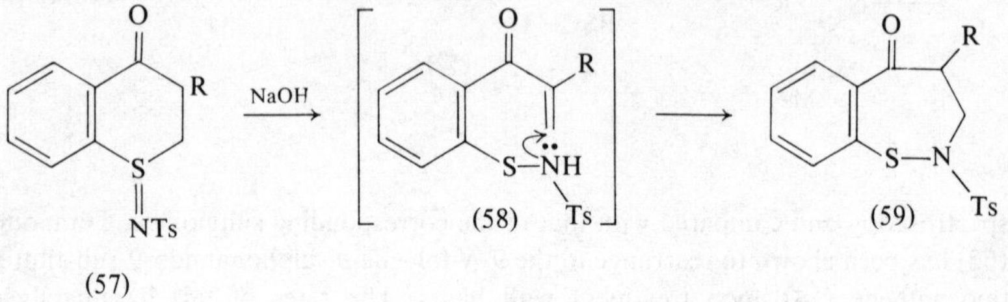

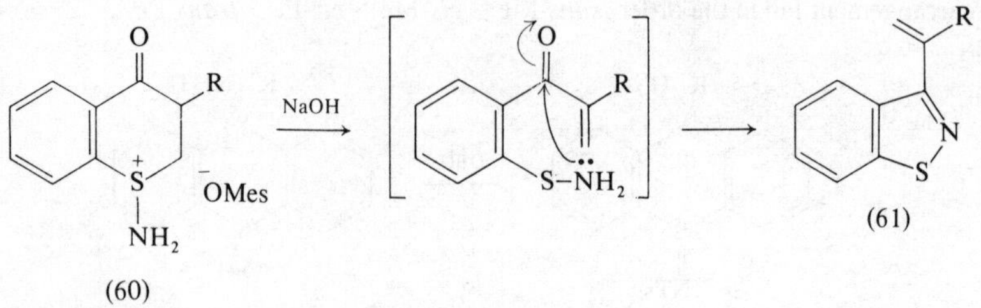

A new application, the transformation of carbonyl compounds into the allylamines (63), has been carried out by treatment of vinylic sulphimides (62) with a strong base, as shown in Scheme 3.[140]

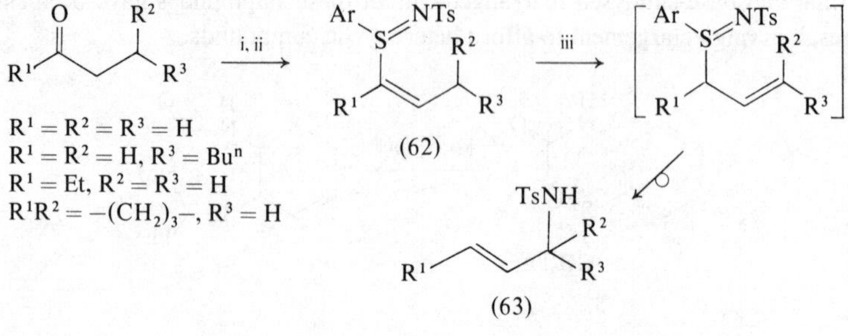

$R^1 = R^2 = R^3 = H$
$R^1 = R^2 = H, R^3 = Bu^n$
$R^1 = Et, R^2 = R^3 = H$
$R^1R^2 = -(CH_2)_3-, R^3 = H$

Reagents: i, ArSH, P_2O_5, CH_2Cl_2; ii, Chloramine-T; iii, NaOEt, EtOH, at 50 °C

Scheme 3

Sulphonamidines. Diaryl disulphides have been found to react with Chloramine-T to afford (64), which is a binary sulphenylating/aminating agent and which gives stereospecific *trans*-adducts (65) upon treatment with olefins.[141]

[139] Y. Tamura, S. M. Bayomi, C. Mukai, M. Ikeda, M. Murase, and M. Kise, *Tetrahedron Lett.*, 1980, **21**, 533.
[140] M. Kakimoto, T. Yamamoto, and M. Okawara, *Tetrahedron Lett.*, 1979, 623.
[141] D. H. R. Barton, M. R. Britten-Kley, and D. Ferreira, *J. Chem. Soc., Perkin Trans. 1*, 1978, 1682.

(64)   (65)

A number of symmetrical and unsymmetrical disulphides afford sulphinimidoyl chlorides (66) in good yields when disulphides are treated with NN-dichloro-amides or NN-dichloro-amines,[142] while treatment of monochloro- or monobromo-acid amides with sulphur and SCl_2 or S_2Cl_2 gives the dichloro-sulphimides $RCON=SCl_2$ ($R = Bu^t$, $ClCH_2$, Ph, or aryl)[140] or sulphur di-imides in high yields.[143] Cycloadditions of N-acyl-S-chloro-sulphimines in the reactions with active-methylene compounds which give N-acyl-sulphenamides and $ClCHXY$ have also been reported.[144] By this process, S-aryl-S-alkylamino-sulphonium methylides p-$R^1C_6H_4S(NHCMe_2R^2)=CR^3R^4$ ($R^1 = Bu^t$, Ph, $PhSO_2$, or Tol; $R^2 = $ Me or CN; $R^3 = R^4 = $ MeCO or CN, or $R^3 = $ CN, $R^4 = CO_2Et$) can be obtained.[139,145,146] The n.m.r. chemical shifts of the compounds (66) and the substituent effects of R^1 have been shown to follow a linear free-energy relationship.[147] Treatment of various mercaptides with NN-dichloroarenesulphonamides has been shown to give the corresponding sulphonyl sulphinamidines (67),[148-154] and their acidities and details of their complex formation have also been reported.[155]

$$RS-NHSO_2C_6H_4X$$
$$\| $$
$$NSO_2C_6H_4X$$

(67)

(66) $R^1 = Bu^t$, $Me_2(CN)C$, $PhSO_2$, or aryl
     $R^2 = $ Bu, $PhCH_2$, Ph, or aryl

Preparation of Azasulphonium Salts. Oxidation of methionine with I_2 to give the cyclic sulphimide is catalysed by general bases, and the rate of the reaction does not give a linear Brönsted plot. The mechanism is proposed to involve the formation

[142] E. S. Levchenko, L. V. Budnik, and T. N. Dubinina, *Zh. Org. Khim.*, 1978, **14**, 1846.
[143] G. S. Borovikova, E. S. Levchenko, and E. M. Dorokhova, *Zh. Org. Khim.*, 1979, **15**, 479.
[144] T. N. Dubinina and E. S. Levchenko, *Zh. Org. Khim.*, 1979, **15**, 743.
[145] E. S. Levchenko and L. V. Budnik, *Zh. Org. Khim.*, 1979, **15**, 748.
[146] L. N. Markovskii, Yu. G. Shermolovich, V. V. Vasil'ev, I. E. Boldeskul, S. S. Trach, and N. S. Zefirov, *Zh. Org. Khim.*, 1978, **14**, 1659.
[147] Yu. P. Egorov and T. G. Zabolotnaya, *Teor. Eksp. Khim.*, 1979, **15**, 265.
[148] R. P. Naumenko, V. I. Sheremet, and M. M. Kremlev, *Vopr. Khim. Khim. Tekhnol.*, 1978, **51**, 57.
[149] R. P. Naumenko, V. I. Sheremet, and M. M. Kremlev, *Vopr. Khim. Khim. Tekhnol.*, 1978, **51**, 3.
[150] A. V. Kharchenko, I. V. Koval, and M. M. Kremlev, *Zh. Org. Khim.*, 1979, **15**, 443.
[151] I. V. Koval, T. G. Oleinik, and M. M. Kremlev, *Zh. Org. Khim.*, 1979, **15**, 2319.
[152] I. V. Koval, A. I. Tarasenko, T. G. Oleinik, and M. M. Kremlev, *Zh. Org. Khim.*, 1979, **15**, 1004.
[153] I. V. Koval, T. G. Oleinik, and M. M. Kremlev, *Vopr. Khim. Khim. Tekhnol.*, 1979, **54**, 24.
[154] S. I. Zlotchenko, M. M. Kremlev, L. P. Yaroshenko, and N. G. Monastyrnaya, *Vopr. Khim. Khim. Tekhnol.*, 1979, **55**, 100.
[155] R. P. Naumenko, V. I. Sheremet, and M. M. Kremlev, *Vopr. Khim. Khim. Tekhnol.*, 1978, **50**, 60.

of a tetra-co-ordinate sulphurane intermediate.[156] L-Methionine gives two diastereo-isomeric isothiazolinium salts.[157] 5-Methyl-1-thia-5-azacyclo-octane (68) generates the dication (69), and hence the radical cation (70) of $R_3\overset{+}{N}-\overset{+}{S}R_2$, upon electrophilic oxidation[158] (Scheme 4). In addition to these azasulphonium salts, several other new 'onium salts have been prepared and characterized by ^1H and ^{19}F n.m.r. spectroscopies or by X-ray crystallography,[159,160] *viz* $[(Me_2N)_xSF_{3-x}]^+$ X^-, $[(Me_2N)_3S]^+$ X^-, and $(Me_3ZN)_3S$ ($X = BF_4^-$ or PF_6^-; $Z = Si$ or C). Optically active alkoxy-aminosulphonium (71) and the related azasulphonium salts (72) have been prepared, the former by methylation of the corresponding sulphinamide. Optically active (71) can transmethylate unsymmetrical tertiary amines or sulphides to give the optically active quaternary ammonium salts and sulphonium salts respectively.[161]

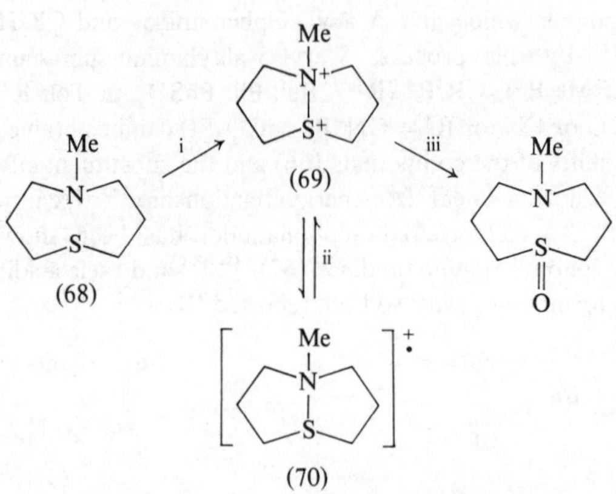

Reagents: i, $NOBF_4$; ii, (70), MeCN; iii, hydrolysis

Scheme 4

(+)-(S)-(71)

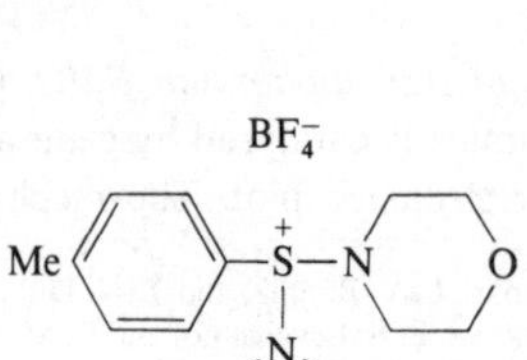

(72)

[156] P. R. Young and L.-S. Hsieh, *J. Am. Chem. Soc.*, 1978, **100**, 7121.
[157] D. Lambeth and D. W. Swank, *J. Org. Chem.*, 1979, **44**, 2632.
[158] W. K. Musker, A. S. Hirschon, and J. T. Doi, *J. Am. Chem. Soc.*, 1978, **100**, 7754.
[159] A. H. Cowley, D. J. Pagel, and M. L. Walker, *J. Am. Chem. Soc.*, 1978, **100**, 7065.
[160] S. Pohl, B. Krebs, U. Seyer, and G. Henkel, *Chem. Ber.*, 1979, **112**, 1751.
[161] K. Okuma, H. Minato, and M. Kobayashi, *Bull. Chem. Soc. Jpn.*, 1980, **53**, 435.

NN-Dialkyl-sulphenamides have been shown to react with t-butyl hypochlorite in the presence of an alcohol to afford a new type of *S*-alkoxy-*NN*-dialkyl-aminosulphonium salt [73; $R^1 = Me$, $R^2 = (CH_2)_5$ or $(CH_2)_2O(CH_2)_2$].[162, 163]

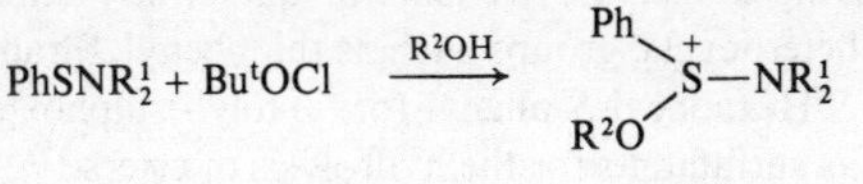

$$PhSNR_2^1 + Bu^tOCl \xrightarrow{R^2OH} \underset{R^2O}{\overset{Ph}{\diagdown}} \overset{+}{S} - NR_2^1$$

(73)

3 Tetra-co-ordinate Sulphur

Sulphoximides.—One of the general procedures is the oxidation of the corresponding sulphimides with $KMnO_4$.[164] In this case, the addition of $MgSO_4$ generally improves the yields of sulphoximides.[165] *N*-Acyl or -sulphonyl substituted sulphimides undergo facile oxidation with H_2O_2[166] or MCPBA[167] under alkaline conditions. By this method, rather reactive *S*-benzyl-sulphoximides, unobtainable with other procedures, can be obtained in good yields, as shown in reaction (4).

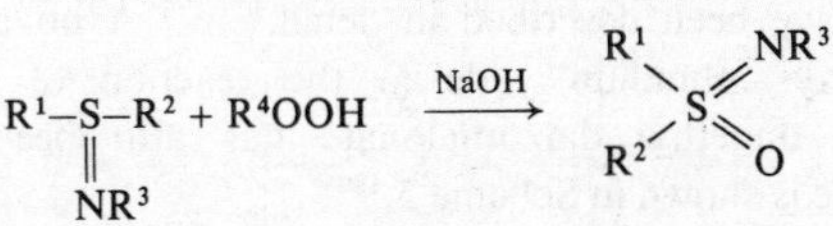

$$R^1-\underset{\underset{NR^3}{\|}}{S}-R^2 + R^4OOH \xrightarrow{NaOH} \underset{R^2}{\overset{R^1}{\diagdown}} \underset{O}{\overset{NR^3}{S}}$$

(4)

(R^1, R^2 = alkyl, aryl, or aralkyl; $R^3 = R^5CO$, R^6SO_2, or Ar)

Aryloxysulphoxonium salts (74), prepared by treating the sulphones with arenediazonium salts, react with primary or secondary amines to give the corresponding sulphoximides. The reaction is considered to proceed through an S_N2 process on the sulphur atom.[168–170] These methods can be used to prepare *N*-alkylated sulphoximides. When the sulphoximide (75; R = OH) is heated at 230 °C for 2 minutes, an intermolecular methyl-transfer reaction can be observed, from the sulphoximide to the carboxylic acid group, to give (75; R = OMe). This

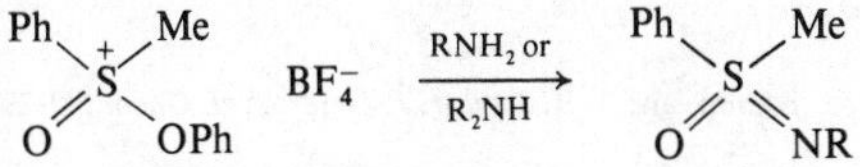

(74)

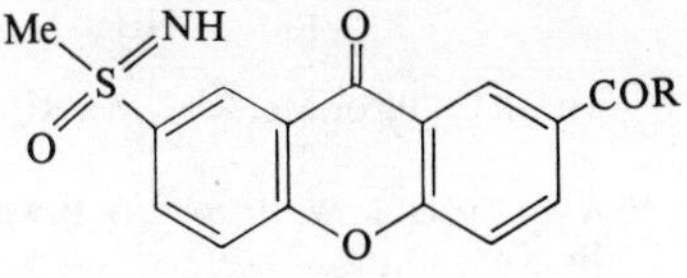

(75)

[162] M. Haake and H. Gebbing, *Synthesis*, 1979, 98.
[163] M. Haake, H. Gebbing, and H. Benack, *Synthesis*, 1979, 97.
[164] D. H. Evans and R. G. Greenwald, U.S. P. 4 113 738 (*Chem. Abstr.*, 1979, **90**, 72 210).
[165] S. Oae, N. Furukawa, T. Akasaka, and M. Moriyama, Japan. Kokai 78 09 717.
[166] C. R. Johnson and R. A. Kirchhoff, *J. Org. Chem.*, 1979, **44**, 2280.
[167] S. H. Huang and D. Swern, *J. Org. Chem.*, 1979, **44**, 2510.
[168] M. Shimagaki, H. Tsuchiya, Y. Ban, and T. Oishi, *Tetrahedron Lett.*, 1978, 3435.
[169] I. W. J. Still and S. Szilagyi, *Synth. Commun.*, 1979, **9**, 923.
[170] G. R. Chalkley, D. J. Snodin, G. Stevens, and M. C. Whiting, *J. Chem. Soc., Perkin Trans. 1*, 1978, 1580.

novel type of methyl-transfer can also be observed in the treatment of methyl phenyl sulphoximide with benzoic acid at 240 °C, which forms methyl benzoate in 15% yield.[171]

Sulphoximides having longer alkyl chains or quaternary ammonium functional groups that contain heterocyclic groups such as thiophenyl, furanyl, or pyridyl have been prepared.[172] S-Hexadecyl-S-phenyl-(or p-tolyl-)sulphoximides have been prepared and tested as surfactants for the hydrolysis of esters.[173]

Optically active N-halogeno-sulphimides, upon hydrolysis under alkaline conditions, give the corresponding sulphoximides. This unusual substitution reaction proceeds with nearly quantitative (96%) retention of configuration.[174] Similar treatment of the N-chloro-SS-diaryl-sulphimides with Chloramine-T in anhydrous acetonitrile, followed by treatment with H_2SO_4, gives the diaryl sulphone di-imines $Ar_2S(=NH)_2$.[175] The reaction of optically active N-chloromethyl-S-phenylsulphoximide with sulphoxides gives optically active α-chloro-sulphoxides directly, but in low optical yields.[176] Upon photolysis or thermolysis in the presence of a radical initiator such as AIBN or benzoyl peroxide, N-chloro- or N-bromo-SS-diaryl-sulphoximides generate the corresponding sulphoximidoyl radicals, which add to olefins or abstract hydrogen from toluene. These homolytic chain reactions have been described in detail.[177,178] A novel preparation of an optically active oxysulphonium ylide in the reaction of an optically active sulphoximide with dimethyl diazomalonate has also been described.[179] The stereochemical cycle is shown in Scheme 5.[180]

A similar deimination of sulphoximides by phenyl(trihalogenomethyl)mercury compounds and subsequent substitution by the carbene species also affords the corresponding oxysulphonium ylides [reaction (5)].[181]

Treatment of various vinylic sulphoximides $R^1SO(=NR^2)CH=CHR^3$ (R^1 = Tol, $PhCH_2$, or naphthyl; R^2 = phthalimido; R^3 = H, and R^1 = Tol, R^2 = phthalimido, R^3 = Me) with 0.5 molar equivalents of (−)-ephedrine results in a Michael addition with asymmetric induction, affording optically active sulphoximides in up to 46% enantiomeric excess.[182]

$$PhHgCX_2Br + \underset{R^2}{\overset{R^1}{\underset{NH}{\overset{O}{S}}}} \longrightarrow \underset{R^2}{\overset{R^1}{\underset{CX_2}{\overset{+O}{S^-}}}} + \underset{R^2}{\overset{R^1}{S}}=O + X_2C=CX_2 \quad (5)$$

(R^1, R^2 = Ph or Me; R^1 = Me, R^2 = Ph)

[171] A. C. Barnes, P. W. Hairsine, D. P. Kay, P. J. Ramm, and J. B. Taylor, *J. Heterocycl. Chem.*, 1979, **16**, 1089.

[172] P. Stoss, G. Satzinger, M. H. Herrmann, and W. Heldt, Ger. Offen. 2 758 613 (*Chem. Abstr.*, 1979, **91**, 107 878).

[173] S. Jugé and G. Meyer, *Tetrahedron*, 1980, **36**, 959.

[174] T. Akasaka, T. Yoshimura, N. Furukawa, and S. Oae, *Chem. Lett.*, 1978, 417.

[175] N. Furukawa, K. Akutagawa, T. Yoshimura, T. Akasaka, and S. Oae, *Synthesis*, 1979, 289.

[176] H. Morita, H. Itoh, N. Furukawa, and S. Oae, *Chem. Lett.*, 1978, 817.

[177] T. Akasaka, N. Furukawa, and S. Oae, *Chem. Lett.*, 1979, 529.

[178] T. Akasaka, N. Furukawa, and S. Oae, *Tetrahedron Lett.*, 1979, 2035.

[179] N. Furukawa, F. Takahashi, T. Yoshimura, and S. Oae, *Tetrahedron Lett.*, 1977, 3633.

[180] N. Furukawa, F. Takahashi, T. Yoshimura, and S. Oae, *Tetrahedron*, 1979, **35**, 317.

[181] J. Iqbal and W. Rahman, *J. Organomet. Chem.*, 1979, **169**, 141.

[182] R. Annunziata and M. Cinquini, *J. Chem. Soc., Perkin Trans. 1*, 1979, 1684.

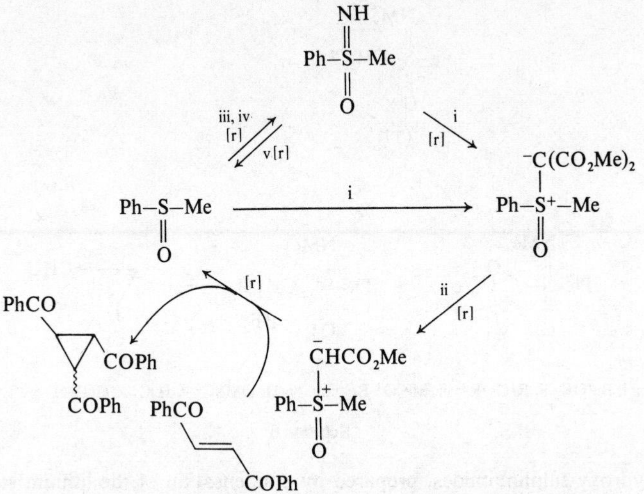

Reagents: i, $N_2C(CO_2Me)_2$, $CuSO_4$, benzene; ii, KOH, aq. MeOH; iii, TsN_3; iv, conc. H_2SO_4; v, $NaNO_2$

Scheme 5

α-Halogeno-sulphoximides are interesting sulphoximide derivatives and have various uses. They are, for example, alkylating reagents like sulphonium salts or S-ylides. Preparation is achieved either by the direct α-halogenation of the *N*-methyl derivatives, by treatment of *N*-methylsulphoximides with t-butyl hypochlorite, or by imination of α-chloro-sulphoxides with *O*-mesitylenesulphonylhydroxylamine.[183] Regioselectivity and stereoselectivity for halogenation have been described.[184] Dialkyl α-halogeno-sulphoximides (76; R^1, R^2 = Et, Pr, or Ph; R^1 = H or Me) have been shown to undergo base-promoted Ramberg–Bäcklund-type rearrangement to afford eventually the olefins. The stereochemistry of the formation of olefins depends upon the substituents; when R = Ph, the *Z*-isomers are predominantly formed, but the *E*-isomers predominate if R is an alkyl group.[185]

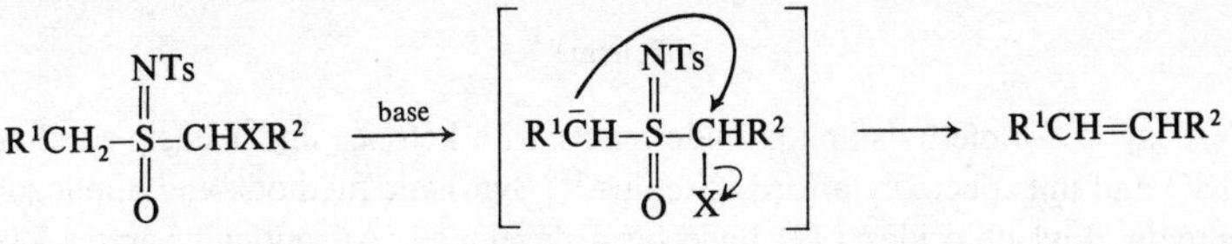

(76)

Similar sulphoximides (77) have also been found to condense with various ketones and aldehydes to give epoxides or simple adducts. The mode of the reaction varies with the type of ketone or aldehyde used (Scheme 6).[186]

[183] C. R. Johnson and C. J. Stark, Jr., *Tetrahedron Lett.*, 1979, 4713.
[184] C. R. Johnson and H. G. Corkins, *J. Org. Chem.*, 1978, **43**, 4136.
[185] C. R. Johnson and H. G. Corkins, *J. Org. Chem.*, 1978, **43**, 4140.
[186] H. G. Corkins, L. Veenstra, and C. R. Johnson, *J. Org. Chem.*, 1978, **43**, 4283.

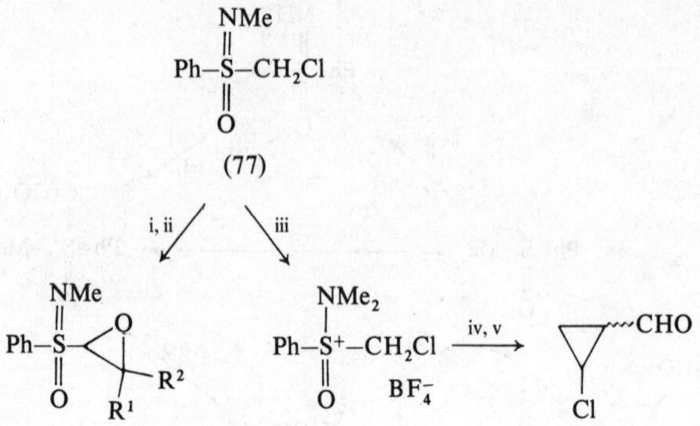

$$\begin{array}{c} NMe \\ \parallel \\ Ph-S-CH_2Cl \\ \parallel \\ O \end{array}$$

(77)

i, ii ↙   ↓ iii

$$\begin{array}{c} NMe \\ \parallel \quad O \\ Ph-S \overset{}{\diagdown} \\ \parallel \quad \diagup R^2 \\ O \quad R^1 \end{array} \qquad \begin{array}{c} NMe_2 \\ \mid \\ Ph-S^+-CH_2Cl \\ \parallel \\ O \quad BF_4^- \end{array} \xrightarrow{\ iv, v\ } \quad \triangleright\!\!\!\sim\!\!\!CHO \\ Cl$$

Reagents: i, Bu^{t}OK; ii, R^{1}COR^2; iii, Me$_3$O$^+$ BF$_4^-$; iv, NaH, DMSO; v, H$_2$C=CHCHO

Scheme 6

β-Hydroxy-sulphoximides, prepared by condensation of the lithium salts of the sulphoximides with aldehydes or ketones, undergo reductive elimination of the sulphoximide groups upon treatment with aluminium amalgam, forming the corresponding alkenes.[187] β-Hydroxy-sulphoximides, after dehydration and treatment of the resulting vinylic sulphoxides with Me$_3$O$^+$ BF$_4^-$, give the sulphoximide salts (78). Michael addition of active methylenic compounds or imines to the compounds (78) and subsequent elimination of NN-dimethylbenzenesulphinamide can give cyclopropanes or aziridines (Scheme 7). An optically active (78), when it reacted with, *e.g.*, methyl cyanoacetate, gave 1-cyano-2-phenylcyclopropane of 25.5% optical purity.[188]

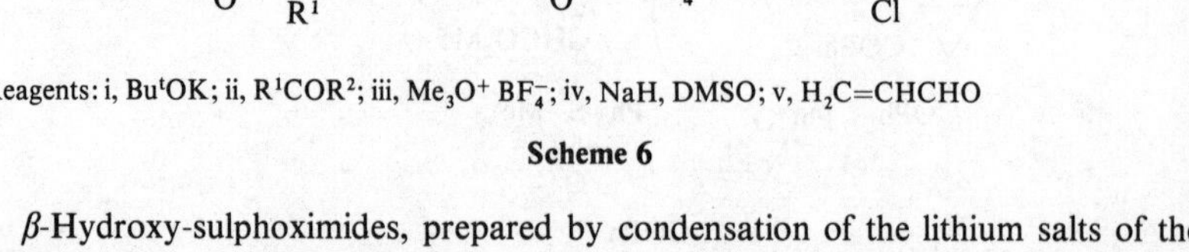

$$\begin{array}{c} R^2 \\ \triangleright\!\!\!-NCMe_2CH_2Ph \\ R^1 \end{array} \xleftarrow{\ i, ii\ } \begin{array}{c} O \\ \parallel \; + \quad R^1 \\ Ph-S-CH=C \\ \mid \qquad\quad R^2 \\ NMe_2 \end{array} \xrightarrow{\ iii, iv\ } \begin{array}{c} R^2 \quad COPh \\ \triangleright\!\!\!< \\ R^1 \quad CO_2Me \end{array}$$

(78)

Reagents: i, NEt$_3$; ii, PhCH$_2$CMe$_2$NH$_2$; iii, NaOMe; iv, PhCOCH$_2$CO$_2$Me

Scheme 7

N-Tosyl-SS-dimethylsulphoximide reacts with ketones under basic conditions in DMSO and unexpectedly affords oxetans.[189] Synthetic methods and applications of sulphoximidoyl chlorides (79) have been developed. An optically active (79) has also been prepared. Since the sulphoximidoyl chloride has a highly electrophilic central sulphur atom, a number of nitrogen- and oxygen-containing nucleophiles have been shown to react with (79) to give the substituted products.[190–192]

[187] C. R. Johnson and R. A. Kirchhoff, *J. Am. Chem. Soc.*, 1979, **101**, 3602.
[188] C. R. Johnson, J. P. Lockard, and E. R. Kennedy, *J. Org. Chem.*, 1980, **45**, 264.
[189] S. C. Welch and A. S. C. P. Rao, *J. Am. Chem. Soc.*, 1979, **101**, 6135.
[190] C. R. Johnson, E. U. Jonsson, and A. Wambsgans, *J. Org. Chem.*, 1979, **44**, 2055.
[191] C. R. Johnson and A. Wambsgans, *J. Org. Chem.*, 1979, **44**, 2278.
[192] C. R. Johnson, E. U. Jonsson, and A. Wambsgans, *J. Org. Chem.*, 1979, **44**, 2061.

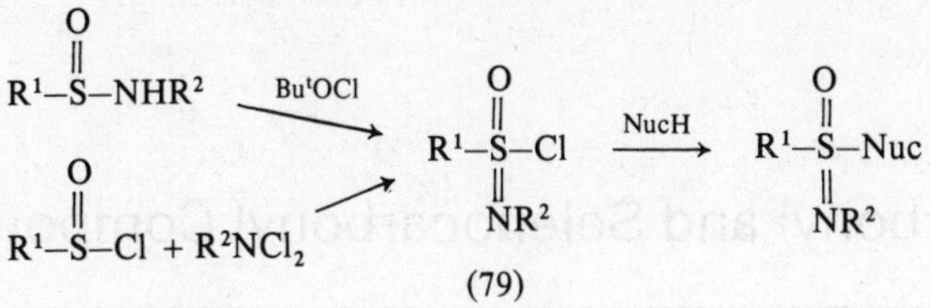

$$R^1-\overset{\overset{\textstyle O}{\|}}{S}-NHR^2 \xrightarrow{\text{Bu}^t\text{OCl}}$$

$$R^1-\overset{\overset{\textstyle O}{\|}}{S}-Cl + R^2NCl_2 \nearrow$$

$$R^1-\overset{\overset{\textstyle O}{\|}}{\underset{\underset{\textstyle NR^2}{\|}}{S}}-Cl \xrightarrow{\text{NucH}} R^1-\overset{\overset{\textstyle O}{\|}}{\underset{\underset{\textstyle NR^2}{\|}}{S}}-Nuc$$

$$(79)$$

The Michael addition of sulphoximides to, *e.g.*, dimethyl acetylenedicarboxylate has been carried out, affording the corresponding adducts. The stereochemical course of the addition has been investigated.[193]

Cyclic oxygen analogues (80) of sulphoximides can also be prepared.[194,195] The biological role of L-methionine sulphoximine and its effect upon glutamine synthetase has been investigated.[196]

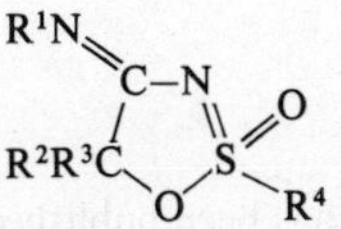

$$\begin{array}{c} R^1N \\ \diagdown \\ R^2R^3C \end{array}\overset{\textstyle C-N}{\diagdown}\begin{array}{c} \diagup O \\ S \\ \diagdown R^4 \end{array}$$

(80) R^1, R^2, R^3, R^4 are alkyl

Sulphodi-imides.—Methods of preparation of a few sulphodi-imides have been reported.[71,72,197] An aza-analogue of sulphurane has also been prepared.[198]

4 Selenium and Tellurium Analogues

Selenium di-imides can be prepared in nearly quantitative yields by treating $RN(SiMe_3)_2$ with SeF_4 or $SeCl_4$. The di-imides react with $RSO_2N=SeCl_2$ to give the unsymmetrical selenium imides $RN=Se=NSO_2R$.[199]

Sulphurane analogues of telluryl compounds are also known to give telluryl-imides similarly. Simple treatment of telluryl or selenyl oxides with primary amines in $CHCl_3$ also yields the corresponding telluryl- or selenyl-imides.[200]

[193] Y. Tamura, S. M. Bayomi, M. Tsunekawa, and M. Ikeda, *Chem. Pharm. Bull.*, 1979, **27**, 2137.
[194] G. L'abbé and A. Verbruggen, *Tetrahedron Lett.*, 1979, 49.
[195] G. L'abbé, C. C. Yu, and S. Toppet, *J. Org. Chem.*, 1979, **44**, 3991.
[196] A. Shrake, E. J. Whiting, Jr., and A. Ginsburg, *J. Biol. Chem.*, 1980, **255**, 581.
[197] W. Heider and O. Glemser, *Chem. Ber.*, 1979, **113**, 237.
[198] L. J. Adzima, C. C. Chiang, I. C Paul, and J. C. Martin, *J. Am. Chem. Soc.*, 1978, **100**, 953.
[199] N. Ya. Derkach, G. G. Barashenko, and E. S. Levchenko, *Zh. Org. Khim.*, 1979, **15**, 1547.
[200] V. I. Naddaka, V. P. Carkin, I. D. Sadekov, V. P. Krasnov, and V. I. Minkin, *Zh. Org. Khim.*, 1979, **15**, 896.

3
Thiocarbonyl and Selenocarbonyl Compounds

BY D. R. HOGG, J. K. LANDQUIST & A. OHNO

PART I: **Thioaldehydes, Thioketones, Thioketens, and their Selenium Analogues**
by A. Ohno

Reviews.—No important review has been published in the past two years.

1 Thioaldehydes and Selenoaldehydes

Synthesis.—No report concerned with the preparation and identification of stable thioaldehydes has appeared; however, pyrrolo[2,1-*b*]thiazole-7- (1) and -5-carbo-selenoaldehydes (2)—(4) have been synthesized by the selenoformylation of pyrrolo[2,1-*b*]thiazoles and characterized by their ^{1}H n.m.r. and u.v. spectra.[1] With the exception of (4), these selenoaldehydes are stable, green, crystalline compounds.

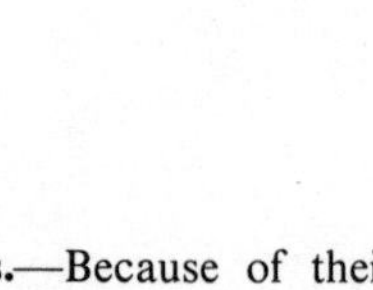

(1)

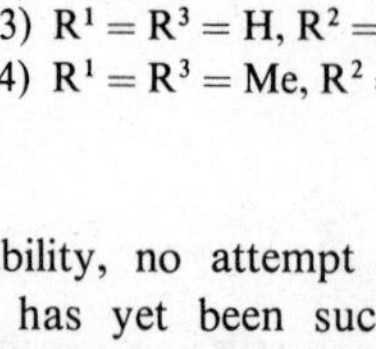

(2) $R^1 = H, R^2 = R^3 = Me$
(3) $R^1 = R^3 = H, R^2 = Me$
(4) $R^1 = R^3 = Me, R^2 = H$

Transient Species.—Because of their instability, no attempt at identifying the intermediacy of thio- or seleno-aldehydes has yet been successful. However, product analyses and mechanistic considerations suggest that thio- or seleno-aldehydes are formed as intermediates in certain reactions. For example, thioformaldehyde has been proposed as a transient species in the reaction of nitroalkanes with sodium methanethiolate, as shown in reactions (1) and (2).[2] The

[1] D. H. Reid, R. G. Webster, and S. McKenzie, *J. Chem. Soc., Perkin Trans. 1*, 1979, 2334.
[2] N. Kornblum, S. C. Carlson, and R. G. Smith, *J. Am. Chem. Soc.*, 1979, **101**, 647.

$$R\cdot + CH_3S^- \longrightarrow RH + \cdot CH_2S^- \qquad (1)$$

$$RNO_2 + \cdot CH_2S^- \longrightarrow RNO_2^{-} + H_2C{=}S \qquad (2)$$

thioformaldehyde that is formed reacts with excess methanethiolate anion to form dimethyl trithiocarbonate. Ozonolysis of benzyl phenylthiolacetate in 1,2-dichloro-ethane affords phenylacetic acid and thiobenzaldehyde (Scheme 1).[3] The latter compound is further oxidized to benzoic acid and, in part, benzaldehyde under the reaction conditions. When 2-azidophenyl 2'-thienyl sulphide is subjected to thermolysis or photolysis, pyrrolo[2,1-*b*]benzothiazole is obtained. The thio-aldehyde (5) has been proposed as an intermediate.[4] Photoisomerization of isothiazol-3(2*H*)-ones yields thiazol-2(3*H*)-ones, probably through a ring-contraction–ring-expansion mechanism, and an α-lactam that is substituted by a thioformyl group, *e.g.* (6)–(10), may be an intermediate.[5]

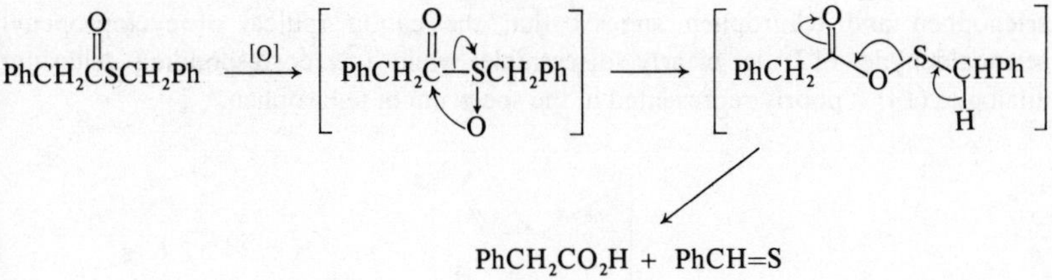

$$PhCH_2CO_2H + PhCH{=}S$$

Scheme 1

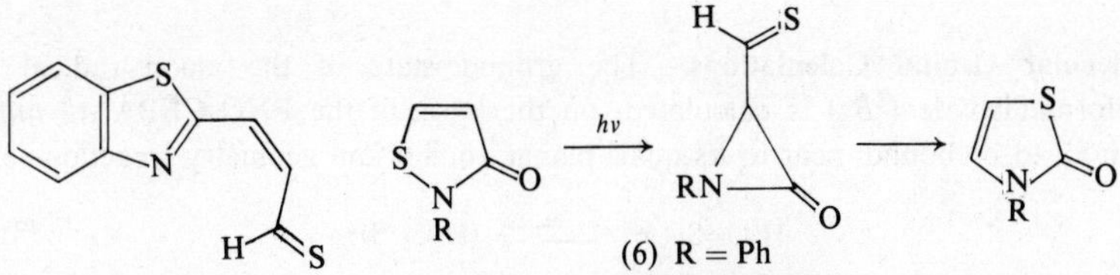

(5)

(6) R = Ph
(7) R = PhCH$_2$
(8) R = ClCH$_2$CH$_2$
(9) R = Bu^t
(10) R = cyclohexyl

2,5-Dihydrothiophenium phenacylide, in refluxing benzene, is decomposed directly or through 2-benzoyl-3,6-dihydro-2*H*-thiopyran into 2-phenylthioglyoxal (11), as shown in Scheme 2.[6] The thioglyoxal is trapped by the thiophenium phenacylide to give 2-benzoyl-5-phenyl-1,3-oxathiole.

The reaction of an aldehyde with sodium hydrogen selenide and an amine hydrochloride generates a selenoaldehyde, which is reduced *in situ* to a symmetric diselenide by sodium borohydride.[7]

[3] H. J. Chaves das Neves and L. S. Godinho, *Tetrahedron*, 1979, **35**, 2053.
[4] J. M. Lindley, O. Meth-Cohn, and H. Suschitzky, *J. Chem. Soc., Perkin Trans. 1*, 1978, 1198.
[5] J. Rokach and P. Hamel, *J. Chem. Soc., Chem. Commun.*, 1979, 786.
[6] U. J. Kempe, T. Kempe, and T. Norin, *J. Chem. Soc., Perkin Trans. 1*, 1978, 1547.
[7] J. W. Lewicki, W. H. H. Günther, and J. Y. C. Chu, *J. Org. Chem.*, 1978, **43**, 2672.

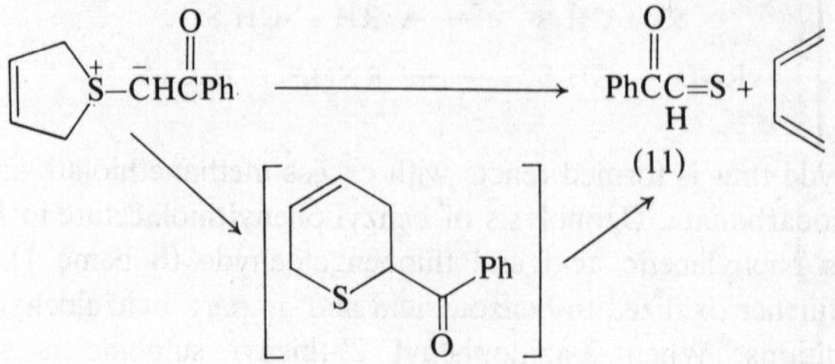

Scheme 2

Mass Spectrometry.—It has been revealed by collisional activation mass spectral studies that protonated thioformaldehyde is about 5—10 kcal mol^{-1} more stable in the structure $H_2C=\overset{+}{S}H$ than in the structure $H_3C-\overset{+}{S}$, which is in reversed order from that of the oxygen analogue.[8] Mass-spectrometric fragmentation studies of selenophen and tellurophen suggest that the cation radical of cyclopropenyl selenoaldehyde (12) is clearly discernible while the corresponding tellurium analogue (13) is poorly represented in the spectrum of tellurophen.[9]

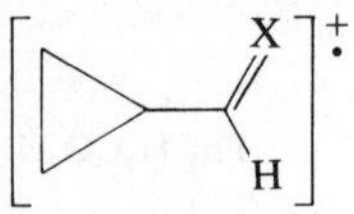

(12) X = Se
(13) X = Te

Molecular Orbital Calculations.—The ground state of the anion radical of thioformaldehyde ($^2B_1$) is calculated, on the basis of the PNO-CEPA *ab initio* method, to be bound, near to its quasi-planar equilibrium geometry [reaction (3)].

$$H_2C=S + e^- \rightleftharpoons (H_2C=S)^{\overline{\cdot}} \qquad (3)$$

The acquisition of an electron should cause only small changes in the H_2C group but a lengthening of the C=S bond.[10]

2 Thioketones

Synthesis.—The highly hindered thioketones (14)—(17) have been prepared [in the yields shown] by the reaction of triphenylphosphorylidene hydrazone with elemental sulphur, by heating the mixture at 130—140 °C under vacuum (1—5 mmHg).[11] The yields are considerably improved when the corresponding hydrazones are allowed to react with disulphur dichloride in the presence of triethylamine in benzene, at 0 °C to room temperature.[12] With the latter method, the yields of (15)—(17) are quantitative. Thioketones (18)—(20) have also been

[8] J. D. Dill and F. W. McLafferty, *J. Am. Chem. Soc.*, 1978, **100**, 2907.
[9] F. Fringuelli and A. Taticchi, *J. Heterocycl. Chem.*, 1978, **15**, 137.
[10] P. Rosmus and H. Bock, *J. Chem. Soc., Chem. Commun.*, 1979, 334.
[11] P. de Mayo, G. L. R. Petrašiūnas, and A. C. Weeden, *Tetrahedron Lett.*, 1978, 4621.
[12] R. Okazaki, K. Inoue, and N. Inamoto, *Tetrahedron Lett.*, 1979, 3673.

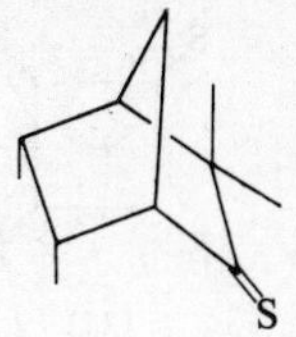

(14) [13%] 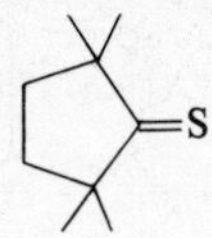   (15) [41%] 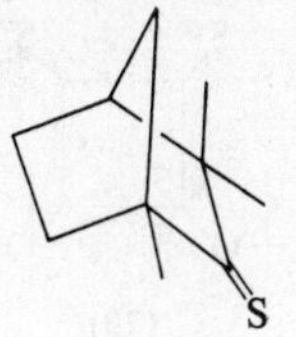   (16) [42%] 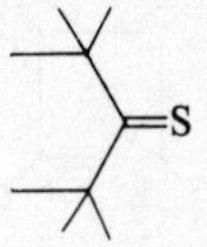   (17) [83%]

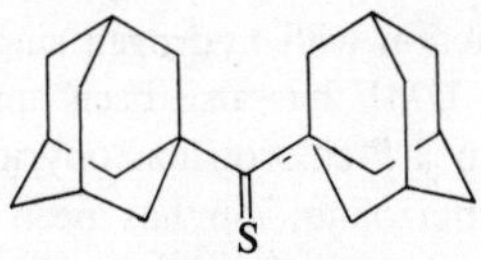

(18) [71%]

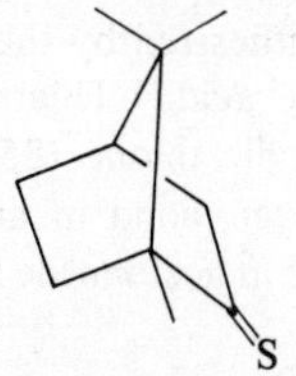

(19) [63%]

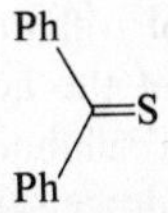

(20) [54%]

prepared by the latter method. $\Delta^{1,4}$-Diene-3-thiones have been prepared from corticosterone-1,4-dien-3-ones and phosphorus pentasulphide.[13] However, the corresponding Δ^4-ene-3-thiones were not isolated when testosterone acetate or androst-4-ene-3,11,17-trione was subjected to the reaction.[13] The enethiones can be trapped *in situ*, giving the expected diene sulphides. Cyclocumarol and phosphorus pentasulphide react to give the three heterocyclic thiones (21)—(23).[13] The more simple enethione (24) has been prepared and found to be stable at −60 °C.[14] The reaction of an aminocyclobutenedione with phosphorus pentasulphide in dichloromethane affords a 1-amino-4-oxo-3-thioxocyclobutene, *e.g.* (25)—(27), instead of its corresponding 3-oxo-4-thioxo-isomer.[15] A number of dianions (28)—(31) of thiosquaric acid have also been prepared by Seitz and his co-workers.[16]

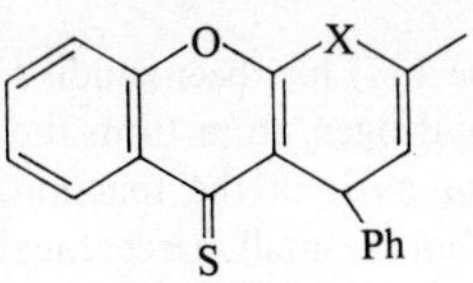

(21) X = S
(22) X = O

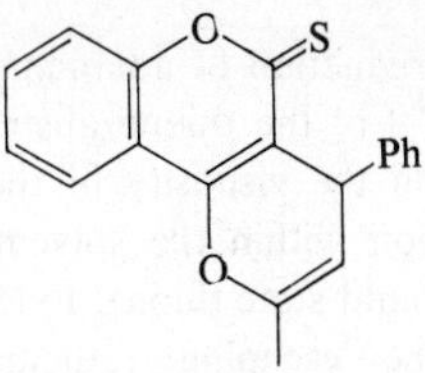

(23)

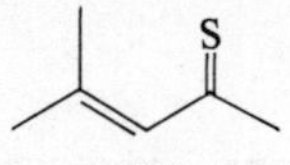

(24)

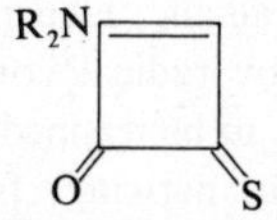

(25) R = Me
(26) $R_2 = -(CH_2)_4-$
(27) $R_2 = -(CH_2)_2O(CH_2)_2-$

[13] D. H. R. Barton, L. S. L. Choi, R. H. Hesse, M. M. Pechet, and C. Wilshire, *J. Chem. Soc., Perkin Trans. 1*, 1979, 1166.
[14] P. Beslin, D. Lagain, and J. Vialle, *Tetrahedron Lett.*, 1979, 2677.
[15] G. Seitz, H. Morck, R. Schmiedel, and R. Sutrisno, *Synthesis*, 1969, 361.
[16] G. Seitz, K. Mann, R. Schmiedel, and R. Matusch, *Chem. Ber.*, 1979, **112**, 990.

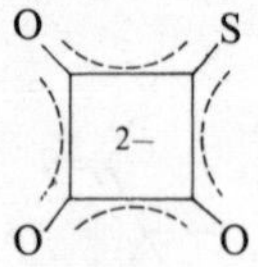

(28) 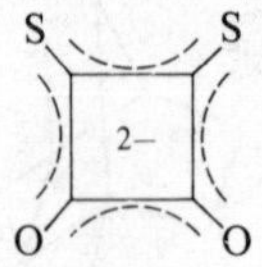   (29) 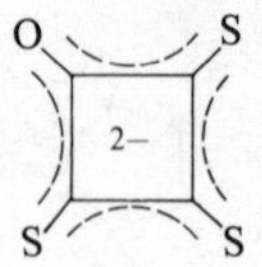   (30) 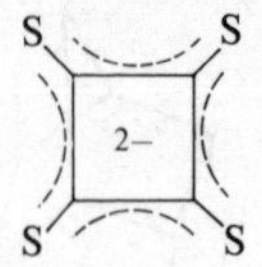   (31)

A modification of the Pinner reaction provides a facile device to prepare β-thioxo-esters: from the corresponding β-enamino-esters in DMF, the thiones (32)—(34) have been synthesized by their reaction with hydrogen sulphide in the presence of trifluoroacetic acid.[17] Thiation in DMF has also been applied to the synthesis of the heterocyclic thione (35) from a 7-chlorotriazolopyridine derivative.[18] The sulphur–nitrogen bond in an isothiazolium ion has been cleaved by sodium hydrogen sulphide in acetonitrile to give the enamino-thione (36).[19]

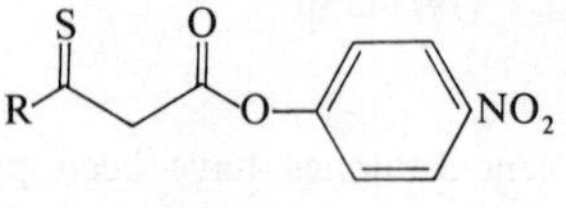

(32) R = Ph [76%]
(33) R = *p*-MeOC$_6$H$_4$ [98%]
(34) R = Me [85%]

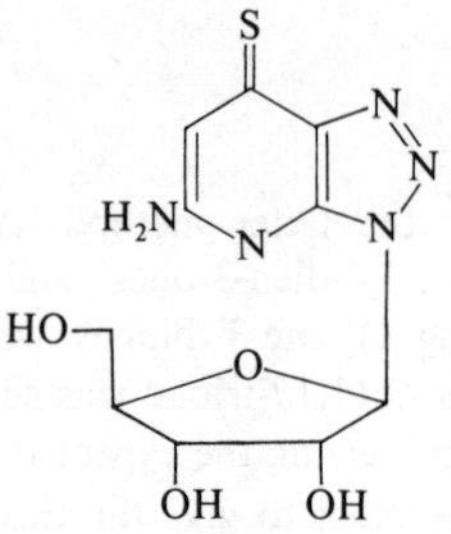

(35)

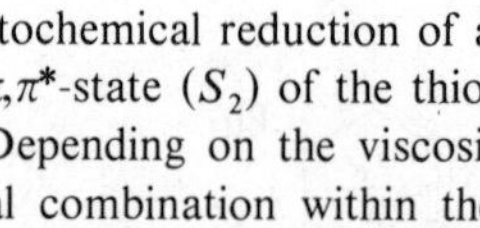

(36)

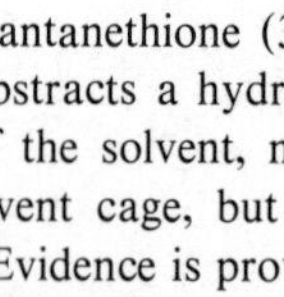

(37)

Reactions.—Photochemical reduction of adamantanethione (37) has been studied in detail.[20] The π,π*-state (S_2) of the thione abstracts a hydrogen atom from the solvent alkane. Depending on the viscosity of the solvent, most of the reaction occurs by radical combination within the solvent cage, but a small percentage escapes, to be trapped by ground-state thione. Evidence is provided that any chain sequence deriving from the escaping radicals must be short ($k = 6.7 \times 10^7$ l mol^{-1} s^{-1} in cyclohexane and 6.3×10^7 l mol^{-1} s^{-1} in cyclopentane). Di-t-butyl thioketone (17) has also been subjected to photoreduction.[21] Only compounds that have an aromatic moiety or an amino-group can reduce the n,π*-state of thione to

[17] F. DiNinno and E. V. Linek, *J. Org. Chem.*, 1979, **44**, 3271.
[18] B. L. Cline, R. P. Panzica, and L. B. Townsend, *J. Org. Chem.*, 1978, **43**, 4910.
[19] A. G. Briggs, J. Czyzewski, and D. H. Reid, *J. Chem. Soc., Perkin Trans. 1*, 1979, 2340.
[20] K. Y. Law and P. de Mayo, *J. Am. Chem. Soc.*, 1979, **101**, 3251.
[21] A. Ohno, M. Uohama, K. Nakamura, and S. Oka, *Bull. Chem. Soc. Jpn.*, 1979, **52**, 1521.

its corresponding thiol. A charge-transfer-type exciplex has been proposed, but without unambiguous proof.

Rajee and Ramamurthy have proposed that singlet oxygen is involved in the photo-oxidation of the thiones (16), (19), (37), and (38)—(40) to the corresponding ketones.[22] On the other hand, Tamagaki and his co-workers have reported that the thioketone (17) is photo-oxidized to di-t-butyl ketone by ground-state oxygen, whereas the photo-oxidation by singlet oxygen affords di-t-butylsulphine.[23]

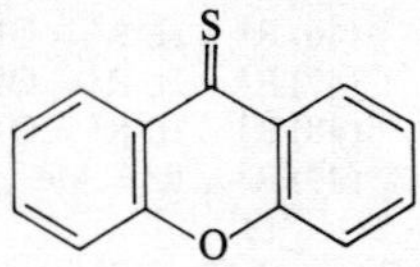

(38) 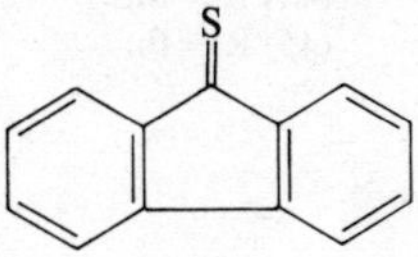   (39) 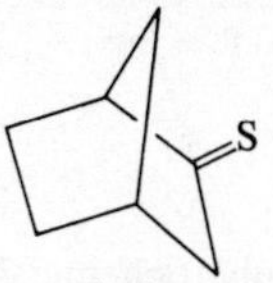   (40)

When the thione (17) is photo-excited to its π,π^*-state, hydrogen-abstraction from alkanes and cycloaddition to olefins takes place, which is in remarkable contrast with the reactivity of (17) in the n,π^*-state.[24]

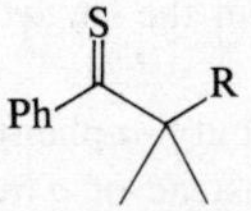

(41) R is a variety of alkyl groups

The mechanism of hydrogen-abstraction of aryl alkyl thioketones (41) to give cyclopentyl thiols has been investigated.[25] By comparison of the rate of hydrogen-abstraction to give a 1,5-biradical with that of product formation, it has been shown that only about one-tenth of the biradicals cyclize. The rest disproportionate to give the original thioketone. The existence of an intermediate and the singlet state as the reactive state have been confirmed. It is also mentioned that hydrogen-abstraction takes place in a similar manner to the alkanone type II singlet process $^1(n,\pi^*)$, although the active species in the photochemistry of thio-ketones is in the π,π^*-state.

When xanthenethione (38) is excited to its n,π^*-state, [2 + 2]cycloaddition proceeds with bis(alkylthio)ethyne. The cyclo-adduct [(42) or (43)] is in thermal equilibrium with its 'open' isomer [(44) or (45)].[26] A similar reaction of (37) with acenaphthylene, indene, or N-phenylmaleimide affords a spirothietan derivative in good yield.[27] Thioxanthenethione in the n,π^*-state cyclo-adds to alkyl- or alkoxy-substituted butatrienes, giving the thietans (46)—(49).[28]

[22] R. Rajee and V. Ramamurthy, *Tetrahedron Lett.*, 1978, 5127.
[23] S. Tamagaki, R. Akatsuka, M. Nakamura, and S. Kozuka, *Tetrahedron Lett.*, 1979, 3665.
[24] R. Rajee and V. Ramamurthy, *Tetrahedron Lett.*, 1978, 3463.
[25] K. W. Ho and P. de Mayo, *J. Am. Chem. Soc.*, 1979, **101**, 5725.
[26] A. C. Brouwer, A. V. E. George, D. Seykens, and H. J. T. Bos, *Tetrahedron Lett.*, 1978, 4839.
[27] H. Gotthardt and S. Nieberl, *Chem. Ber.*, 1978, **111**, 1471.
[28] R. G. Visser and H. J. T. Bos, *Tetrahedron Lett.*, 1979, 4857.

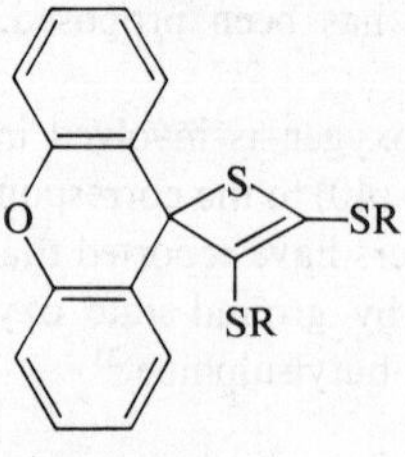

(42) R = Me
(43) R = Bu^t

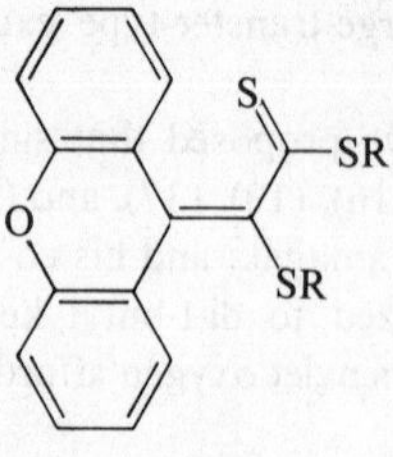

(44) R = Me
(45) R = Bu^t

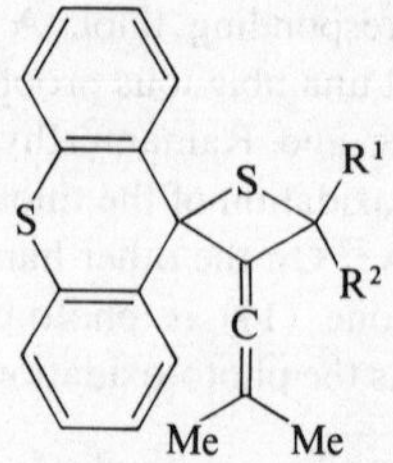

(46) R^1 = H, R^2 = OMe
(47) R^1 = H, R^2 = OBu^t
(48) R^1 = H, R^2 = OPh
(49) R^1 = R^2 = Me

The photochemical formation of thietan from (−)-3-menthyl methacrylate and xanthenethione (38) or 4,4′-dimethoxythiobenzophenone proceeds with three to four-and-a-half times higher optical purity through the triplet n,π^*-state (17—18% e.e.) of the thione than through the singlet π,π^*-state (4—6% e.e.).[29] It has been proposed that the carbon–carbon single bond in the biradical triplet intermediate allows the closure to the four-membered ring to occur preferentially, by the pathway of lowest energy, and thus increases the selectivity of the reaction. The synchronous reaction pathway from the singlet π,π^*-state does not involve such a rotation.

NN-Di-isopropyl- and *NN*-di-s-butyl-α-phenylthioxoacetamides, (50) and (51), undergo γ-hydrogen-abstraction, instead of δ-hydrogen-abstraction, on irradiation, to give the corresponding β-lactams. The same reaction with (52) affords a disulphide (53).[30] The selective abstraction of γ-hydrogen has been interpreted in terms of an electron-transfer mechanism.

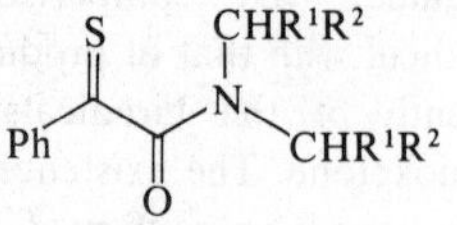

(50) R^1 = R^2 = Me
(51) R^1 = Me, R^2 = Et
(52) R^1 = H, R^2 = Me

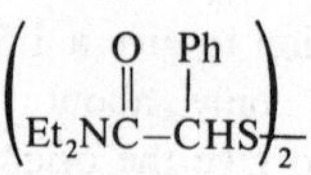

(53)

Thioacetylacetone in the enol form can be tautomerized to the enethiol form on irradiation with light of wavelength 353 nm, at −178 °C.[31] The interconversion between enol and enethiol forms is reversible.

Photo-irradiation (with light of wavelength longer than 520 nm) of a mixture of thiobenzophenone, benzenethiol, and a cobalt(III) complex in acetonitrile results in the generation of hydrogen gas, as shown in reaction (4).[32] The yield is better than 300%, based on the amount of thioketone that is consumed or cobalt(III) that is reduced. Thiobenzophenone (20) plays a catalytic role.

[29] H. Gotthardt and W. Lenz, *Tetrahedron Lett.*, 1979, 2879.
[30] H. Aoyama, S. Suzuki, T. Hasegawa, and Y. Omote, *J. Chem. Soc., Chem. Commun.*, 1979, 899.
[31] L. Carlsen and F. Duus, *J. Chem. Soc., Perkin Trans. 2*, 1979, 1533.
[32] S. Oishi and K. Nozaki, *Chem. Lett.*, 1979, 549.

$$[(PhS)Co^{III}(dmg)_2(py)] + PhSH \xrightarrow[Ar_2 C=S]{h\nu(>520 \text{ nm})} [Co(dmg)_2(py)] + PhSSPh + H_2\uparrow \quad (4)$$

Reactions of thioketones with nucleophilic reagents have been studied to some extent. Thiobenzophenone (20) and its 4,4'-substituted derivatives react with cyclopentadienyliron dicarbonyl anion and cyclopentadienyl-molybdenum or -tungsten tricarbonyl anion at room temperature to give fulvenes.[33] Phase-transfer catalysis conditions improve the yields. An electron-transfer mechanism has been proposed for this desulphurization and coupling reaction. An electron-transfer mechanism has also been proposed for the reduction of thiopivalophenone (54) by 1-benzyl-1,4-dihydronicotinamide in acetonitrile.[34] 2,3-Diphenylcyclopropene-thione (55) reacts with *N*-methylarylamines to give the bicyclic thioamides (56).[35] The formation of a thioketen as an intermediate has been suggested. A similar reaction scheme has also been proposed for the reaction of (55) with 3,4-dihydroisoquinoline.[36]

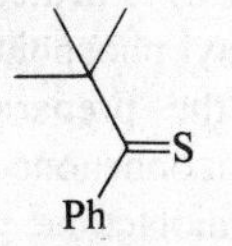

(54) 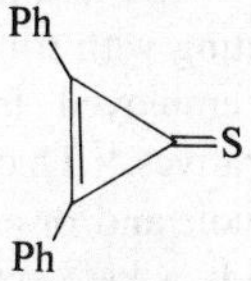 (55) 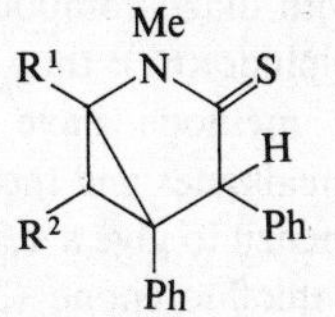 (56)

Reactions of dialkyl thioketones and cycloalkanethiones with trialkyl phosphites give dialkyl 1-alkylthio- or 1-mercapto-dialkyl- and -cycloalkyl-phosphonates, respectively.[37,38] Initial attack by the phosphorus atom takes place on the thiocarbonyl-carbon atom to form a betaine, which is converted into product by an alkyl migration or by a proton migration. The reaction of 4*H*-pyran-4-thiones with ammonia gives 1*H*-pyridine-4-thiones or bis-(4-pyridyl) disulphides, whereas that of 3-methoxy-2-methyl-4*H*-pyran-4-thione with guanidine affords 1,3-thiazine derivatives. 4*H*-Pyran-4-thiones react with hydrazine to give pyridazines, pyrazoles, and 1-amino-1*H*-pyridine-4-thiones. 4*H*-Pyran-4-thiones having a hydroxy-group in the β-position give the corresponding pyridazine derivatives.[39]

Active-methylene compounds are converted into the corresponding thioketones by successive reaction with thionyl chloride and triphenylphosphine.[40] The thioketones thus prepared are too unstable to be isolated, and they undergo dimerization to give the 1,3-dithietans (57)—(59). However, the intermediacy of thioketones has been confirmed by u.v. spectrometry. The transient formation of thioketone and then its dimerization or a Diels–Alder reaction with an olefin has

[33] H. Alper and H.-N. Paik, *J. Am. Chem. Soc.*, 1978, **100**, 508.
[34] A. Ohno, S. Yasui, K. Nakamura, and S. Oka, *Bull. Chem. Soc. Jpn.*, 1978, **51**, 290.
[35] T. Eicher, J. L. Weber, and J. Kopf, *Liebigs Ann. Chem.*, 1978, 1222.
[36] T. Eicher and D. Krause, *Tetrahedron Lett.*, 1979, 1213.
[37] S. Yoneda, T. Kawase, and Z. Yoshida, *J. Org. Chem.*, 1978, **43**, 1980.
[38] T. Kawase, S. Yoneda, and Z. Yoshida, *Bull. Chem. Soc. Jpn.*, 1979, **52**, 3342.
[39] H. Besso, K. Imafuku, and H. Matsumura, *Bull. Chem. Soc. Jpn.*, 1978, **51**, 179.
[40] K. Oka, *J. Org. Chem.*, 1979, **44**, 1736.

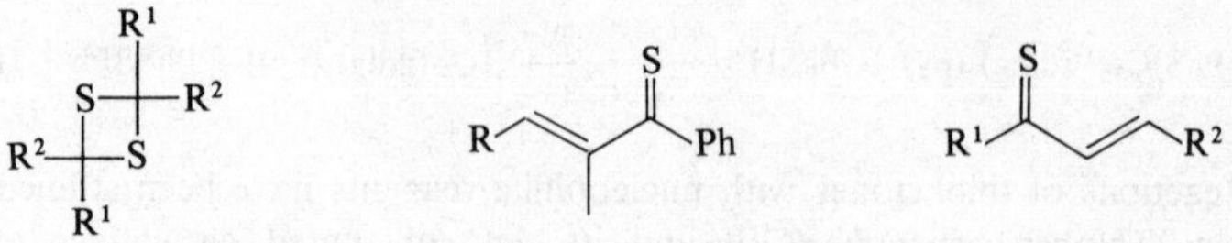

(57) $R^1 = R^2 = CO_2Et$
(58) $R^1 = PhCO, R^2 = Ph$
(59) $R^1 = Ph, R^2 = CO_2Et$

(60) $R = p\text{-}MeOC_6H_4$
(61) $R = p\text{-}ClC_6H_4$

(62) $R^1 = R^2 = Ph$
(63) $R^1 = Ph, R^2 = Me$

also been reported for the $\alpha\beta$-unsaturated compounds (60)—(63).[41,42] Thiophen derivatives are prepared from diethyl acetylsuccinate through diethyl thio-acetylsuccinate, followed by its enethiolization and cyclization.[43] The addition of thioketones to unsaturated compounds is a versatile method for the preparation of heterocycles: treatment of thiobenzophenone (20) and its 4,4'-dimethoxy-derivative with diphenylketen gives the corresponding β-thiolactones (64) and (65), respectively.[44] Highly crowded olefins have been synthesized by the reaction of thioketones with diazo-compounds.[45,46] The adduct (66) is heated, as the solid, to give the episulphide (67); this, on heating with trimethyl phosphite, gives the olefin (68). Similar methods have been employed for the preparation of various adamantylidenealkanes and their derivatives.[47] Thiobenzophenone (20) is allowed to react with azibenzil to give a 1,3-oxathiole and/or a β-thiolactone, whereas the same reaction with thiofluorenone (39) yields a keto-episulphide.[48] 1,3-Oxathioles have also been synthesized from monothiobenzils and diazomethanes.[49]

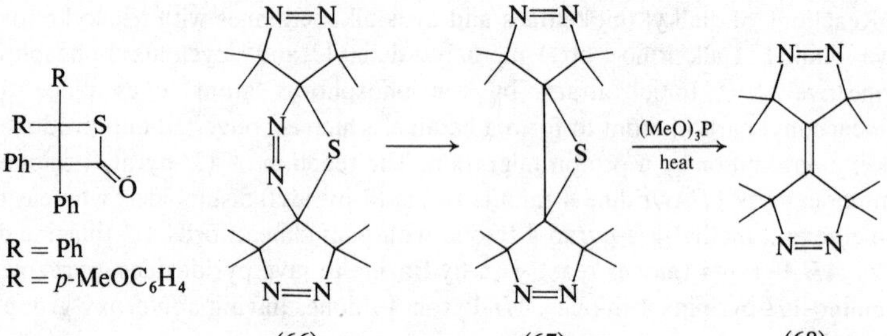

(64) R = Ph
(65) $R = p\text{-}MeOC_6H_4$

(66)   (67)   (68)

At this point it is appropriate to summarize the chemistry of thioquinones. Thermolysis of *p*-quino-bis(benzo-1,3-dithiole) *S*-oxide affords cyclic tri- and tetra-sulphides in about 30% combined yield. The intermediacy of *o*-dithioquinone (69) and species derived from it by desulphurization [*e.g.* (70) and (71)] has been

[41] J.-P. Guémas, A. Reliquet, F. Reliquet, and H. Quiniou, *C. R. Hebd. Seances Acad. Sci., Ser. C*, 1979, **288**, 89.
[42] T. Karakasa and S. Motoki, *J. Org. Chem.*, 1978, **43**, 4147.
[43] F. Duus, *J. Chem. Soc., Perkin Trans. 1*, 1978, 293.
[44] H. Kohn, P. Charumilind, and Y. Gopichand, *J. Org. Chem.*, 1978, **43**, 4961.
[45] R. J. Bushby, M. D. Pollard, and W. S. McDonald, *Tetrahedron Lett.*, 1978, 3851.
[46] R. J. Bushby and M. D. Pollard, *J. Chem. Soc., Perkin Trans. 1*, 1979, 2401.
[47] F. Cordt, R. M. Frank, and D. Lenoir, *Tetrahedron Lett.*, 1979, 505.
[48] S. Mataka, S. Ishi-i, and M. Tashiro, *J. Org. Chem.*, 1978, **43**, 3730.
[49] C. Bak and K. Praefcke, *Chem. Ber.*, 1979, **112**, 2744.

(69)   (70)   (71)

suggested.[50] *o*-Thiobenzoquinone methide is generated, by photochemical desulphonylation, from 3*H*-1,2-benzodithiole 2,2-dioxide.[51] The methide undergoes [4 + 2]cycloaddition with *N*-phenylmaleimide. The cycloadditions of *o*-thioquinone methide with olefins are also reported in several papers.[52-54] The enol acetate of 4-thiochromanone undergoes ring-opening on photolysis. A novel 1,5-acyl migration from oxygen to sulphur in the intermediate *o*-thioquinone methide (72) has been suggested.[55] Nitrogen analogues of *o*-thioquinone methide are also generated by photolysis of *N*-phenyl-2-benzothiazolinone[56] and by the reaction of phenothiazine derivatives with trialkyl phosphites.[57]

(72)   (73)   (74)

Spectroscopy.—Photoelectron spectroscopy and electronic absorption and emission spectra indicate that there exists a strong interaction between the two thiocarbonyl groups in (73), whereas the carbonyl and thiocarbonyl groups in (74) interact only weakly.[58] U.v., i.r., and ^{13}C n.m.r. spectra have been reported for compounds (28)—(31).[16] The relative stabilities of tautomers of β-thioxoketones have been studied on the basis of u.v. and n.m.r. spectroscopies and of molecular orbital calculations.[59] It is concluded that when the thioxoketone is substituted by at least one vinyl group, the enol form is more stable than the enethiol form, whereas methyl and/or phenyl substituents stabilize the enethiol form more than the enol form. E.s.r. spectroscopy has revealed that thiobenzoylphenylmethylene has the triplet biradical structure (75) and that the carbon–carbon bond in (75) has greater double-bond character than the carbon–sulphur bond.[60]

It has been suggested, from *X*-ray crystallography, that there is little interaction between the two sulphur atoms of (3-methyl-4-phenyl-Δ^4-thiazolin-2-ylidene)thioacetophenone (76).[61]

[50] M. Sato, M. V. Lakshmikantham, M. C. Cava, and A. F. Garito, *J. Org. Chem.*, 1978, **43**, 2084.
[51] A. G. Hortmann, A. J. Aron, and A. K. Bhattacharya, *J. Org. Chem.*, 1978, **43**, 3374.
[52] R. Okazaki, F. Ishii, and N. Inamoto, *Bull. Chem. Soc. Jpn.*, 1978, **51**, 309.
[53] R. Okazaki, K.-T. Kang, K. Sunagawa, and N. Inamoto, *Chem. Lett.*, 1978, 55.
[54] R. Okazaki, F. Ishii, K. Sunagawa, and N. Inamoto, *Chem. Lett.*, 1978, 51.
[55] I. W. J. Still and T. S. Leong, *Tetrahedron Lett.*, 1979, 3613.
[56] L. R. Sousa and J. G. Bucher, III, *Tetrahedron Lett.*, 1978, 2267.
[57] J. I. G. Cadogan and N. J. Tweddle, *J. Chem. Soc., Perkin Trans. 1*, 1979, 1278.
[58] K. N. Tantry, P. K. Basu, V. Ramamurthy, C. N. R. Rao, E. A. Seddon, and J. C. Green, *Tetrahedron Lett.*, 1979, 4787.
[59] L. Carlsen and F. Duus, *J. Am. Chem. Soc.*, 1979, **100**, 281.
[60] H. Murai, T. Torres, and O. P. Strausz, *J. Am. Chem. Soc.*, 1979, **101**, 3976.
[61] R. J. S. Beer, D. McMonagle, M. S. S. Siddiqui, A. Hordvik, and K. Jynge, *Tetrahedron*, 1979, **35**, 1199.

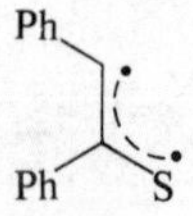

(75) 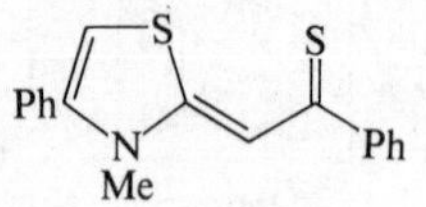   (76) 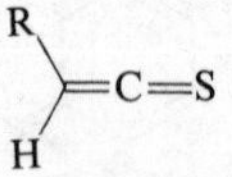   (77) R = Me
                                               (78) R = NO$_2$

3 Thioketens and Selenoketens

Synthesis.—4- or 5-Substituted isothiazoles give thioketens on gas-phase thermolysis at about 600 °C.[62] Chemical trapping of an intermediate and measurement of activation parameters have revealed the reaction mechanism. Dimethylamine was used to trap the thioketens (77) and (78). Thermolysis has also been applied to the synthesis of the thioketen (71), as shown in Scheme 3,[63] the formation of which was established by photoelectron spectroscopy.

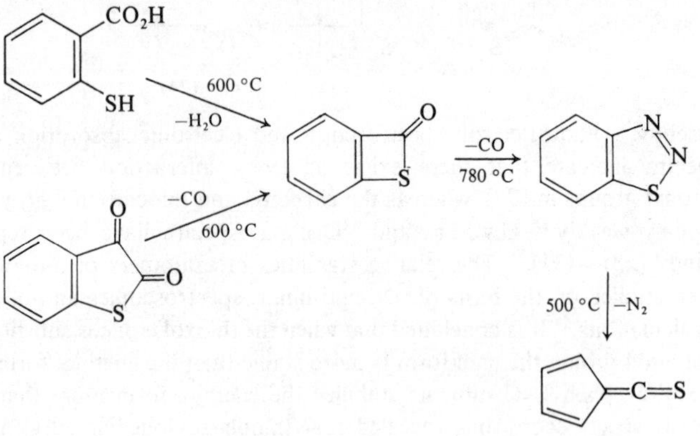

Scheme 3

Photochemical elimination of nitrogen from a bicyclic 1,2,3-thiadiazole results in the formation of a thioketen, *e.g.* (79), instead of a thiiren (see Scheme 4),[64] which is in contrast to the reaction of monocyclic thiadiazoles. When 1,2,3-selenadiazoles are subjected to thermolysis at 500—600 °C, they give the selenoketens (80)—(82).[65] The selenoketens have been trapped and characterized at −196 °C.

The reaction of a ketone with a thioalkynolate gives a β-thionolactone. The intermediacy of a thioketen has been suggested.[66]

[62] G. E. Castillo and H. E. Bertorello, *J. Chem. Soc., Perkin Trans. 1*, 1978, 325.
[63] R. Schulz and A. Schweig, *Tetrahedron Lett.*, 1979, 59.
[64] U. Timm, H. Bühl, and H. Meier, *J. Heterocycl. Chem.*, 1978, **15**, 697.
[65] A. Holm, C. Berg, C. Bjerre, B. Bak, and H. Svanholt, *J. Chem. Soc., Chem. Commun.*, 1979, 99.
[66] N. Miyaura, T. Yanagi, and A. Suzuki, *Chem. Lett.*, 1979, 535.

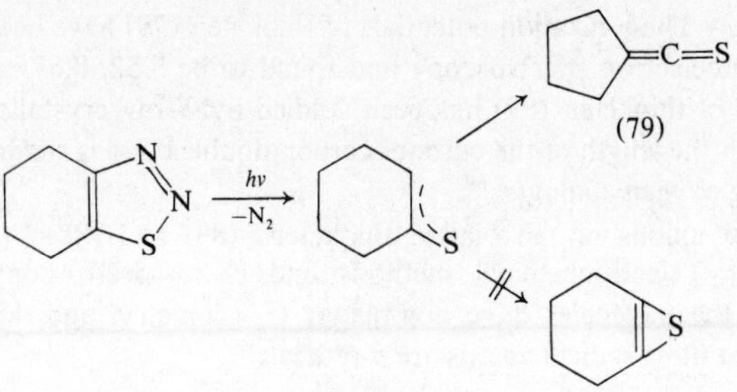

Scheme 4

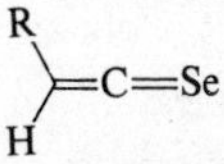

(80) R = H
(81) R = Me
(82) R = Bu^t

Reactions.—Thioketen (83) reacts with azomethines to give β-thiolactams. It has been suggested that the dipolar species (84) is formed, instead of the classical dipolar species (85), in the rate-determining step of the cycloaddition.[67] Bis-(trifluoromethyl)thioketen (86) adds to azomethines, isothiocyanates, and azides to form 1,3,5-dithiazines and thiazetidines, 1,3-dithietans, and Δ^3-1,2,3,4-thia-triazolines, respectively.[68] The thioketen (86) also undergoes [4 + 2]cycloadditions with dienes.[68]

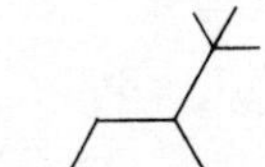

(83)

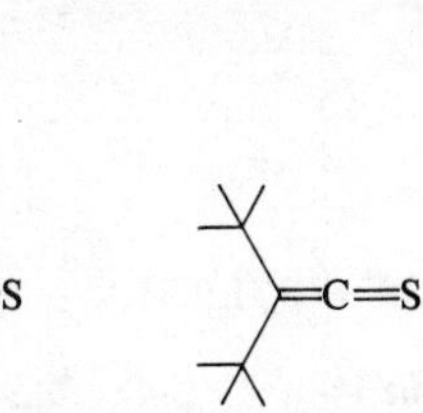

(84)

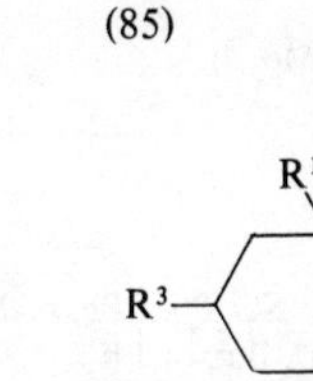

(85)

(CF$_3$)$_2$C=C=S

(86)

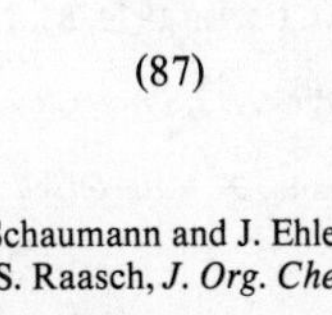

(87)

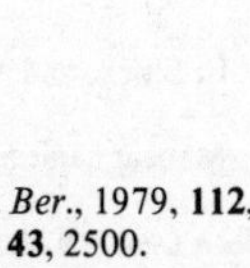

(88)

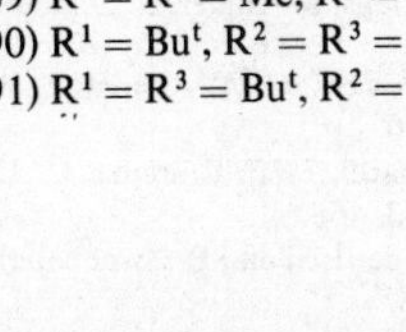

(89) R^1 = R^2 = Me, R^3 = H
(90) R^1 = Bu^t, R^2 = R^3 = H
(91) R^1 = R^3 = Bu^t, R^2 = H

[67] E. Schaumann and J. Ehlers, *Chem. Ber.*, 1979, **112**, 1000.
[68] M. S. Raasch, *J. Org. Chem.*, 1978, **43**, 2500.

Spectroscopy.—The ionization potentials of thioketen (79) have been measured by means of photoelectron spectroscopy and found to be 8.52, 8.67, and 11.14 eV.[63] The structure of thioketen (87) has been studied by X-ray crystallography, and it was found that the length of the carbon–carbon double bond is similar to that in the corresponding oxygen analogue.[69]

The radical anions of the dialkyl thioketens (83) and (88)—(91) have been investigated by electrochemical methods and e.s.r. spectroscopy. The results indicate that the molecules have non-planar C_s geometry, and this supports the suggestion that these radical anions are σ-radicals.[70]

PART II: Sulphines and Sulphenes *by D. R. Hogg*

1 Sulphines

Sulphine, $H_2C=S=O$, generated by vacuum pyrolysis of 1,3-dithietan 1-oxide (see Vol. 4, p. 136) or allyl methyl sulphoxide, has been trapped in an argon matrix at 18 K and its i.r. spectrum studied.[1] A complete vibrational assignment is proposed. The region of low ionization energy of the p.e. spectra of $(p\text{-}XC_6H_4)_2C=S=O$ (X = H, OMe, Me, I, Br, Cl, or NO$_2$) has been assigned on the basis of perturbation M.O. theory arguments coupled with CNDO/2 calculations and substituent effects.[2] The HOMO has π-symmetry, unlike that for the corresponding thioketones, which is mainly sulphur lone-pair.

Preparation of sulphines by the alkylidenation of sulphur dioxide, using α-silyl carbanions (see Vol. 5, p. 133), has been extended by the development of an alternative synthesis of the carbanions by addition of organolithium compounds to vinylsilanes,[3] as shown in Scheme 1.

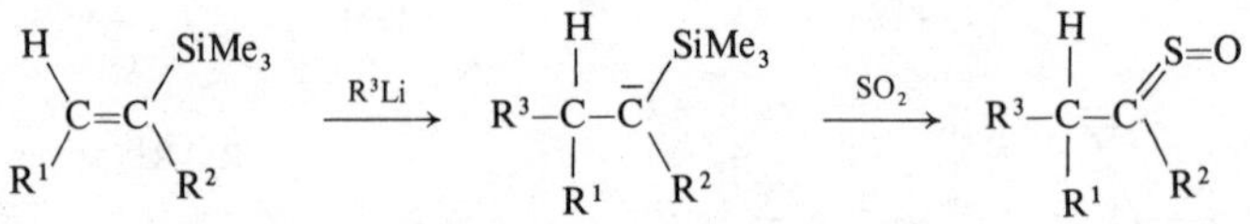

$$R^1 = H, R^2 = SO_2Ph, SPh, \text{ or Ph; } R^1R^2 = C_6H_4(CH_2)_2$$
$$R^3 = Me, Bu^n, Bu^t, \text{ or Ph}$$

Scheme 1

[69] E. Schaumann, S. Harto, and G. Adiwidjaja, *Chem. Ber.*, 1979, **112**, 2698.
[70] C.-P. Klags, S. Köhler, E. Schaumann, W. Schmüser, and J. Voss, *J. Phys. Chem.*, 1979, **83**, 738.

[1] D. E. Powers, C. A. Arrington, W. C. Harris, E. Block, and V. F. Kalasinsky, *J. Phys. Chem.*, 1979, **83**, 1890.
[2] F. Bernardi, F. P. Colonna, C. Distefano, G. Maccagnani, and G. Spunta, *Z. Naturforsch.*, *Teil A*, 1978, **33**, 468.
[3] M. van der Leij and B. Zwanenburg, *Tetrahedron Lett.*, 1978, 3383.

The $\alpha\beta$-unsaturated sulphine (1) was obtained as a transient intermediate by oxidation of the dimer of 2-benzylidene-1-thiotetralone with MCPBA, followed by pyrolysis of the resulting product in boiling xylene.[4] The vinylsulphines (2) were obtained by rearrangement of the vinylsulphinylcarbenes (3) that were generated by photolysis of 5-alkylsulphinyl-3,3-dimethyl-3H-pyrazoles or by spontaneous ring-opening of the corresponding sulphinylcyclopropenes.[5] Allyl groups migrated without rearrangement. The sulphines (2) were trapped by addition to dimethyl-diazomethane (see also Chap. 2, Pt I, p. 82).

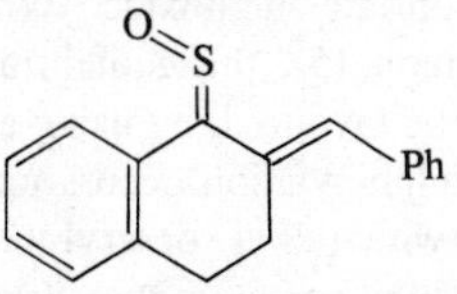

The principal component of the lachrymatory factor of the onion (*Allium cepa*) has been identified by flash vacuum pyrolysis (FVP)/microwave techniques as (Z)-propanethial S-oxide, $CH_3CH_2CH{=}S{=}O$.[6] It functions as a 1,3-dipole and as a dipolarophile, and it dimerizes in cold benzene to give[7] principally *trans*-3,4-diethyl-1,2-dithietan 1,1-dioxide rather than 2,4-diethyl-1,3-dithietan 1,3-dioxide, as previously proposed. Formation of this sulphine from its cellular precursor, *trans*-(+)-S-(prop-1-enyl)-L-cysteine sulphoxide, is suggested[6] to involve the formation of a Schiff base and the elimination of (E)-prop-1-enesulphenic acid, which then rearranges to the sulphine (Scheme 2).

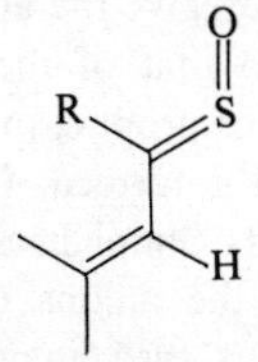

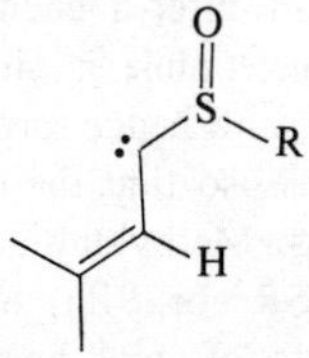

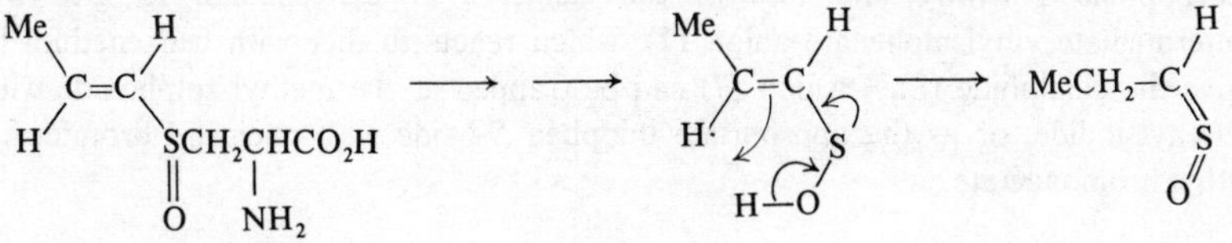

Scheme 2

t-Butyl vinyl sulphoxide and t-butyl prop-1-enyl sulphoxide were shown to give similarly the corresponding sulphines on FVP at 250 °C, and the n.m.r spectra of the products were compared.[6] In the latter case, the intermediate sulphenic acid was trapped with methyl propiolate.[7]

The various reactions of nucleophiles with sulphines continue to be explored; Scheme 3 provides a summary. Functionalized carbanions, *e.g.* XCH_2^- (X = CN, SOMe, or $SO_2C_6H_4Me$-*p*) or $\overline{C}HRCONEt_2$, proceed by the more usual thiophilic

[4] T. Karakasa and S. Motoki, *Tetrahedron Lett.*, 1979, 3961.
[5] M. Franck-Neumann and J. J. Lohmann, *Tetrahedron Lett.*, 1979, 2397.
[6] E. Block, R. E. Penn, and L. K. Revelle, *J. Am. Chem. Soc.*, 1979, **101**, 2200; E. Block, L. K. Revelle, and A. A. Bazzi, *Tetrahedron Lett.*, 1980, **21**, 1277.
[7] E. Block, A. A. Bazzi, and L. K. Revelle, *J. Am. Chem. Soc.*, 1980, **102**, 2490.

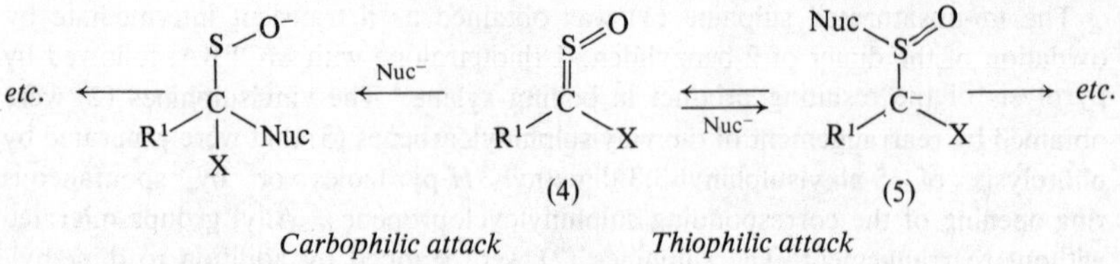

(4)   (5)

Carbophilic attack   *Thiophilic attack*

Scheme 3

mechanism; after quenching, they give the appropriate sulphoxide.[8] When the pK_a of the nucleophile is similar to that of the anion (5), the equilibrium may be displaced, and hence formation of product may be favoured, by using an excess of nucleophile, so that the dianion is formed. Desulphinylation occurs under certain conditions. Methyl-lithium reacts thiophilically with (alkyl- or aryl-thio)sulphines (4; $X = SR^2$ or SPh) to give the anions of dithioacetal monoxides; these are equivalents of acyl anions, and they may be allowed to react with suitable electrophiles.[9] The dithioacetal monoxides give the corresponding aldehydes[9] and the unsymmetrical disulphides[10] R^2SSMe or PhSSMe, in excellent yields, on acidolytic cleavage. With a suitable leaving-group on the α-carbon atom and an absence of α-hydrogen atoms (loss of which would give a vinylsulphenate), the anion (5) could be stabilized by elimination. Thus (4; $X = SMe$; $R^1 = CMe_2SO_2C_6H_4Me$-p) reacts thiophilically with Grignard reagents, carbanions, and alcohols in the presence of base, to give $\alpha\beta$-unsaturated sulphoxides or sulphinate esters [6; $X = SMe$; $Y = Ph$, CH_2Ph, CH_2SO_2Ph, $CH(CN)Ph$, OMe, or OEt].[11]

α-Oxo-carbanions, *e.g.* 2-lithiocyclohexanone and lithiated α-tetralone, have a low tendency to thiophilic reaction, and they do not react with di-p-tolylsulphine. They react with p-tolyl(phenylthio)sulphine (4; $R^1 = C_6H_4Me$-p, $X = SPh$) by a carbophilic pathway and loss of the elements of benzenethiol to give the intermediate vinylsulphenate anion (7), which reacts further with benzenethiol to give the disulphide (8).[12] Anion (7) can be trapped as the methyl sulphoxide with methyl iodide, or as the appropriate thiophen S-oxide with phenacyl bromide or ethyl bromoacetate.

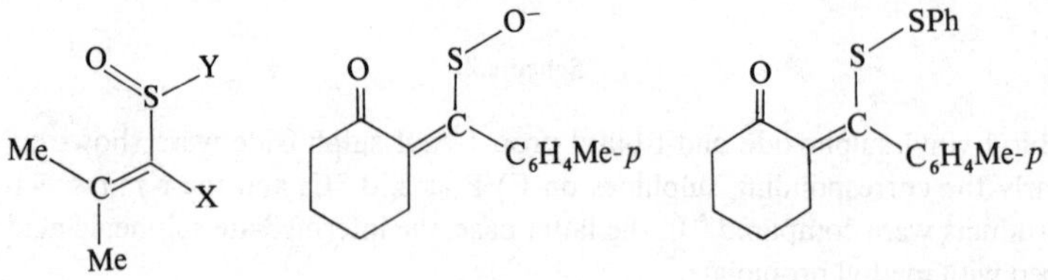

(6)   (7)   (8)

Di-t-butylsulphine has been suggested to undergo a one-electron transfer from t-butylmagnesium chloride to give eventually t-butyl alcohol and di-t-butyl

[8] J. A. Loontjes, M. van der Leij, and B. Zwanenburg, *Recl. Trav. Chim. Pays-Bas*, 1980, **99**, 39.
[9] G. E. Veenstra and B. Zwanenburg, *Tetrahedron*, 1978, **34**, 1585.
[10] B. Zwanenburg and P. Kielbasinski, *Tetrahedron*, 1979, **35**, 169.
[11] M. van der Leij, H. J. M. Strijtveen, and B. Zwanenburg, *Recl. Trav. Chim. Pays-Bas*, 1980. **99**, 45.
[12] M. van der Leij and B. Zwanenburg, *Recl. Trav. Chim. Pays-Bas*, 1980, **99**, 49.

thioketone.[13] With primary and secondary alkylmagnesium halides, an initial thiophilic reaction has been proposed to occur. The anion (5; $R^1 = X = Bu^t$; Nuc = Me, Et, $PhCH_2$, or allyl) from the primary Grignard reagent was suggested to undergo removal of a proton followed by simultaneous ring-closure and extrusion of oxygen, to give the thiiran (9). In contrast, the anion (5; $R^1 = X = Bu^t$, Nuc = Pr^i or cyclohexyl) from the secondary Grignard has been proposed to react by a one-electron transfer followed by loss of oxygen, to give the sulphide (10).[13]

(9)   (10)   (11)

Sulphines react with diphenylnitrilimine, $Ph\overset{+}{C}=N\overset{-}{N}Ph$, to give the corresponding 1,3,4-thiadiazoline 1-oxides in good yields.[14] As with other 1,3-dipoles, the regiochemistry resembles that of the corresponding diaryl thioketones (see Vol. 5, p. 136), but product equilibration *via* ring-opening leads to loss of the steric integrity of the sulphine during addition.

Co-ordination compounds of platinum that involve sulphines as ligands have been prepared.[15] Geometrical isomers of thione *S*-imides (11), which are the nitrogen analogues of sulphines, have been reported,[16] and thiosulphines, $R^1R^2C=S=S$, have been postulated as new reactive intermediates.[17]

2 Sulphenes

Further work on the base-catalysed hydrolysis of 2,4-dinitrophenyl arylmethane-sulphonates by the *E*lcBi mechanism (see Vol. 4, p. 139) has revealed a change in the rate-limiting step in concentrated pyridine buffers from the proton-transfer stage to the breakdown of an encounter complex of the sulphene with 2,4-dinitrophen-oxide ion. Studies of substituent effects in these reactions have been extended.[18]

Sulphenes have been postulated as intermediates in the substitution reactions of α-disulphones, which follow an elimination/addition mechanism with all except the most weakly basic strong nucleophiles.[19] Alkyl sulphinyl sulphones follow the direct substitution mechanism.[19] Although perfluorodisulphene has not yet been prepared from the sulphonyl chloride or by fluorination of disulphene, it is obtainable by oxidation (with CrO_3 and HNO_3) of tetrafluoro-1,3-dithietan. The perchloro-compound is obtained similarly (by $KMnO_4$ plus HOAc).[20]

[13] A. Ohno, M. Uohama, K. Nakamura, and S. Oka, *J. Org. Chem.*, 1979, **44**, 2244.
[14] B. F. Bonini, G. Maccagnani, G. Mazzanti, L. Thijs, G. E. Veenstra, and B. Zwanenburg, *J. Chem. Soc., Perkin Trans. 1*, 1978, 1218.
[15] J. W. Gosselink, G. van Koten, K. Vrieze, B. Zwanenburg, and B. H. M. Lammerink, *J. Organomet. Chem.*, 1979, **179**, 411; J. W. Gosselink, A. M. F. Brouwers, G. van Koten, and K. Vrieze, *J. Chem. Soc., Chem. Commun.*, 1979, 1045.
[16] A. Tangerman and B. Zwanenburg, *Recl. Trav. Chim. Pays-Bas*, 1979, **98**, 127.
[17] A. Senning, *Angew. Chem., Int. Ed. Engl.*, 1979, **18**, 941.
[18] S. Thea and A. Williams, *J. Chem. Soc., Chem. Commun.*, 1979, 715; S. Thea, M. G. Harum, and A. Williams, *ibid.*, 1979, 717.
[19] J. L. Kice, O. Farning, and S.-M. Woo, *Phosphorus Sulfur*, 1979, **7**, 47.
[20] R. Seelinger and W. Sundermeyer, *Angew. Chem., Int. Ed. Engl.*, 1980, **19**, 223.

Sulphenes having at least one aryl substituent are known to react with sulphur dioxide to give the appropriate carbonyl compounds. Sulphur dioxide is not an efficient sulphene trap, and the yields of carbonyl compounds are not high. ^{18}O-Labelling shows[21] that the oxygen of the carbonyl group does not arise exclusively from the sulphur dioxide, and hence that the mechanism is complex. Exchange of sulphur dioxide with the sulphene could not be detected.[21] CNDO/B calculations of the potential-energy surface for the combination of carbene and sulphur dioxide to give sulphene, and its cyclization to the α-sultine (12), have been reported.[22] A cyclic sulphoxylic ester (13) has been suggested as an intermediate.

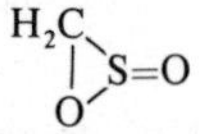

$$(12) \qquad (13) \qquad (14)$$

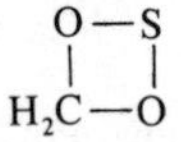

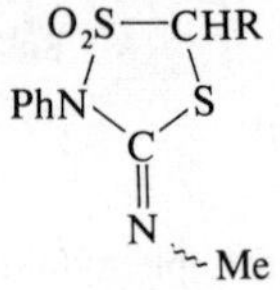

Further cycloadditions have been reported. 4-Methyl-5-(phenylimino)-1,2,3,4-thiatriazoline acts as a masked 1,3-dipole and reacts with sulphenes, with loss of nitrogen, to give[23] the sultams (14). α-Alkoxycarbonyl-sulphenes undergo cyclo-addition with isobutenylmorpholine, but give uncyclized enamine derivatives (15) with cyclopent- and cyclohex-enylmorpholines.[24] Trimethylsilylsulphene reacts similarly with isobutenylmorpholine and also forms a cyclo-adduct with 1,1-diethoxyethane.[25] α-Keto-sulphenes react with 3-substituted 2-phenylazirines by a concerted process to give[26] the oxathiazolines (16). 1,4-Cycloaddition reactions of sulphene to propenones (giving 1,2-oxathiin, 2,2-dioxides),[27] to 3-aminomethylene-4-piperidones (giving 4-amino-3,4,5,6,7,8-hexahydro-1,2-oxathiino[5,6-c]pyridine 2,2-dioxides),[28] and to the corresponding quinolones[28] have been reported. Similar additions to 5-aminomethylene-1,5,6,7-tetrahydroindol-4-ones gave the new tricyclic heterocyclic system 3,4,5,6-tetrahydro-7H-1,2-oxathiino[6,5-e]indole.[29] Ad-

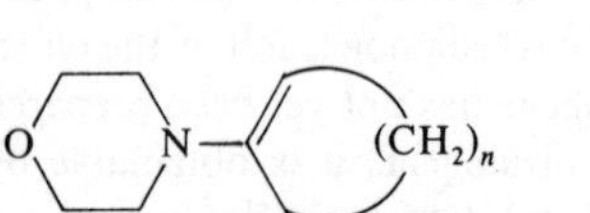

$$(15) \qquad (16) \qquad (17)$$

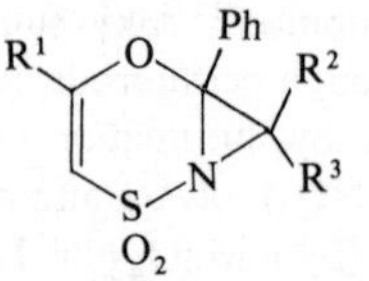

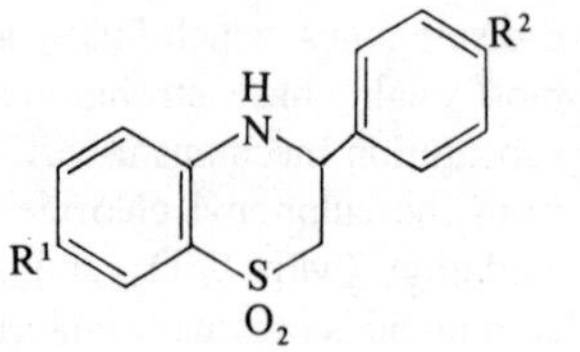

[21] J. F. King and M. Aslam, *Can. J. Chem.*, 1979, **57**, 3278.

[22] L. Carlsen and J. P. Snyder, *J. Org. Chem.*, 1978, **43**, 2216.

[23] G. L'abbé, A. Timmerman, C. Martens, and S. Toppet, *J. Org. Chem.*, 1978, **43**, 4951.

[24] A. Etienne and B. Desmazieres, *J. Chem. Res. (S)*, 1978, 484.

[25] A. G. Shipov, A. V. Kisin, and Yu, I. Baukov, *Zh. Obshch. Khim.*, 1979, **49**, 1170.

[26] O. Tsuge and M. Noguchi, *Heterocycles*, 1978, **9**, 423.

[27] F. Evangelisti, P. Schenone, and A. Bargagna, *J. Heterocycl. Chem.*, 1979, **16**, 217; A. Bargagna, P. Schenone, F. Bondavalli, and M. Longbardi, *ibid.*, 1980, **17**, 33.

[28] L. Mosti, P. Schenone, and G. Menozzi, *J. Heterocycl. Chem.*, 1979, **16**, 177.

[29] L. Mosti, P. Schenone, and G. Menozzi, *J. Heterocycl. Chem.*, 1979, **16**, 913.

dition of sulphene to the Schiff base p-$R^1C_6H_4N=CHC_6H_4R^2$-p is suggested[30] to form a 1,2-thiazetidine, which then undergoes migration of SO_2 and proton transfer to give the benzothiazine (17).

Syntheses of $[(\eta^5\text{-}C_5H_5)W(CO)_3SO_2CH_2D]$[31] and $[(Ph_3P)Mn(CO)_4SO_2CH_3]$[32] that involve the insertion of sulphene into the metal–deuterium and the metal–hydrogen bond have been reported.

PART III: **Thioureas, Thiosemicarbazides, Thioamides, Thiono- and Dithio-carboxylic Acids, and their Selenium Analogues** *by J. K. Landquist*

Reviews.—Two reference books[1,2] that deal with topics covered in this chapter have been published, and there have been reviews of the chemistry of aromatic thioureas[3] and of the applications in organic synthesis of thioamides and thioureas,[4] and of thiophosgene.[5]

1 Thioureas and Selenoureas

Synthesis.—Diarylthioureas have been made from arylamines and either bis-(alkoxythiocarbonyl) disulphides[6] or CS_2 and an alumina catalyst.[7] The standard reaction of amines with isothiocyanates was used in the preparation of adamantyl-[8,9] and glycosyl-[10,11] thioureas. Benzoyl isothiocyanate, when treated with sterically hindered amines, *e.g.* Pr^i_2NH or dicyclohexylamine, gives

[30] M. Rai, S. Kumar, K. Krishan, and A. Singh, *Chem. Ind. (London)*, 1979, 26.
[31] I.-P. Lorenz, *Angew. Chem., Int. Ed. Engl.*, 1978, **17**, 285.
[32] I.-P. Lorenz, A. Baur, and K. Hintzer, *J. Organomet. Chem.*, 1979, **182**, 375.

[1] F. Duus, in 'Comprehensive Organic Chemistry,' Vol. 3, ed. D. N. Jones, Pergamon Press, Oxford, 1979.
[2] S. Scheithauer and R. Mayer, 'Topics in Sulfur Chemistry, Vol. 4, Thio- and dithio-carboxylic acids and their derivatives,' Georg Thieme Verlag, Stuttgart, 1979.
[3] L. Grehn, *Acta Univ. Ups., Abstr. Uppsala Diss. Fac. Sci.*, 1979, 509.
[4] H. Singh, *J. Indian Chem. Soc.*, 1979, **56**, 545.
[5] S. Sharma, *Synthesis*, 1978, 803.
[6] P. P. Gnatyuk, V. A. Malii, G. A. Pestova, and S. V. Tsarenko, USSR P. 694 502/1979 (*Chem. Abstr.*, 1980, **92**, 76 149).
[7] M. K. Gadzhiev and Sh. Sh. Barabadze, *Izv. Akad. Nauk Gruz. SSR, Ser. Khim.*, 1979, **5**, 128 (*Chem. Abstr.*, 1980, **92**, 76 073).
[8] A. Kreutzberger and A. Tantawy, *Arch. Pharm. (Weinheim, Ger.)*, 1978, **311**, 770.
[9] J. W. Tilley, P. Levitan, and M. J. Kramer, *J. Med. Chem.*, 1979, **22**, 1009.
[10] H. Takahashi, N. Nimura, and H. Ogura, *Chem. Pharm. Bull.*, 1979, **27**, 1130.
[11] H. Takahashi, K. Takeda, N. Nimura, and H. Ogura, *Chem. Pharm. Bull.*, 1979, **27**, 1137.

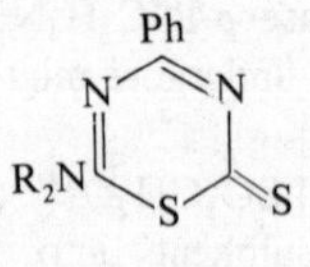

$$Ph$$

(1)

1,3,5-thiadiazine-2-thiones (1) instead of acylthioureas, possibly because the acylthiourea is attacked by benzoyl isothiocyanate with loss of benzoic acid.[12]

A general method for preparing 1,1-disubstituted thioureas in 50—80% yield employs the combined reagent triphenylphosphine–thiocyanogen (Scheme 1).[13, 14]

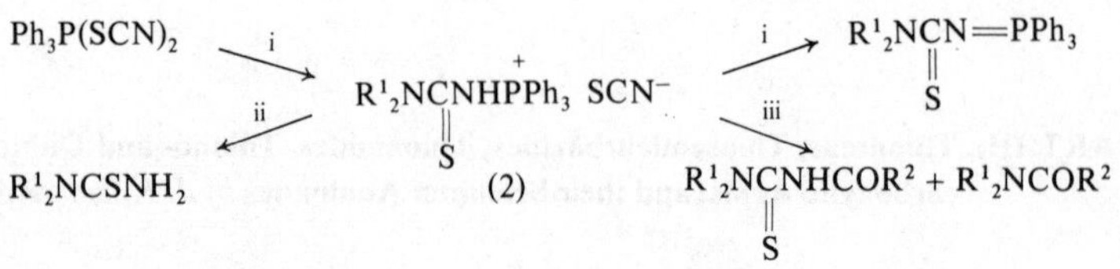

Reagents: i, R^1_2NH; ii, H_2O; iii, R^2CO_2H

Scheme 1

The intermediate phosphonium thiocyanate (2) gives phosphinimines with an excess of amine, and acylthioureas and amides when treated with carboxylic acids. A Beckmann rearrangement of amidoximes $RC(NOH)NH_2$ with $PhSO_2Cl$ in pyridine, followed by treatment with H_2S, gives thioureas $RNHCSNH_2$ in 55—76% yields.[15]

The radical phenylation of thiourea or of NN'-diphenylthiourea to give the S-phenyl derivatives proceeds better with nitrosoacetanilides than with phenyl-azotriphenylmethane as the source of radicals.[16, 17]

$$H_2NCSNRCSNH_2 \qquad\qquad PhCONHCONHCSNH_2$$

(3)   (4)

Addition of H_2S to alkyldicyanamides affords 1-alkyl-1-cyanothioureas, and then (with NEt_3 catalysis) 3-alkyl-dithiobiurets (3).[18] 3-Alkyl-2-thiobiurets are made similarly from 1-alkyl-1-cyanoureas. 1-Benzoyl-2-thiobiuret is obtained from benzoyl isothiocyanate and urea,[19, 20] and it rearranges to 1-benzoyl-4-thiobiuret (4)

[12] H. Hartmann, L. Beyer, and E. Hoyer, *J. Prakt. Chem.*, 1978, **320**, 647.
[13] Y. Tamura, M. Adachi, T. Kawasaki, and Y. Kita, *Tetrahedron Lett.*, 1978, 1753.
[14] Y. Tamura, T. Kawasaki, M. Adachi, and Y. Kita, *Chem. Pharm. Bull.*, 1979, **27**, 1636.
[15] M. V. Arnaudov, *Dokl. Bolg. Akad. Nauk*, 1978, **31**, 309 (*Chem. Abstr.*, 1978, **89**, 197 131).
[16] B. V. Kopylova, I. I. Kandror, and R. Kh. Freidlina, *Dokl. Akad. Nauk SSSR*, 1978, **243**, 1197 (*Chem. Abstr.*, 1979, **90**, 103 583).
[17] B. V. Kopylova, I. I. Kandror, and R. Kh. Freidlina, *Izv. Akad. Nauk SSSR, Ser. Khim.*, 1979, 1138 (*Chem. Abstr.*, 1979, **91**, 91 320).
[18] P. H. Benders and P. A. E. van Erkelens, *Synthesis*, 1978, 775.
[19] M. N. Basyouni and A. M. A. El-Khamry, *Chem. Ind. (London)*, 1978, 670.
[20] M. N. Basyouni and A. M. A. El-Khamry, *Bull. Chem. Soc. Jpn.*, 1979, **52**, 3728.

when heated in water.[21] Acid hydrolysis affords a convenient preparation of monothiobiuret. Thiocarbamoylcarbodi-imides are made by treatment of 1,1,5-trisubstituted dithiobiurets with cyanuric chloride and NEt_3.[22] They readily dimerize to 1,3,5-thiadiazines unless they are stabilized by bulky substituents, and with water, alcohols, and amines they provide monothiobiurets, O-alkyl-thioisobiurets, and thiocarbamoylguanidines, respectively. The thiocarbamidic sulphur atom of thioureas has been protected by t-butylation before reaction with, *e.g.*, iso-thiocyanates. The protecting group is readily removed with warm dilute HCl.[23]

Among the more unusual thiourea derivatives are crown ethers containing thiourea links[24] and N-(phenylthiocarbamoyl)ethyleneimine oligomers, which are selective adsorbents for Cu^{2+} and Hg^{2+} ions.[25] S-Amino-isothioureas are obtained by hydrolysis of 5-arylsulphonylimino-1,3,2,4-dithiadiazolidine 3-oxides, which are made from 4-alkyl-5-arylsulphonylimino-1,2,3,4-thiatriazolines and N-sulphinyl-amines as shown in Scheme 2.[26] Phosphorylated isothioureas $R^1NHC(SR^2)=NP-(X)R^3R^4$ have been prepared from $R^1NHC(SR^2)=NH$ and $ClP(X)R^3R^4$ (X = O or S),[27] and N-(trimethylsilyl)thioureas were obtained from Me_3SiCl and mono-, di-, or tri-substituted thioureas with NEt_3 or BuLi.[28] The reaction of halogenodiorganyl-boranes with N-(trimethylsilyl)thioureas affords thioureido- or isothioureido-boranes, with displacement of the trimethylsilyl group.[29]

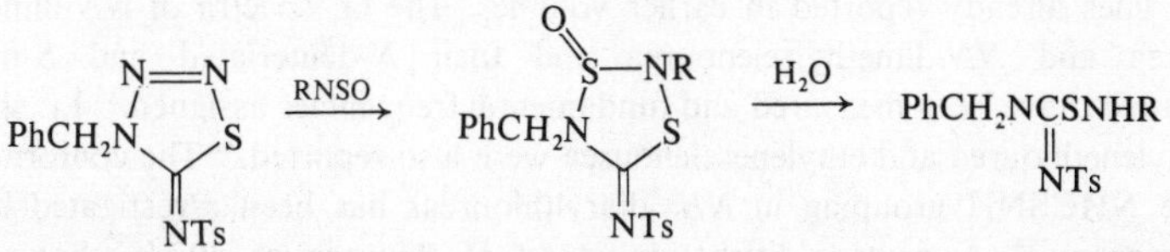

Scheme 2

Persubstituted guanidines, *e.g.* pentamethylguanidine, react with activated isothiocyanates to give stable dipolar compounds (5), which may be protonated to give thiocarbamoylguanidinium salts (6). The analogous dipole (7), from CS_2 and pentamethylguanidine, undergoes a [2 + 2] cycloreversion to MeNCS and tetramethylthiourea on heating.[30] Similar compounds made from tetramethyl-isothiourea and isothiocyanates give methyl NN-dimethyldithiocarbamate by [2 + 2] cycloreversion.[31]

[21] D. L. Klayman and J. P. Scovill, *J. Org. Chem.*, 1979, **44**, 630.

[22] J. Goerdeler and R. Losch, *Chem. Ber.*, 1980, **113**, 79.

[23] M. H. Damle, *Indian J. Chem., Sect. B*, 1978, **16**, 396 (*Chem. Abstr.*, 1978, **89**, 146 401).

[24] A. V. Bogatsky, N. G. Lukyanenko, and T. I. Kirichenko, *Tetrahedron Lett.*, 1980, **21**, 313.

[25] H. Tsukube, T. Araki, H. Inoue, and A. Nakamura, *J. Polym. Sci., Polym. Lett. Ed.*, 1979, **17**, 437 (*Chem. Abstr.*, 1979, **91**, 75 052).

[26] G. L'abbé, A. Van Asch, J. P. Declercq, G. Germain, and M. Van Meerssche, *Bull. Soc. Chim. Belg.*, 1978, **87**, 285.

[27] L. Ya. Bogel'fer, V. N. Zontova, V. V. Negrebetskii, A. F. Grapov, S. F. Dymova, and N. N. Mel'nikov, *Zh. Obshch. Khim.*, 1978, **48**, 1729 (*Chem. Abstr.*, 1978, **89**, 163 666).

[28] W. Walter, H. Kubel, and H-W. Luke, *Liebigs Ann. Chem.*, 1979, 263.

[29] W. Maringgele, *Z. Naturforsch., Teil B*, 1980, **35**, 164.

[30] E. Schaumann, E. Kausch, and E. Rossmanith, *Liebigs Ann. Chem.*, 1978, 1543.

[31] E. Schaumann and E. Kausch, *Liebigs Ann. Chem.*, 1978, 1560.

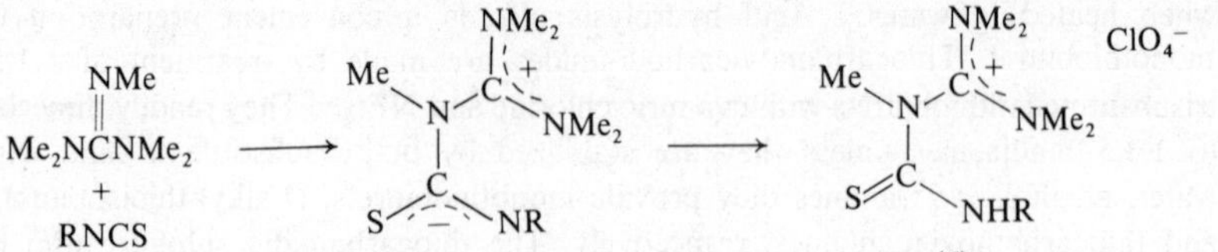

(5)

(6)

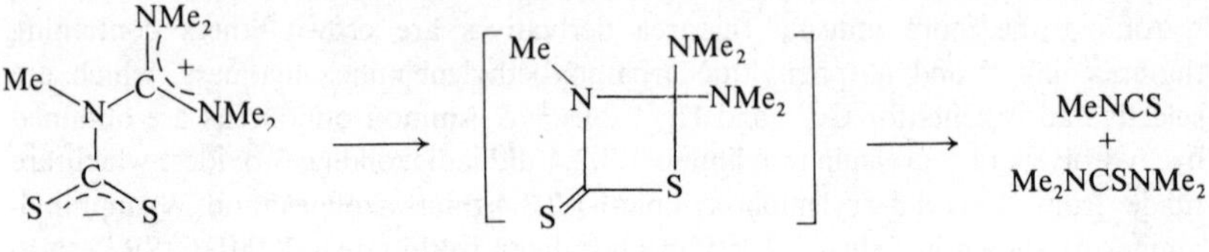

(7)

An improved method for the preparation of selenoureas from S-methyl-isothioureas and H_2Se has been described.[32]

Physical Properties.—Physicochemical investigation of thioureas has continued on many lines already reported in earlier volumes. The i.r. spectra of NN-dimethyl-thiourea and NN-dimethylselenourea and their N-deuteriated and S-methyl derivatives have been measured and fundamental frequencies assigned.[33] I.r. spectra of ethylenethiourea and ethyleneselenourea were also reported.[34] The conformation of the NHCSNH grouping in NN'-diarylthioureas has been investigated by i.r. spectroscopy.[35] A neutron-diffraction study of thiourea at 293 K showed the molecule to be planar, with one hydrogen atom ~0.07 Å out of plane, this being associated with strong hydrogen-bonding that links molecules in chains.[36] The bond-electron distribution has been investigated, using a refined model.[37] The binding energies of sulphur $2p$, nitrogen $1s$, and carbon $1s$ electrons in some alkylated thioureas and related thiocarbonyl compounds have been determined by X-ray p.e. spectroscopy.[38] The positive charge in isothiouronium salts is concentrated on carbon and nitrogen.

The electronic and ^1H n.m.r. spectra of monothiobiuret and dithiobiuret have been determined, and the results of quantum-mechanical calculations were used in a discussion of the electronic structures, p.e. and electronic spectra, and conformational stability.[39] The structures of some (trimethylsilyl)thioureas were determined

[32] V. I. Cohen, *Synthesis*, 1980, 60.
[33] K. Dwarakanath and D. N. Sathyanarayana, *Indian J. Pure Appl. Phys.*, 1979, **17**, 171 (*Chem. Abstr.*, 1979, **91**, 107 351).
[34] K. Dwarakanath and D. N. Sathyanarayana, *Indian J. Chem., Sect. A*, 1979, **18**, 302 (*Chem. Abstr.*, 1980, **92**, 31 344).
[35] B. Galabov, G. Vassilev, N. Neykova, and A. Galabov, *J. Mol. Struct.*, 1978, **44**, 15.
[36] D. Mullen, G. Heger, and W. Treutmann, *Z. Kristallogr., Kristallgeom., Kristallphys., Kristall-chem.*, 1979, **148**, 95 (*Chem. Abstr.*, 1979, **90**, 178 467).
[37] D. Mullen and E. Hellner, *Acta Crystallogr., Sect. B*, 1978, **34**, 2789.
[38] R. Szargan, R. Scheibe, L. Beyer, Ya. V. Salyn, and V. I. Nefedov, *Tetrahedron*, 1979, **35**, 59.
[39] V. C. Kyothi Bhasu, D. N. Sathyanarayana, and C. C. Patel, *Proc. Indian Acad. Sci., Sect. A*, 1979, **88**, 91 (*Chem. Abstr.*, 1979, **91**, 65 395).

by i.r. and ^{1}H n.m.r. spectrometry and the barriers to rotation about C—N bonds were measured.[28] Further work on dynamic ^{1}H and ^{13}C effects in tetramethylthiourea, arising from restricted rotation about C—N bonds, has been reported.[40]

X-Ray diffraction analysis of the 1:1 complex of thiourea with bis-[2-(*o*-methoxyphenoxy)ethoxyethyl] ether showed that the hydrogen atoms of the amino-groups are hydrogen-bonded to all the oxygen atoms, showing bifurcated hydrogen bonds.[41]

Reactions.—*Oxidation.* Thioureas and alkyl derivatives are oxidized very rapidly by ICl in the presence of NaHCO$_3$ [reaction (1)]. Titration of the liberated iodine

$$RNHCSNHR + 8\,ICl + 5\,H_2O \longrightarrow RNHCONHR + 4\,I_2 + H_2SO_4 + 8\,HCl \quad (1)$$

(by ICl$_3$, to regenerate ICl) gives an analytical method for thioureas or their precursors (*i.e.* amines or isothiocyanates).[42] Thiourea, ethylenethiourea, and *N*-phenylthiourea are oxidized to the corresponding formamidinesulphinic acids by microsomal amine oxidase from pigs' liver. Formamidinesulphenic acids were identified as intermediates.[43] Oxidation of mixtures of *NN'*-diarylthiourea and thiosemicarbazide with H$_2$O$_2$ affords 3-amino-4-aryl-5-arylamino-1,2,4-triazoles.[44]

Alkylation. The heat of reaction of tetramethylthiourea with methyl iodide in MeCN has been measured, and the kinetics of the reaction were studied in various solvents.[45] The reaction is second-order when it occurs in polar solvents, but in non-polar solvents the kinetics are complicated by the formation of aggregates of ions. The kinetics of the reaction of thiourea with chloroacetic acid and bromoacetic acid have also been measured.[46] The reaction of tetra(hydroxymethyl)phosphonium chloride with thiourea or phenylthiourea occurs through a methylol group, with subsequent loss of formaldehyde, to give, *e.g.*, PhNHC-(S)NHCH$_2$P(O)(CH$_2$OH)$_2$.[47] Condensation of thiourea with aromatic aldehydes and triphenyl phosphite, followed by hydrolysis and *S*-alkylation, affords 1-(*S*-alkylisothioureido)benzylphosphonic acids (8), which are intermediates for the preparation of 1-guanidinobenzylphosphonic acids.[48]

$$\underset{\displaystyle (8)}{\overset{\displaystyle SR^2}{\underset{\displaystyle PO_3H_2}{R^1C_6H_4CHNHC\!=\!NH}}} \qquad \underset{(9)}{R^1R^2P(X)N\!=\!\!=\!C(SMe)NHCOR^3}$$

[40] F. A. L. Anet and M. Ghiaci, *J. Am. Chem. Soc.*, 1979, **101**, 6857.

[41] I-H. Suh and W. Saenger, *Angew. Chem.*, 1978, **90**, 565.

[42] K. V. Verma, *Bull. Chem. Soc. Jpn.*, 1979, **52**, 2155.

[43] D. M. Ziegler, *Biochem. Soc. Trans.*, 1978, **6**, 94.

[44] K. N. Rajasekharan, *Indian J. Chem.*, *Sect. B*, 1979, **17**, 69 (*Chem. Abstr.*, 1980, **92**, 22 475).

[45] Y. Kondo, S. Hirano, and N. Tokura, *J. Chem. Soc.*, *Perkin Trans. 2*, 1979, 1738.

[46] J. Kaválek, S. El-Bahaie, and V. Štěrba, *Collect. Czech. Chem. Commun.*, 1980, **45**, 263.

[47] S. Bhatnagar, J. C. Gupta, K. Lal, and H. L. Bhatnagar, *Indian J. Chem.*, *Sect. A*, 1978, **16**, 356 (*Chem. Abstr.*, 1978, **89**, 146 994).

[48] J. Oleksyszyn and R. Tyka, *Pol. J. Chem.*, 1978, **52**, 1949 (*Chem. Abstr.*, 1979, **90**, 104 066).

Acyl Derivatives. Acylation of N^1-phosphorylated or -thiophosphorylated *S*-methylisothiourea occurs at N-2, giving (9).[49] Arylamines are converted into aryl isothiocyanates by condensation with benzoyl isothiocyanate and thermal decomposition of the benzoylthiourea in boiling *o*-dichlorobenzene. The yield improves on dilution.[50]

Additions to Multiple Bonds. In most of the reactions reported recently, the addition of a thiourea to a multiple bond has been followed by cyclization at another reactive centre. An exception is the reaction of *S*-benzylisothiourea with aroylaryl-acetylenes, which gives α-aroyl-β-(benzylmercapto)styrenes (Scheme 3), and not the expected 4,6-diaryl-2-(benzylthio)pyrimidines.[51] Cycloaddition of thiourea to divinyl sulphide in the presence of acids gives salts of 6-amino-2,4-dimethyl-1,3,5-dithiazine.[52] Thiourea condenses with chalcone to give 3,6-dihydro-2-oxo-4,6-diphenyl-1,3-thiazine (10),[53] but with other $\alpha\beta$-unsaturated ketones, *e.g.* 2-methylene-3-oxoquinuclidine[54] and cyclohex-2-enones,[55] only the nitrogen atoms are involved, giving the hexahydropyrimidines (11) and (12) respectively. Addition of thiourea to the benzoquinones (13) occurs through the sulphur atom, subsequent cyclization giving benzoxathioles (14).[56] Condensation of dimethyl acetylene-dicarboxylate with 1-methyl- and 1-ethyl-thiourea gives the isomeric 1,3-thiazines (15) and (16) in the ratio 43:38 (R = Me) and 54:22 (R = Et).[57] Mono- and di-substituted selenoureas condense with dimethyl acetylenedicarboxylate, giving the 1,3-selenazines (17; R^1 = H, R^2 = benzyl or Ar) and (17; R^1 = R^2 = Me or Ar).[58]

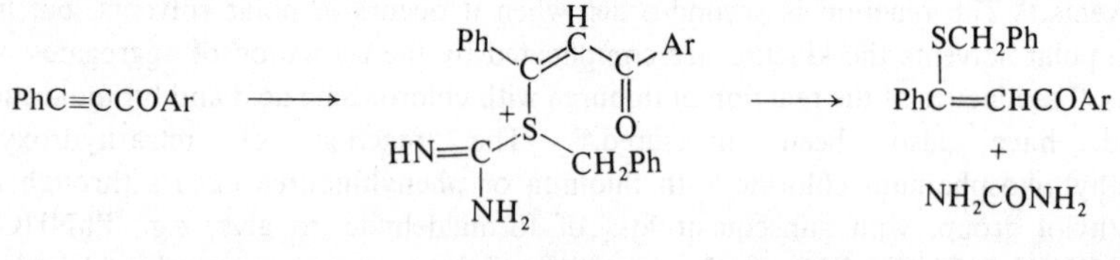

PhC≡CCOAr ⟶ [structure: Ph–C=C(H)–C(=O)–Ar with HN=C(S)(NH₂) and CH₂Ph] ⟶ PhC(SCH₂Ph)=CHCOAr + NH₂CONH₂

Scheme 3

Carbonyl sulphide reacts with *S*-benzylisothiourea, giving 2-imino-4-oxo-6-thioxohexahydro-1,3,5-triazine (18; R = H), and with *S*-benzyl-*N*-methyliso-thiourea to give the dimethyl derivative (18; R = Me).[59] Chlorosulphonyl iso-

[49] A. F. Grapov, A. F. Vasil'ev, V. N. Zontova, V. V. Galushina, and N. N. Mel'nikov, *Zh. Obshch. Khim.*, 1979, **49**, 2474 (*Chem. Abstr.*, 1980, **92**, 110 135).
[50] S. Rajappa, T. G. Rajagopalan, R. Sreenivasan, and S. Kanal, *J. Chem. Soc., Perkin Trans. 1*, 1979, 2001.
[51] F. G. Baddar, F. H. Hajjar, and N. R. El-Rayyes, *J. Heterocycl. Chem.*, 1978, **15**, 105.
[52] B. A. Trofimov, G. M. Gavrilova, G. A. Kalabin, V. V. Bairov, and S. V. Amosova, *Khim. Geterotsikl. Soedin.*, 1979, 1466.
[53] D. N. Dhar, A. K. Singh, and H. C. Misra, *Indian J. Chem., Sect. B*, 1979, **17**, 25 (*Chem. Abstr.*, 1980, **92**, 6483).
[54] V. A. Bondarenko, E. E. Mikhlina, T. Ya. Filipenko, K. F. Turchin, Yu. N. Sheinker, and L. N. Yakhontov, *Khim. Geterotsikl. Soedin.*, 1979, 1393.
[55] W. Wendelin and W. Kern, *Monatsh. Chem.*, 1979, **110**, 1345.
[56] K. Srihari and V. Sundaramurthy, *Indian J. Chem., Sect. B*, 1979, **18**, 80 (*Chem. Abstr.*, 1980, **92**, 41 484).
[57] L. I. Giannola, S. Palazzo, P. Agozzino, L. Lamartina, and L. Ceraulo, *J. Chem. Soc., Perkin Trans. 1*, 1978, 1428.
[58] A. Shafiee, F. Assadi, and V. I. Cohen, *J. Heterocycl. Chem.*, 1978, **15**, 39.
[59] J. M. Parnandiwar, *Indian J. Chem., Sect. B*, 1978, **16**, 927 (*Chem. Abstr.*, 1979, **91**, 20 456).

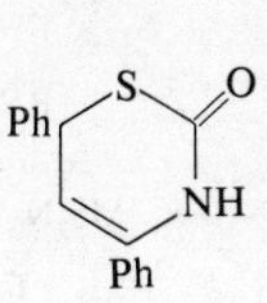

(10)

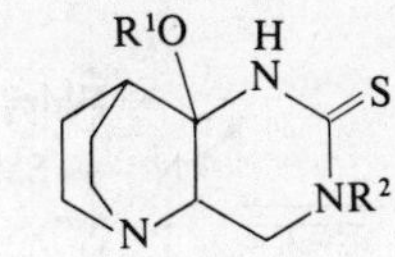

(11)

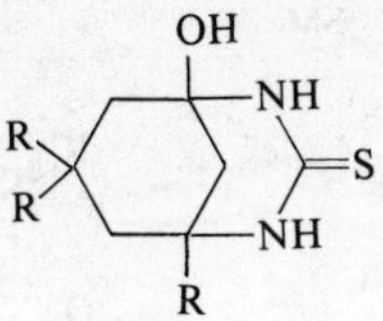

(12)

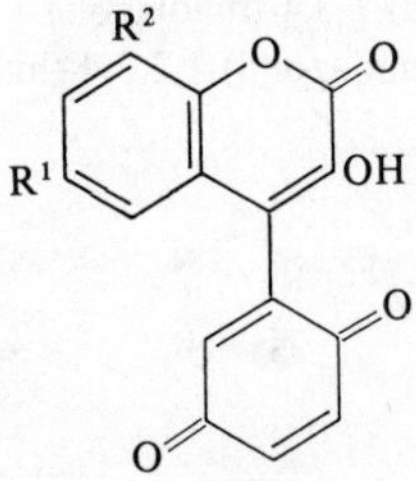

(13)

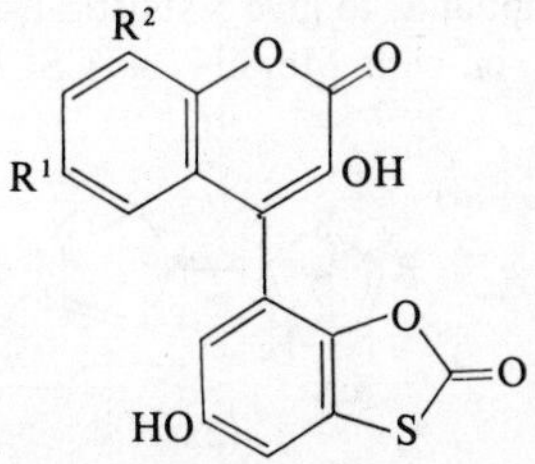

(14)

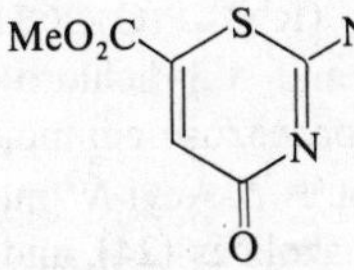

(15)

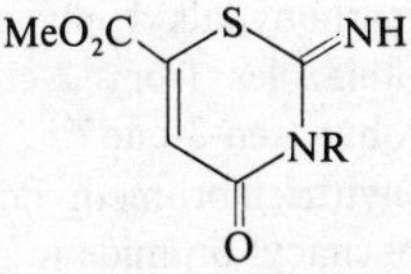

(16)

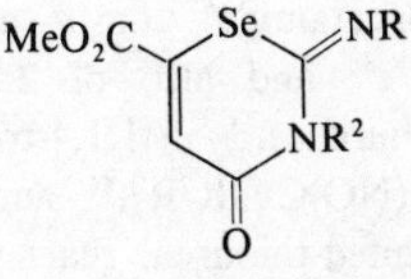

(17)

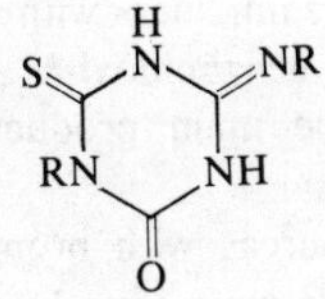

(18)

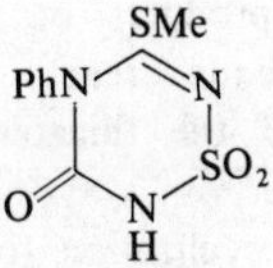

(19)

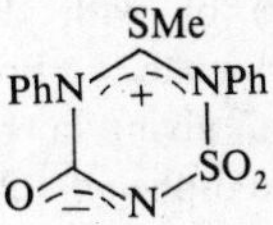

(20)

cyanate condenses with *S*-methyl-*N*-phenylthiourea and with *S*-methyl-*NN'*-diphenylthiourea to give the thiatriazines (19) and (20) respectively.[60] Per-substituted *S*-methylisothioureas add to sulphonyl isocyanates to give dipolar compounds (21), which are stable if $R^1 = Me$ but which undergo [2 + 2] cycloreversion to isocyanates and *N*-sulphonyl-*S*-methyl-isothioureas if R^1 is more bulky. Benzoyl and thiobenzoyl isocyanates behave similarly.[61]

The reactive thiourea *NN'*-thiocarbonyldi-imidazole reacts with aliphatic

[60] S. Karady, J. S. Amato, D. Dortmund, A. A. Patchett, R. A. Reamer, R. J. Tull, and L. M. Weinstock, *Heterocycles*, 1979, **12**, 1199.
[61] E. Schaumann, E. Kausch, J-P. Imbert, and G. Adiwidjaja, *Chem. Ber.*, 1978, **111**, 1475.

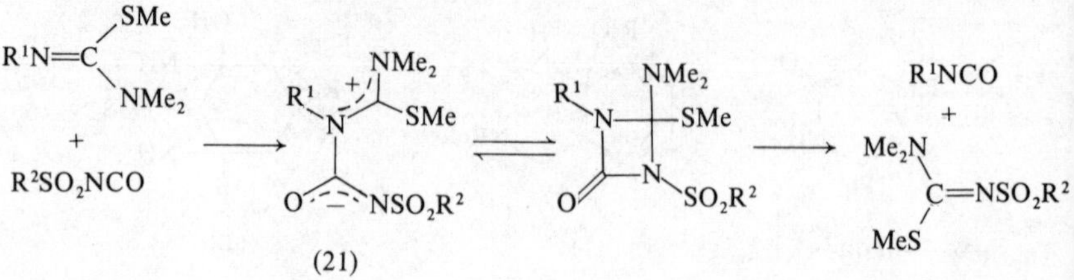

(21)

diazo-compounds to give 5-substituted-2-(1-imidazolyl)-1,3,4-thiadiazoles (22), and with HN_3 or with Me_3SiN_3 + $CSCl_2$, giving 5-(1-imidazolyl)-1,2,3,4-thiatriazole (23).[62]

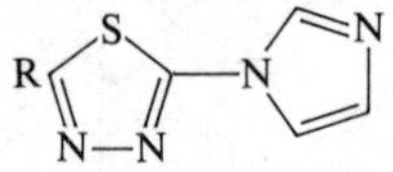

(22) 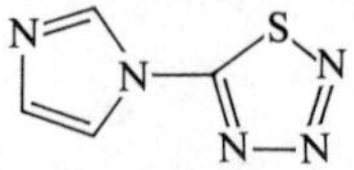   (23)

Formation of Thiazoles. The widely exploited synthetic applications of thioureas to give thiazole derivatives include the preparation of 2-imino-4-thiazolidinones from chloroacetamide[63] or α-(alkoxycarbonyl)alkyl phosphates $(R^1O)_2P(O)OCHR^2\text{-}CO_2R^3$;[64] and also of 2-aminothiazoles from 2-chloro- and 1,3-dichloro-2,3-epoxybutanes,[65] 1,1,4-trichlorobut-1-en-3-one,[66] β-chloronitroso-compounds $R^1CH(NO)CClR^2R^3$,[67] and phenyl(trichloromethyl)carbinol.[68] N-Acyl-N'-mono-substituted thioureas react with phenacyl bromide to give thiazolines (24), and not 5-acylthiazoles, as previously reported. The latter compounds, *e.g.* (25), are obtained by the reaction of phenacyl bromide with the O-methyl-N'-phenyl derivatives of the acylthiourea (from methyl benzimidate and phenyl isothiocyanate) or with the addition products of N-arylbenzamidines with phenyl isothiocyanate.[69] Bromonitromethane reacts with N-benzimidoyl-N'-phenyl-thiourea, giving a poor yield of the thiazole (26), the main product being 5-anilino-3-phenyl-1,2,4-thiadiazole.[70]

Application of the Hugershoff cyclization (of arylthioureas with bromine) to pyrrolylthioureas[71] and to N-acyl-N'-(3-thienyl)thioureas[72] gave pyrrolothiazoles

[62] A. Martvoň, L. Floch, and S. Sekretár, *Tetrahedron*, 1978, **34**, 453.

[63] L. I. Mizrakh, L. Yu. Polonskaya, B. I. Bryantsev, and T. M. Ivanova, *Zh. Org. Khim.*, 1978, **14**, 1553 (*Chem. Abstr.*, 1978, **89**, 163 482).

[64] L. I. Mizrakh, L. Yu. Polonskaya, B. I. Bryantsev, and T. N. Doronchenkova, *Zh. Obshch. Khim.*, 1978, **48**, 568 (*Chem. Abstr.*, 1978, **89**, 43 255).

[65] A. A. Durgaryan, G. E. Esayan, and R. H. Arakelyan, *Arm. Khim. Zh.*, 1979, **32**, 29 (*Chem. Abstr.*, 1979, **91**, 123 664).

[66] A. N. Mirskova, G. G. Levkovskaya, I. D. Kalikhman, and M. G. Voronkov, *Zh. Org. Khim.*, 1979, **15**, 2301 (*Chem. Abstr.*, 1980, **92**, 128 792).

[67] J. Beger, C. Thielemann, and P. D. Thong, *J. Prakt. Chem.*, 1979, **321**, 249.

[68] W. Reeve and W. R. Coley, *Can. J. Chem.*, 1979, **57**, 444.

[69] S. Rajappa, M. D. Nair, B. G. Advani, R. Sreenivasan, and J. A. Desai, *J. Chem. Soc., Perkin Trans. I*, 1979, 1762.

[70] S. Rajappa and B. G. Advani, *Indian J. Chem., Sect. B*, 1978, **16**, 749 (*Chem. Abstr.*, 1979, **90**, 152 059).

[71] L. Grehn, *Chem. Scr.*, 1979, **13**, 78 (*Chem. Abstr.*, 1980, **92**, 6454).

[72] L. Grehn, *J. Heterocycl. Chem.*, 1978, **15**, 81.

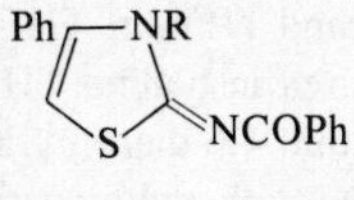

(24)

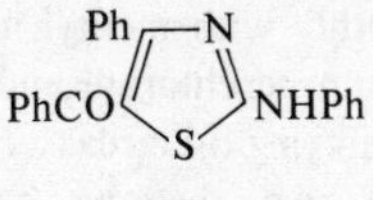

(25)

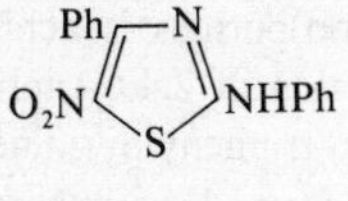

(26)

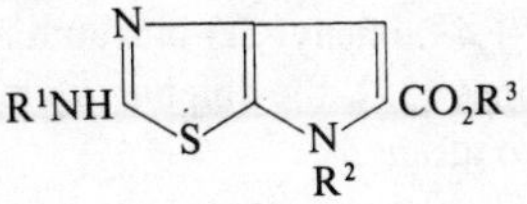

(27)

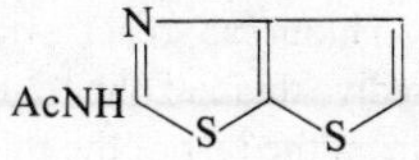

(28)

(27) and thienothiazoles (28), but the analogous 2-thienylthioureas were brominated at positions 3 or 3 and 5 in the thiophen ring, and they did not cyclize.

Miscellaneous Reactions. Tetramethylthiourea forms 1:1 adducts with arsenic trihalides and arylarsenic dihalides.[73] The nitrosation of dimethylamine by sodium nitrite at pH 4 is catalysed by thiourea, less efficiently by *NN'*-dimethylthiourea, and more efficiently by tetramethylthiourea.[74] Base-catalysed condensation of thiourea with phthalaldehyde gives *N*-thiocarbamoyl-1,3-dihydroxyisoindoline (29)

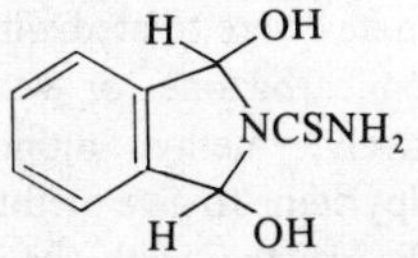

(29)

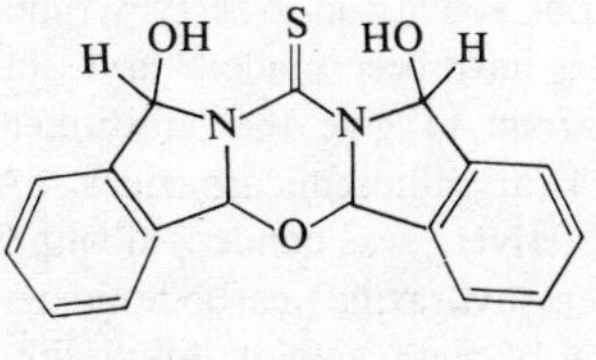

(30)

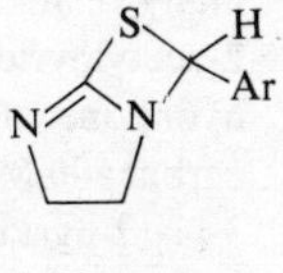

(31)

and the 1:2-adduct (30).[75] Ethylenethiourea reacts with aromatic aldehydes and BF₃ etherate to give 2-aryl-1,3-thiazetidines (31).[76] 2*H*-1,3-Oxazetes are obtained by lithiation of acylthioureas and treatment with methyl iodide (Scheme 4).[77]

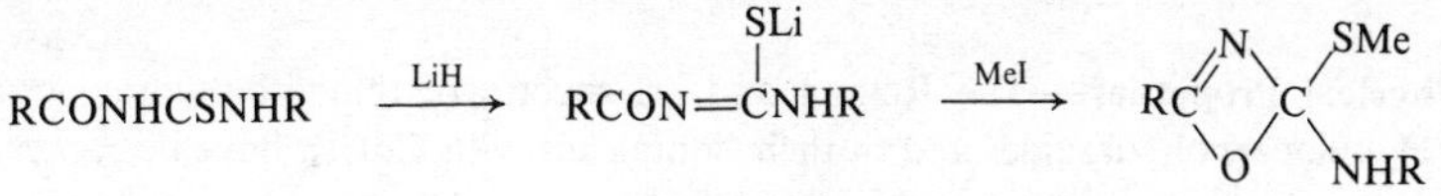

Scheme 4

The cyclodesulphurization of *N*-(*o*-aminoaryl)- or *N*-(6-aminopyrimidin-5-yl)-*N'*-glycosylthioureas to give (glycosylamino)benzimidazoles and (glycosyl-

[73] D. J. Williams and K. J. Wynne, *Inorg. Chem.*, 1978, **17**, 1108.

[74] M. Masui, C. Ueda, T. Yasuoka, and H. Ohmori, *Chem. Pharm. Bull.*, 1979, **27**, 1274.

[75] R. D. Reynolds, D. F. Guanci, C. B. Neynaber, and R. J. Conboy, *J. Org. Chem.*, 1978, **43**, 3838.

[76] M. Yokoyama and H. Monma, *Tetrahedron Lett.*, 1980, **21**, 293.

[77] P. Kristian, P. Kutschy, and M. Dzurilla, *Collect. Czech. Chem. Commun.*, 1979, **44**, 1324.

amino)purines is achieved smoothly with methyl iodide and NEt_3 in THF.[78] The reaction of 2,4,6-triphenylpyrylium perchlorate and thiourea in boiling THF gives 2,4,6-triphenylpyridine; with S-alkylisothioureas, 2-alkylthio-4,6-diarylpyrimidines are formed with elimination of acetophenone, and NN'-diphenylthiourea gives 1,2,4,6-tetraphenylpyridinium perchlorate.[79] Some reactions of thio-bis(formamidines) to give isothioureido-pyrimidines and -1,3,5-triazines have been described.[80,81] Thioureas react with 1-chloro-1,4-diphenyl-2,3-diazabutadiene to give, unexpectedly, thiadiazoline-2-carboxamidines (32), the sulphur atom becoming separated from the rest of the thiourea by two atoms.[82]

(32)   (33)   (34)

2 Thiosemicarbazides, Dithiocarbazates, and Thiocarbohydrazides

Synthesis.—Many syntheses have been directed to the preparation of thiosemicarbazones for testing as possible therapeutic agents.[83-85] Following the discovery of antimalarial activity in 2-acetylpyridine 4-phenylthiosemicarbazone, many related substances have been made;[83] aryl isothiocyanates were treated with 2-acetylpyridine hydrazone to give the substituted thiosemicarbazone, or with hydrazine to give 4-(aryl)thiosemicarbazides. Alternatively, methyl dithiocarbazate, $MeSC(=S)NHNH_2$, was condensed with 2-acetylpyridine to give methyl 3-[1-(2-pyridyl)ethylidene]hydrazine carbodithioate, which reacted with arylamines with elimination of methanethiol. Alkyl dithiocarbazates from n-propyl to n-nonyl have been prepared for the first time, by alkylation of hydrazinium dithiocarbazate, $NH_2NHC(=S)SH\cdot NH_2NH_2$, and corresponding alkyldithiocarbamoyl derivatives of picolino-, nicotino-, and isonicotino-hydrazides were made.[86] 2-Acetoxyvinyl isothiocyanate and hydrazine react at room temperature to give 4-(β-acetoxyvinyl)thiosemicarbazide (33), but dihydro-1,2,4-triazine-3-thione (34) is obtained at higher temperatures.[87]

Physical Properties.—The Raman and i.r. spectra of thiocarbohydrazide and of [$^2H_6$]thiocarbohydrazide, and of their complexes with $CdCl_2$, have been measured,

[78] H. Takahashi, N. Nimura, N. Obata, H. Sakai, and H. Ogura, *Chem. Pharm. Bull.*, 1979, **27**, 1153.
[79] E. A. Zvezdina, M. P. Zhdanova, and G. N. Dorofeenko, *Khim. Geterotsikl. Soedin.*, 1979, 324.
[80] R. Evers, *Z. Chem.*, 1979, **19**, 250.
[81] R. Evers and E. Fischer, *Z. Chem.*, 1979, **19**, 290.
[82] W. T. Flowers, S. F. Moss, J. F. Robinson, D. R. Taylor, A. E. Tipping, and M. J. Haley, *J. Chem. Soc., Chem. Commun.*, 1979, 149.
[83] D. L. Klayman, J. F. Bartosevich, T. S. Griffin, C. J. Mason, and J. P. Scovill, *J. Med. Chem.*, 1979, **22**, 855.
[84] A. Wahab, *Arzneim.-Forsch*, 1979, **29**, 466.
[85] A. Andreani, D. Bonazzi, V. Cavrini, R. Gatti, G. Giovanninetti, M. Rambaldi, L. Franchi, A. Nanetti, and M. A. Guarda, *Farmaco., Ed. Sci.*, 1978, **33**, 754.
[86] S. Kubota, M. Uda, Y. Mori, F. Kametani, and H. Terada, *J. Med. Chem.*, 1978, **21**, 591.
[87] A. W. Faull and R. Hull, *J. Chem. Res. (S)*, 1979, 240.

calculations of force constants were carried out, and vibrational assignments were made.[88] The ring–chain tautomerism of acetone thiosemicarbazone and of its N^2- and N^4-methyl derivatives [reaction (2)] was investigated by n.m.r. spectroscopy. The cyclic form is favoured in deuterio-TFA and the acyclic form in $(^2H_6)DMSO$, but the 2,4-dimethyl derivative is cyclic in both solvents.[89]

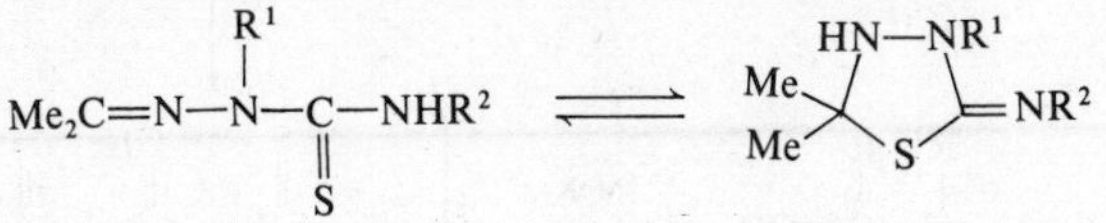

$$Me_2C{=}N{-}\overset{\overset{\displaystyle R^1}{|}}{N}{-}\overset{\overset{\displaystyle \|}{C}}{\underset{\displaystyle S}{}}{-}NHR^2 \rightleftharpoons \underset{Me}{\overset{Me}{{>}}}\overset{HN{-}NR^1}{\underset{S}{\diagup}}{=}NR^2 \qquad (2)$$

Several conformational studies have been reported.[90–92] The *trans,cis* conformer (35) of *S*-methyl dithiocarbazate was isolated by low-temperature crystallization and was subjected to *X*-ray crystal analysis.[92] The known solid conformer (36) is the *cis,trans* one.

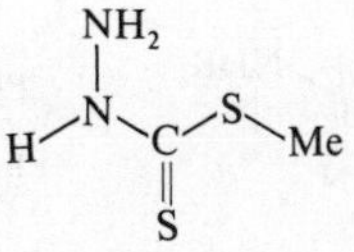

(35) *trans,cis* 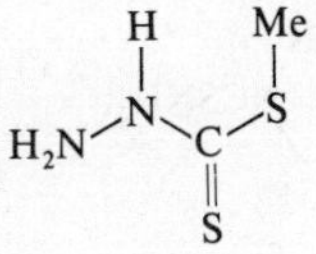   (36) *cis,trans*

Reactions.—Successive alkylations of the two sulphur atoms in potassium β-toluene-*p*-sulphonyldithiocarbazate, $TsNHNHCS_2^- K^+$, with different alkyl groups, to give unsymmetrical dithiocarbonate toluene-*p*-sulphonylhydrazones, are stereoselective, the last alkyl group introduced going *cis* to the toluene-*p*-sulphonylimino-group.[93] The dianions derived from these compounds, *e.g.* with BuLi, are also alkylated stereoselectively. Thermolysis of the sodium salts of *S*-alkyl *S'*-methyl dithiocarbonate toluene-*p*-sulphonylhydrazones, followed by methylation, affords a convenient synthesis of 1,1-bis(methylthio)buta-1,3-dienes *via* a [2,3]sigmatropic rearrangement of a dithiocarbene,[94] as shown in Scheme 5.

Condensation of thiosemicarbazide with 2-methylene-3-oxoquinuclidine affords the thiocarbamoylpyrazoline (37).[54] 2,4,6-Triphenylpyrylium perchlorate reacts with thiosemicarbazide, giving 1-thiocarbamido-2,4,6-triphenylpyridinium perchlorate.[79] Desyl bromide, PhCOCHBrPh, reacts with 4-alkyl-thiosemi-carbazides $NH_2NHCSNHR$ to give 2-alkylamino-1,3,4-thiadiazines (38), 3-amino-thiazolone-2-alkylimides (39), or 3-alkylthiazolone-2-hydrazones (40), depending on the substituents present.[95] Similar isomeric products may be prepared selectively

[88] K. Dwarakanath, D. N. Sathyanarayana, and K. Volka, *Bull. Soc. Chim. Belg.*, 1978, **87**, 667, 677.
[89] M. Uda and S. Kubota, *J. Heterocycl. Chem.*, 1979, **16**, 1273.
[90] C. Gors, F. Baert, J. P. Henichart, and R. Houssin, *J. Mol. Struct.*, 1979, **55**, 223.
[91] V. V. Dunina, E. I. Kazakova, V. M. Potapov, and E. G. Rukhadze, *Zh. Org. Khim.*, 1978, **14**, 2075 (*Chem. Abstr.*, 1979, **90**, 71 645).
[92] R. Mattes and H. Weber, *J. Chem. Soc., Dalton Trans.*, 1980, 423.
[93] T. Nakai and K. Mikami, *Chem. Lett.*, 1979, 465.
[94] T. Nakai and K. Mikami, *Chem. Lett.*, 1978, 1243.
[95] E. Bulka, W. D. Pfeiffer, C. Tröltsch, E. Dilk, H. Gärtner, and D. Daniel, *Collect. Czech. Chem. Commun.*, 1978, **43**, 1227.

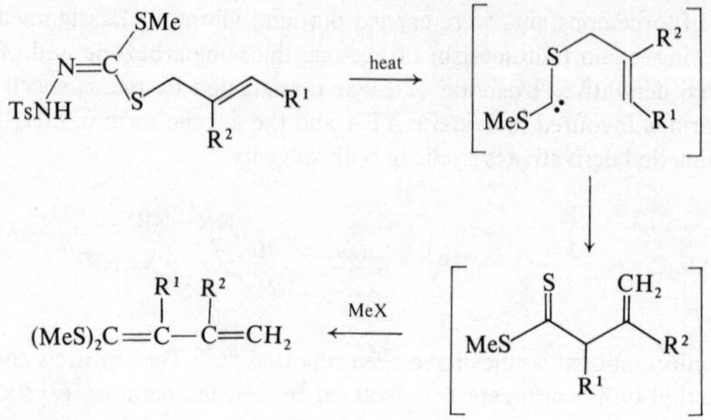

Scheme 5

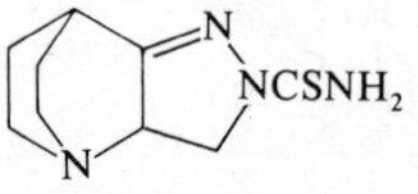

(37) 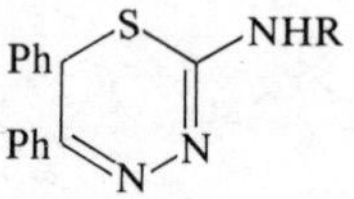   (38) 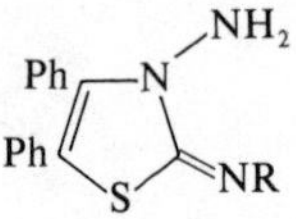   (39)

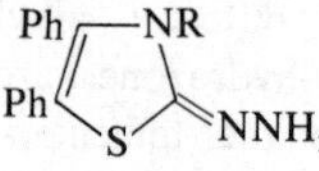

(40) 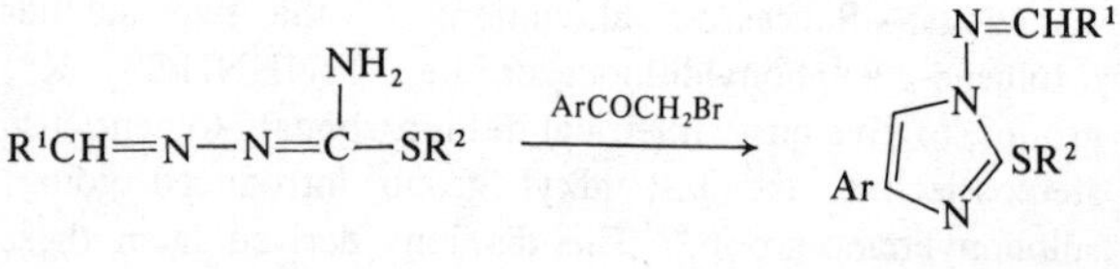   (41)

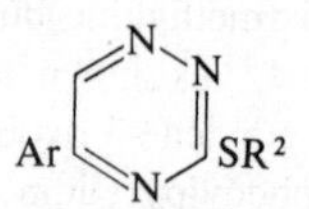

(42) 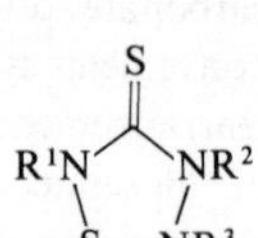   (43)

from thiosemicarbazide and ethyl 4-chloroacetoacetate by varying the acidity, solvent polarity, and temperature.[96] Isothiosemicarbazones of aromatic aldehydes and ketones react with phenacyl bromides to give 2-mercaptoimidazole derivatives (41), but isothiosemicarbazones of aliphatic aldehydes give poor yields of (41), and 3,5-disubstituted 1,2,4-triazines (42) are produced.[97] 1,2,4-Trisubstituted thiosemicarbazides react with thionyl chloride and pyridine, giving 1,2,3,5-thiatriazolidine-4-thione 1-oxides (43).[98]

[96] E. Campaigne and T. P. Selby, *J. Heterocycl. Chem.*, 1978, **15**, 401.
[97] C. Yamazaki, *Bull. Chem. Soc. Jpn.*, 1978, **51**, 1846.
[98] S. D. Ziman, *J. Heterocycl. Chem.*, 1979, **16**, 895.

Cyclization of 1-alkylthiocarbamoyl-3-thiosemicarbazides, RNHCSNHNH-CSNH$_2$, in alkaline media gives both 4-alkyl-5-amino-1,2,4-triazoline-3-thiones (44), by loss of H$_2$S, and 4-alkyl-1,2,4-triazolidine-3,5-dithiones (45), by loss of NH$_3$. 1-Alkoxythiocarbonyl-3-thiosemicarbazides lose H$_2$S under weakly alkaline conditions, giving 5-alkoxy-1,2,4-triazoline-3-thiones (46), but in strong alkali they eliminate alcohol, giving (45). 1-Alkoxycarbonyl-3-thiosemicarbazides cyclize to 5-thiono-1,2,4-triazolidin-3-ones in both weak and strong alkali.[99] Other cyclizations of (acyl)thiosemicarbazides[100] and of *S*-alkyl *N*-acyldithiocarbazates[101] have been reported. Cyclodesulphurization of 1-(*o*-aminobenzoyl)thiosemicarbazides with DCCI gives benzotriazepin-5-ones (47).[102] Rate constants for the ring-closure of 1-methylisatin 3-thiosemicarbazones to indolotriazines have been determined.[103]

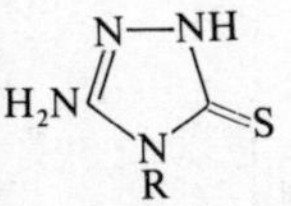

(44) 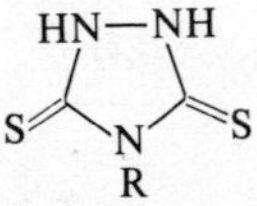　　　　　(45) 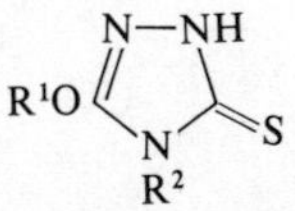　　　　　(46) 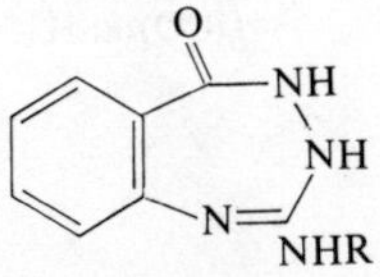　　　　　(47)

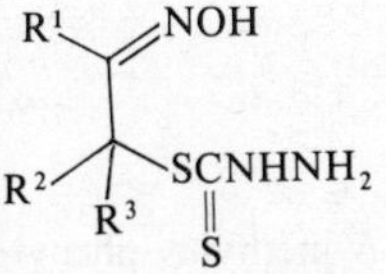

(48) 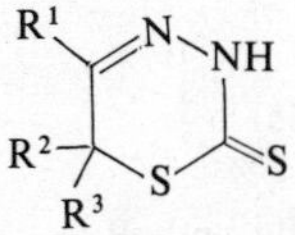　　　　　(49) 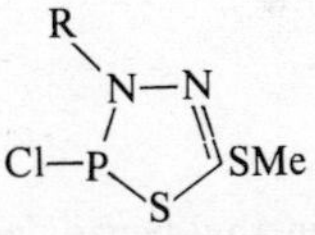　　　　(50) R = Ac or Ph 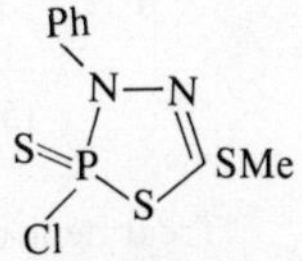　　　(51)

Salts of dithiocarbazic acid, NH$_2$NHCS$_2^-$ M$^+$, react with β-chloro-nitroso-compounds R^{1}CH(NO)CR^{2}R^{3}Cl, giving β-oximinoalkyl esters (48), which readily cyclize to 6*H*-1,3,4-thiadiazine-2-thiones (49).[104] Methyl *N*-phenyldithiocarbazate reacts with PCl$_3$ and PSCl$_3$ to give the phosphorus heterocycles (50) and (51), in which the chlorine atoms are labile and may be replaced by nucleophiles. Methyl *N*-acetyldithiocarbazate reacts similarly with PCl$_3$, but with POCl$_3$ or PSCl$_3$ it is dehydrated to give 2-methyl-5-methylthio-1,2,4-thiadiazolium chloride.[105]

3 Thioamides and Selenoamides

Synthesis.—The addition of H$_2$S to nitriles to give primary thioamides is conveniently performed in the presence of a phase-transfer catalyst in dilute aqueous Na$_2$S and a liquid organic phase.[106] Other examples of this reaction employ

[99] H. W. Altland and P. A. Graham, *J. Heterocycl. Chem.*, 1978, **15**, 377.
[100] M. Balogh, I. Hermecz, Z. Mészáros, and L. Pusztay, *J. Heterocycl. Chem.*, 1980, **17**, 175.
[101] G. Mazzone and F. Bonina, *Farmaco, Ed. Sci.*, 1978, **33**, 438.
[102] A. Mohsen, M. E. Omar, and F. A. Ashour, *J. Heterocycl. Chem.*, 1979, **16**, 1435.
[103] A. B. Tomchin and U. Lepp, *Zh. Org. Khim.*, 1978, **14**, 1544 (*Chem. Abstr.*, 1979, **90**, 6362).
[104] J. Beger, C. Thielemann, and P. D. Thong, *J. Prakt. Chem.*, 1979, **321**, 959.
[105] N. I. Shvetsov-Shilovskii, D. P. Nesterenko, and A. A. Stepanova, *Zh. Obshch. Khim.*, 1979, **49**, 1896 (*Chem. Abstr.*, 1980, **92**, 41 846).
[106] L. Cassar, S. Panossian, and C. Giordano, *Synthesis*, 1978, 917.

NH_3, NEt_3, or pyridine as the basic catalyst.[107-110] Nitriles are also converted into thioamide derivatives in the presence of the Lewis acids $SnCl_4$ and AsF_5, using $AcSH^{111}$ and Et_4N^+ SH^{-112} as sources of SH^-. Nitriles react with *OO*-dialkyl *S*-hydrogen phosphorodithioates, giving unstable adducts (52), which react further with the dialkyl phosphorodithioate to give thioamides, phosphorylated thioamides (53), and sulphides.[113] Benzonitrile and *OO*-dialkyl *O*-hydrogen phosphoro-thionates give similar products in low yield.[114] Isocyanides react with $R_2^1P(S)SH$ (R^1 = alkoxy or Ph), giving *S*-thiophosphorylated thioformamides (54), which rearrange to the *N*-isomers (55).[115]

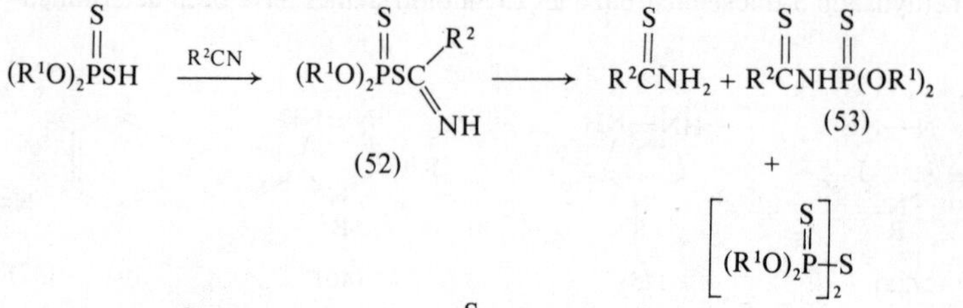

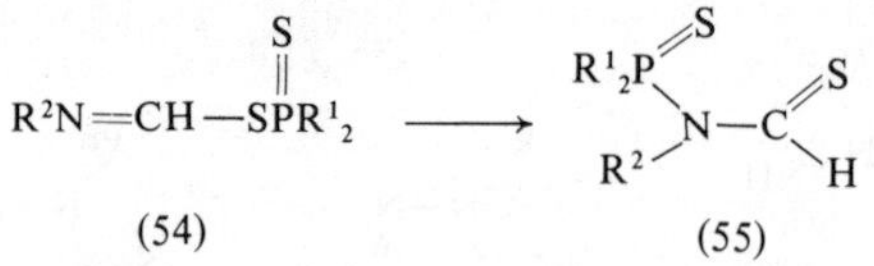

Treatment of *N*-methyl-benzoylacetanilide with P_4S_{10} gave *N*-methyl-*N*-phenyl-β-mercaptothiocinnamamide (56), which was isolated as its palladium complex.[116] Further examples of the thionation of carboxamides by P_4S_{10}[117,118] and by *p*-methoxyphenylthionophosphine sulphide dimer[119,120] have been reported. The latter reagent caused the thioamides (57) to cyclize to 6-ethylthieno[2,3-*d*]-[1,3]thiazine-4-thiones (58) when the reaction time was prolonged.[119] Carbox-

[107] L. N. Kulaeva, A. D. Grabenko, and P. S. Pel'kis, *Khim. Geterotsikl. Soedin.*, 1978, 909.
[108] S. N. Banerjee and C. Ressler, *Int. J. Pept. Protein Res.*, 1979, **14**, 234.
[109] T. Jaworski and J. Terpinski, *Pol. J. Chem.*, 1978, **52**, 2067 (*Chem. Abstr.*, 1979, **90**, 54 437).
[110] V. A. Pechenyuk, L. B. Dashkevich, S. A. Chistyakova, and E. N. Kuvaeva, *Zh. Org. Khim.*, 1978, **14**, 745 (*Chem. Abstr.*, 1978, **89**, 42 389).
[111] T. A. Tember-Kovaleva, R. A. Slavinskaya, and T. N. Sumarokova, *Zh. Obshch. Khim.*, 1978, **48**, 1556 (*Chem. Abstr.*, 1978, **89**, 172 822).
[112] L. Kolditz and I. Beierlein, *Z. Chem.*, 1978, **18**, 452.
[113] A. N. Pudovik, R. A. Cherkasov, M. G. Zimin, and N. G. Zabirov, *Zh. Obshch. Khim.*, 1978, **48**, 926 (*Chem. Abstr.*, 1978, **89**, 107 958); *Izv. Akad. Nauk SSSR, Ser. Khim.*, 1979, 861 (*Chem. Abstr.*, 1979, **91**, 56 575).
[114] M. G. Zimin, N. G. Zabirov, and A. N. Pudovik, *Zh. Obshch. Khim.*, 1979, **49**, 1164 (*Chem. Abstr.*, 1979, **91**, 74 316).
[115] M. G. Zimin, N. G. Zabirov, V. I. Nikitina, and A. N. Pudovik, *Zh. Obshch. Khim.*, 1979, **49**, 2651 (*Chem. Abstr.*, 1980, **92**, 129 027).
[116] S. Kitagawa and H. Tanaka, *Chem. Pharm. Bull.*, 1978, **26**, 3028.
[117] J. S. Davidson, *Synthesis*, 1979, 359.
[118] J. K. Schneider, P. Hofstetter, E. Pretsch, D. Ammann, and W. Simon, *Helv. Chim. Acta*, 1980, **63**, 217.
[119] K. Clausen and S. O. Lawesson, *Nouv. J. Chim.*, 1980, **4**, 43.
[120] H. Fritz, P. Hug, S. O. Lawesson, E. Logemann, B. S. Pedersen, H. Sauter, S. Scheibye, and T. Winkler, *Bull. Soc. Chim. Belg.*, 1978, **87**, 525.

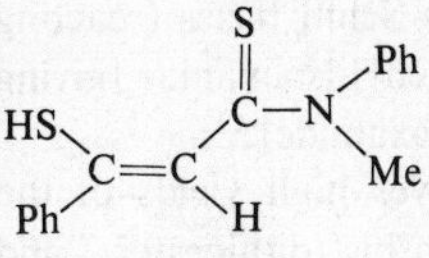

(56) 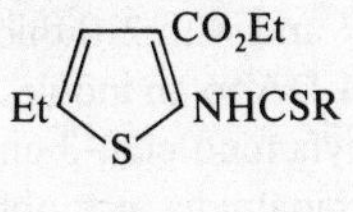 (57) 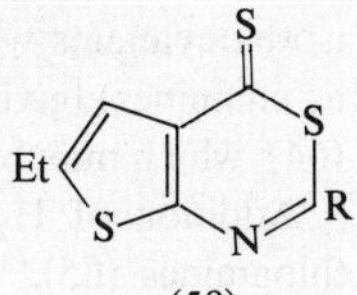 (58)

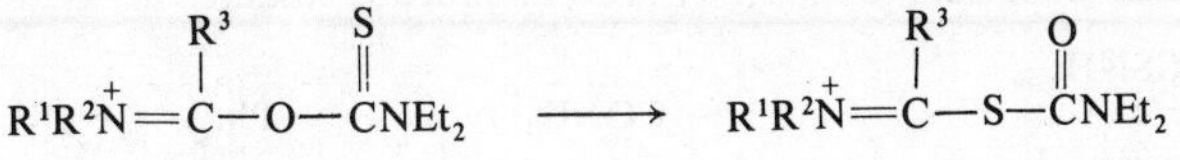

(59) (60)

amides are also converted into thioamides by the organophosphorus compounds $(Me_2N)_xP(S)Cl_{3-x}$, but not by $(Me_2N)_3PS$.[121] Diethylthiocarbamoyl chloride converts primary carboxamides and thioamides into nitriles, and tertiary amides into thioamides. The reaction mechanism that has been proposed involves the formation of an iminium salt (59), which rearranges to (60). Elimination of COS and $Et_2NH \cdot HCl$ from either (59) or (60) (if $R^1 = R^2 = H$) will give a nitrile, and elimination of Et_2NCOCl from (60) gives a thioamide.[122] Chloroiminium salts (Vilsmeier reagents) react reversibly with thioamides with interchange of sulphur and chlorine [reaction (3)]; if the equilibrium is favourable, the new thioamide may be isolated after hydrolysis of the reaction mixture.[123]

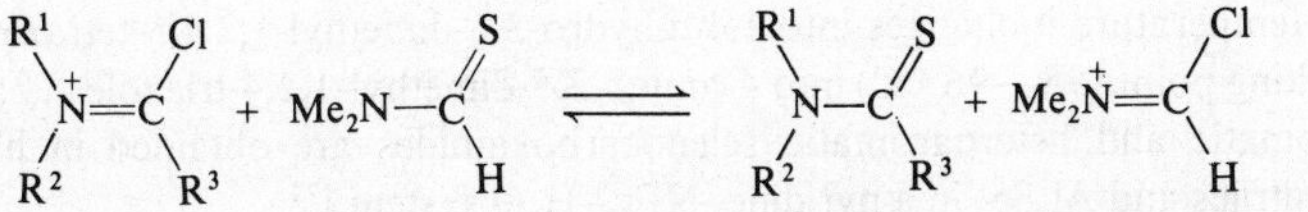

(3)

The reaction of Grignard reagents with triphenylphosphine thiocyanate, $Ph_3P(SCN)_2$, is a convenient method for preparing *N*-unsubstituted thioamides in fair yield.[124] The application of the Willgerodt–Kindler reaction to aromatic aldehydes and primary aliphatic amines to yield $ArCSNHR$ has been reported.[125,126] Thioamides are not obtained with aromatic amines. Nitrobenzyl halides react with sulphur and secondary amines in a dipolar aprotic solvent at 20 °C to give nitrothiobenzamides, but reduction of the nitro-group also occurs at higher temperatures.[127]

A thiocarbamoyl group may be introduced directly into anisole and *o*- or *m*-dimethoxybenzenes by treatment with KSCN in PPA at 60 °C or with concentrated H_2SO_4 at room temperature to give the thioanisamides (61) in *ca.* 50% yield.[128] Substituted thiocarbamoyl groups are introduced by the reaction of

[121] B. S. Pedersen and S. O. Lawesson, *Bull. Soc. Chim. Belg.*, 1977, **86**, 693.
[122] M. Ogata and H. Matsumoto, *Heterocycles*, 1978, **11**, 139.
[123] M. Helbert, J. P. Renou, and M. L. Martin, *Tetrahedron*, 1979, **35**, 1087.
[124] Y. Tamura, T. Kawasaki, M. Adachi, and Y. Kita, *Synthesis*, 1979, 887.
[125] R. C. Moreau, P. Loiseau, J. Bernard, F. Sébastien, and R. Leroy, *Eur. J. Med. Chem.*, 1979, **14**, 317.
[126] R. C. Moreau and P. Loiseau, *Ann. Pharm. Fr.*, 1978, **36**, 269.
[127] R. Mayer, H. Viola, J. Reichert, and W. Krause, *J. Prakt. Chem.*, 1978, **320**, 313.
[128] S. Sastry and N. A. Kudav, *Indian J. Chem., Sect. B*, 1979, **18**, 455 (*Chem. Abstr.*, 1980, **92**, 128 540).

isothiocyanates with benzoylacetamide [giving (62)],[129] with Schiff bases (reacting as enamines) [giving (63)],[130,131] and with 2-(lithiomethyl)phenyl isocyanide [giving (64), which may be cyclized by LDA to an indole-3-thiocarboxamide].[132]

Addition of H_2S to 1-(dialkylamino)pent-3-en-1-ynes gives high yields of the thioamides (65).[133] Polymeric thioamides are obtained from bis-(dithioesters) and diamines by uncatalysed polycondensation in benzene–DMF at room temperature.[134] Monothio-diaroylamines, $RC(O)NHC(S)R$, are made by hydrolysis of the dithiazolium salts (66), *e.g.* by NEt_3 in aqueous ethanol.[135]

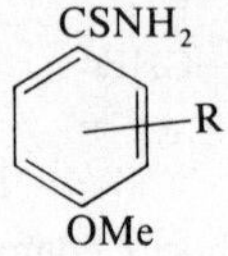

(61) 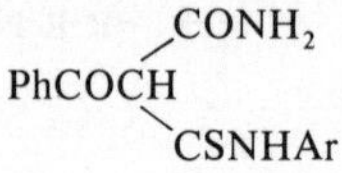   (62) 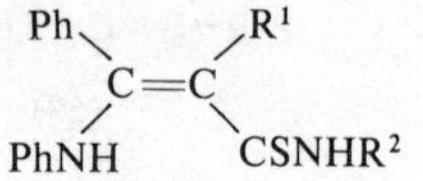   (63)

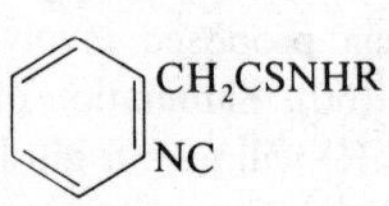

(64)   (65)   (66)

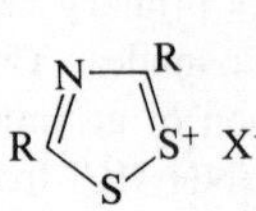

The preparation of thioacethydrazide, $MeCSNHNH_2$, has now been accomplished, both from the known nickel complex and from carboxymethyl dithioacetate [$MeC(=S)SCH_2CO_2H$] and hydrazine.[136] It is stable at $-30\ ^\circ C$, but at room temperature it changes into 1,4-dihydro-3,6-dimethyl-1,2,4,5-tetrazine and at its melting point (95—96 $^\circ C$) into 4-amino-3,5-dimethyl-1,2,4-triazole.

Aromatic and heteroaromatic selenocarboxamides are obtained in high yield from nitriles and Al_2Se_3 in a pyridine–NEt_3–H_2O system.[137]

Physical Properties.—In studies of the secondary thioamide function, the i.r. and, where possible, the Raman spectra of a series of *N*-methyl-thioamides XCSNHMe (X = CSNHMe, $CSNH_2$, $CONH_2$, or CO_2K) and of *N*- and *C*-deuteriated products were determined, vibrational analyses were proposed, and the results were compared with data for other *N*-methyl-thioamides.[138,139] The ^{13}C n.m.r. spectra of a series of tertiary thioamides of formic, acetic, trifluoroacetic, propionic, and

[129] R. G. Dubenko, E. F. Gorbenko, and P. S. Pel'kis, *Zh. Org. Khim.*, 1979, **15**, 1483 (*Chem. Abstr.*, 1979, **91**, 174 975).

[130] W. Zankowska-Jasinska and H. Borowiec, *Pol. J. Chem.*, 1978, **52**, 1155 (*Chem. Abstr.*, 1978, **89**, 163 536).

[131] W. Zankowska-Jasinska and H. Borowiec, *Pol. J. Chem.*, 1978, **52**, 1683 (*Chem. Abstr.*, 1979, **90**, 54 632).

[132] Y. Ito, K. Kobayashi, and T. Saegusa, *Tetrahedron Lett.*, 1979, 1039.

[133] S. E. Tolchinskii, I. A. Maretina, and A. A. Petrov, *Zh. Org. Chem.*, 1979, **15**, 650 (*Chem. Abstr.*, 1979, **91**, 19 860).

[134] G. Levesque and J. C. Gressier, *J. Polym. Sci., Polym. Lett. Ed.*, 1979, **17**, 281 (*Chem. Abstr.*, 1979, **91**, 21 171).

[135] J. Liebscher and H. Hartmann, Ger. (East) P. 135 901 (*Chem. Abstr.*, 1979, **91**, 193 019).

[136] K. A. Jensen and E. Larsen, *Acta Chem. Scand., Ser. A*, 1979, **33**, 137.

[137] V. I. Cohen, *Synthesis*, 1978, 668.

[138] H. O. Desseyn, A. J. Aarts, E. Esmans, and M. A. Herman, *Spectrochim. Acta, Part A*, 1979, **35**, 1203.

[139] H. O. Desseyn, A. J. Aarts, and M. A. Herman, *Spectrochim. Acta, Part A*, 1980, **36**, 59.

butyric acids were completely assigned, and a linear relationship was found between the ^{13}C shifts of C=S in ethyl-thioamides and C=O in ethyl-amides.[120]

The stereochemistry of methyl *N*-methylthiobenzimidates, *e.g.* (67) and (68), that have substituents at one or more of the 2-, 4-, and 6-positions has been investigated.

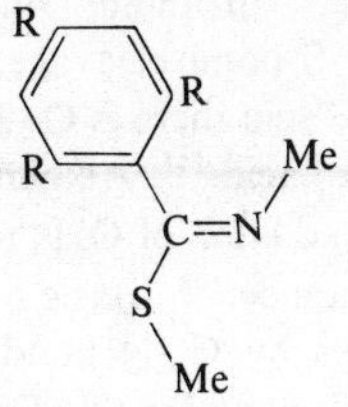

(E)

(67)

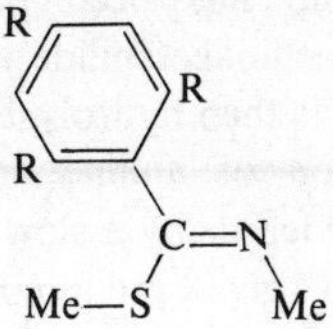

(Z)

(68)

The ^{13}C n.m.r. spectra of E/Z mixtures of twelve compounds were assigned and the angles of twist about the aryl–imino-carbon bond were determined.[140] Subsequently, the pure (E)- and (Z)-isomers were obtained by preparative t.l.c. at 258 K or by stereospecific syntheses from (E)- and (Z)-thioamides. The chemical shifts of the protons in the methyl groups were measured and the $(E):(Z)$ ratios and the barriers to isomerization were determined by ^1H n.m.r. spectroscopy.[141] The conformational states of some *NN*-di(primary alkyl)-amides and -thioamides were studied by dynamic ^1H n.m.r. spectroscopy, and the barriers to the interconversion of enantiomers that were thus found were in good agreement with molecular mechanics calculations.[142] I.r. and n.m.r. data for optically active thioamides $R^1CSNR^2CHMeR^3$ indicate that there are mixtures of (E) and (Z) conformers,[143] and i.r. and ^{31}P n.m.r. studies of phosphorylated thiobenzamides provide evidence of a tautomeric equilibrium [reaction (4)], the amide form being favoured in more

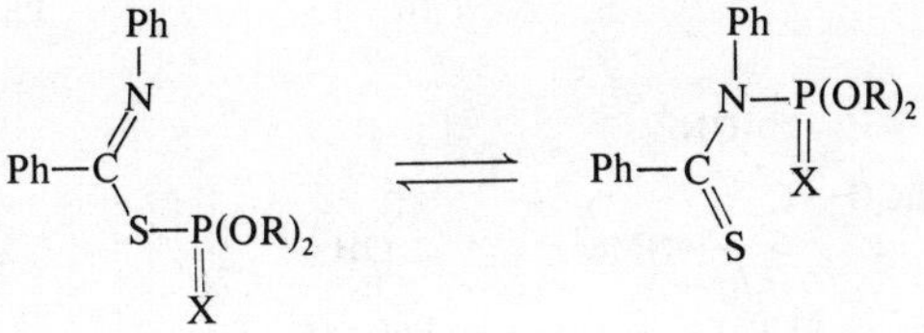

(4)

polar solvents.[144] The heats of vaporization of *NN*-dimethylthiobenzamide and of methyl *N*-methylthiobenzimidate, and the heats of methylation of these compounds with $MeOSO_2F$, show an enthalpy difference that favours the thioamide structure.[145]

[140] W. Walter, W. Ruback, and C. O. Meese, *Org. Magn. Reson.*, 1978, **11**, 612.
[141] W. Walter, W. Ruback, and C. O. Meese, *Chem. Ber.*, 1980, **113**, 171.
[142] U. Berg, M. Grimaud, and J. Sandström, *Nouv. J. Chim.*, 1979, **3**, 175.
[143] V. V. Dunina, V. M. Potapov, E. G. Rukhadze, and E. I. Kazakova, *Zh. Org. Khim.*, 1978, **14**, 2064 (*Chem. Abstr.*, 1979, **90**, 86 614).
[144] M. G. Zimin, M. M. Afanas'ev, and A. N. Pudovik, *Zh. Obshch. Khim.*, 1979, **49**, 2621 (*Chem. Abstr.*, 1980, **92**, 110 306).
[145] P. Beak, J-K. Lee, and J. M. Ziegler, *J. Org. Chem.*, 1978, **43**, 1536.

The i.r. spectra of selenoacetamide and the *N*-deuteriated species have been determined and fundamental frequencies assigned.[146]

Reactions.—*Hydrolysis.* In several kinetic studies of the hydrolysis of thioamides, the formation of a tetrahedral intermediate by addition of H_3O^+ or OH^- to the thiocarbonyl group has been postulated.[147–149] In dilute sulphuric acid the intermediate from thioacetanilide undergoes C—S bond cleavage, giving H_2S and acetanilide, which is then hydrolysed, but in 48% acid there is C—N bond cleavage, with the formation of aniline and thiolacetic acid.[147] Alkaline hydrolysis of thiobenzamides proceeds by a slow nucleophilic addition of OH^- to C=S and a fast decomposition to give predominantly the amide.[148] Basic methanolysis of *N*-aryl-*N*-(methyl)thiobenzamides, however, goes by C—N bond cleavage of the intermediate, with liberation of the *N*-methylarylamine.[150] Thioanilides, when treated with Na_2PdCl_4, give complexes which are desulphurized by wet DMSO, giving anilides.[151]

Oxidation. The rates of formation of *S*-oxides of substituted *N*-(methyl)-thiobenzanilides by oxidation with H_2O_2 in aqueous dioxan have been measured and the effects of substituents were analysed. The results suggest that the thioanilides have the (*E*) configuration (69), and not the (*Z*)-form (70).[152] Oxidation

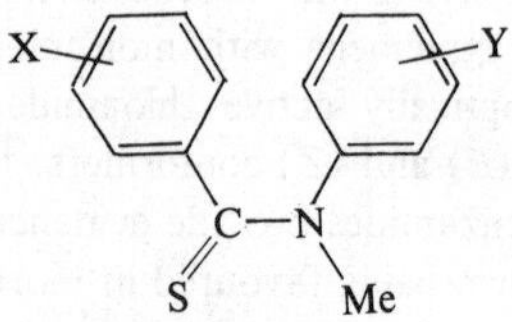

(69)

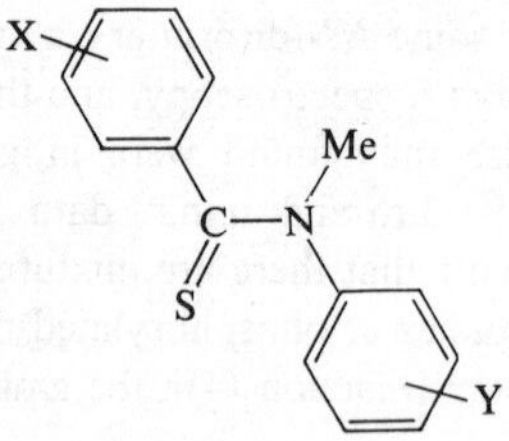

(70)

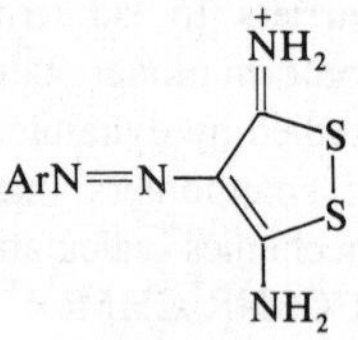

(71)

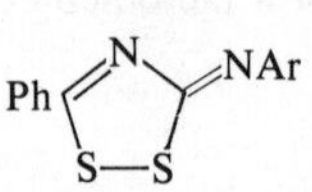

(72)

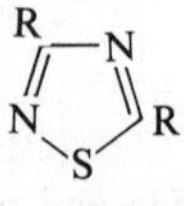

(73)

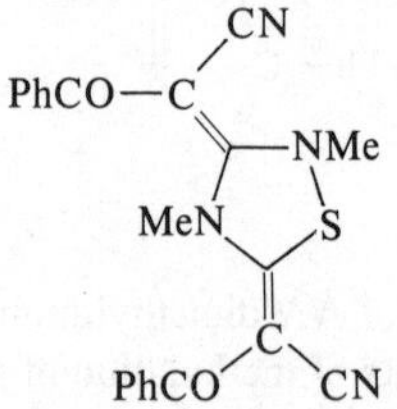

(74)

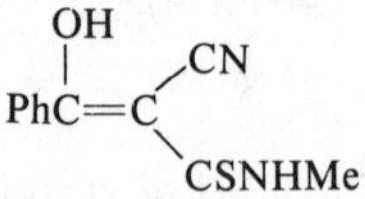

(75)

[146] K. R. Gayathri-Devi, D. N. Sathyanarayana, and K. Volka, *Spectrochim. Acta, Part A*, 1978, **34**, 1137.

[147] J. T. Edward and S. C. Wong, *J. Am. Chem. Soc.*, 1979, **101**, 1807.

[148] J. Mollin and P. Bouchalová, *Collect. Czech. Chem. Commun.*, 1978, **43**, 2283.

[149] R. Ahmad and A. A. Khan, *Indian J. Chem., Sect. A*, 1979, **18**, 264 (*Chem. Abstr.*, 1980, **92**, 21 725).

[150] T. J. Broxton, L. W. Deady, and J. E. Rowe, *Aust. J. Chem.*, 1978, **31**, 1731.

[151] H. Alper and J. K. Currie, *J. Organomet. Chem.*, 1979, **168**, 369.

[152] W. Walter and O. H. Bauer, *Liebigs Ann. Chem.*, 1979, 248.

by bromine of the arylhydrazones of dithiomesoxalic acid diamide (from arenediazonium salts and dithiomalonamide) gives 3,5-diamino-4-arylazo-1,2-dithiolium salts (71),[107] and similar oxidation of $PhCH_2SC(Ph){=}NCSNHAr$ gives the dithiazoles (72), with elimination of the benzyl group.[153] Oxidation of thiobenzamides with nitrous acid in concentrated hydrochloric acid gives the corresponding benzonitriles and 1,2,4-thiadiazoles (73). 2,6-Dichlorothiobenzamide, however, gives 2,6-dichlorophenyl isothiocyanate as the main product (33%).[154] The 1,2,4-thiadiazolidine (74) is obtained by oxidation of (75) with I_2 and NEt_3.[155]

Alkylation and Arylation. The S-methylation of thioamides by diazomethane is catalysed by silica gel.[156] S-Phenyl derivatives of thiobenzanilides are obtained in high yield by radical arylation with nitrosoacetanilide or phenylazotriphenylmethane.[157,158] Thiols are made from alkyl (or cycloalkyl, or aralkyl) halides by their reaction with dimethylthioformamide and treatment of the imidium ester with methanol; other methods of hydrolysis lead to side-reactions, giving impure products.[159] Mono- and bi-protic thiocarboxamides condense with ethyl bromocyanoacetate and one equivalent of NaOEt with spontaneous separation of sulphur and the formation of β-amino-α-cyano-acrylates, as shown in reaction (5).[160] The

$$R^1C{=}S + BrCH \begin{matrix} CN \\ \diagup \\ \diagdown \\ CO_2Et \end{matrix} \quad\longrightarrow\quad \left[\begin{matrix} R^1C{-}S{-}CH{-}CN \\ \| \quad\quad | \\ R^2N \quad\quad CO_2Et \end{matrix} \right] \quad\longrightarrow\quad R^1C{=}C \begin{matrix} CN \\ \diagup \\ \diagdown \\ CO_2Et \end{matrix} \quad (5)$$

(with NHR^2 below $R^1C{=}S$, and R^2NH below the product $R^1C{=}C$)

similar sulphide contractions *via* alkylative coupling described previously[161] (see Vol. 2, p. 235) employed trialkyl phosphite or triphenylphosphine to remove sulphur from the postulated episulphide intermediate. Biprotic thiocarboxamides behave as sources of H_2S when treated with α-halogenated ketones, esters, or nitriles and ethoxide ion under non-hydrolytic conditions, and give nitriles and symmetrical sulphides derived from the halogenated compound.[162] Thioacetamide has similarly been used to provide sulphur in the synthesis of many-membered cyclic sulphides (thiophanes).[163]

1,4-Benzoquinone and thioacetamide react by Michael addition at the sulphur atom; with an excess of quinone, oxidation and cyclization occur, to give (76). At lower quinone ratios, acetonitrile is eliminated from the Michael adduct to give

[153] R. Rai and V. K. Verma, *Indian J. Chem., Sect. B*, 1979, **18**, 284 (*Chem. Abstr.*, 1980, **92**, 94 310).
[154] M. Badahir, S. Nitz, H. Parlar, and F. Korte, *Z. Naturforsch., Teil. B*, 1979, **34**, 768.
[155] H. Kunzek, E. Nesener, and J. Voigt, *Z. Chem.*, 1978, **18**, 172.
[156] H. Nishiyama, H. Nagase, and K. Ohno, *Tetrahedron Lett.*, 1979, 4671.
[157] R. Kh. Freidlina, I. I. Kandror, and I. O. Bragina, *Izv. Akad. Nauk SSSR, Ser. Khim.*, 1979, 1165 (*Chem. Abstr.*, 1979, **91**, 56 580).
[158] I. I. Kandror, I. O. Bragina, and R. Kh. Freidlina, *Dokl. Akad. Nauk SSSR*, 1979, **249**, 867 (*Chem. Abstr.*, 1980, **92**, 163 665).
[159] K. Hattori, T. Takido, and K. Itabashi, *Nippon Kagaku Kaishi*, 1979, 105 (*Chem. Abstr.*, 1979, **90**, 137 380).
[160] H. Singh and C. S. Gandhi, *J. Chem. Res. (S)*, 1978, 407; *Synth. Commun.*, 1978, **8**, 469.
[161] M. Roth, P. Dubs, E. Götschi, and A. Eschenmoser, *Helv. Chim. Acta*, 1971, **54**, 710.
[162] H. Singh and C. S. Gandhi, *Synth. Commun.*, 1979, **9**, 569.
[163] E. Hammerschmidt, W. Bieber, and F. Vögtle, *Chem. Ber.*, 1978, **111**, 2445.

2,5-dihydroxythiophenol, which may then be oxidized to the disulphide. Sterically hindered quinones act only as oxidizing agents.[164]

Keten *SS*-acetals are obtained from thioamides that have at least one hydrogen atom α- to the thiocarbonyl group by *S*-methylation and treatment of the imidium ester with propane-1,3-dithiol.[165] Treatment of the imidium esters from *NN*-disubstituted thioamides with 4-amino-4-methylpentan-2-ol gives 2-substituted 4,6,6-trimethyl-5,6-dihydro-1,3-oxazines,[166] and with sodium azide they give alkylidenetriazenes (1,2,3-triazabutadienes) (77),[167] not dihydrothia(S^{IV})triazoles (78), as previously claimed[168] (see Vol. 3, p. 399). Structure (77) was demonstrated by *X*-ray studies.

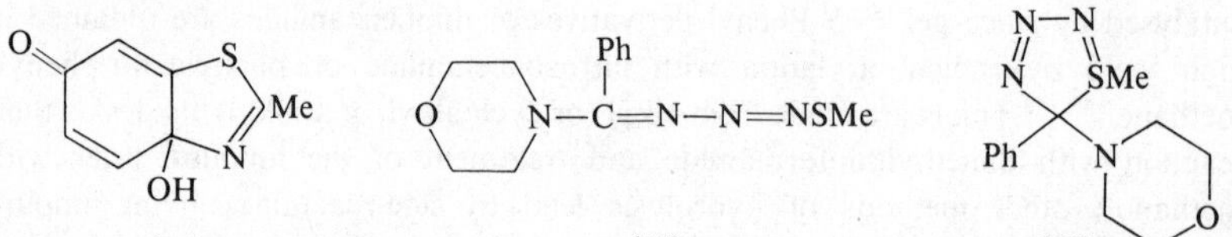

(76)   (77)   (78)

Nucleophilic Reactions of the α-Carbon Atom. Thioacetamide condenses with dimethylformamide dimethyl acetal, giving *trans*-3-(dimethylamino)thioacryl-amide.[169] The Japp–Klingemann reaction with 3-(phenylthiocarbamoyl)-pentane-2,4-dione effects acetyl cleavage, giving 2-methyl-1-(phenylthiocarbam-oyl)glyoxal arylhydrazones,[170] and with (arylthiocarbamoyl)benzoylacetamides (62) it displaces the carboxamido-group.[129] Araldehydes condense with cyanothio-acetamide to give 3-aryl-2-cyano-thioacrylamides (79), which dimerize reversibly to thiopyran derivatives (80).[171]

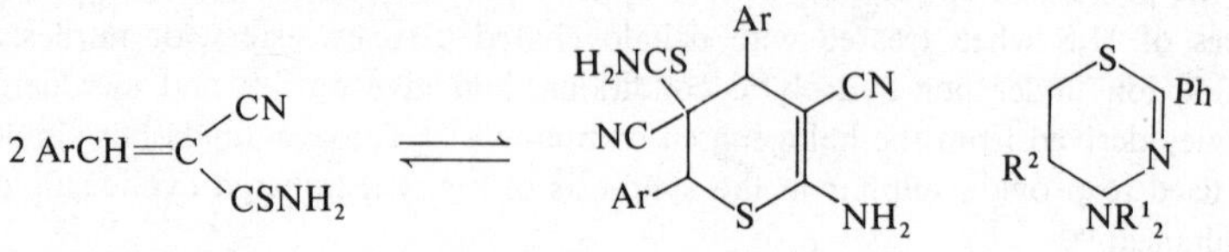

(79)   (80)   (81)

Diels–Alder and Similar Reactions. Thiobenzoylformamidines, $PhC(S)N=CH-NR^1_2$, behave as dienes, reacting with dienophiles to give dihydro-1,3-thiazines (81).[172] On the other hand, cyanothioformamides, and especially their *N*-acyl

[164] V. Horak and W. B. Manning, *J. Org. Chem.*, 1979, **44**, 120.
[165] T. Harada, Y. Tamaru, and Z. Yoshida, *Tetrahedron Lett.*, 1979, 3525.
[166] T. Harada, Y. Tamaru, and Z. Yoshida, *Chem. Lett.*, 1979, 1353.
[167] G. L'abbé, A. Willocx, J. P. Declercq, G. Germain, and M. Van Meerssche, *Bull. Soc. Chim. Belg.*, 1979, **88**, 107.
[168] S. I. Mathew and F. Stansfield, *J. Chem. Soc., Perkin Trans. 1*, 1974, 540.
[169] S. Rajappa and B. G. Advani, *Indian J. Chem., Sect. B*, 1978, **16**, 819 (*Chem. Abstr.*, 1979, **90**, 86 694).
[170] F. A. Amer, A. H. Harhash, and M. A. Awad, *Z. Naturforsch., Teil. B*, 1978, **33**, 660.
[171] J. S. A. Brunskill, A. De, and D. F. Ewing, *J. Chem. Soc., Perkin Trans. 1*, 1978, 629.
[172] J. C. Meslin and H. Quiniou, *Bull. Soc. Chim. Fr., Part 2*, 1979, 347.

derivatives, are dienophiles which add dienes across the C=S bond[173] and which also undergo cycloaddition with 1,3-dipolar compounds such as diazoalkanes, diphenylnitrilimine, benzonitrile oxide, and phenyl azide to give heterocyclic adducts or their decomposition products.[174] Acylated cyanothioformanilides add to methylenecyclohexane or to β-pinene by the ene-reaction, giving unsaturated sulphides, *e.g.* (82).[175]

Additions to αβ-Unsaturated Thioamides. Thiourea and toluene-*p*-sulphonic acid add to phenylthiopropiolanilide, PhC≡CCSNHPh, to give an isothiouronium toluene-*p*-sulphonate, from which *N*-phenyl-β-mercaptothiocinnamamide is obtained by hydrolysis.[176] Organolithium compounds R^1Li and Grignard reagents react with αβ-unsaturated thioamides by 1,4-addition. The lithium enolate anion can be captured by its reaction with electrophiles; *e.g.*, with PhSSPh to give (83).[177,178] Soft nucleophiles, *e.g.* sodiomalonic ester, are unreactive or give intractable mixtures when treated with αβ-unsaturated thioamides, but in the presence of a Pd^{II} salt they give α-palladiated β-substituted thioamides, which are more stable than oxa-(π-allyl)palladium complexes.[179] The palladium may be removed by hydrogenation. Thiobenzamides are recovered unchanged after treatment with two equivalents of an arylmagnesium bromide, but with four equivalents of Grignard reagent the intermediate $Ar^1C(=NMgBr)SMgBr$ is attacked; it gives a thiobenzophenone $Ar^1C(=S)Ar^2$ and a benzophenone imide $Ar^1C(=NH)Ar^2$. The ratio of the products depends on the nature of Ar^1 and Ar^2. *N*-Substituted thiobenzamides are unreactive.[180]

(82)   (83)   (84)

Displacement of Sulphur. The Wittig reaction of resonance-stabilized phosphorus ylides with thioacyl-urethanes gives acylated enamines, *e.g.* (84), from which β-keto-esters are obtained by hydrolysis.[181] Hydrazine displaces sulphur in thioanilides, giving amidrazones,[117] but with two equivalents of arylselenoamides it gives 2,5-diaryl-1,3,4-selenadiazoles.[182] Selenoamides react with an excess of hydrazine to give 1,2,4,5-tetrazines (see also Chap. 6, p. 294). 1,2-Bis(thio-benzamido)ethane, when treated with NEt_3 and HgO, cyclizes to 2-phenyl-

[173] K. Friedrich and M. Zamkanei, *Chem. Ber.*, 1979, **112**, 1867.
[174] K. Friedrich and M. Zamkanei, *Chem. Ber.*, 1979, **112**, 1873.
[175] K. Friedrich and M. Zamkanei, *Chem. Ber.*, 1979, **112**, 1916.
[176] S. Kitagawa and H. Tanaka, *Chem. Pharm. Bull.*, 1978, **26**, 1021.
[177] Y. Tamaru, T. Harada, H. Iwamoto, and Z. Yoshida, *J. Am. Chem. Soc.*, 1978, **100**, 5221.
[178] Y. Tamaru, T. Harada, and Z. Yoshida, *J. Am. Chem. Soc.*, 1979, **101**, 1316.
[179] Y. Tamaru, M. Kagotani, and Z. Yoshida, *J. Org. Chem.*, 1979, **44**, 2816.
[180] T. Karakasa, T. Hanzawa, and S. Motoki, *Bull. Chem. Soc. Jpn.*, 1979, **52**, 3469.
[181] A. Gossauer, F. Roessler, H. Zilch, and L. Ernst, *Liebigs Ann. Chem.*, 1979, 1309.
[182] V. I. Cohen, *J. Heterocycl. Chem.*, 1979, **16**, 365.

1-thiobenzoyl-2-imidazoline, and with ethylenediamine and HgO it gives two molecules of 2-phenylimidazoline.[183]

Synthesis of Thiazole Derivatives. N-Thiobenzoyl-α-amino-acids, when treated with TFAA, give 5-(2'-benzamidoacyloxy)-2-phenylthiazoles (85; X = OCOCHRNHCOPh), half of the starting material serving to acylate the hydroxythiazole. The corresponding amides give 2-phenyl-5-(trifluoro-acetamido)thiazoles (85; X = NHCOCF$_3$), and thiobenzoyl-leucine methyl ester is converted into benzoyl-leucine methyl ester.[184] N-Formyl- and N-thioformyl-sarcosine N-methylthioamide cyclize to 5-methylamino-1,3-thiazolium salts (86) and the isomeric 4-mercapto-1,3-imidazolium salts (87) when TFA is present. The latter give the meso-ionic compound anhydro-1,3-dimethyl-4-mercaptoimidazolium hydroxide when treated with a base.[185] Meso-ionic anhydro-4-hydroxythiazolium hydroxides (88) are obtained from *gem*-dicyano-epoxides (89) and N-mono-substituted thioamides,[186] a side-reaction giving some thiazolidinone (90).

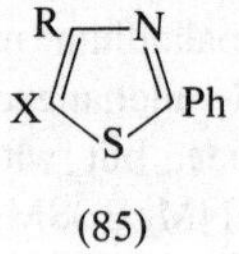

(85) 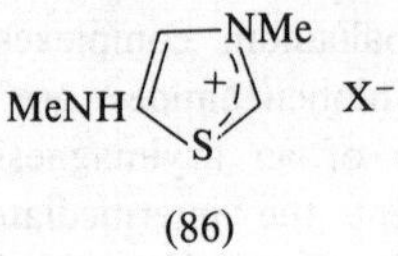   (86) 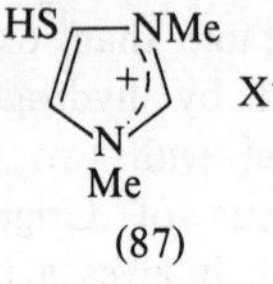   (87)

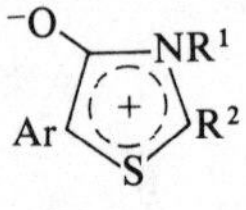

(88) 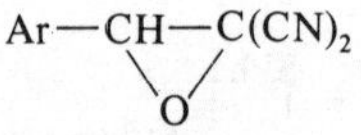   (89) 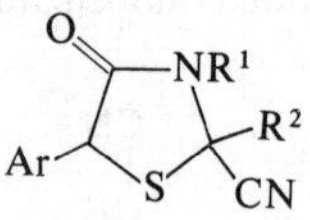   (90)

The Hantzsch thiazole synthesis from thioamides and α-halogeno-ketones has been applied to NN-(dimethyl)dithio-oxamide to give 2-(dimethylthio-carbamoyl)thiazoles,[187] but in some cases it took another course. Thus, 1-alkylthiocarbamoyl-1-cyanoacetophenones (91) react with bromo-ketones with elimination of alkylamine, giving 2-methylidene-1,3-oxathioles (92).[188] The thio-anilide of acetoacetic acid reacts with 2-bromo-1-tetralone to give the spiro-

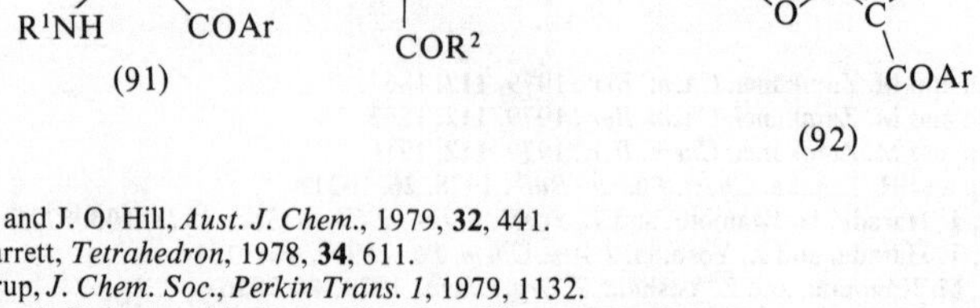

(91)   + MeCHBr   →   (92)

[183] G. Ewin and J. O. Hill, *Aust. J. Chem.*, 1979, **32**, 441.
[184] G. C. Barrett, *Tetrahedron*, 1978, **34**, 611.
[185] M. Begtrup, *J. Chem. Soc., Perkin Trans. 1*, 1979, 1132.
[186] M. Baudy, A. Robert, and A. Foucaud, *J. Org. Chem.*, 1978, **43**, 3732.
[187] H. U. Kibbel, E. Peters, and M. Michalik, *Z. Chem.*, 1979, **19**, 19.
[188] H. Dehne and P. Krey, *Z. Chem.*, 1979, **19**, 251.

compound (93), and the corresponding *N*-ethylthioamide gives the analogous ethylimino-compound, but the major product in this case is the benzindole derivative (94).[189] (See also Chap. 6, p. 285.)

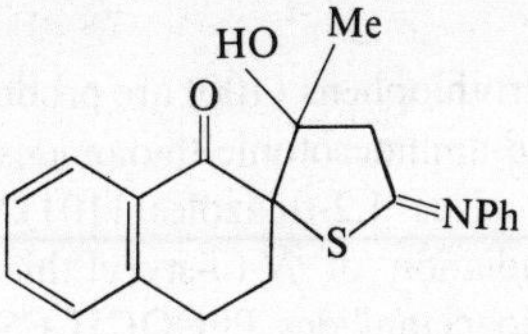

(93) 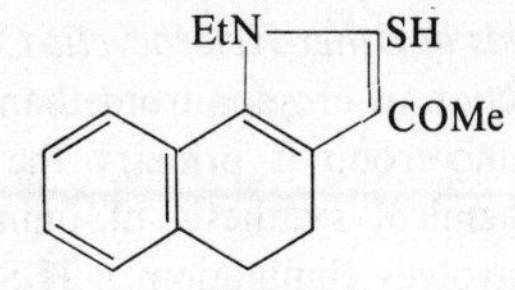   (94)

The α-thiocarbamoylcarbodi-imides (95), which are obtained by the reaction of sterically hindered isothiocyanates R²NCS with 3-dialkylamino-2,2-dimethyl-2*H*-azirines[190] (see Vol. 5, pp. 141, 158), are in equilibrium with the dipolar species (96). This last compound is not detectable by spectroscopy, but its presence may be demonstrated by its protonation or by its reduction with NaBH₄ to give stable thiazolidine derivatives. The carbodi-imides (95; R¹ = Me) rearrange on heating, to give the dipolar imidazoles (97), in which some participation of the acyclic structure (98) is suggested by the low barrier to rotation of the dimethylamino-group (from n.m.r. studies) and by a further isomerization to give (99).[191] This behaviour is closely paralleled by the properties of the adducts of 3-dialkylamino-2,2-dimethyl-2*H*-azirines with carbon disulphide, which exist in the solid state as heterocyclic dipolar compounds and as α-(dialkylthiocarbamoyl)isopropyl isothiocyanates when melted.[192, 193]

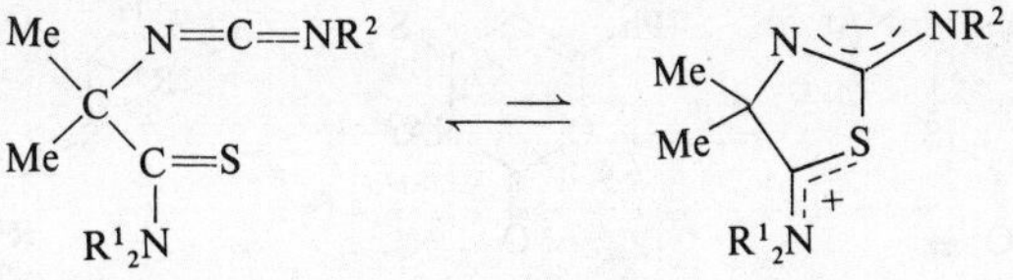

(95)   (96)

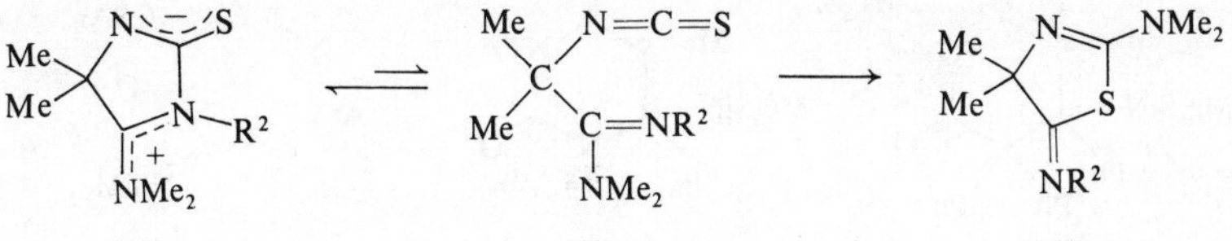

(97)   (98)   (99)

When pentafluorophenylthioureas or pentafluorophenyl thio- or dithio-carbamates are heated in DMF, they cyclize to give 4,5,6,7-tetrafluorobenzo-

[189] E. H. Marten and C. A. Maggiulli, *J. Heterocycl. Chem.*, 1978, **15**, 1277.
[190] E. Schaumann, E. Kausch, and W. Walter, *Chem. Ber.*, 1977, **110**, 820.
[191] E. Schaumann, H. Behr, and G. Adiwidjaja, *Liebigs Ann. Chem.*, 1979, 1322.
[192] S. Chaloupka, H. Heimgartner, H. Schmid, H. Link, P. Schönholzer, and K. Bernauer, *Helv. Chim. Acta*, 1976, **59**, 2566.
[193] E. Schaumann, E. Kausch, S. Grabley, and H. Behr, *Chem. Ber.*, 1978, **111**, 1486.

thiazoles. Thionation of acyl derivatives of pentafluoroaniline with P_4S_{10} in dioxan at 90 °C causes cyclization to tetrafluorobenzothiazoles, but N^1-acyl-N^2-(penta-fluorophenyl)hydrazines can be thionated to thiohydrazides, which cyclize to 5,6,7,8-tetrafluorobenzo-1,3,4-thiadiazines when heated in DMF.[194]

Synthesis of Other Heterocyclic Compounds. Nitrothiophens (100) are produced by the reaction of bromonitromethane with α-acyl-β-aminocrotonic thioamides, but if the amino-group is primary the reaction also gives 1,2-thiazoles (101).[195] The photochemical synthesis of quinolines by irradiation of N-(o-styryl)thioamides (102) involves elimination of H_2S.[196] Benzoylthioacetanilides $PhCOCH_2CSNHAr$ and phosgene give 4-arylimino-2-oxo-6-phenyl-1,3-oxathiins (103), which rearrange in acid solution to 3-aryl-2,3-dihydro-2-oxo-6-phenyl-1,3-oxazine-4-thiones (104).[197] A general route to $6H$-1,3,5-oxathiazines (105) by α-thioamidoalkylation of aldehydes R^4CHO with $R^1C(S)NHCH(OR^3)R^2$ has been described.[198] The reaction of N-methyl-N-(trimethylsilyl)thiobenzamide with diphenylketen or tosyl isocyanate gives the heterocyclic compounds (106) and (107) respectively.[199] Thiobenzoyl isocyanates give 2:1 adducts with trimethylsilyl cyanide which have been formulated as 1,3,5-thiadiazepine derivatives (108).[200]

(100) 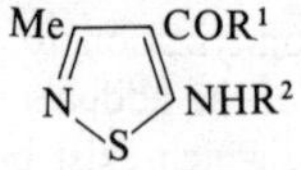   (101) 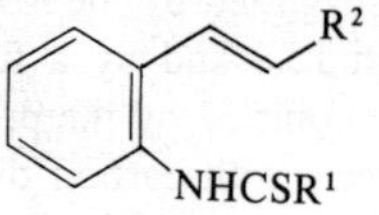   (102)

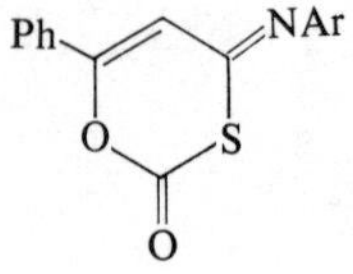

(103) 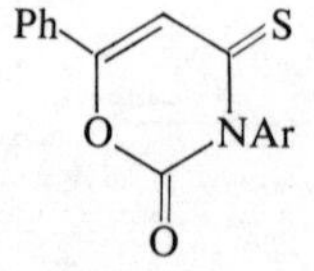   (104) 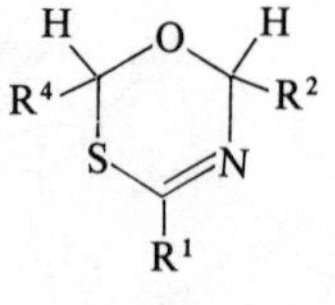   (105)

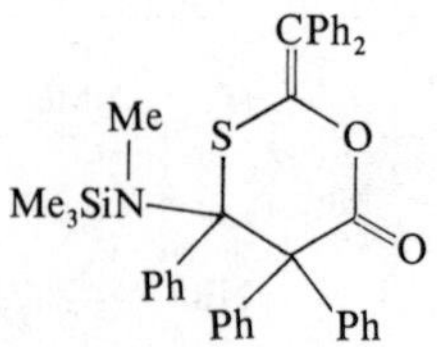

(106) 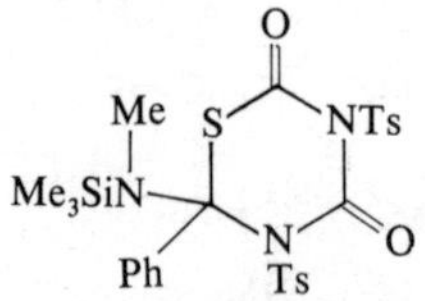   (107) 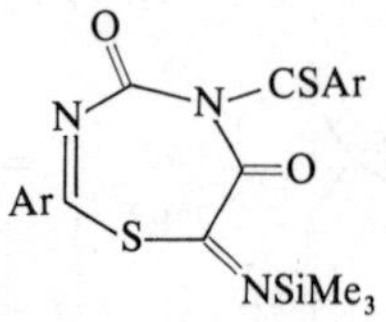   (108)

[194] F. E. Herkes, *J. Fluorine Chem.*, 1979, **13**, 1.

[195] S. Rajappa and R. Sreenivasan, *Indian J. Chem., Sect. B*, 1978, **16**, 752 (*Chem. Abstr.*, 1979, **90**, 151 899).

[196] P. de Mayo, L. K. Sydnes, and G. Wenska, *J. Chem. Soc., Chem. Commun.*, 1979, 499.

[197] W. Zankowska-Jasinska, Z. Kamela, and U. Zieba, *Pol. J. Chem.*, 1979, **53**, 1251 (*Chem. Abstr.*, 1980, **92**, 6480).

[198] C. Giordano, A. Belli, R. Erbea, and S. Panossian, *Synthesis*, 1979, 801.

[199] W. Walter, H. W. Lueke, and T. Proll, *J. Organomet. Chem.*, 1979, **173**, 33.

[200] O. Tsuge and S. Urano, *Heterocycles*, 1979, **12**, 1319.

N-Aryl-*N*-methyl-*N'*-thioacylhydrazines, $RCH_2C(S)NHNMePh$, cyclize to bis-(1-methyl-2-indolyl) disulphides (109) and salts of 2-aminoindoline (110) when $POCl_3$ or PCl_3 is present.[201] Some cycloadditions of thioacylhydrazines give 1,3,4-thiadiazole derivatives: addition of thioaroylhydrazines to methyl propiolate or dimethyl acetylenedicarboxylate gives (111; $R^1 = H$ or CO_2Me),[202] and *N*-alkyl-*N*-thioaroylhydrazines react with aryldibromodiazabutadienes $ArCH=N-N=CBr_2$ to give the 1,3,4-thiadiazolium bromides (112), which are converted into meso-ionic 1,3,4-thiadiazolidine-2-benzylidenehydrazinides (113) by treatment with anhydrous NH_3 or NEt_3.[203]

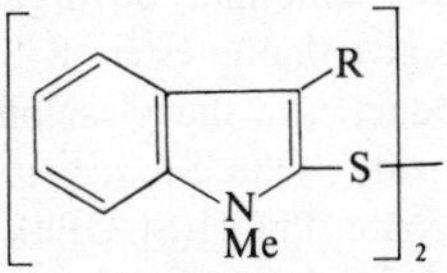

(109) 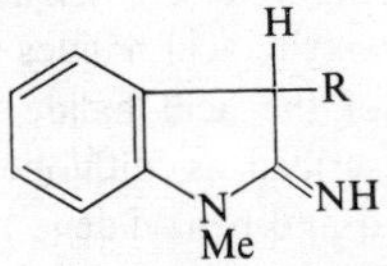   (110) 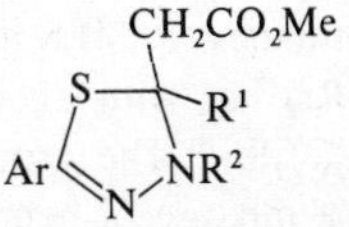   (111)

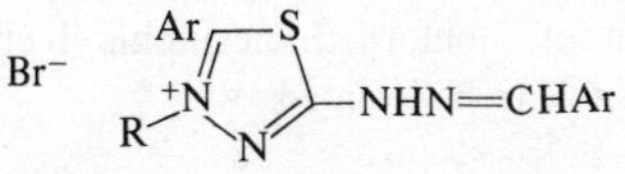

(112) 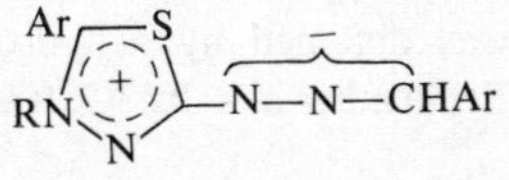   (113)

Miscellaneous. Unstable 2,4,6-triphenyl-1-thioacylaminopyridinium salts are obtained from thioacylhydrazines and 2,4,6-triphenylpyrylium tetrafluoroborate. They decompose on heating to give nitriles, triphenylpyridine, and sulphur.[204] Radical anions of *NN*-dialkyl-benzamides, -thiobenzamides, and -selenobenzamides, generated by electrochemical reduction in anhydrous DMF, were studied by e.p.r. spectroscopy. The $CSNR_2$ and $CSeNR_2$ groups withdraw much spin density from the aromatic ring, and are comparable with the nitro-group in this respect. The new thioamides and selenoamides have been made from the amides and P_4S_{10} or P_4Se_{10} in dry pyridine.[205]

4 Thiono- and Dithio-carboxylic Acids, their Derivatives, and their Selenium Analogues

Synthesis of Derivatives of Thionocarboxylic Acids.—Thiono-esters are obtained by the reaction of dialkoxycarbonium ions with Na_2S or anhydrous NaHS in acetonitrile (Scheme 6).[206] Cyclic carbonium salts from 1,2- or 1,3-diols give monothiono-esters of these diols. The tetrahedral intermediates in this process may be alkylated to give

[201] A. N. Kost, G. A. Golubeva, and A. G. Popova, *Khim. Geterotsikl. Soedin.*, 1979, 344.
[202] N. D. Heindel, G. Friedrich, and M. C. Tsai, *J. Heterocycl. Chem.*, 1980, **17**, 191.
[203] E. Cawkill, W. D. Ollis, C. A. Ramsden, and G. P. Rowson, *J. Chem. Soc., Perkin Trans. 1*, 1979, 724.
[204] A. R. Katritzky, J. Lewis, and P-L. Nie, *J. Chem. Soc., Perkin Trans. 1*, 1979, 446.
[205] J. Voss and F. R. Bruhn, *Liebigs Ann. Chem.*, 1979, 1931.
[206] F. Khouri and M. K. Kaloustian, *Tetrahedron Lett.*, 1978, 5067.

$$\overset{+}{PhC}(OR)_2 \ BF_4^- \xrightarrow{\ Na_2S\ } PhC(SNa)(OR)_2 \longrightarrow PhCSOR$$

Scheme 6

monothio-orthoesters.[207] $\alpha\beta$-Unsaturated thiono-esters have been made by thiolysis of imidates from cinnamic, crotonic, and fumaric acids at low temperatures (Scheme 7). At higher temperatures there is also addition of H_2S at the β-position, giving (114), which may be converted into the unsaturated thiono-ester by heating.[208] The reaction of alkyl halides with the potassium or mercury(II) salt of thiolacetic acid gives only thiolesters.[209] Equimolecular amounts of organo-silyl-thianes $(R_3^1Si)_2S$ and carboxylic acid halides react to give thiono-esters $R_3^1SiO-C(S)R^2$, but with an excess of the acid halide the product is a diacyl sulphide $(R^2CO)_2S$.[210] The product described as 'dithiobenzoyl oxide', $(PhCS)_2O$,[211] is, in fact, a mixture of benzoic acid and benzylidene thiobenzoate, $PhCH(SCOPh)_2$.[212] The kinetics of the thermal decomposition of 5-phenyl-1,2,3,4-thiatriazole in the temperature range 52.2—61.5 °C to give benzonitrile, sulphur, and nitrogen suggest that thiobenzoyl azide is an intermediate in the process. At ~100 °C the decomposition gives phenyl isothiocyanate and nitrogen.[213] Oxythioacetyl chloride (115) was obtained by dehydrochlorination of 1-chloroethanesulphinyl chloride with NEt_3 in ether at −30 °C. It was too unstable to be isolated.[214]

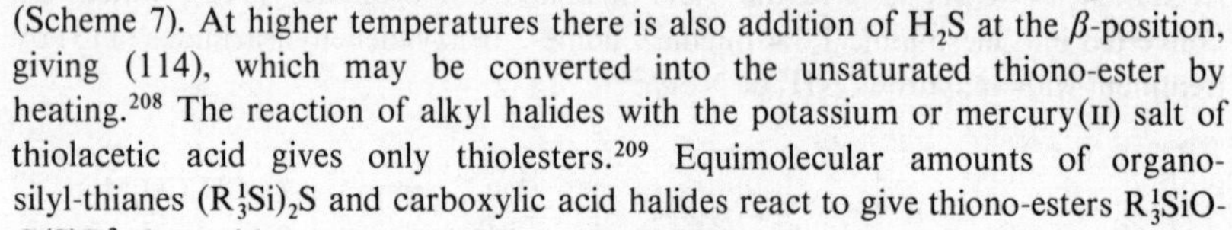

Scheme 7

Potassium 1,2-diseleno-oxalate has been prepared by saponification of diphenyl oxalate by alcoholic K_2Se. It is very sensitive to oxygen. The ion was shown by *X*-ray analysis to have the *trans*-planar conformation (116).[215]

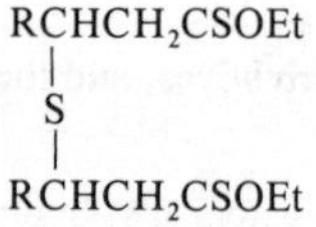

(114) 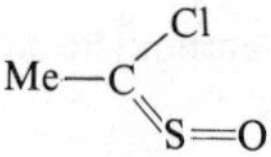   (115) 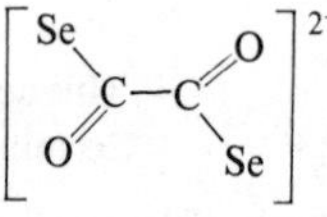   (116)

[207] F. Khouri and M. K. Kaloustian, *J. Am. Chem. Soc.*, 1979, **101**, 2249.

[208] R. Hoffmann and K. Hartke, *Chem. Ber.*, 1980, **113**, 919.

[209] S. Uemura, N. Watanabe, A. Toshimitsu, and M. Okano, *Bull. Chem. Soc. Jpn.*, 1978, **51**, 1818.

[210] E. P. Lebedev, V. A. Baburina, and M. D. Mizhiritskii, *Zh. Obshch. Khim.*, 1979, **49**, 1081 (*Chem. Abstr.*, 1979, **91**, 39 577).

[211] D. T. Lewis, *J. Chem. Soc.*, 1940, 831.

[212] A. Senning and K. Schank, *Acta Chem. Scand.*, *Ser. B*, 1978, **32**, 780.

[213] A. Holm, L. Carlsen, and E. Larsen, *J. Org. Chem.*, 1978, **43**, 4816.

[214] L. A. Carpino and J. R. Williams, *J. Org. Chem.*, 1979, **44**, 1177.

[215] C. Matz and R. Mattes, *Z. Zaturforsch.*, *Teil. B*, 1978, **33**, 461.

Synthesis of Dithiocarboxylic Acids and their Derivatives.—The preparation of thioterephthalic acid from 1,4-di(chloromethyl)benzene, sulphur, and sodium methoxide (see Vol. 5, p. 179) has been reported by another group.[216] In a similar reaction, the addition of chloroacetonitrile to sulphur and triethylamine in DMF, followed by alkylation, gave methyl and ethyl cyanodithioformates. These were not isolated, but they condensed when water was added, giving E/Z mixtures of 1,2-di(alkylmercapto)-1,2-dicyanoethylenes.[217] Ethylene dithioacetals from aromatic aldehydes are decomposed by sodium hydride in DMF containing HMPA, with elimination of ethylene and the liberation of an aryldithiocarboxylate anion; this may be alkylated to give the dithioester (Scheme 8).[218]

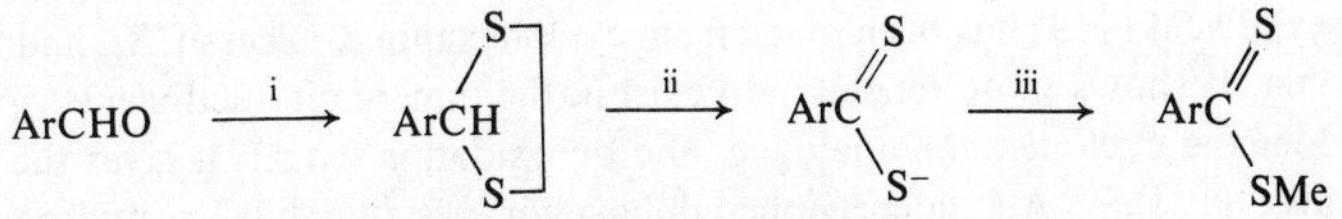

Reagents: i, $(CH_2SH)_2$, H^+; ii, NaH, DMF, HMPA; iii, MeI

Scheme 8

Methyl dithiocinnamate was obtained in low yield from styrylmagnesium bromide by reaction with CS_2 and methylation with MeI, and in better yield by low-temperature thiolysis of the thioimidate (117). At room temperature the thiolysis was complicated by addition of H_2S to give the saturated sulphide analogous to (114). The dithiocinnamate dimerizes reversibly, by [4 + 2]cyclo-addition, to form a product that was assigned the structure (118) on n.m.r. evidence.

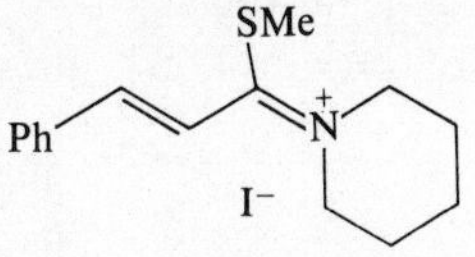

(117) 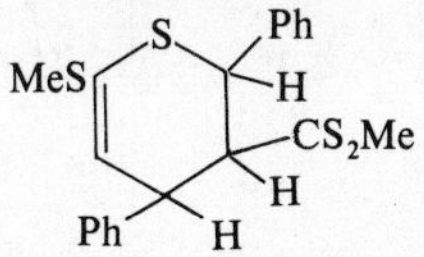   (118)

Another stereoisomer is obtained by heating the dimer with methyl dithio-cinnamate at 125 °C. Methyl dithiocrotonate could only be isolated as its dimer.[208] Another approach to unsaturated dithioesters employs the addition of an unsaturated Grignard reagent to phenyl isothiocyanate, followed by methylation and thiolysis of the resulting *S*-methyl-*N*-phenylimidate. With allylic Grignard reagents, inversion occurs through attack by the isothiocyanate at the allylic position. Double bonds in the $\beta\gamma$-position can be moved into conjugation with the thiocarbonyl group by NEt_3 catalysis.[219] The synthesis of dithioesters from Grignard reagents and CS_2 is improved by the addition of CuBr, *i.e.* by the use of an organocopper reagent.[220]

[216] J. M. Fabre, E. Torreilles, and L. Giral, *Tetrahedron Lett.*, 1978, 3703.
[217] R. Mayer, W. Thiel, and H. Viola, *Z. Chem.*, 1979, **19**, 56.
[218] N. C. Gonnella, M. V. Lakshmikantham, and M. P. Cava, *Synth. Commun.*, 1979, **9**, 17.
[219] P. Gosselin, S. Masson, and A. Thuillier, *Tetrahedron Lett.*, 1978, 2715.
[220] H. Westmijze, H. Kleijn, J. Meijer, and P. Vermeer, *Synthesis*, 1979, 432.

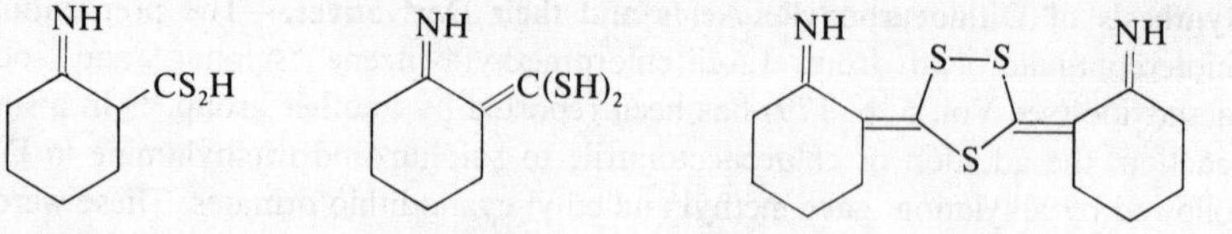

(119)　　　　　　(120)　　　　　　　　　　(121)

The synthesis of dithiocarboxylates by addition of CS_2 to enamines or activated methylene groups is further exemplified in the pyrimidine series[221-223] and in the preparation of pyridinium and quinolinium zwitterions from *N*-methyl anhydro-bases of 2-methyl-pyridines and -quinolines.[224,225] 2-Iminocyclohexanedithio-carboxylic acid (119) has been made from cyclohexanone, excess CS_2, and aqueous ammónia. It shows more tendency to exist in the dimercaptomethylene form (120) than does the cyclopentane analogue, and on oxidation with I_2 it gives the trithiole derivative (121).[226] Alkylidenetriphenylphosphoranes that have a hydrogen atom attached to the negative carbon atom add to CS_2 in a 2:1 ratio, giving phosphonium salts (122) in which the dithiocarboxylic acid component can be alkylated successively on both sulphur atoms to yield first 2-(triphenylphosphor-anylidene)alkanedithioates (123) and then phosphonium salts (124) that give keten dithioacetals (125) and triphenylphosphine oxide upon alkaline hydrolysis (Scheme 9). Compounds (123) do not undergo the Wittig reaction. The adducts from CS_2 and other alkylidenephosphoranes are betaines (126), from which dithiocarboxylic

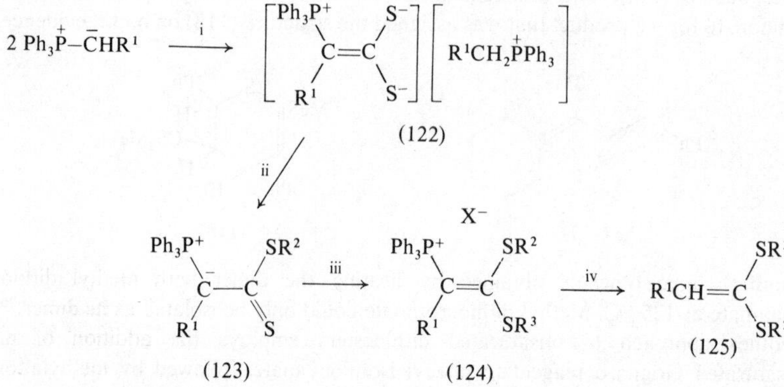

(122)

(123)　　　　　　(124)　　　　　　　　(125)

Reagents: i, CS_2; ii, alkylation, R^2X; iii, alkylation, R^3X; iv, H_2O, OH^-

Scheme 9

[221] H. Okuda, Y. Tominaga, Y. Matsuda, and G. Kobayashi, *Heterocycles*, 1979, **12**, 485.

[222] Y. Tominaga, H. Okuda, Y. Mitsuomi, Y. Matsuda, G. Kobayashi, and K. Sakemi, *Heterocycles*, 1979, **12**, 503.

[223] Y. Tominaga, T. Machida, H. Okuda, Y. Matsuda, and G. Kobayashi, *Yakugaku Zasshi*, 1979, **99**, 515.

[224] W. O. Foye, Y. J. Lee, K. A. Shah, and J. M. Kauffman, *J. Pharm. Sci.*, 1978, **67**, 962.

[225] W. O. Foye and J. M. Kauffman, *J. Pharm. Sci.*, 1979, **68**, 336.

[226] T. Takeshima, A. Yano, N. Fukada, Y. Hirose, and M. Muraoka, *J. Chem. Res. (S)*, 1979, 140.

esters are obtained by alkylation to the phosphonium salts (127) and removal of triphenylphosphine by electrolytic reduction.[227]

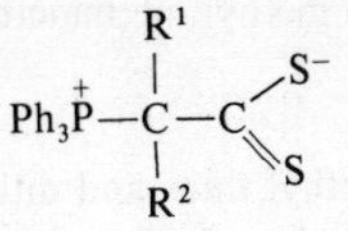

(126) 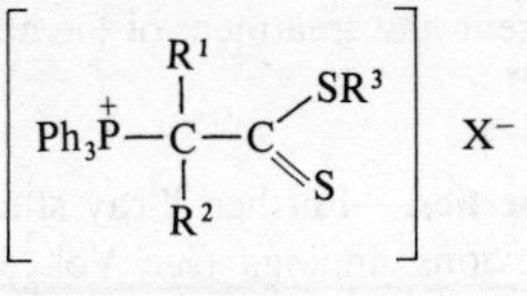   (127)

Methyl and ethyl dithioacetates, when heated with trimethyl phosphite, undergo a thio-Claisen condensation with loss of H_2S, giving *cis*- and *trans*-β-(alkylthio)dithiocrotonic esters.[228] Alkylation of potassium 1,1-dithio-oxalate in water gives exclusively the *S*-monoester. A second alkyl group may be introduced, located on oxygen, by alkylating the sodium salt of the monoester in HMPA.[229] Dimethyl trithio-oxalate has been made by alkylating potassium trithio-oxalate in MeOH.[230] Irradiation of xanthenethione with bis-(t-butylthio)ethyne gives both an $\alpha\beta$-unsaturated dithioester (128) and a thiet (129), which equilibrate in solution.[231] With bis(methylthio)ethyne only the ester was obtained[232] (see Vol. 4, p. 181).

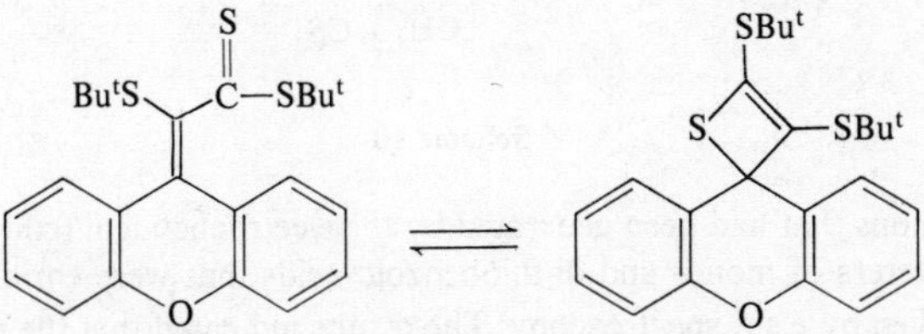

(128)   (129)

On standing in a refrigerator, dithioacetic acid forms crystals containing two trimers (1,3,5-trithian derivatives) as well as a trace of the previously reported tetramethylhexathia-adamantane. The isomers, which were not separated, regenerate monomer on distillation or chromatography.[233] Convenient methods for the preparation of anhydrous sodium dithiocarboxylates[234] and other alkali-metal salts[235] have been described. Exchange of sulphur between ^{35}S and methyl dithiobenzoate or methyl cyclohexanedithiocarboxylate requires a basic solvent, *e.g.* pyridine, and occurs at the $C=S$ and not the SMe position.[236]

[227] H. J. Bestmann, R. Engler, H. Hartung, and K. Roth, *Chem. Ber.*, 1979, **112**, 28.
[228] Z. Yoshida, S. Yoneda, T. Kawase, and M. Inaba, *Tetrahedron Lett.*, 1978, 1285.
[229] J. Voss and H. Gunther, *Synthesis*, 1978, 849.
[230] U. Niemer, K. Mennemann, and R. Mattes, *Chem. Ber.*, 1978, **111**, 2113.
[231] A. C. Brouwer, A. V. E. George, D. Seykens, and H. J. T. Bos, *Tetrahedron Lett.*, 1978, 4839.
[232] A. C. Brouwer and H. J. T. Bos, *Tetrahedron Lett.*, 1976, 209.
[233] G. Levesque, A. Mahjoub, and A. Thuillier, *Tetrahedron Lett.*, 1978, 3847.
[234] S. Kato, K. Ito, R. Hattori, M. Mizuta, and T. Katada, *Z. Naturforsch., Teil. B*, 1978, **33**, 976.
[235] S. Kato, S. Yamada, H. Goto, K. Terashime, M. Mizuta, and T. Katada, *Z. Naturforsch., Teil. B*, 1980, **35**, 458.
[236] L. Lešetický and F. Picha, *Collect. Czech. Chem. Commun.*, 1978, **43**, 2403.

Alkyl trithioperesters [alkyl thiocarbonyldisulphides, $R^1C(=S)SSR^2$] were obtained by the reaction of t-butylsulphenyl iodide with a metal (preferably Pb) salt of a dithio-acid,[237] but they are made more conveniently by the addition of CS_2 to a Grignard reagent and treatment of the adduct with methyl methanethiolsulphonate, $MeSO_2SMe$.[238]

Physical Properties.—Further X-ray studies of methyl thio- and dithio-esters have confirmed previous findings (see Vol. 5, p. 180) that the ester group is almost planar, with a Z conformation.[230,239] Kinetic studies of the gas-phase pyrolysis of dithioacetates have shown that the reaction is a concerted unimolecular process, consistent with the mechanism shown in Scheme 10.[240]

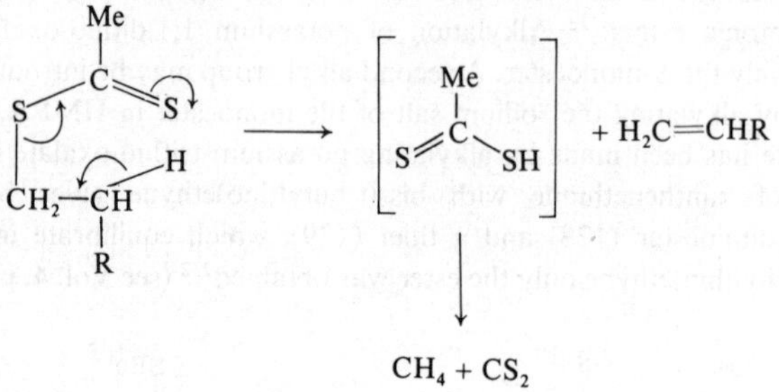

CH$_4$ + CS$_2$

Scheme 10

Radical anions that had been generated by the electrochemical reduction of some ($^2H_3$)methyl esters of mono- and di-thiobenzoic acids that were enriched in ^{13}C or ^{33}S were studied by e.s.r. spectroscopy. The results indicated that the radical anions are benzyl radicals, and not thiyl radicals.[241]

Reactions.—2-Iminocyclopentanedithiocarboxylic acid adds to Schiff bases at room temperature to form esters (130); these cyclize to 1,3-thiazines (131), with loss of arylamine, on warming.[242] Similar 1,3-thiazines are readily obtained from cyclohexanone, CS_2, and ammonia.[226] The addition of dithio-acids to dienes goes by a concerted reaction at $\sim 50\ ^\circ C$ and by a free-radical mechanism at higher temperatures, the products being allyl or vinyl dithioesters and/or dithiolans, e.g. (132).[243,244]

A dication (133), containing a tetrathian ring, has been made by electrochemical oxidation of sodium 4-diethylaminodithiobenzoate, and also by oxidation with

[237] D. F. Aycock and G. R. Jurch Jr., *J. Org. Chem.*, 1979, **44**, 569.
[238] R. S. Sukhai and L. Brandsma, *Synthesis*, 1979, 971.
[239] U. Niemer and R. Mattes, *Chem. Ber.*, 1978, **111**, 2118.
[240] N. Al-Awadi, D. B. Bigley, and R. E. Gabbott, *J. Chem. Soc., Perkin Trans. 2*, 1978, 1223.
[241] C-P. Klages, W-D. Malmberg, and J. Voss, *J. Chem. Res. (S)*, 1979, 160.
[242] N. Fukada, M. Kato, H. Seki, M. Kawana, and T. Takeshima, *J. Chem. Soc., Perkin Trans. 1*, 1978, 558.
[243] G. Levesque, A. Mahjoub, and A. Thuillier, *Colloq. Int. CNRS*, 1977, **278**, 365 (*Chem. Abstr.*, 1979, **91**, 174 492).
[244] J. C. Gressier, G. Levesque, A. Mahjoub, and A. Thuillier, *Bull. Soc. Chim. Fr., Part 2*, 1979, 355.

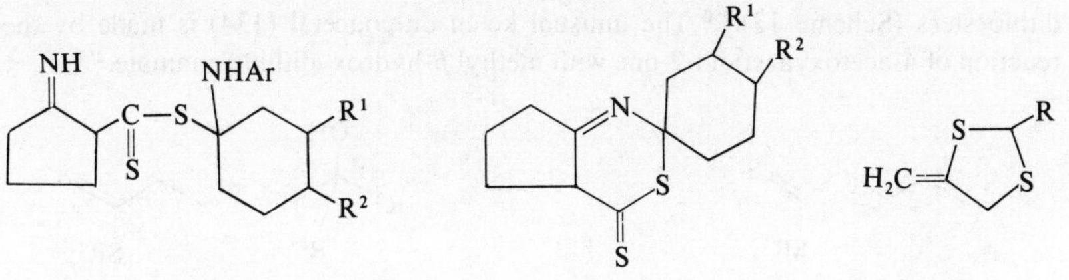

(130)   (131)   (132)

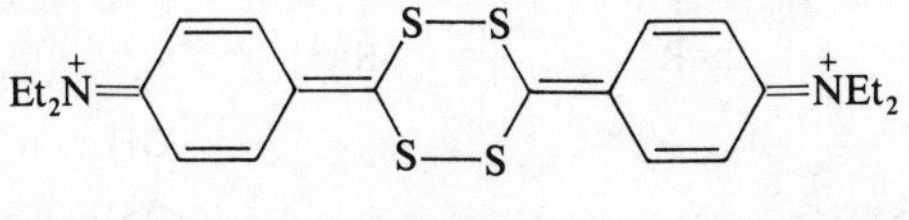

(133)

iodine.[245] Electroreductive alkylation of dithioesters in the presence of alkyl iodides gives dithioacetals; keten dithioacetals also are obtained from alkyl dithio-propionates.[246]

At −78 °C, *O*-ethyl thioesters undergo nucleophilic addition of alkyl-lithium at the C=S group. The products may be alkylated to give monothioacetals, or converted into thioenolates by warming to room temperature with LDA (Scheme 11).[247] Anions generated from dithioesters with LDA have been employed as

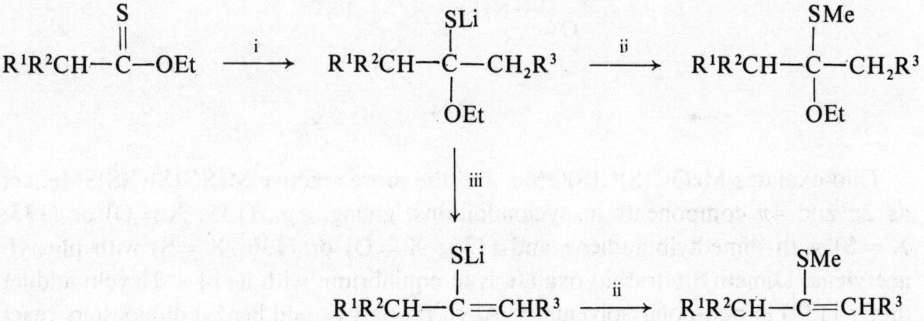

Reagents: i, R³CH₂Li; ii, MeI; iii, LDA

Scheme 11

equivalents of acyl anions in reaction with electrophiles[248] (see also Chap. 2, Pt I, p. 95) and as intermediates for preparing keten dithioacetals.[249] The methyl and isopropyl esters of pent-4-enedithioic acid, when treated successively with potassium hydride and s-butyl-lithium, give dianions of penta-2,4-dienedithioate esters. These react with carbonyl compounds at the 5-position, affording anions that may be alkylated to give keten dithioacetals or quenched with water to give substituted

[245] G. Cauquis and A. Deronzier, *J. Chem. Soc., Chem. Commun.*, 1978, 809.
[246] L. Kistenbrügger and J. Voss, *Liebigs Ann. Chem.*, 1980, 472.
[247] L. Narasimhan, R. Sanitra, and J. S. Swenton, *J. Chem. Soc., Chem. Commun.*, 1978, 719.
[248] A. I. Meyers, T. A. Tait, and D. L. Comins, *Tetrahedron Lett.*, 1978, 4657.
[249] B. Cazes and S. Julia, *Tetrahedron Lett.*, 1978, 4065.

dithioesters (Scheme 12).[250] The unusual keten dithioacetal (134) is made by the reaction of 4-acetoxyazetidin-2-one with methyl β-hydroxydithiocinnamate.[251]

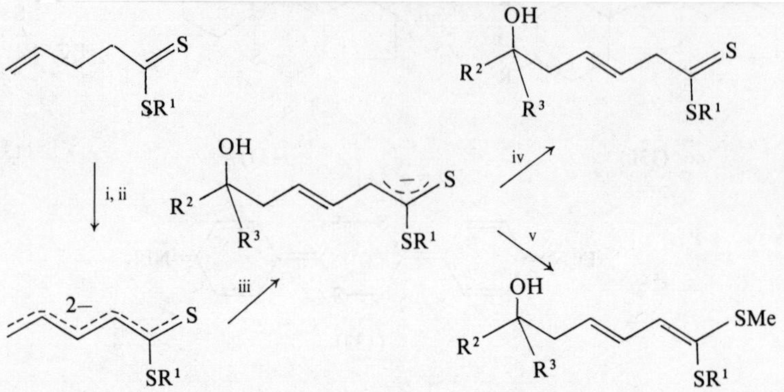

Reagents: i, KH, at 0 °C; ii, Bu^{s}Li, at −78 °C, TMEDA; iii, R^{2}COR^3; iv, H$_2$O; v, MeI

Scheme 12

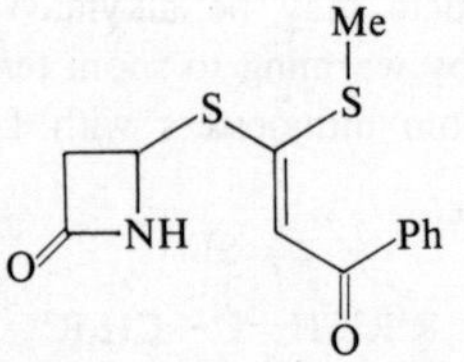

(134)

Thio-oxalates MeOC(S)C(S)OMe, and the more reactive MeSC(S)C(S)SMe, act as 2π and 4π components in cycloadditions, giving, *e.g.*, (135; X = O) or (135; X = S) with dimethylbutadiene and (136; X = O) or (136; X = S) with phenyl-acetylene. Dimethyl tetrathio-oxalate is in equilibrium with its [4 + 2]cyclo-adduct dimer (137) in non-polar solvents.[252] Allyl, propargyl, and benzyl dithioesters react with dimethyl acetylenedicarboxylate to form 1,3-dithioles, *e.g.* (138), with migration of the allyl, propargyl, or benzyl group.[253,254] Thietanones (139) are obtained by the addition of diphenylketen to dithiobenzoic esters.[255] Some further meso-ionic 1,3-dithiolones (140; R = Me or thioalkyl) (see Vol. 4, p. 185) have been made by dehydration of α-carboxybenzyl dithioesters RC(S)SCHPhCO$_2$H, and their cycloaddition reactions have been investigated.[256]

Acyl thioacyl sulphides, R^{1}C(S)SC(O)R^2, disproportionate in the presence of

[250] M. Pohmakotr and D. Seebach, *Tetrahedron Lett.*, 1979, 2271.
[251] M. Shibuya and S. Kubota, *Heterocycles*, 1979, **12**, 1315.
[252] K. Hartke, T. Kissel, J. Quante, and G. Henssen, *Angew. Chem., Int. Ed. Engl.*, 1978, **17**, 953.
[253] V. N. Drozd and O. A. Popova, *Tetrahedron Lett.*, 1979, 4491.
[254] V. N. Drozd, *Zh. Org. Khim.*, 1979, **15**, 1106 (*Chem. Abstr.*, 1979, **91**, 193 229).
[255] V. N. Drozd and O. A. Popova, *Zh. Org. Khim.*, 1979, **15**, 2602 (*Chem. Abstr.*, 1980, **92**, 163 777).
[256] H. Gotthardt, O. M. Huss, and C. M. Weisshuhn, *Chem. Ber.*, 1979, **112**, 1650.

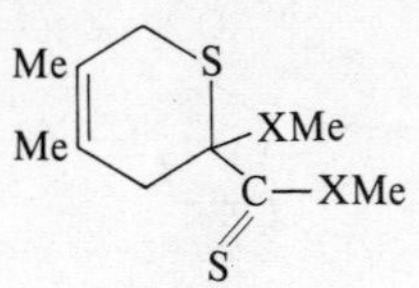

(135) 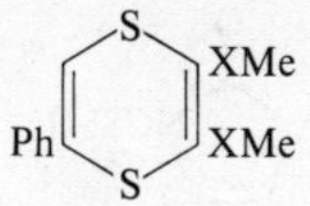   (136) 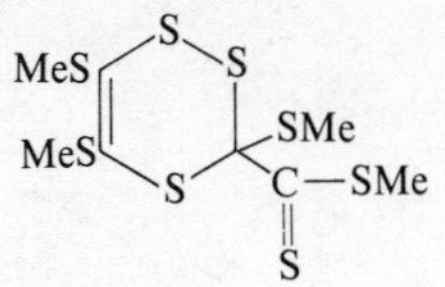   (137)

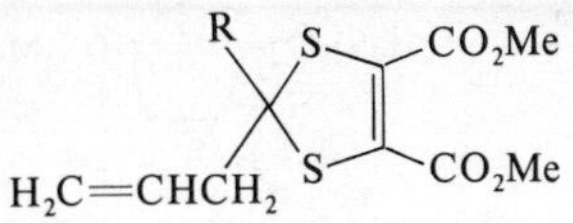

(138) 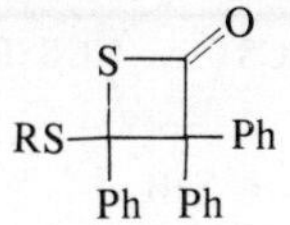   (139) 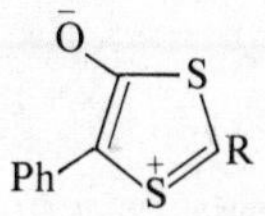   (140)

lithium alkoxides or lithium mercaptides to give bis(thioacyl) sulphides and bis(acyl) sulphides.[257]

Aliphatic selone esters $RC(Se)OEt$ react with a stoicheiometric amount (0.5 mol) of hydrazine to give hydrazones $RC(OEt)=NN=C(OEt)R$,[182] and with semi-carbazides and thiosemicarbazides they give 3-hydroxy- or 3-mercapto-1,2,4-triazoles.[258]

5 Thio- and Dithio-carbamates

Synthesis.—The chief method for the preparation of dithiocarbamates remains the reaction of CS_2 with amines, sometimes in the presence of an added base, and alkylation of the dithiocarbamate ion if an ester is required. The heat of formation of diethylammonium diethyldithiocarbamate, made by this method, has been determined.[259] Polymeric dithiocarbamates are made from CS_2 with amino-celluloses[260] or linear poly(iminoethylene).[261] Amino-acids, when treated with CS_2 and alkali, and then with ethyl bromide, give dithiocarbamic esters, $EtSCSNHCHRCO_2H$, which are sufficiently volatile for mass spectrometry by the direct-inlet probe, without the need to esterify the carboxyl groups.[262] Dithiocarbamates from amino-acids have been incorporated into macrocyclic compounds which are multidentate ligands.[263] Cyclic amides, including 2-pyrrolidone, 2-piperidone, imidazolidin-2-one, and 2,5-dioxopiperazine, give 1-dithiocarboxylic acids and/or 1,3-bis-dithiocarboxylic acids, depending on the reaction conditions (Scheme 13), but imido-groups that are between two carbonyl groups are not attacked by CS_2.[264] 2-Aminopyridine methiodide reacts with CS_2 and NaOH to

[257] S. Kato, K. Sugino, M. Mizuta, and T. Katada, *Angew. Chem.*, 1978, **90**, 718.

[258] V. I. Cohen, *J. Heterocycl. Chem.*, 1978, **15**, 237.

[259] K. J. Cavell, J. O. Hill, and R. J. Magee, *Thermochim. Acta*, 1979, **33**, 377.

[260] S. Imai, M. Muroi, A. Hamaguchi, R. Matsushita, and M. Koyama, *Anal. Chim. Acta*, 1980, **113**, 139.

[261] K. J. Ivin and E. D. Lillie, *Makromol. Chem.*, 1978, **179**, 591.

[262] J. Szafranek, G. Blotny, and P. Vouros, *Tetrahedron*, 1978, **34**, 2763.

[263] G. Blotny and J. F. Biernat, *Pol. J. Chem.*, 1979, **53**, 689 (*Chem. Abstr.*, 1979, **91**, 123 725).

[264] T. Takeshima, M. Ikeda, M. Yokoyama, N. Fukada, and M. Muraoka, *J. Chem. Soc., Perkin Trans. 1*, 1979, 692.

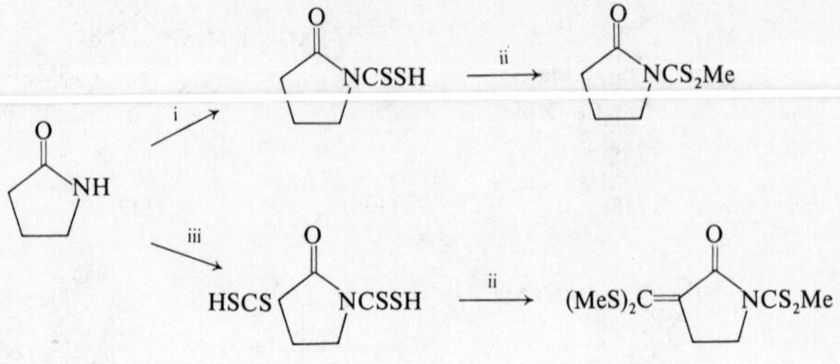

Reagents: i, CS$_2$, KOH; ii, MeI; iii, CS$_2$, NaH

Scheme 13

give a dithiocarbamate (141), in which methylation gives successively a methyl ester (142) and a quaternized SS-dimethyl N-(2-pyridyl)dithiocarbonimidate (143).

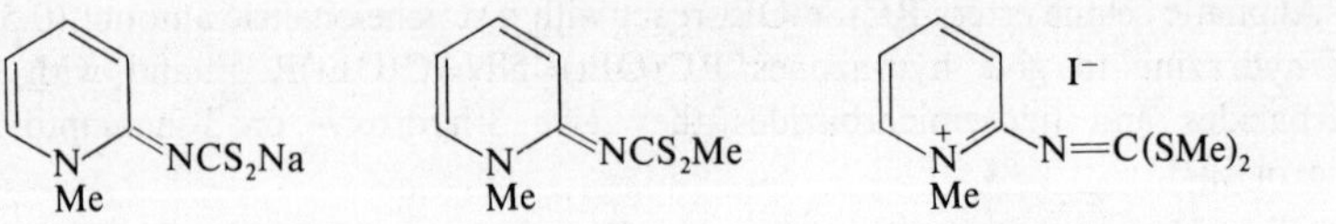

Similar products are obtained from the methiodides of 2-aminopyrimidine and 2-aminothiazole, but 4-aminopyridine methiodide gives 1-methyl-4-thiopyridone.[265] Some pyridyl-alkyl esters of dithiocarbamates have been made by treatment of the sodium salts with 2-(chloromethyl)pyridine or 2-vinylpyridine,[266] and 6-chloro-3-pyridyl NN-dimethyldithiocarbamate by reaction with diazotized 3-amino-6-chloropyridine.[267]

A second route to dithiocarbamic esters uses the reaction of isothiocyanates with thiols.[266,268] Cysteine and similar compounds containing mercapto- and amino-groups react with isothiocyanates through the mercapto-group at pH 4—6, giving dithiocarbamates, but, in alkaline solution (pH ~ 9), the reaction occurs at the amino-group.[269] The reaction of amines with aryloxythiocarbonyl chlorides, ArOCSCl, gave thiocarbamic esters of phenols.[270]

γ-Aryltrithioallophanic acids, ArNHC(S)NHCSSH, are prepared by thiolysis of their allylbenzyl esters, which are made from aryl isothiocyanates and PhCH$_2$SC-(=NH)SCH$_2$CH=CH$_2$.[271] Thiolysis of 2-alkoxycarbonyl-3-imino-5-methylthio-1,2,4-thiadiazolines gives N^1-alkoxycarbonyl-N^2-[(methylthio)thiocarbonyl]guanidines with loss of sulphur (Scheme 14).[272] Thiocarbamate esters that are unsubsti-

[265] K. Mizuyama, Y. Tominaga, Y. Matsuda, and G. Kobayashi, *Chem. Pharm. Bull.*, 1979, **27**, 2879.
[266] J. Berner and K. Kirmse, *Z. Chem.*, 1978, **18**, 381.
[267] K. Krowicki, *Pol. J. Chem.*, 1979, **53**, 889 (*Chem. Abstr.*, 1979, **91**, 140 689).
[268] F. Pavanetto, M. Mazza, L. Montanari, and T. Modena, *Farmaco, Ed. Sci.*, 1979, **34**, 808.
[269] D. Podhradský, Ľ. Drobnica, and P. Kristian, *Experientia*, 1979, **35**, 154.
[270] V. Konečný, S. Varkonda, and N. Kuličková, *Collect. Czech. Chem. Commun.*, 1979, **44**, 918.
[271] M. H. Damle, *Indian J. Chem., Sect. B*, 1979, **17**, 17 (*Chem. Abstr.*, 1980, **92**, 41 527).
[272] T. Fuchigami and T. Nonaka, *Chem. Lett.*, 1979, 829.

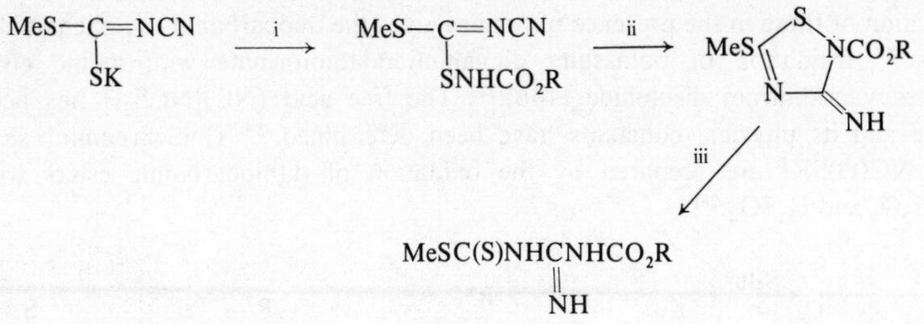

MeS—C=NCN $\xrightarrow{\text{i}}$ MeS—C=NCN $\xrightarrow{\text{ii}}$ (structure)
|
SK   SNHCO$_2$R

iii

MeSC(S)NHCNHCO$_2$R
‖
NH

Reagents: i, ClNHCO$_2$R; ii, H$^+$; iii, H$_2$S

Scheme 14

tuted on nitrogen are obtained, with other products, by prolonged treatment of alkyl or aryl thiocyanates with $(Pr^iO)_2P(S)SH$ at 20 °C.[273]

Sodium dialkyldithiocarbamates react with perfluoropropene both by substitution and by addition, to give $CF_3CF=CFSC(S)NR_2$ [(E) and (Z) forms] and $CF_3CHFCF_2SC(S)NR_2$, respectively. Pyrolysis of the latter compounds gives quantitative yields of dialkylthiocarbamoyl fluorides.[274] Treatment of N-trimethyl-silyl-amines with an excess of thiophosgene gives good yields of thiocarbamoyl chlorides of excellent purity. Bis(thiocarbamoyl) chlorides, $X(NRCSCl)_2$, are obtained from NN'-bis(trimethylsilyl)diamines.[275]

Physical Properties.—The i.r. spectra of dithiocarbamate and diselenocarbamate ions have been compared.[276] In continuation of earlier work,[277] a dynamic ^1H n.m.r. investigation of restricted rotation at the N—C—X bond in $H_2C=CMeCH_2$-N(R)—CXY (X = O, Y = SEt; or X = S, Y = OPh) has been made.[278] Numerous dithiocarbamates, mostly metal salts and complexes, have been studied by ^{13}C n.m.r. spectroscopy, and the chemical shifts of carbon in the NCS_2 fragment have been correlated with the degree of π-bonding.[279] An X-ray crystallographic study of the fungicide Denmert (144) showed the pyridyl and benzyl groups to be in a *syn* configuration, but n.m.r. evidence showed that the *syn*- and *anti*-forms are in equilibrium in solution.[280]

Reactions.—Electrolytic oxidation of dialkylamines and CS_2 in an undivided cell, with solvent extraction of the product, provides a simple preparative method for bis(dialkylthiocarbamoyl) disulphides (thiuram disulphides),[281] and electrolytic

[273] A. N. Pudovik, R. A. Cherkasov, M. G. Zimin, and R. M. Kamalov, *Zh. Obshch. Khim.*, 1978, **48**, 2645 (*Chem. Abstr.*, 1979, **90**, 137 443).
[274] T. Kitazume, S. Sasaki, and N. Ishikawa, *J. Fluorine Chem.*, 1978, **12**, 193.
[275] S. J. i Skorini and A. Senning, *Tetrahedron*, 1980, **36**, 539.
[276] K. N. Tantry and M. L. Shankaranarayana, *Indian J. Chem., Sect. A*, 1979, **18**, 347 (*Chem. Abstr.*, 1980, **92**, 31 345).
[277] E. Kleinpeter, R. Widera, and M. Mühlstädt, *J. Prakt. Chem.*, 1977, **319**, 133; *Z. Chem.*, 1977, **17**, 298.
[278] E. Kleinpeter, R. Widera, and M. Mühlstädt, *J. Prakt. Chem.*, 1978, **320**, 279.
[279] H. L. M. Van Gaal, J. W. Diesveld, F. W. Pijpers, and J. G. M. Van der Linden, *Inorg. Chem.*, 1979, **18**, 3251.
[280] S. Tanaka, T. Kato, S. Yamamoto, H. Yoshida, Z. Taira, and W. H. Watson, *Agric. Biol. Chem.*, 1978, **42**, 287.
[281] S. Torii, H. Tanaka, and K. Misima, *Bull. Chem. Soc. Jpn.*, 1978, **51**, 1575.

oxidation of these in the presence of amines gives the thiocarbamoylsulphenamides (145).[282] Oxidation of potassium dicyanimidodithioformate with iodine gives *N*-tetracyanothiuram disulphide (146).[283] The free acid $(NC)_2NCS_2H$ has been made and its physical constants have been determined.[284] Thiolcarbamic esters $R^1R^2NC(O)SR^3$ are prepared by the oxidation of dithiocarbamic esters with $K_2Cr_2O_7$ and H_2SO_4.[285]

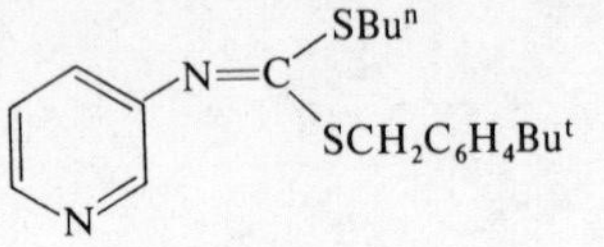

(144) 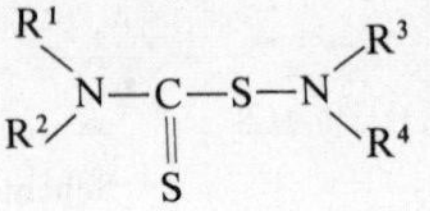   (145) 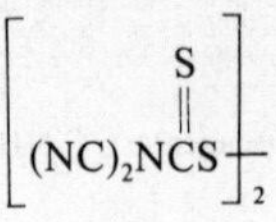   (146)

Pyrolysis of *S*-ethyl *N*-disubstituted dithiocarbamates gives ethylene, CS_2, and secondary amines and/or a trisubstituted thiourea and ethanethiol.[286] *NNSS'*-Tetrasubstituted dithiocarbamidium iodides, obtained by *S*-alkylating dithiocarbamates, are in equilibrium with their precursors, and this equilibrium has been used in the conversion of allylic dithiocarbamates, particularly allylic 1-pyrrolidine-carbodithioates (147), into allylic iodides by treatment with methyl iodide; bis(methylthio)methylenepyrrolidinium iodide (148) is insoluble, and it precipitates from the reaction mixture (Scheme 15).[287] *NNSS'*-Tetramethyldithiocarbamidium iodide (149) reacts with hydroxylamine or semicarbazide with displacement of the dimethylamino-group, but with other nucleophiles, *e.g.* thiosemicarbazide or

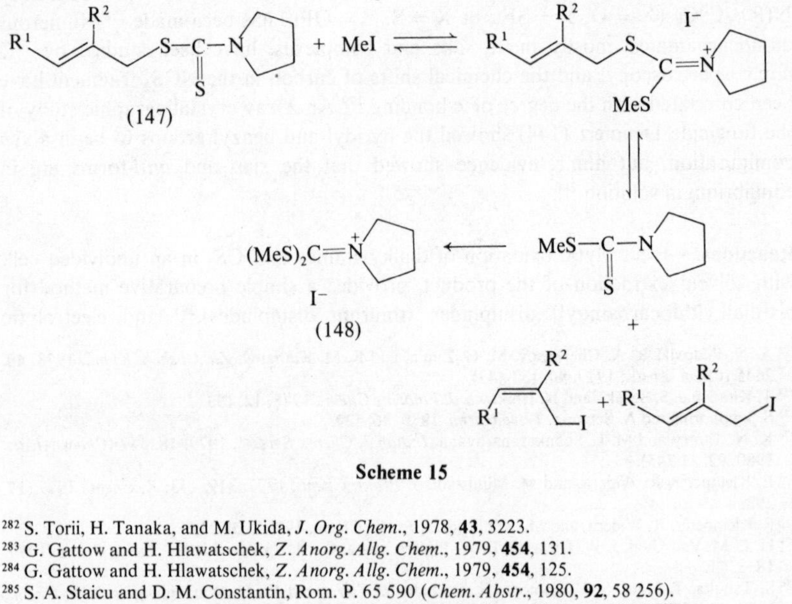

Scheme 15

[282] S. Torii, H. Tanaka, and M. Ukida, *J. Org. Chem.*, 1978, **43**, 3223.
[283] G. Gattow and H. Hlawatschek, *Z. Anorg. Allg. Chem.*, 1979, **454**, 131.
[284] G. Gattow and H. Hlawatschek, *Z. Anorg. Allg. Chem.*, 1979, **454**, 125.
[285] S. A. Staicu and D. M. Constantin, Rom. P. 65 590 (*Chem. Abstr.*, 1980, **92**, 58 256).
[286] M. S. Chande, *J. Indian Chem. Soc.*, 1979, **56**, 386 (*Chem. Abstr.*, 1979, **91**, 192 553).
[287] A. Sakurai, T. Hayashi, I. Hori, Y. Jindo, and T. Oishi, *Synthesis*, 1978, 370.

amines, one or both of the methylthio-groups are replaced.[288] The morpholine analogue of (149) loses morpholine on thiolysis, giving dimethyl trithiocarbonate, but its other reactions include addition of cyanide ion to give (150) and replacement of the methylthio-groups by morpholine, giving the guanidinium salt (151).[289]

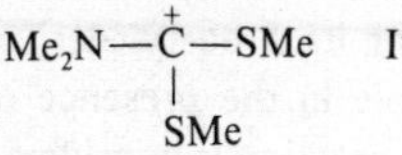

$Me_2N-\overset{+}{\underset{|}{C}}-SMe \quad I^-$
                    $|$
                  SMe

(149)

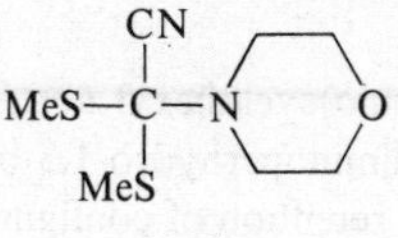

(150)

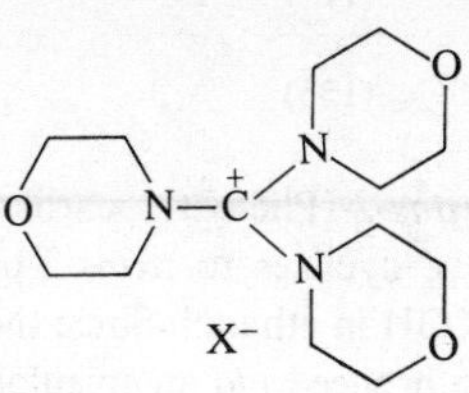

(151)

S-Methyl dithiocarbamates attack the reactive methylene groups of cyano-acetamide, cyanoacetic ester, or bis(phenylsulphonyl)methane with loss of methanethiol and formation of the thioamides $R^1R^2CHC(S)NR^3R^4$.[290, 291] Diazomethane and phenyldiazomethane, which is more reactive, insert into dialkyl-aminomethyl esters of dialkyldithiocarbamic acids (152; $X = NR^3R^4$) or ethyl xanthic acid (152; $X = OEt$), giving esters (153; $R^5 = H$ or Ph, $X = NR^3R^4$ or OEt) that have an extended alkylene chain.[292]

$R^1R^2NCH_2SC\overset{\displaystyle S}{\underset{\displaystyle X}{}}$

(152)

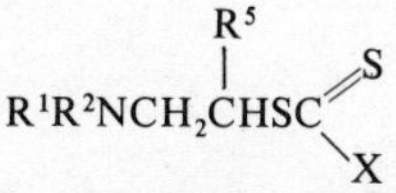

$R^1R^2NCH_2CHSC\overset{\displaystyle S}{\underset{\displaystyle X}{}}$ with R^5 above

(153)

$PhNC(S)OMe$
    $|$
  $COCl$

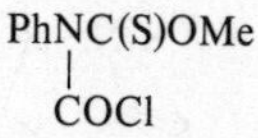

(154)

Alkali-metal salts of *O*-methyl *N*-phenylthiocarbamate are acylated on nitrogen by $COCl_2$, giving the carbamoyl chloride (154), but they react with benzene-sulphonyl chloride to give the disulphide $[PhN=C(OMe)S\text{+}]_2$, presumably through *S*-acylation of the ambident anion and decomposition of the acyl derivative.[293] Treatment of *O*-ethyl *N*-monosubstituted thiocarbamates with butyl-lithium gives lithiated derivatives, which decompose to give isothiocyanates, the reaction being accelerated by CS_2.[294] Acylation of methyl *N*-aryldithiocarbamates with propio-lactone in Ac_2O gives 4-oxo-2-thioxotetrahydro-1,3-thiazines (155) in low yield.[295] 2-Dialkylamino-1,3-oxathiol-2-ylium salts (156) are conveniently made by cyclizing 2-oxoalkyl dithiocarbamates (157) with dimethyl sulphate at 100—120 °C.[296]

[288] S. A. Okecha, *Chem. Ind.* (*London*), 1979, 526.

[289] F. Stansfield and M. D. Coomassie, *J. Chem. Soc., Perkin Trans. 1*, 1979, 2708.

[290] I. M. Bazavova, R. G. Dubenko, and P. S. Pel'kis, *Zh. Org. Khim.*, 1978, **14**, 195 (*Chem. Abstr.*, 1978, **88**, 152 374).

[291] I. M. Bazavova, E. F. Gorbenko, R. G. Dubenko, and P. S. Pel'kis, *Zh. Org. Khim.*, 1979, **15**, 1213 (*Chem. Abstr.*, 1979, **91**, 211 048).

[292] Y. Akasaka, K. Suzuki, T. Morimoto, and M. Sekiya, *Chem. Pharm. Bull.*, 1979, **27**, 2327.

[293] W. Mack, *Chem. Ber.*, 1980, **113**, 165.

[294] T. Fujinami, M. Ashida, and S. Sakai, *Nippon Kagaku Kaishi*, 1978, 773 (*Chem. Abstr.*, 1978, **89**, 108 570).

[295] W. Hanefeld, *Arch. Pharm.* (*Weinheim, Ger.*), 1979, **312**, 635.

[296] Y. Ueno and M. Okawara, *Synthesis*, 1979, 182.

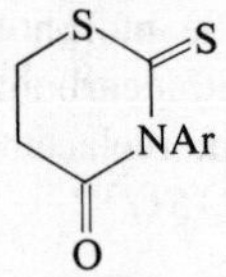

(155) 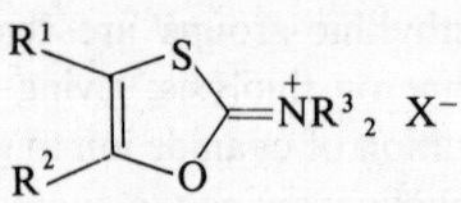   (156) 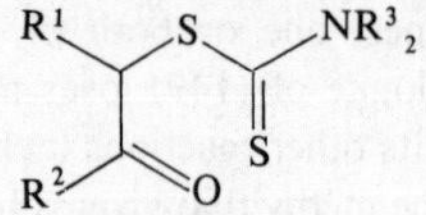   (157)

trans-2-(Phenylthiocarbamoylthio)cyclohexyl acetate (158) or its *S*-acetyl derivative cyclizes to *trans*-2-phenyliminoperhydro-1,3-benzodithiole in the presence of KOH in ethanol. Since there is retention of configuration, the reaction is considered to proceed *via* an episulphonium ion intermediate, as shown in Scheme 16. The *cis* analogues do not cyclize.[297]

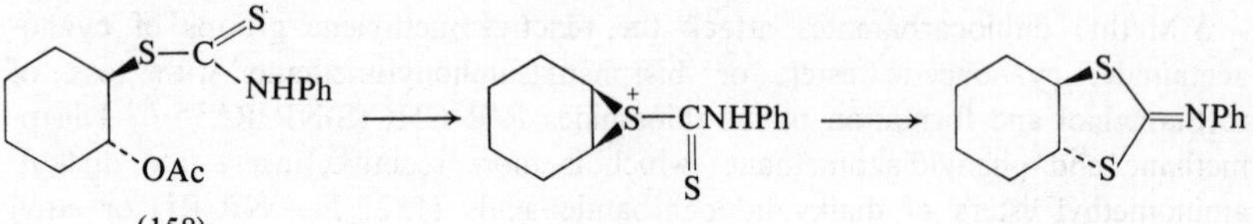

(158)

Scheme 16

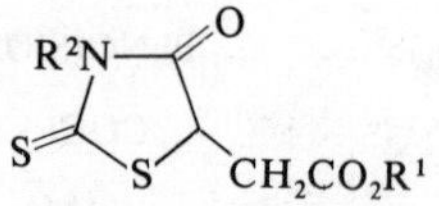

(159) R¹ = Me or Et 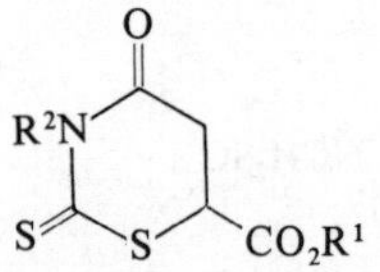   (160) R¹ = Me or Et

Several reactions leading to thiazole derivatives have been reported. Primary amines react with CS_2 and methyl or ethyl maleate to give rhodaninylacetates (159), and not the expected 1,3-thiazines (160).[298] The potassium salts of cyanimidodithiocarbonates, when alkylated by γ-bromocrotonic esters or γ-bromocrotononitrile, give 4-aminothiazoles through attack of the cyano-group on the activated methylene group [reaction (6)]. Salts of N^1N^1-disubstituted N^2-cyano-

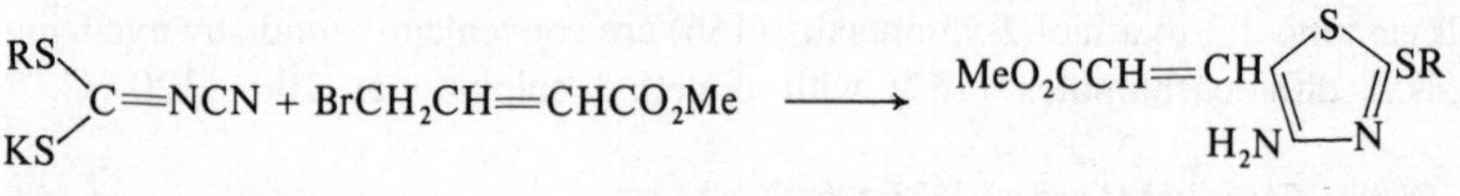

$$\begin{matrix} RS \\ \diagdown \\ KS \end{matrix} C{=}NCN + BrCH_2CH{=}CHCO_2Me \longrightarrow MeO_2CCH{=}CH \underset{H_2N}{\overset{S}{\diagup}} SR \quad (6)$$

thiourea undergo a similar ring-closure, but N^1-monosubstituted N^2-cyanothioureas cyclize to thiazolidines (161) by an internal Michael addition.[299] Some S^1-alkyl S^2-benzyl *N*-cyanoimidodithiocarbonates are cyclized by base to the analogous 4-aminothiazoles.[300] *S*-Methyl *N*-acyldithiocarbamates may be alkylated

[297] R. C. Foster and L. N. Owen, *J. Chem. Soc., Perkin Trans. 1*, 1978, 1208.
[298] W. Hanefeld and W. Hinz, *Arch. Pharm. (Weinheim, Ger.)*, 1980, **313**, 20.
[299] D. Wobig, *Liebigs Ann. Chem.*, 1978, 1118.
[300] P. Lipka and D. Wobig, *Liebigs Ann. Chem.*, 1979, 757.

to give *N*-acyliminodithiocarbonic acid diesters (162), and when these have a reactive methylene group, *e.g.* (162; $R^1 = Ph$, $R^2 = PhCH_2$ or $PhCOCH_2$), they cyclize to thiazoles (163; R = Ph or PhCO). Mono- and di-substitution of the diesters (162) with amines gives thioureas, guanidines, or heterocyclic compounds.[301]

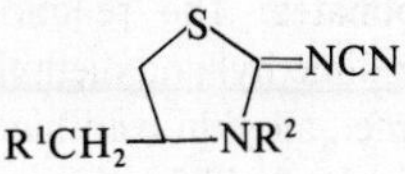

(161) 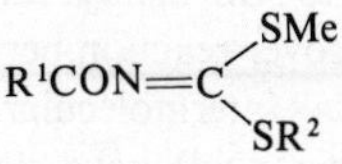   (162) 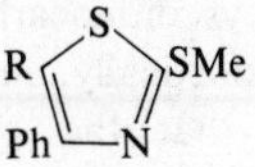   (163)

Sodium *NN*-diethyldithiocarbamate reacts with $TeCl_4$ to give $Te(Et_2NCS_2)_3Cl$, which has a pentagonal-bipyramidal structure.[302] Some other organotellurium dithiocarbamates $R^1_2Te(S_2CNR^2_2)_2$ have been reported.[303]

The hydroxylamine derivative (164) is of interest since it undergoes a hetero-Cope rearrangement through the intermediate (165), which has been isolated (Scheme 17).[304]

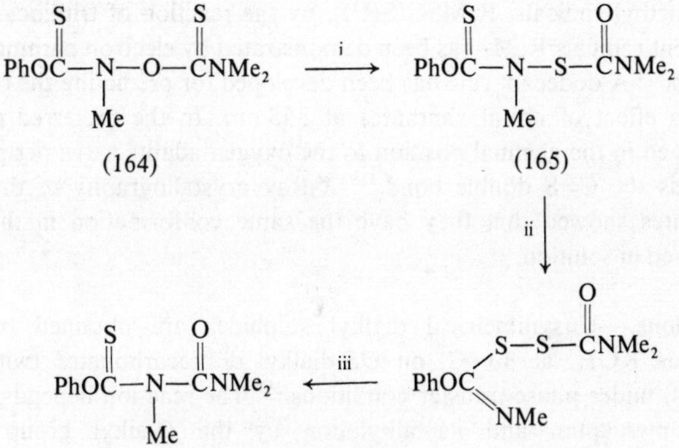

Reagents: i, MeOH, at 25 °C; ii, MeOH, at 55 °C; iii, desulphurize

Scheme 17

6 Xanthates and Trithiocarbonates

Preparation.—The reaction of alcohols with CS_2 under strongly alkaline conditions remains the standard method for preparing xanthates (*O*-alkyl dithiocarbonates). Alkylation of xanthates to *OS*-dialkyl dithiocarbonates is conveniently done by a phase-transfer process, using an alkyl halide or an alkyl methanesulphonate,[305] and

[301] M. Augustin, M. Richter, and S. Salas, *J. Prakt. Chem.*, 1980, **322**, 55.
[302] W. Schnabel, K. Von Deuten, and G. Klar, *Chem.-Ztg.*, 1979, **103**, 231.
[303] T. N. Srivastava, R. C. Srivastava, and A. Bhargava, *Indian J. Chem.*, Sect. A, 1979, **18**, 236 (*Chem. Abstr.*, 1980, **92**, 163 801).
[304] A. J. Lawson, *J. Chem. Soc., Chem. Commun.*, 1979, 456.
[305] I. Degani and R. Fochi, *Synthesis*, 1978, 365.

it may also be performed by 1-alkyl-2,4,6-triphenylpyridinium perchlorates.[306] S-Methyl dithiocarbonates of 6-O-tritylamylose and 6-O-tritylcellulose with a high degree of substitution were obtained by treating the O-trityl polysaccharide, dissolved in DMSO, successively with the methylsulphinyl anion, with CS_2, and with methyl iodide. Oxidation of the unmethylated xanthate with nitrous acid gave dithiobis(thioformates).[307] Boron trifluoride etherate catalyses the rearrangement of OS-dialkyl dithiocarbonates to SS-dialkyl dithiocarbonates. The reaction was studied kinetically, and a crossover reaction between the dimethyl and diethyl esters showed that the process is intermolecular.[308] Some alkylthioxanthic acids (mono-esters of trithiocarbonic acid) have been made by acidification of their potassium salts.[309] Aryl chlorodithiocarbonates, $ArSC(S)Cl$, are obtained in ~20% yield from arenediazonium chlorides and CS_2, with catalysis by copper or CuCl.[310]

Physical Properties.—The i.r. spectra and normal vibrations of O-methyl thiocarbonate, S-methyl dithiocarbonate, N-methylthiocarbamate, methyl trithio-carbonate, and similar ions have been related and compared.[311] Band assignments were made for the i.r. spectra of monoalkyl trithiocarbonates and for the Raman spectrum of methyl trithiocarbonate.[312] The mass-spectral fragmentation patterns of dialkyl trithiocarbonates have been studied.[313] The formation of tris(organo-thiyl)methyl radicals, $R_nM\dot{S}C(SR^1)_2$, by the reaction of trithiocarbonates with the transient radicals $R_nM\cdot$ has been demonstrated by electron paramagnetic resonance studies.[314] A dodecant rule has been developed for predicting the relative sign of the Cotton effect of chiral xanthates at 355 nm. In the preferred conformation the hydrogen in the geminal position to the oxygen adapts a *syn* periplanar orientation towards the C=S double bond.[315] X-Ray crystallography of three carbohydrate xanthates showed that they have the same conformation in the crystal as that preferred in solution.[316]

Reactions.—Unsymmetrical dialkyl sulphides are obtained by the action of aqueous KOH, at 80 °C, on OS-dialkyl dithiocarbonates (with different alkyl groups) under phase-transfer conditions.[317] The reaction depends on the liberation of a mercaptan and its alkylation by the O-alkyl group of the diester. Ethoxythiocarbonyl derivatives of amino-acids are made by acylating the amino-acids with the stable reagent bis(ethoxythiocarbonyl) sulphide, $(EtOCS)_2S$. This is prepared by an improved method from potassium ethylxanthate (2 mol) and ethyl chloroformate (1 mol). Esters of amino-acids are acylated by carboxymethyl

[306] A. R. Katritzky, U. Gruntz, N. Mongelli, and M. C. Rezende, *J. Chem. Soc., Chem. Commun.*, 1978, 133; *J. Chem. Soc., Perkin Trans. 1*, 1979, 1953.
[307] N. W. H. Cheetham, *Carbohydr. Res.*, 1978, **65**, 144.
[308] K. Komaki, T. Kawata, K. Harano, and T. Taguchi, *Chem. Pharm. Bull.*, 1978, **26**, 3807.
[309] D. Hoerner and G. Gattow, *Z. Anorg. Allg. Chem.*, 1978, **442**, 195.
[310] H. Viola and R. Mayer, *Z. Chem.*, 1979, **19**, 289.
[311] K. N. Tantry and M. L. Shankaranarayana, *Proc. Indian Acad. Sci., Sect. A*, 1979, **88**, 457 (*Chem. Abstr.*, 1980, **92**, 118 832).
[312] D. Hoerner and G. Gattow, *Z. Anorg. Allg. Chem.*, 1978, **442**, 204.
[313] D. Hoerner and G. Gattow, *Z. Anorg. Allg. Chem.*, 1978, **444**, 117.
[314] D. Forrest and K. U. Ingold, *J. Am. Chem. Soc.*, 1978, **100**, 3868.
[315] H. Paulsen, B. Elvers, H. Redlich, E. Schüttpelz, and G. Snatzke, *Chem. Ber.*, 1979, **112**, 3842.
[316] P. Luger, B. Elvers, and H. Paulsen, *Chem. Ber.*, 1979, **112**, 3855.
[317] I. Degani, R. Fochi, and V. Regondi, *Synthesis*, 1979, 178.

O-ethylxanthate, $EtOC(=S)SCH_2CO_2H$.[318]   *O*-Ethyl S-cyanomethyl dithio-carbonate condenses with ketones, under basic conditions, to give alk-2-ene-nitriles,[319] as shown in Scheme 18.

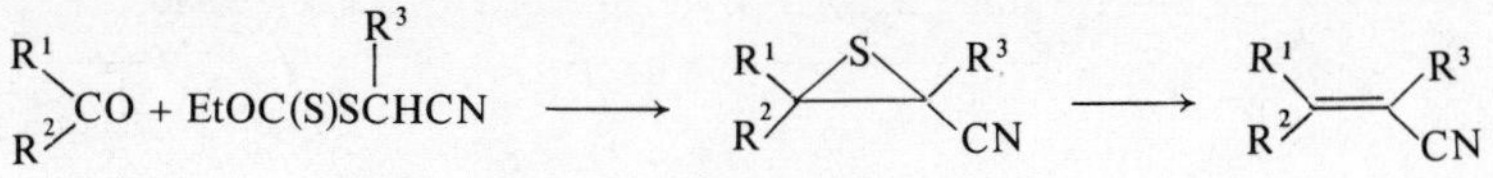

Scheme 18

Thermolysis of bis-(*S*-methyl dithiocarbonates) of 1,2- and 1,3-diols (166; $n = 0$ or 1) gives cyclic *OS*-di- or tri-methylene *SS*-dimethyl trithio-orthocarbonates (167; $n = 0$ or 1), the C–S bond being formed preferentially at a primary carbon atom.[320] This reaction has been applied in the carbohydrate field.[321,322] *S*-Methyl dithiocarbonates derived from suitably protected monosaccharides or amino-glycosides are reduced to deoxy-sugars by tributyltin hydride or tributyltin deuteride, the reaction being highly stereoselective.[323–327] Vicinal diols are converted into olefins in high yield by the reaction of their bis-(*S*-methyl dithiocarbonate) derivatives with tri-n-butylstannane.[328] Rearranged olefins are obtained from cyclohexyl xanthates, *e.g.* (168), with elimination of the xanthate residue, by heating with $[RhCl(PPh_3)_3]$ or its Ir, Pd, or Pt analogue.[329]

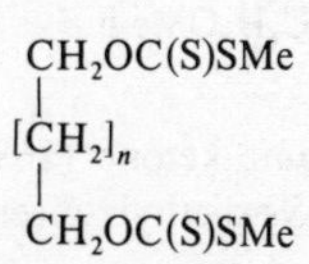

(166)

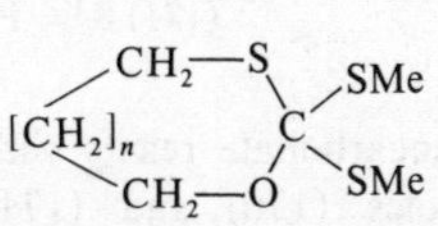

(167)

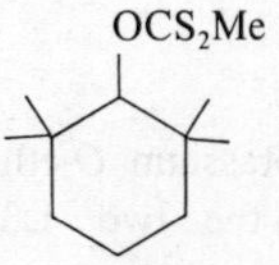

(168)

1,3-Dithiole-2-thiones are obtained in good yield by alkylating sodium t-butyl trithiocarbonate with propargyl halides and heating the unsymmetrical ester with TFA in acetic acid (Scheme 19).[330]

Alkyl (and aryl) chlorodithioformates, RSC(S)Cl, react with potassium toluene-*p*-thiosulphonate to give RSC(S)SSTs, RSC(S)Ts, and [RSC(S)S]$_2$, none of the expected primary product, *i.e.* RSC(S)STs, being isolated.[331]

[318] G. Barany, B. W. Fulpius, and T. P. King, *J. Org. Chem.*, 1978, **43**, 2930.

[319] K. Tanaka, N. Ono, Y. Kubo, and A. Kaji, *Synthesis*, 1979, 890.

[320] A. Faure and G. Descotes, *Synthesis*, 1978, 286.

[321] G. Descotes and A. Faure, *Synthesis*, 1976, 649.

[322] A. Faure, B. Kryczka, and G. Descotes, *Carbohydr. Res.*, 1979, **74**,127.

[323] D. H. R. Barton and S. W. McCombie, *J. Chem. Soc., Perkin Trans. 1*, 1975, 1574.

[324] T. Hayashi, T. Iwaoka, N. Takeda, and E. Ohki, *Chem. Pharm. Bull.*, 1978, **26**, 1786.

[325] J. J. Patroni and R. V. Stick, *J. Chem. Soc., Chem. Commun.*, 1978, 449.

[326] J. J. Patroni and R. V. Stick, *Aust. J. Chem.*, 1979, **32**, 411.

[327] E. M. Acton, R. N. Goerner, H. S. Uh, K. J. Ryan, D. W. Henry, C. E. Cass, and G. A. Lepage, *J. Med. Chem.*, 1979, **22**, 518.

[328] A. G. M. Barrett, D. H. R. Barton, and R. Bielski, *J. Chem. Soc., Perkin Trans. 1*, 1979, 2378.

[329] R. Hamilton, T. R. B. Mitchell, J. J. Rooney, and M. A. McKervey, *J. Chem. Soc., Chem. Commun.*, 1979, 731.

[330] N. F. Haley, *Tetrahedron Lett.*, 1978, 5161.

[331] H. C. Hansen and A. Senning, *J. Chem. Soc., Chem. Commun.*, 1979, 1135.

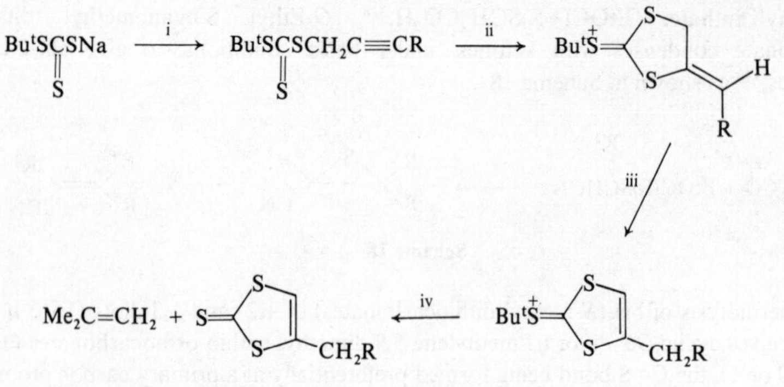

Reagents: i, BrCH$_2$C≡CR; ii, TFA, at 25 °C; iii, 100 °C; iv, HOAc, at 100 °C

Scheme 19

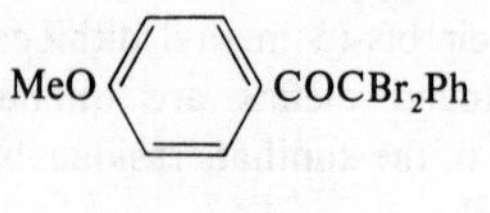

(169)

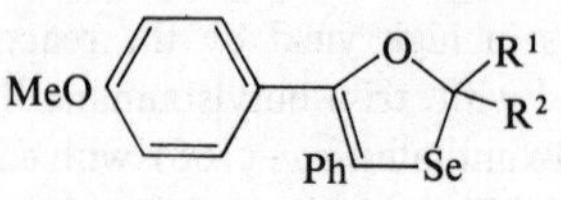

(170) R^{1}R^2 = O
(171) R^1 = Ph, R^2 = COC$_6$H$_4$OMe-*p*

Potassium *O*-ethyl diselenocarbonate reacts with the dibromo-ketone (169) to give the two 1,3-oxaselenoles (170) and (171) and *O,Se*-diethyl diseleno-carbonate.[332]

[332] C. Bak, K. Praefcke, and L. Henriksen, *Chem.-Ztg.*, 1978, **102**, 361.

4

Small-ring Compounds of Sulphur and Selenium

BY C. G. VENIER

1 Thiirans (Episulphides)

Reviews—A book on thiirans[1] and a review covering the reactivity of thiirans[2] have been published by the same authors. Thiirens,[3] thiiranium ions,[4] and thiirenium ions[5] have been reviewed. Small rings are among the compounds covered in a book on reactive species[6] and a review on the pyrolysis of sulphones.[7] Two reviews on heterocyclic analogues of methylenecyclopropanes have appeared,[8,9] and a survey of the recent literature in the area of heterocyclic compounds includes three- and four-membered rings that contain chalcogens.[10] For coverage of biologically active steroidal episulphides and naturally occurring episulphides, see the bibliography on page 232.

Formation.—The full paper showing that the compound previously reported to be 2,3-dibenzoyl-2,3-diphenylthiiran is actually 2,4,5-triphenyl-1,3-oxathiole has appeared[11] (see Vol. 1, p. 115 and Vol. 2, p. 102).

The preparation of thiirans by the reaction of diazo-compounds with thiocarbonyls has received widespread use.[12–22] In particular, a number of preparations of

[1] A. V. Fokin and A. F. Kolomiets, 'Khimiya Tiiranov' ('Chemistry of Thiiranes'), Nauka, Moscow, 1978.
[2] A. V. Fokin and A. F. Kolomiets, *Russ. Chem. Rev. (Engl. Transl.)*, 1976, **45**, 25.
[3] M. Torres, E. M. Lown, and O. P. Strausz, *Heterocycles*, 1978, **11**, 697.
[4] (a) V. N. Gogte and H. M. Modak, in 'New Trends in Heterocyclic Chemistry,' ed. R. B. Mitra, Elsevier, Amsterdam, 1979, p. 142; (b) W. A. Smit, N. S. Zefirov, I. V. Bodrikov, and M. Z. Krimer, *Acc. Chem. Res.*, 1979, **12**, 282.
[5] G. Capozzi, V. Lucchini, and G. Modena, *Rev. Chem. Intermed.*, 1979, **2**, 347.
[6] E. Block, 'Reactions of Organosulfur Compounds,' Academic Press, New York, 1978.
[7] F. Vogtle and L. Rossa, *Angew. Chem., Int. Ed. Engl.*, 1979, **18**, 515.
[8] G. L'abbé, in 'Topics in Organic Sulphur Chemistry,' ed. M. Tisler, University Press, Ljubljana, Yugoslavia, 1978, Ch. 6.
[9] G. L'abbé, *Angew. Chem., Int. Ed. Engl.*, 1980, **19**, 276.
[10] A. R. Katritzky and P. M. Jones, *Adv. Heterocycl. Chem.*, 1979, **25**, 303.
[11] U. J. Kempe, T. Kempe, and T. Norin, *J. Chem. Soc., Perkin Trans. 1*, 1978, 1547.
[12] F. M. E. Abdel-Megeid, A. A. Elbarbary, and F. A. Gad, *Pol. J. Chem.*, 1979, **53**, 1877.
[13] M. O-oka, A. Kitamura, R. Okazaki, and N. Inamoto, *Bull. Chem. Soc. Jpn.*, 1978, **51**, 301.
[14] K. Friedrich and M. Zamkanei, *Chem. Ber.*, 1979, **112**, 1873.
[15] M. Tashiro, S. Mataka, and S. Ishi-i, *Heterocycles*, 1979, **12**, 184.
[16] S. Mataka, S. Ishi-i, and M. Tashiro, *J. Org. Chem.*, 1978, **43**, 3730.
[17] M. S. Raasch, *J. Org. Chem.*, 1979, **44**, 632.
[18] M. Atzmueller and F. Vogtle, *Chem. Ber.*, 1979, **112**, 138.
[19] D. H. R. Barton, L. S. L. Choi, R. H. Hesse, and M. M. Pechet, *J. Chem. Soc., Perkin Trans. 1*, 1979, 1166.

highly hindered thiirans and olefins have used this procedure; *e.g.*, of (1).[12] Tetrasubstituted alkenes are also prepared *via* 1,3,4-thiadiazolines and hence episulphides[23,24] (see Vol. 5, p. 188). The flash pyrolysis of 1,3-oxathiolan-5-ones (2), related to the pyrolysis of thiadiazoline by loss of carbon dioxide rather than nitrogen, also gives thiirans stereospecifically.[25,26]

The conversion of oxirans into thiirans by the action of sulphur nucleophiles has been further exploited, using thiourea[27-32] or a thiocyanate.[33-35] Treatment of alkenes with (3) or (4), followed by reduction with LiAlH$_4$, gives thiirans in good yields in a process that provides an alternative method for the conversion of alkenes into episulphides.[36] Vicinal iodo-thiocyanates are converted into thiirans when treated with base.[37]

Conversions of aldehydes and ketones into thiirans have been reported, using the α-anions of alkyl thiolbenzoates (5),[38] the dianions of α-mercapto-esters,[39] and, in

(1) $n = 1$ or 0 (*i.e.* a π-bond)

(2)   (3)   (4)

(5)   (6)

[20] W. Friedrichsen, M. Betz, E. Bueldt, H. J. Jurgens, R. Schmidt, I. Schwarz, and K. Visser, *Liebigs Ann. Chem.*, 1978, 440.

[21] M. D. Bachi, O. Goldberg, and A. Gross, *Tetrahedron Lett.*, 1978, 4167.

[22] M. A. F. Elkaschef, F. M. E. Abdel-Megeid, and A. A. Elbarbary, *Acta Chim. Acad. Sci. Hung.*, 1977, 157.

[23] D. J. Humphreys, C. E. Newall, G. H. Phillipps, and G. A. Smith, *J. Chem. Soc., Perkin Trans. 1*, 1978, 45.

[24] A. Krebs and W. Rueger, *Tetrahedron Lett.*, 1979, 1305.

[25] T. B. Cameron and H. W. Pinnick, *J. Am. Chem. Soc.*, 1979, **101**, 4755.

[26] T. B. Cameron and H. W. Pinnick, *J. Am. Chem. Soc.*, 1980, **102**, 744.

[27] H. Prinzbach, H. P. Boehm, S. Kagabu, V. Wessely, and H. V. Rivera, *Tetrahedron Lett.*, 1978, 1243.

[28] F. Haviv and B. Belleau, *Can. J. Chem.*, 1978, **56**, 2677.

[29] B. Kraska and L. Mester, *Tetrahedron Lett.*, 1978, 4583.

[30] I. L. Kuranova, E. V. Snetkova, and V. A. Gindin, *Vestn. Leningr. Univ., Fiz. Khim.*, 1978, 140 (*Chem. Abstr.*, 1978, **89**, 107 190).

[31] A. Hasegawa, Y. Kawai, H. Kasugai, and M. Kiso, *Carbohydr. Res.*, 1978, **63**, 131.

[32] T. Takigawa, M. Tominaga, K. Mori, and M. Matsui, *Agric. Biol. Chem.*, 1979, **43**, 1591.

[33] M. Goguelin and M. Sepulchre, *Makromol. Chem.*, 1979, **180**, 1215.

[34] I. M. Abdullabekov, F. Kh. Agaev, A. L. Shabanov, and M. M. Movsumzade, *Dokl. Akad. Nauk Az. SSR*, 1978, 34 (*Chem. Abstr.*, 1979, **90**, 6157).

[35] C. A. Kingsbury, D. L. Durham, and R. Hutton, *J. Org. Chem.*, 1978, **43**, 4696.

[36] M. U. Bombala and S. V. Ley, *J. Chem. Soc., Perkin Trans. 1*, 1979, 3013.

[37] R. C. Cambie, H. H. Lee, P. S. Rutledge, and P. D. Woodgate, *J. Chem. Soc., Perkin Trans. 1*, 1979, 765.

[38] D. B. Reitz, P. Beak, R. F. Farney, and L. S. Helmick, *J. Am. Chem. Soc.*, 1978, **100**, 5428.

[39] K. Tanaka, N. Yamagishi, R. Tanikaga, and A. Kaji, *Bull. Chem. Soc. Jpn.*, 1979, **52**, 3619.

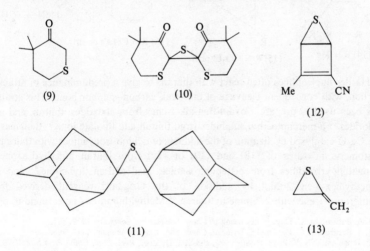

(7)   (8)

an optically active form, using (6) in the presence of (7)[40] (see Vol. 4, p. 188). Also, derivatives (8) are converted into thiirans by sequential treatment with NaBH$_4$ and NaH.[41]

Diazotization of cysteine and its derivatives and of penicillamine and its derivatives yields thiirancarboxylic acids and esters.[42]

Direct addition of various forms of elemental sulphur to alkenes has been reported: of sulphur atoms (produced by photolysis of COS[43] or by laser irradiation of S$_8$[44]) and of cyclo-octasulphur (involving irradiation[45] or activation by amines[46,47]).

Treatment of (9) with thionyl chloride gave a 6% yield of (10), which could be desulphurized to the corresponding olefin with triphenylphosphine.[48]

Adamantylideneadamantene episulphide (11) was prepared by demethylation of the corresponding methyl episulphonium salt with chloride or bromide ion.[49]

Di-t-butylsulphine reacts with primary Grignard reagents to give thiirans.[50] The

(9)   (10)   (12)

(11)   (13)

[40] K. Soai and T. Mukaiyama, *Bull. Chem. Soc. Jpn.*, 1979, **52**, 3371.
[41] V. Calo, L. Lopez, and G. Pesce, *Gazz. Chim. Ital.*, 1979, **109**, 703.
[42] C. D. Maycock and R. J. Stoodley, *J. Chem. Soc., Perkin Trans. 1*, 1979, 1852.
[43] A. G. Sherwood, I. Safarik, B. Verkoczy, G. Almadi, H. A. Wiebe, and O. P. Strausz, *J. Am. Chem. Soc.*, 1979, **101**, 3000.
[44] D. R. Betteridge and J. T. Yardley, *Chem. Phys. Lett.*, 1979, **62**, 570.
[45] S. Inoue, T. Tezuka, and S. Oae, *Phosphorus Sulfur*, 1978, **4**, 219.
[46] J. Emsley, D. W. Griffiths, and G. J. J. Jayne, *J. Chem. Soc., Perkin Trans. 1*, 1979, 228.
[47] H. Kutlu, *Doga*, 1979, **3**, 71 (*Chem. Abstr.*, 1979, **91**, 211 321).
[48] U. Luhmann and W. Luettke, *Chem. Ber.*, 1978, **111**, 3246.
[49] J. Bolster and R. M. Kellogg, *J. Chem. Soc., Chem. Commun.*, 1978, 630.
[50] A. Ohno, M. Uohama, K. Nakamura, and S. Oka, *J. Org. Chem.*, 1979, **44**, 2244.

'Dewar' thiophen (12) has been isolated, following irradiation of the parent thiophen.[51,52]

Allene episulphide (13) has been prepared, by flash vacuum pyrolysis, independently by two groups seeking cyclopropanethione.[53,54]

For spectroscopic information, see the bibliography on p. 232.

Reactions.—Desulphurization of thiirans to alkenes remains an important strategy for the preparation of hindered alkenes, *e.g.* (14),[24] and is the most typical thiiran reaction. Phosphines and phosphites are the reagents that are most often used.[12–14,17,19,20,21,24,48] Lithium reagents desulphurize *trans*-2,3-diphenylthiiran to (*E*)-stilbene, while the *cis*-isomer is converted into a 1 : 1 mixture of (*Z*)-stilbene and the two isomers of α-mercaptostilbene.[55] 2-Phenylthiiran is desulphurized by α-metallated isocyanides.[56] Thermal extrusions of sulphur include the conversion of (15) into (16),[27] amongst others.[15,18,22] Desulphurizations using zinc and acetic acid,[23] molybdenum[57] or palladium[58] complexes, and the oxaziridine (17)[59] have been reported.

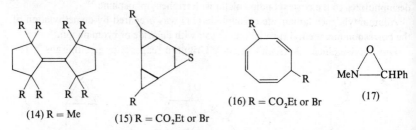

(14) R = Me   (15) R = CO$_2$Et or Br   (16) R = CO$_2$Et or Br   (17)

Harder nucleophiles often react with thiirans to give a predominance of attack at carbon, with consequent cleavage of the weak carbon–sulphur bond. This strategy has been used to prepare 7-oxa-4-thia-alk-1-enes from alkoxides, thiiran, and allyl chlorides,[60] β-mercaptoethyl sulphides from thiiran and thiolate ions,[61] thiazines by the Cu$_2$O-catalysed cyclization of the adducts of α-lithio-isocyanides with thiirans,[56] tautomeric mixtures of (18) and (19) or (20) from thiiran and hydrazones,[62] *N*-methylpyrrolidines from bis(epithio)linoleic acid and methylamine,[63] and 2,4-dimethylhex-5-ene-2-thiol by addition of the Grignard reagent derived from 3-chlorobut-1-ene with 2,2-dimethylthiiran.[64] Methylthiiran has been found to open

[51] J. A. Barltrop, A. C. Day, and E. Irving, *J. Chem. Soc., Chem. Commun.*, 1979, 881.
[52] J. A. Barltrop, A. C. Day, and E. Irving, *J. Chem. Soc., Chem. Commun.*, 1979, 966.
[53] E. Jongejan, T. S. V. Buys, H. Steinberg, and T. J. De Boer, *Recl. Trav. Chim. Pays-Bas*, 1978, **97**, 214.
[54] E. Block, R. E. Penn, M. D. Ennis, T. A. Owens, and S. L. Yu, *J. Am. Chem. Soc.*, 1978, **100**, 7436.
[55] B. F. Bonini, G. Maccagnani, G. Mazzanti, and P. Piccinelli, *Tetrahedron Lett.*, 1979, 3987.
[56] U. Schollkopf, R. Jentsch, and K. Madawinata, *Liebigs Ann. Chem.*, 1979, 451.
[57] M. W. Bishop, J. Chatt, and J. R. Dilworth, *J. Chem. Soc., Dalton Trans.*, 1979, 1.
[58] A. L. Batch, L. S. Benner, and M. M. Olmstead, *Inorg. Chem.*, 1979, **18**, 2996.
[59] Y. Hata and M. Watanabe, *J. Org. Chem.*, 1980, **45**, 1691.
[60] A. A. Rodin, K. A. V'yunov, A. I. Ginak, and E. G. Sochilin, *Zh. Org. Khim.*, 1979, **15**, (a) p. 1272; (b) p. 2252.
[61] B. Bannister, *J. Chem. Soc., Perkin Trans. 1*, 1978, 967.
[62] S. M. Shevchenko and A. A. Potekhin, *Khim Geterotsikl. Soedin.*, 1978, 1569.
[63] I. L. Kuranova and E. V. Snetkova, *Zh. Org. Khim.*, 1978, **14**, 2460.
[64] P. K. Claus, W. Rieder, and F. W. Vierhapper, *Monatsh. Chem.*, 1978, **109**, 609.

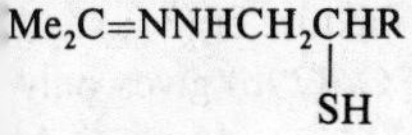

$$Me_2C=NNHCH_2CHR$$

(18)

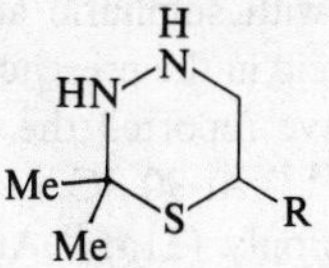

(19)

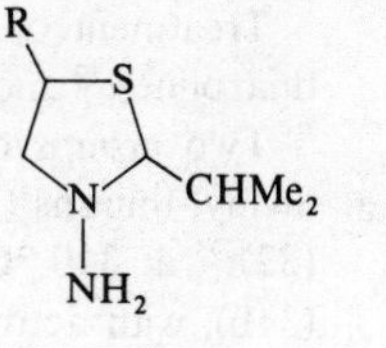

(20)

faster than methyloxiran with butylamine[65] or alkoxides,[60] even though it is less strained.

Electrophilic opening of thiirans in molecules possessing an accessible double bond leads to dihalogenated cyclic sulphides; *e.g.*, (21) gives (22) with chlorine or (23) with bromine. The intermediacy of (24) establishes that the initial, and exclusive, attack by halogen is at sulphur.[66] Treatment of (21) with SCl_2 gives (25), probably by a similar route.[66]

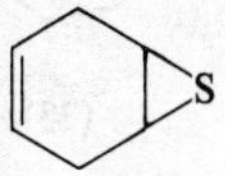

(21)

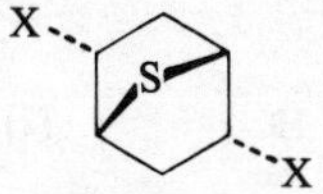

(22) X = Cl
(23) X = Br

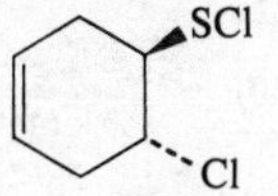

(24)

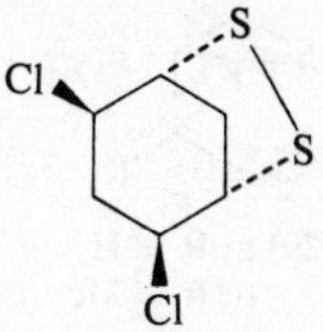

(25)

Bis(epithio)cyclohexa-1,4-diene (26) gives the *cis*-1,4-sulphur-substituted derivative (27) upon treatment with acetyl chloride, but the *cis*-1,3-sulphur-substituted derivative (28) on ring-opening with mercury(II) acetate.[28] Thiiran plus acetyl bromide gives 2-bromoethyl thiolacetate,[67] while carboxylic acid anhydrides give mixtures of 2-acyloxy-1-acylthio- and 1-acyloxy-2-acylthio-derivatives with n-alkyl-thiirans.[68] A preliminary report shows that phenyl acetate reacts with thiiran to give a low yield of 2-phenoxyethyl thiolacetate.[69] The rapid reaction of methylthiiran with HCl can be used as a non-basic scavenger for HCl that is produced in reactions.[70] Bis(epithio)linoleic acid gives tetrahydrothiophens with acids.[71]

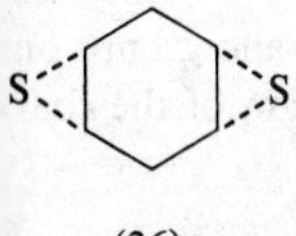

(26)

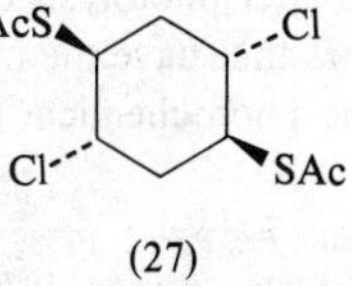

(27)

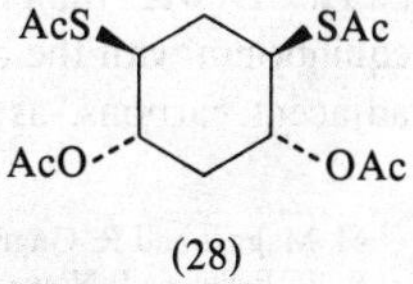

(28)

[65] H. Kakiuchi, T. Iijima, and H. Horie, *Tetrahedron*, 1979, **35**, 303.
[66] P. H. McCabe and A. Stewart, *J. Chem. Soc., Chem. Commun*, 1980, 100.
[67] L. Anoardi, R. Fornasier, D. Sostero, and U. Tonellato, *Gazz. Chim. Ital.*, 1978, **108**, 707.
[68] Y. Taguchi and Y. Suhara, *Yukagaku*, 1979, **28**, 270 (*Chem. Abstr.*, 1979, **91**, 56 719).
[69] K. Funahashi, *Chem. Lett.*, 1978, 1043.
[70] A. F. Kolomiets, A. V. Fokin, and A. A. Krolevets, *Izv. Akad. Nauk SSSR, Ser. Khim.*, 1978, 1468.
[71] I. L. Kuranova and E. V. Snetkova, *Zh. Org. Khim.*, 1978, **14**, 2165.

Treatment of episulphides with sulphuric acid in the presence of nitriles gives thiazolines,[72] and with acetic acid in the presence of imines gives thiazolidines.[73]

Two groups of workers have reported the preparation and thermolysis of the divinyl-thiirans (29) and (30).[74,75] At 90 °C, (29a) gives a 1:1 mixture of (31a) and (32);[75] at 340 °C, (29a) gives only (31a).[74] At 110 and 300 °C, (29b) gives only (31b), with activation parameters $\Delta H^{\ddagger} = 105$ kJ mol^{-1} and $\Delta S^{\ddagger} = -46$ e.u.[74] At 100 °C, (30) gives only (32), which is slowly converted into (33);[75] at 420 °C, (30) gives a mixture of products, (31a) (29%) and (34) (15%) being the only sulphur-containing ones.[74] The direct production of (31) is a Cope rearrangement, while all other products probably arise from processes involving cleavage of carbon–sulphur bonds to form radicals. At 400 °C, *cis*- and *trans*-2-phenyl-3-vinylthiiran are in equilibrium, and they react *via* a radical pathway to give (35), (36), and (37).[74]

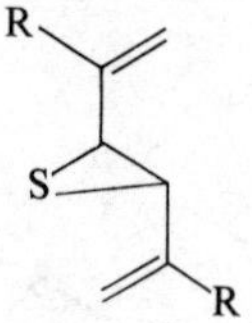

(29) a; R = H
     b; R = Me

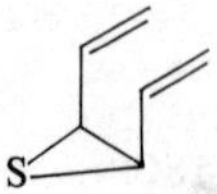

(30)

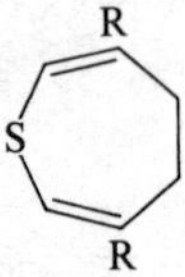

(31) a; R = H
      b; R = Me

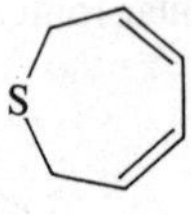

(32)

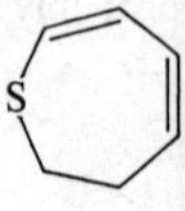

(33)

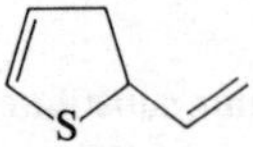

(34)

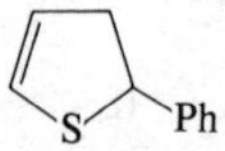

(35)

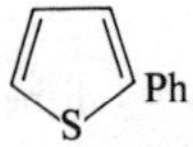

(36)

(37)

Cycloaddition reactions starting from 2-iminothiirans have been reviewed,[8,9] new results including the addition of isocyanides to give 2,4-bis(imino)thietans[76] and a demonstration that (38) can react through all three combinations of terminal atoms in the intermediate (39) to give (40), (41), and (42) (Scheme 1).[77]

Diethylgermylene reacts with thiiran to give (43), probably *via* the intermediacy of 2,2-diethyl-2-germathietan and 3-germapentane-3-thione,[78] and a germaphosphine (44) is trapped by thiiran to give (45).[79]

The 'Dewar' thiophen (12), isolated after photolysis of the parent,[52] is in thermal equilibrium with the other three isomers that have the methyl and cyano-groups on adjacent carbons, as evidenced by the photochemical interconversion of the four

[72] M. Maguet and R. Guglielmetti, *Bull Soc. Chim. Fr., Part 2*, 1978, 539.
[73] S. W. Pelletier, J. Nowacki, and N. V. Mody, *Synth. Commun.*, 1979, **9**, 201.
[74] J. C. Pommelet and J. Chuche, *J. Chem. Res. (S)*, 1979, 56.
[75] M. P. Schneider and M. Schnaithmann, *J. Am. Chem. Soc.*, 1979, **101**, 254.
[76] G. L'abbé, J.-P. Dekerk, J.-P. Declercq, G. Germain, and M. Van Meerssche, *Tetrahedron Lett.*, 1979, 3213.
[77] G. L'abbé, J.-P. Dekerk, S. Toppet, J.-P. Declercq, G. Germain, and M. Van Meerssche, *Tetrahedron Lett.*, 1979, 1819.
[78] J. Barrau, M. Bouchaut, H. Lavayssiere, G. Dousse, and J. Satgé, *Helv. Chim. Acta*, 1979, **62**, 152.
[79] C. Couret, J. Satgé, J. D. Andriamizaka, and J. Escudié, *J. Organomet. Chem.*, 1978, **157**, C35.

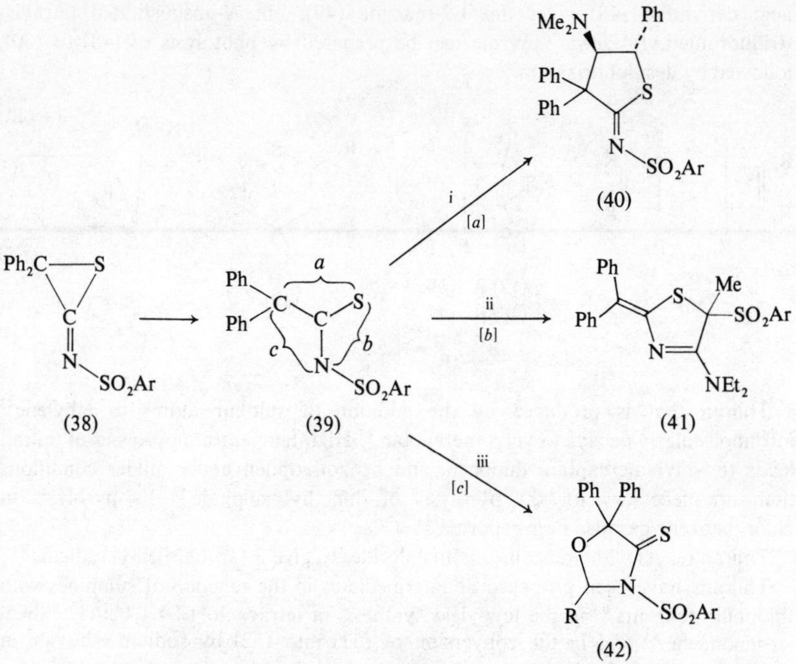

Reagents: i, PhCH=CHNMe$_2$; ii, MeC≡CNEt$_2$; iii, RCHO

Scheme 1

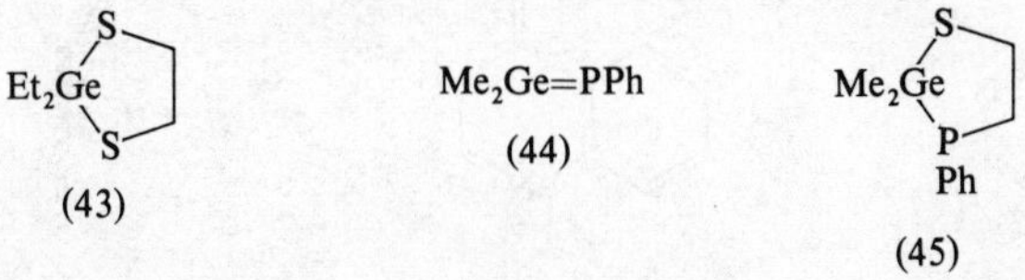

isomers. Neither of the thiophen isomers with the methyl group and the cyano-group on carbon atoms which are not formally bonded in the 'Dewar' structure is produced from (12).[51] This led the investigators to postulate interconversion by 'sulphur walk' around the periphery of the cyclobutene.

Further reactions of tetrakis(trifluoromethyl)-'Dewar' thiophen have been reported. The half-life for reversion to the normal thiophen is 5.1 h at 160 °C, in benzene. The isomerization occurs at room temperature in triphenylphosphine; in DMSO, a dimer (46) precipitates immediately.[80] Diels–Alder reactions with butadienes, furan, pyrrole, and cyclopentadiene give the expected products.[81] The furan adduct is converted into an oxahomocubane by photolysis, after desulphurization. Heating of the azide adduct (47) leads, after desulphurization, to pyrroles through a thiet intermediate.[81] Under the same conditions, the hydrazoic

[80] Y. Kobayashi, I. Kumadaki, A. Ohsawa, Y. Sekine, and A. Ando, *Heterocycles*, 1977, **6**, 1587.
[81] Y. Kobayashi, A. Ando, K. Kawada, I. Kumadaki, and A. Ohsawa, *Heterocycles*, 1979, **12**, 211.

acid derivative (48) gives the 1,2-thiazine (49), but *N*-unsubstituted tetrakis-(trifluoromethyl)-'Dewar' pyrrole can be prepared by photolysis of (48) to (50), followed by desulphurization.[82]

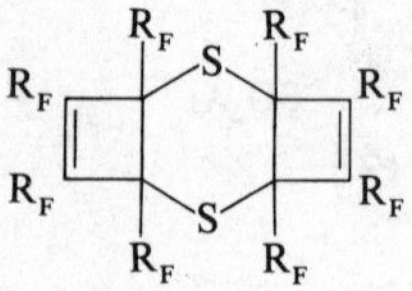

(46)

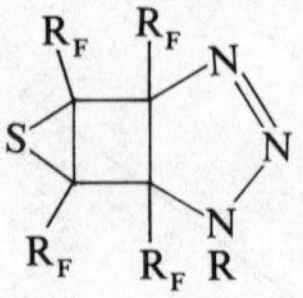

(47) R = alkyl
(48) R = H

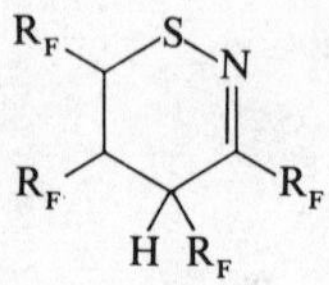

(49)

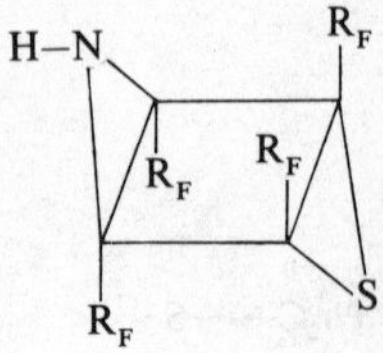

(50)

$(R_F = CF_3)$

Thiiran that is produced by the addition of sulphur atoms to ethylene[43] intramolecularly decays to vinyl mercaptan.[43] High-temperature pyrolysis of thiiran leads to vinyl mercaptan, thiophen, and benzothiophen under milder conditions than are necessary for the pyrolysis of dimethyl sulphide.[83] Its pyrolysis in chlorobenzene has also been reported.[84]

Thiiran reacts with arenediazonium chlorides to give 2-(arylthio)ethyl radicals.[85]

Thiirans have been proposed as intermediates in the reaction of sulphines with thiophilic reagents,[86] in the low-yield synthesis of tetracyclo[12.4.1.1^{3,12}.1^{5,10}]henicosanonaene,[87] and in the conversion of (51) into (52) by sodium ethoxide in DMF[88] or by concentrated sulphuric acid.[89]

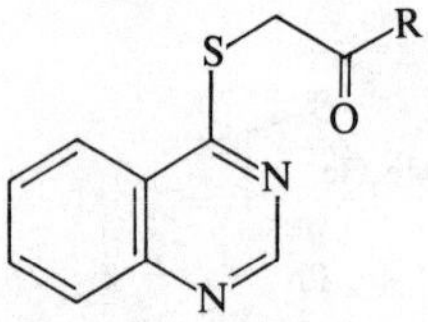

(51)

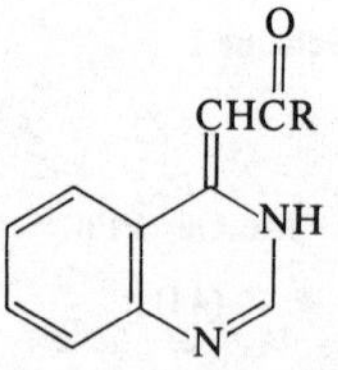

(52)

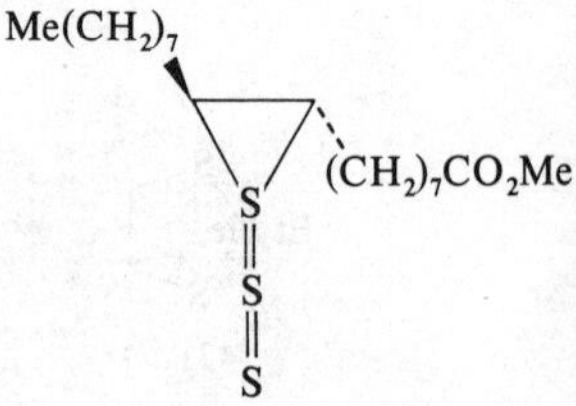

(54)

$R_2\overset{-}{C}-\overset{+}{X}$
(53) X = PPh$_3$ or SPh$_2$

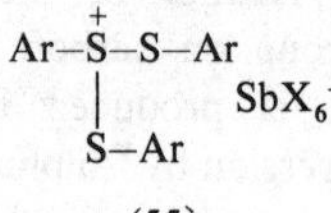

$Ar-\overset{+}{S}-S-Ar$
|
S$-$Ar

SbX$_6^-$

(55)

[82] Y. Kobayashi, A. Ando, and I. Kumadaki, *J. Chem. Soc., Chem. Commun.*, 1978, 509.
[83] E. N. Deryagina, E. N. Sukhomazova, O. B. Bannikova, and M. G. Voronkov, *Izv. Akad. Nauk SSSR, Ser. Khim.*, 1979, 2103.
[84] E. N. Sukhomazova, E. N. Deryagina, and M. G. Voronkov, *Zh. Org. Khim.*, 1980, **16**, 405.
[85] R. Kh. Freidlina, B. V. Kopylova, and L. V. Yashkina, *Tezisy Dokl. Nauchn. Sess. Khim. Tekhnol. Org. Soedin. Sery Sernistykh Neftei, 14th*, 1975 (publ. 1976), 114 (*Chem. Abstr.*, 1978, **89**, 129 149).
[86] J. A. Loontjes, M. van der Leij, and B. Zwanenburg, *Recl. Trav. Chim. Pays-Bas*, 1980, **99**, 39.
[87] W. Wagemann, M. Iyoda, H. M. Deger, J. Sombroek, and E. Vogel, *Angew. Chem., Int. Ed. Engl.*, 1978, **17**, 956.
[88] H. Singh, S. S. Narula, and C. S. Gandhi, *J. Chem. Res. (S)*, 1978, 324.
[89] H. Singh, C. S. Gandhi, and M. S. Bal, *Heterocycles*, 1980, **14**, 3.

2 Thiiran Oxides (Episulphoxides and Episulphones)

Lithium reagents desulphurize *trans*-2,3-diphenylthiiran 1-oxide, but give about equal yields of *cis*-stilbene and the lithium salt of stilbene-α-sulphenic acid with *cis*-2,3-diphenylthiiran 1-oxide.[55] Decomposition of the *trans*-isomer in the presence of ylides (53) leads to small yields of sulphines.[90] Pyridine *N*-oxide is deoxygenated by SO.[91] In all these cases, kinetic measurements show that decomposition of the thiiran 1-oxide occurs independently of the substrate.

An episulphoxide intermediate has been proposed in the thiophilic reactions of sulphines[86] (see Chap. 3, Pt II, p. 162). The necessity of a 'pseudopericyclic' process in the isomerization of tetrakis(trifluoromethyl)-'Dewar' thiophen *S*-oxide has been questioned.[92] Calculations of the electronic transitions of thiiran and methylthiiran oxides have been related to their u.v. and c.d. spectra.[93]

Episulphones are treated in a review of the pyrolysis of sulphones.[7] A detailed analysis of the vibrational spectra of sulphones includes thiiran 1,1-dioxide.[94]

A molecular formula (54) appears in *Chemical Abstracts* as a product proposed to arise from the treatment of methyl oleate or methyl elaidate with sulphur at 140—160 °C.[95] No details are available to the Reporter.

3 Thiiranium Ions (Episulphonium Ions)

Two reviews of thiiranium ions have appeared.[4] The main conclusion of reference 4*b* is that episulphonium ions are generally not involved in reactions unless their counter-ions are very poorly nucleophilic. Additional studies supporting this point have appeared. The existence of two Ad_E-2 mechanisms has been established,[96,97] additional examples of the reactions of nucleophiles with pre-formed thiiranium ions are reported,[98,99] and a new method for the preparation of thiiranium ions by the reaction of alkenes with aryl bis(arylthio)sulphonium ions (55) has been described.[100,101]

While there are normally two reactions of thiiranium ions with nucleophiles, namely ring-opening (with cleavage of a carbon–sulphur bond) or desulphurization to an alkene, a third reaction, *i.e.* dealkylation to (56), was reported for (57) when chloride or bromide ion was used as the nucleophile. When fluoride, iodide, azide, or

[90] B. F. Bonini, G. Maccagnani, G. Mazzanti, P. Pedrini, and P. Piccinelli, *J. Chem. Soc., Perkin Trans. 1*, 1979, 1720.

[91] B. F. Bonini, G. Maccagnani, G. Mazzanti, and P. Pedrini, *Tetrahedron Lett.*, 1979, 1799.

[92] J. P. Snyder and T. A. Halgren, *J. Am. Chem. Soc.*, 1980, **102**, 2861.

[93] G. L. Bendazzoli, P. Palmieri, G. Gottarelli, I. Moretti, and G. Torre, *Gazz. Chim. Ital.*, 1979, **109**, 19.

[94] J. Brunvoll and I. Hargittai, *Acta Chim. Acad. Sci. Hung.*, 1978, **96**, 337.

[95] D. Rankov, A. Popov, and S. Ivanov, *Symp. Pap., IUPAC Int. Symp. Chem. Nat. Prod., 11th*, 1978, **2**, 223 (*Chem. Abstr.*, 1980, **92**, 94 151).

[96] I. V. Bodrikov, A. V. Borisov, L. V. Chumakov, N. S. Zefirov, and W. A. Smit, *Tetrahedron Lett.*, 1980, **21**, 115.

[97] I. V. Bodrikov, T. S. Ganzhenko, F. M. Sokova, and N. S. Zefirov, *Zh. Org. Khim.*, 1980, **16**, 246.

[98] A. S. Gybin, M. Z. Krimer, W. A. Smit, V. S. Bogdanov, and E. A. Vorob'eva, *Izv. Akad. Nauk SSSR, Ser. Khim.*, 1979, 563.

[99] A. S. Gybin, W. A. Smit, and M. Z. Krimer, *Izv. Akad. Nauk SSSR, Ser. Khim.*, 1979, 927.

[100] A. S. Gybin, W. A. Smit, V. S. Bogdanov, M. Z. Krimer, and J. B. Kalyan, *Tetrahedron Lett.*, 1980, **21**, 383.

[101] A. S. Gybin, W. A. Smit, V. S. Bogdanov, and M. Z. Krimer, *Izv. Akad. Nauk SSSR, Ser. Khim.*, 1979, 1663.

methylthiolate ions or methyl-lithium were used, the alkene (58) was formed.[49] The episulphonium ion (59) has been observed.[102] The use of thiiranium ions in synthesis has received recent attention.[103-106] New calculations of the structures of thiiranium ions have appeared.[107]

(56)   (57)   (58)   (59)

(Ad = 2-adamantylidene)

Seleniranium[108] and selenirenium[109] ions have been reported. Thiirenium ions have been reviewed[5] and a crystal structure has been reported.[110]

4 Thiirens and Thiiren Oxides

A review has appeared.[3] The preparation and observation by a second research group[111] of thiirens from 1,2,3-thiadiazoles by irradiation at low temperature (8 K) (see Vol. 5, p. 197) has engendered a spirited discussion of its i.r. spectrum.[112,113] *Ab initio* calculations of the structure and the i.r. spectrum of thiiren lead to the conclusion that thiiren is properly anti-aromatic, but cast doubt on the existence of a disputed band at 660 cm^{-1}.[114] Thiirens have also been prepared by irradiation of vinylene trithiocarbonates at low temperatures[115] and of 1,2,3-thiadiazoles in the gas phase.[116] Both 4- and 5-methyl-1,2,3-thiadiazoles give (60) in the presence of hexafluorobut-2-yne.[116] Similarly, (61) and (62) give the same ratio of isomeric 2- or 3-methyltriphenylthiophens upon thermolysis in the presence of diphenyl-acetylene.[117,118]

A ^{13}C-labelling experiment has established firmly that (63) does *not* give (64).[119] Chemical labelling of benzothiadiazole (65) has established that, in the presence of

[102] J. Ohishi, K. Tsuneoka, S. Ikegami, and S. Akaboshi, *J. Org. Chem.*, 1978, **43**, 4013.
[103] W. A. J. de Loos, A. J. W. van den Berg-van Kuijk, H. M. van Iersel, J. W. de Haan, and H. M. Buck, *Recl. Trav. Chim. Pays-Bas*, 1980, **99**, 53.
[104] T. Terasawa and T. Okada, *Heterocycles*, 1980, **14**, 99.
[105] F. Cooke, R. Moerck, J. Schwindeman, and P. Magnus, *J. Org. Chem.*, 1980, **45**, 1046.
[106] B. M. Trost, W. C. Vladuchick, and A. J. Bridges, *J. Am. Chem. Soc.*, 1980, **102**, 3548.
[107] (a) M. J. S. Dewar and G. P. Ford, *J. Am. Chem. Soc.*, 1979, **101**, 783; (b) H. Lischka and H. J. Kohler, *Chem. Phys. Lett.*, 1979, **63**, 326.
[108] (a) S. Raucher, *Tetrahedron Lett.*, 1978, 2261; (b) N. Miyoshi, Y. Ohno, K. Kondo, S. Murai, and N. Sonoda, *Chem. Lett.*, 1979, 1309; (c) K. C. Nicolaou, R. L. Magolda, W. J. Sipio, W. E. Barnette, Z. Lysenko, and M. M. Joullié, *J. Am. Chem. Soc.*, 1980, **102**, 3784.
[109] C. N. Filer, D. Ahern, R. Fazio, and E. J. Shelton, *J. Org. Chem.*, 1980, **45**, 1313.
[110] R. Destro, T. Pilati, and M. Simonetta, *Nouv. J. Chim.*, 1979, **3**, 533.
[111] M. Torres, A. Clement, J. E. Bertie, H. E. Gunning, and O. P. Strausz, *J. Org. Chem.*, 1978, **43**, 2490.
[112] M. Torres, I. Safarik, A. Clement, J. E. Bertie, and O. P. Strausz, *Nouv. J. Chim.*, 1979, **3**, 365.
[113] A. Krantz and J. Laureni, *J. Org. Chem.*, 1979, **44**, 2730.
[114] B. A. Hess, jun., L. J. Schaad, and C. S. Ewig, *J. Am. Chem. Soc.*, 1980, **102**, 2507.
[115] M. Torres, A. Clement, H. E. Gunning, and O. P. Strausz, *Nouv. J. Chim.*, 1979, **3**, 149.
[116] J. Font, M. Torres, H. E. Gunning, and O. P. Strausz, *J. Org. Chem.*, 1978, **43**, 2487.
[117] E. Schaumann, J. Ehlers, W. R. Forster, and G. Adiwidjaja, *Chem. Ber.*, 1979, **112**, 1769.
[118] E. Schaumann, J. Ehlers, and H. Mrotzek, *Liebigs Ann. Chem.*, 1979, 1734.
[119] U. Timm, H. Buhl, and H. Meier, *J. Heterocycl. Chem.*, 1978, **15**, 697.

a hydrogen donor, (65) must go through a symmetric intermediate in some cases.[120] The effect is more noticeable in photolysis than in thermolysis. An intermediate, most probably benzothiiren, has been observed in an argon matrix at 8 K.[111] The photolysis of (66) to (67) is best rationalized by the intermediacy of (68).[121]

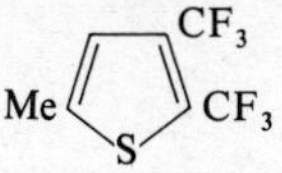

(60)

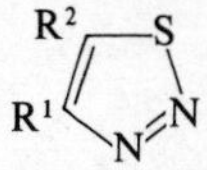

(61) R¹ = Ph, R² = Me
(62) R¹ = Me, R² = Ph

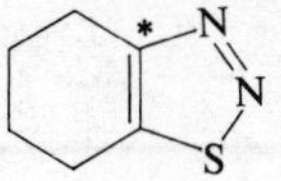

(63)

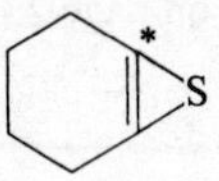

(64)

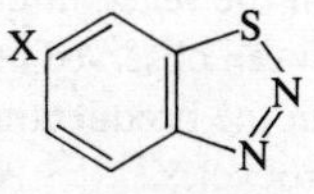

(65)

X = MeO, Cl, Me,
or MeO₂C

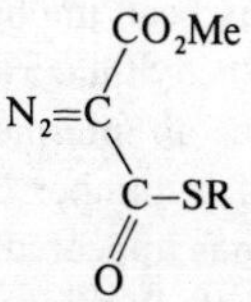

(66)

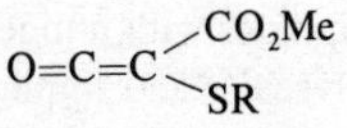

(67)

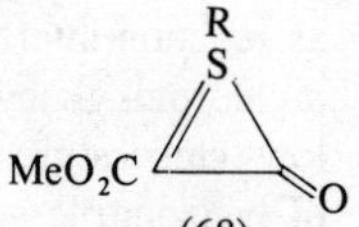

(68)

A full paper detailing the preparation and chemical properties of diphenylthiiren 1-oxide has appeared.[122] The reactions of thiiren 1,1-dioxides with α-metallated isocyanides[123, 124] and lithium azide[125] have been reported. The latter leads to a new heterocyclic system (69). Dimethyl- and diphenyl-thiiren 1,1-dioxides are reported to react with the 1,3-dipoles (70) and (71) to give (72) or (73) respectively.[124] In the case of (71) and diphenylthiiren 1,1-dioxide, (74) is also formed.[124] The Diels–Alder reaction with a diene-amine, but not the structure of the product, is mentioned in the same report.

Extensive calculations and discussions of the comparative conjugative and

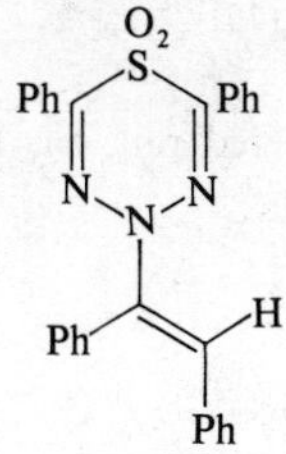

(69)

PhC≡NX

(70) X = NPh
(71) X = CHPh

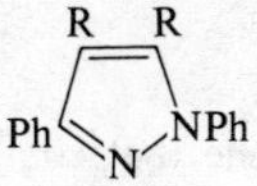

(72)

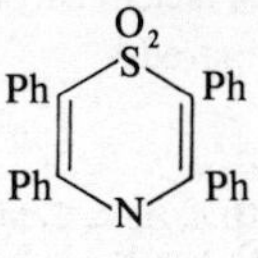

(74)

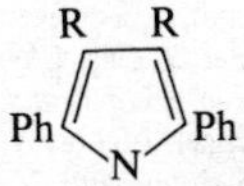

(73)

[120] R. C. White, J. Scoby, and T. D. Roberts, *Tetrahedron Lett.*, 1979, 2785.
[121] V. Georgian, S. K. Boyer, and B. Edwards, *J. Org. Chem.*, 1980, **45**, 1686.
[122] L. A. Carpino and H.-W. Chen, *J. Am. Chem. Soc.*, 1979, **101**, 390.
[123] Y. Yoshida, M. Komatsu, Y. Ohshiro, and T. Agawa, *J. Org. Chem.*, 1979, **44**, 830.
[124] Y. Yoshida, M. Uesaka, M. Komatsu, Y. Ohshiro, and T. Agawa, *Heterocycles*, 1980, **14**, 91.
[125] B. B. Jarvis, G. P. Stahly, and H. L. Ammon, *Tetrahedron Lett.*, 1978, 3781.

inductive effects of C=S, C=O, S=O, and SO_2, with relation to the structure and
properties of (75), have been accomplished.[126] In the case of all except the thiiren
1,1-dioxide, the calculated energy levels have been compared to those determined
by p.e. spectroscopy.[127]

Some of the recent papers on thiirenium ions are listed in the bibliography
on page 232.

5 Three-membered Rings with More than One Heteroatom

Oxathiiran, oxathiiran 2-oxide, and oxathiiren[128] and oxathiiran 1-oxide[129] have
been treated theoretically. An oxathiiran[130] and an oxathiiren[131] have been proposed
as reaction intermediates. The first dithiiran has been implicated in the reaction of
α-chloroalkyl acyl disulphides with amines.[132] Phenylthiazirine has been observed in
low-temperature photolyses[133] and implicated as an intermediate in the production
of benzonitrile sulphide from the photolysis of a number of heterocycles.[134, 135] A
book treating thiazirines amongst other topics has appeared.[136] Dithiaziridines have
been proposed to exist in the reactions of N-thiosulphinyl compounds.[137]
2-Imino-oxadithiirans have been proposed as first intermediates in the oxidation of
N-thiosulphinyl compounds,[138] and a dithiaziridine S-oxide in the decomposition of
(76).[138] A compound of the structure (77) has been proposed as a product of the
reaction of methyl oleate or methyl elaidate with sulphur at 140—160 °C.[95] No
details are available to the Reporter.

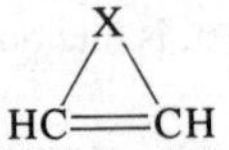

(75) X = CH_2, C=O, C=S,
      S=O, or SO_2
                        (76) 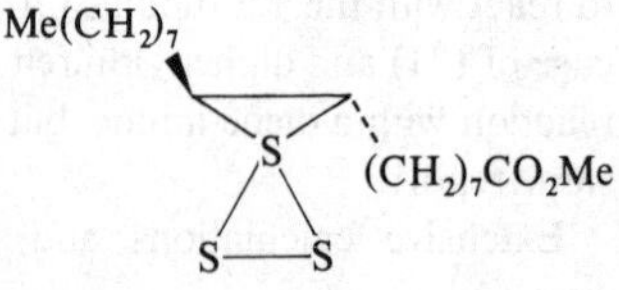   (77) 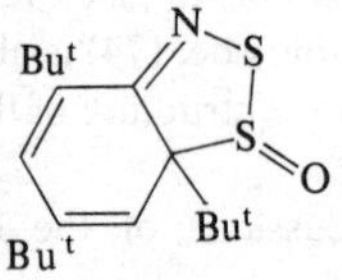

A selenacarbaborane that contains a B—C—Se three-membered ring has been
prepared.[139]

---

[126] H. L. Hase, C. Mueller, and A. Schweig, *Tetrahedron*, 1978, **34**, 2983.
[127] C. Mueller, A. Schweig, and H. Vermeer, *J. Am. Chem. Soc.*, 1978, **100**, 8056.
[128] L. Carlsen and J. P. Snyder, *J. Org. Chem.*, 1978, **43**, 2216.
[129] L. Carlsen, *J. Chem. Soc., Perkin Trans. 2*, 1980, 188.
[130] K. Praefcke and C. Weichsel, *Liebigs Ann. Chem.*, 1980, 333.
[131] C. F. Ebbinghaus, P. Morrissey, and R. L. Rosati, *J. Org. Chem.*, 1979, **44**, 4697.
[132] A. Senning, *Angew. Chem., Int. Ed. Engl.*, 1979, **18**, 941.
[133] A. Holm, N. Harrit, and I. Trabjerg, *J. Chem. Soc., Perkin Trans. 1*, 1978, 746.
[134] A. Holm and N. H. Toubro, *J. Chem. Soc., Perkin Trans. 1*, 1978, 1445.
[135] A. Holm, J. J. Christiansen, and C. Lohse, *J. Chem. Soc., Perkin Trans. 1*, 1979, 960.
[136] A. Holm, 'On the Trail of Nitrile Sulfides, Thioacylnitrenes and Thiazirines', Oersted Institute,
      Copenhagen University, Denmark, 1978.
[137] Y. Inagaki, R. Okazaki, and N. Inamoto, *Bull. Chem. Soc. Jpn.*, 1979, **52**, 2002.
[138] Y. Inagaki, R. Okazaki, and N. Inamoto, *Bull. Chem. Soc. Jpn.*, 1979, **52**, 3615.
[139] G. D. Friesen, A. Barriola, and L. J. Todd, *Chem. Ind. (London)*, 1978, 631.

6 Thietans

Formation.—A number of thietans have been prepared in high yield from 1,3-bis(benzenesulphonates) by treatment with sulphide ion in DMSO.[140] The photochemical addition of (78) and (79) to cyclic alkenes gives spiro-thietans.[141]

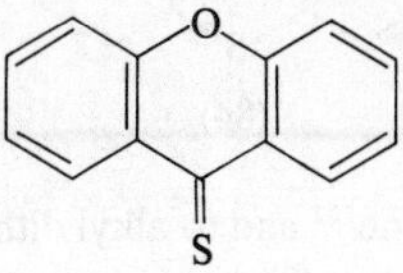

(78)

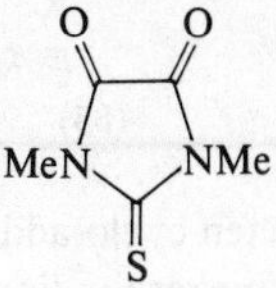

(79)

Di-t-butyl thioketone photochemically adds to alkoxy- and cyano-substituted alkenes to give thietans, but abstracts hydrogen from simple alkenes.[142] Thiobenzophenone reacts with acrylonitriles to give thietans.[143]

Chloromethyl-oxirans react with barium hydroxide that is saturated with H_2S to give 3-hydroxy-thietans.[144] Aldehyde–dimedone adducts give thietans (80) when treated with SCl_2.[145] Ring-contraction of a dithiolan with tris(diethylamino)phosphine gives a 2-propyl-thietan.[146] Treatment of (81) with triphenylphosphine and diethyl azodicarboxylate gives (82).[147]

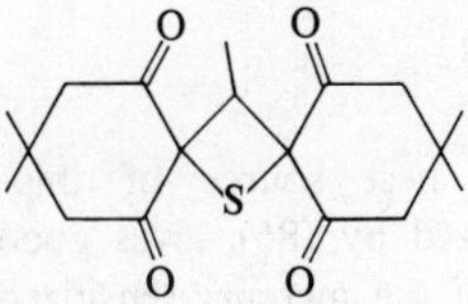

(80)

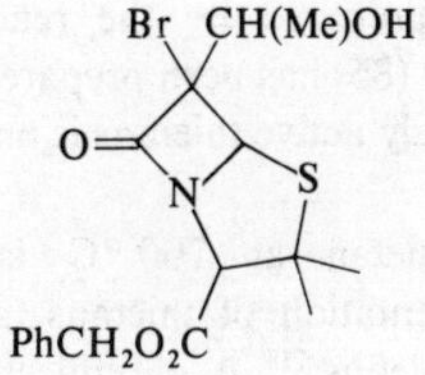

(81)

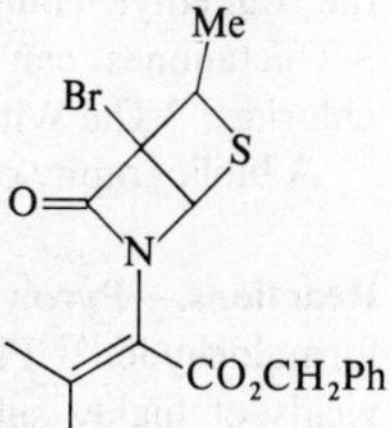

(82)

Methylenethietans, *e.g.* (83), are prepared from quaternized 4-methyl-5-(2-chloroethyl)-1,3-thiazoles by treatment with base.[148] Thioxanthione adds photochemically to the terminal double bonds of hexa-2,3,4-trienes to give 3-allenyl-thietans, *e.g.* (84).[149]

The preparation and structure determination of a 2,4-bis(imino)thietan[76] and a tris(imino)thietan[150] have been reported.

[140] M. Buza, K. K. Andersen, and M. D. Pazdon, *J. Org. Chem.*, 1978, **43**, 3827.
[141] H. Gotthardt and S. Nieberl, *Chem. Ber.*, 1978, **111**, 1471.
[142] R. Rajee and V. Ramamurthy, *Tetrahedron Lett.*, 1978, 3463.
[143] C. C. Leung, *Diss. Abstr. Int. B*, 1979, **39**, 5938.
[144] F. S. Abbott and K. Haya, *Can. J. Chem.*, 1978, **56**, 71.
[145] S. Ito and J. Mori, *Bull. Chem. Soc. Jpn.*, 1978, **51**, 3403.
[146] D. R. Crump, *Tetrahedron Lett.*, 1978, 5233.
[147] F. DiNinno, *J. Am. Chem. Soc.*, 1978, **100**, 3251.
[148] H.-J. Federsel, J. Bergman, and U. Stenhede, *Heterocycles*, 1979, **12**, 751.
[149] R. G. Visser and H. J. T. Bos, *Tetrahedron Lett.*, 1979, 4857.
[150] G. L'abbé, L. Huybrechts, J.-P. Declercq, G. Germain, and M. Van Meerssche, *J. Chem. Soc., Chem. Commun.*, 1979, 160.

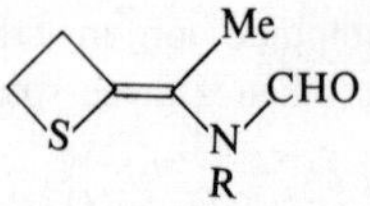

(83) 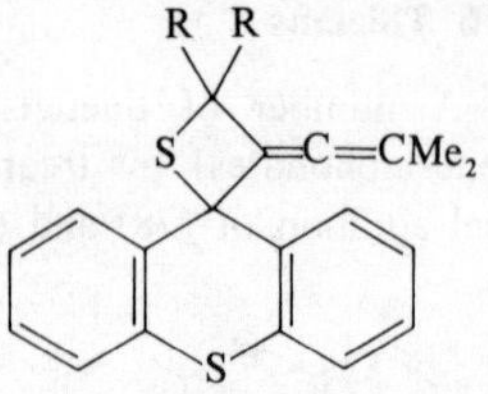   (84)

Diphenylketen cyclo-adds to thiobenzophenone[151] and to alkyl dithiobenzoates[152] to give the corresponding 2-oxothietans (β-thio-β-lactones), but with *p*-anisyl β-styryl thioketone to give only the six-membered-ring adduct.[153] Azibenzil reacts with thiobenzophenone to give mostly oxathiolan, but a small amount of a tetraphenylthietanone is also formed. In the case of di-(*p*-anisyl) thioketone, a 44% yield of the 2-thietanone is reported, and has been attributed to the trapping by thioketone of di-(*p*-anisyl)keten that arises from the azibenzil.[16] However, no structure proof for the thietanone is provided, and the melting point of the di-(*p*-anisyl) derivative is substantially different from that reported in reference 151 for the 2-thietanone.

The di-, tri-, and tetra-methyl-3-thietanones have been prepared by the ring-closure of α,α'-dibromo-ketones with sulphide ion.[154] The stereochemistry of the dimethyl compounds was determined by reduction to the thietanols.[154] 3-Thietanones can also be prepared by the reaction of ketones with thionyl chloride.[155] The Wittig reagent (85) has been prepared.[156]

A bibliography on biologicaly active thietans is on p. 232.

Reactions.—Pyrolysis of thietan at 700 °C is the best source of thio-formaldehyde.[157] Photofragmentation of thietans, sensitized by (86), gives good yields of highly substituted olefins.[158] A careful study of the mercury-sensitized decomposition of thietan suggests the trimethylenethiyl biradical as the first intermediate.[159] Photolysis of (87) gives (88).[160]

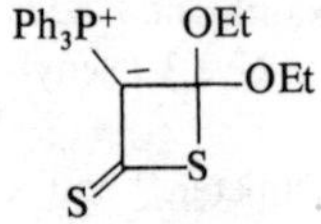

(85)   (87) 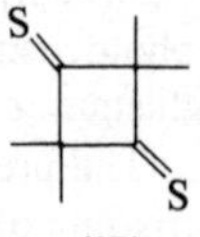   (88)

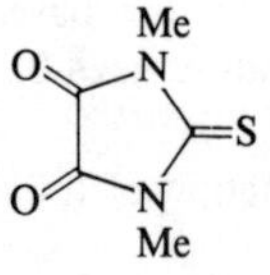

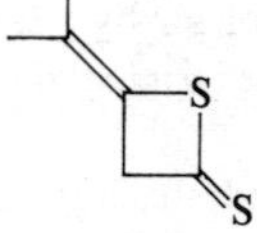

(86)

[151] H. Kohn, P. Charumilind, and Y. Gopichand, *J. Org. Chem.*, 1978, **43**, 4961.
[152] V. N. Drozd and O. A. Popova, *Zh. Org. Khim.*, 1979, **15**, 2602.
[153] T. Karakasa, H. Yamaguchi, and S. Motoki, *J. Org. Chem.*, 1980, **45**, 927.
[154] B. Fohlisch and W. Gottstein, *Liebigs Ann. Chem.*, 1979, 1768.
[155] J. S. Pizey and K. Symeonides, *Phosphorus Sulfur*, 1980, **8**, 1.
[156] H. J. Bestmann and R. W. Saalfrank, *J. Chem. Res. (S)*, 1979, 313.
[157] R. H. Judge and G. W. King, *J. Mol. Spectrosc.*, 1979, **74**, 175.
[158] H. Gotthardt and S. Nieberl, *Liebigs Ann. Chem.*, 1979, 866.
[159] D. R. Dice and R. P. Steer, *Can. J. Chem.*, 1978, **56**, 114.
[160] K. Muthuramu and V. Ramamurthy, *J. Chem. Soc., Chem. Commun.*, 1980, 243.

Oxidation of simple thietans with peroxy-acids gives monoxides[140, 145] and dioxides.[144] However, oxidation of tetraphenyl-2-thietanone (89) with *m*-chloroperoxybenzoic acid gives (90), which is a stable carboxylic–sulphinic acid anhydride.[161] Thietan dioxides are formed in the oxidation of thietans by NaMnO$_4$.[140] 2-Arylsulphonyl-3-aryl-oxaziridines oxidize sulphides (including thietans) to sulphoxides.[162] 3-Hydroxythietan can be oxidized to 3-thietanone, without oxidation at sulphur, with DMSO and benzoic anhydride.[163]

Thionyl chloride reacts with 2-phenyl-3-hydroxythietan to give 3-phenylallyl chloride rather than the expected 2-phenyl-3-chlorothietan.[144] Of the 3-chloro-thiacycloalkanes, 3-chlorothietan is the most reactive in both S_N1 and S_N2 reactions.[164] A substantial amount of ring-opening to allyl phenyl disulphide occurs in the reaction of 3-chlorothietan with thiophenoxide ion.[164] Attack at sulphur by the sulphur nucleophiles has also been suggested in the reaction of 3-thietanones with hydrogen sulphide ion;[163] α-mercapto-ketones and their di- and tri-sulphide derivatives are the only products, with no evidence of sulphur on the α'-position being found.[154]

Tris(imino)thietan (91) reacts with enamines to give 2-aminopyrroles,[165] with ynamines to give (92),[165] with hydrazoic acid to give the tetrazole (93),[166] with NaN$_3$ to give (94),[166] and with MeOH, Et$_2$NH, and EtSH to give the derivatives (95).[166] Pyrolysis of the sodium salt of the tosylhydrazone of 3-thietanone gives methylenethiiran.[54]

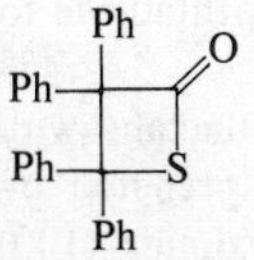

(89) 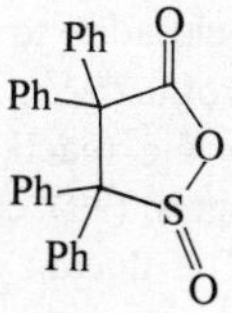   (90) 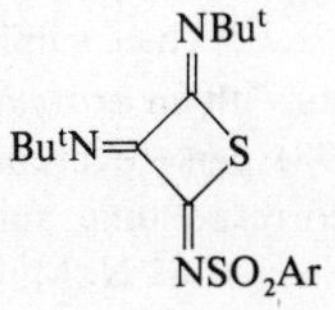   (91) 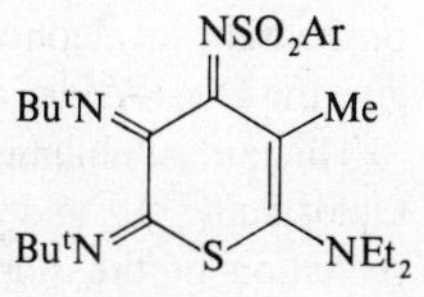   (92)

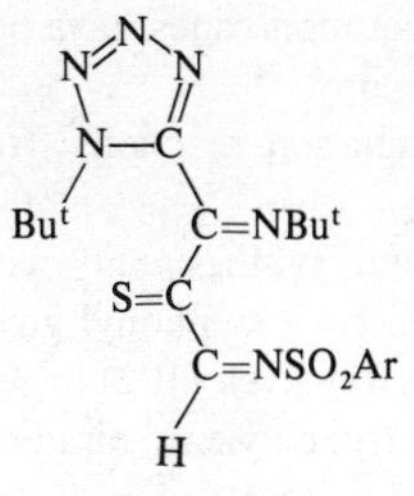

(93) 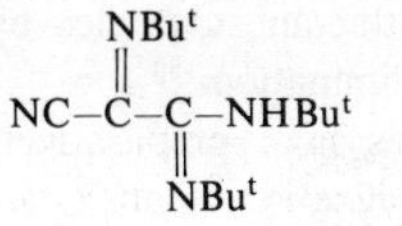   (94) 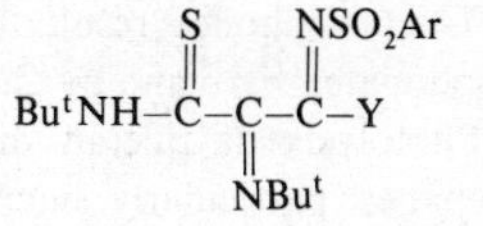   (95) Y = MeO, Et$_2$N, or EtS

[161] H. Kohn, P. Charumilind, and S. H. Simonsen, *J. Am. Chem. Soc.*, 1979, **101**, 5431.
[162] F. A. Davis, R. Jenkins, jun., and S. G. Yocklovich, *Tetrahedron Lett.*, 1978, 5171.
[163] B. Fohlisch and B. Czauderna, *Phosphorus Sulfur*, 1978, **4**, 167.
[164] H. Morita and S. Oae, *Heterocycles*, 1977, **6**, 1593.
[165] G. L'abbé, L. Huybrechts, S. Toppet, J.-P. Declercq, G. Germain, and M. Van Meerssche, *Bull. Soc. Chim. Belg.*, 1978, **87**, 893.
[166] G. L'abbé, L. Huybrechts, S. Toppet, J.-P. Declercq, G. Germain, and M. Van Meerssche, *Bull Soc. Chim. Belg.*, 1979, **88**, 297.

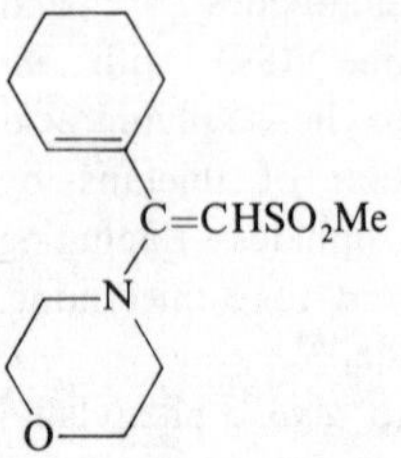

(96) 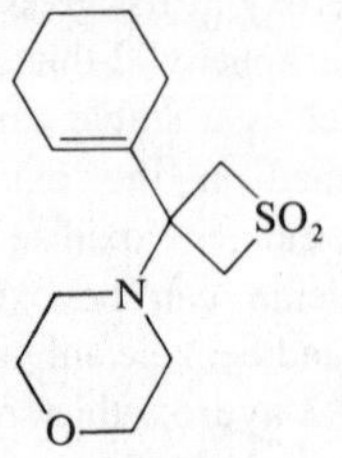   (97) 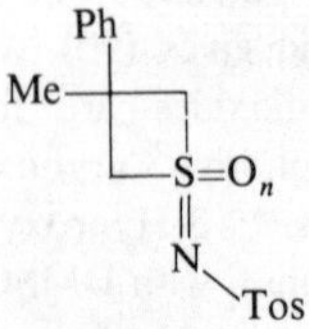   (98) $n = 0$ (100)
                                               (99) $n = 1$

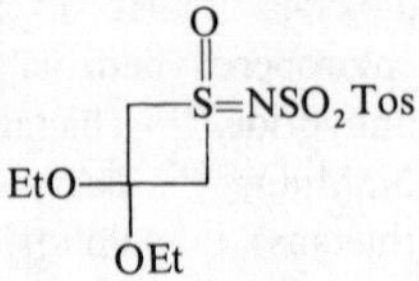

Neighbouring-group participation by β-sulphur, through thietanium ions, has been demonstrated.[167]

7 Thietan Oxides and Imides

Formation.—Preparation of thietan oxides by oxidation of thietans is covered in the previous section. Cycloaddition of sulphenes to electron-rich alkenes, the other important method of preparation of episulphones, has been further used for the preparation of α-chloro- and β-amino-,[168] α-sila- and β-amino-,[169] and α-alkoxycarbonyl- and β-amino-thietan dioxides.[170] The sulphonyl enamine (96) can be converted into the thietan dioxide (97) by treatment with acid, a procedure that usually hydrolyses enamines to ketones.[171] Regeneration of (96) from (97) occurs on heating in alcoholic base.[171] Thiet sulphone cyclo-adds to cyclopentadiene to give the Diels–Alder adducts with an *exo/endo* ratio of 0.25.[172]

Thietan sulphilimines (98) can be prepared by the reaction of thietans with Chloramine T, and the corresponding sulphoximines (99) can be prepared by oxidation of the sulphilimines with $NaMnO_4$.[140] The thietan sulphoximine (100) results from the cycloaddition of an iminosulphene to 1,1-diethoxyethene.[173]

Reactions.—2,2,4,4-Tetra-acylthietan monoxides are photochemically converted into cyclopropanes.[145] The conformational aspects of thietan monoxides have been assessed, using lanthanide shift reagents and n.m.r. spectroscopy.[174]

Thiet sulphones result from thietan sulphones by elimination reactions: from sulphonates[144, 175] and by Cope elimination.[168]

Photolysis of thietan dioxides is a preparatively useful synthesis of cyclopropanes, particularly since stabilization of an α-carbanion by a sulphonyl group allows wide latitude for manipulation.[176] Thietan dioxide (101) gives (102) in 95% yield upon photolysis.[145] Trimethylene sulphone gives trimethylene biradicals

[167] E. L. Eliel, W. H. Pearson, L. M. Jewell, and A. G. Abatjoglou, *Tetrahedron Lett.*, 1980, **21**, 331.
[168] W. Ried and H. Bopp, *Chem. Ber.*, 1978, **111**, 1527.
[169] A. G. Shipov, A. V. Kisin, and Yu. I. Baukov, *Zh. Obshch. Khim.*, 1979, **49**, 1170.
[170] A. Etienne and B. Desmazieres, *J. Chem. Res. (S)*, 1978, 484.
[171] R. A. Ferri, G. Pitacco, and E. Valentin, *Tetrahedron*, 1978, **34**, 2537.
[172] H. D. Martin, R. Iden and H. J. Schiwek, *Tetrahedron Lett.*, 1978, 3337.
[173] C. R. Johnson, E. U. Jonsson, and C. C. Bacon, *J. Org. Chem.*, 1979, **44**, 2055.
[174] D. J. H. Smith, J. D. Finlay, C. R. Hall, and J. J. Uebel, *J. Org. Chem.*, 1979, **44**, 4757.
[175] P. Dalla Croce, P. Del Buttero, S. Maiorana, and R. Vistocco, *J. Heterocycl. Chem.*, 1978, **15**, 515.
[176] J. D. Finlay, D. J. H. Smith, and T. Durst, *Synthesis*, 1978, 579.

during vacuum-u.v. photolysis.[177] The hexamethyl-3-methylenethietan dioxide (103) gives methylenecyclopropanes with less deuterium scrambling than required of conventional trimethylenemethane intermediates.[178]

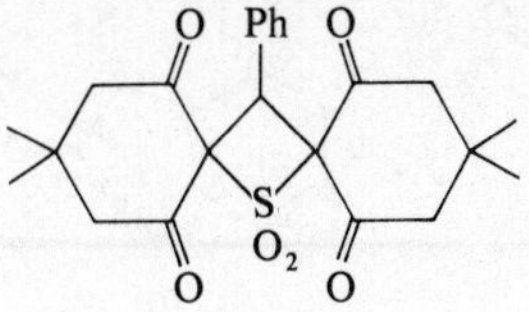

(101) 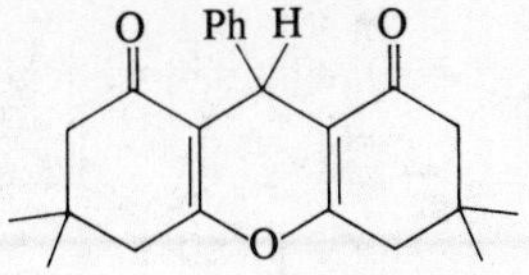   (102) 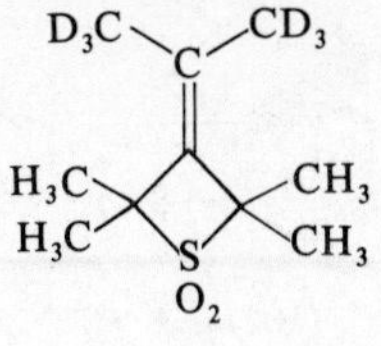   (103)

2-Phenyl-3-hydroxythietan 1,1-dioxide gives benzyl methyl sulphone with hydroxide ion, but benzyl methyl ketone with sulphuric or phosphoric acids.[144]

8 Thiet and its Derivatives

Thiets have been reviewed.[179]

Formation.—Phenyl-, 2-naphthyl-, 2-thienyl-, and 2-furanyl-thiets are prepared from aryl methyl ketones, dimethylamine, and methanesulphonyl chloride.[180] X-Ray crystal structures establish that the compounds arising from the cyclo-addition of alkynes to thioxanthones are thiets, and not spiro-dihydropyrans, as originally proposed.[181] Additional work shows that alkynyl sulphides also give thiets with thioxanthone, and that the adduct (104) is in rapid equilibrium with the open form (105) in solution, although they can be isolated separately.[182,183] Thiet (106) was isolated after thermolysis of a 'Dewar' thiophen–azide adduct.[81]

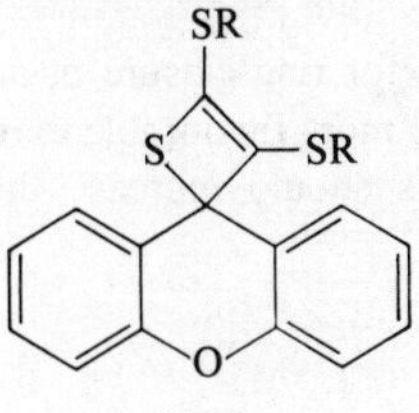

(104) 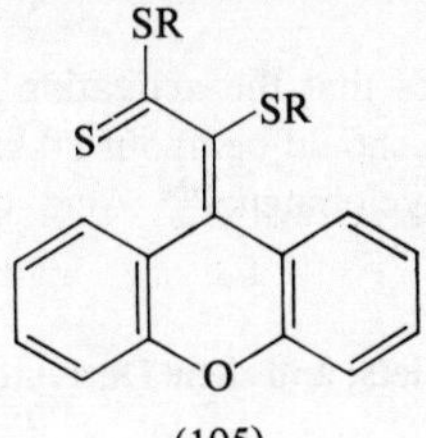   (105) 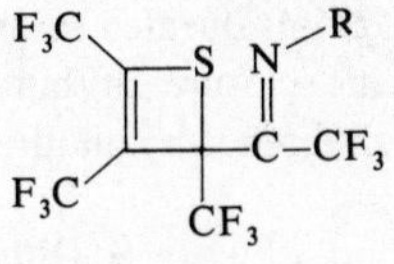   (106)

A potentially useful reaction for the large-scale preparation of benzothiet (107) has been reported. It involves the thermolysis of (108) at 700 °C and is estimated to

[177] A. A. Scala and I. Colon, *J. Phys. Chem.*, 1979, **83**, 2025.
[178] R. J. Bushby and M. D. Pollard, *Tetrahedron Lett.*, 1978, 3855.
[179] D. C. Dittmer, in 'New Trends in Heterocyclic Chemistry', ed. R. B. Mitra, Elsevier, Amsterdam, 1979.
[180] B. H. Patwardhan, E. J. Parker, and D. C. Dittmer, *Phosphorus Sulfur*, 1979, **7**, 5.
[181] H. Gotthardt and O. M. Huss, *Tetrahedron Lett.*, 1978, 3617; *cf.* H. Gotthardt and S. Nieberl, *ibid.*, 1976, 3563.
[182] A. C. Brouwer, A. V. E. George, D. Seykens, and H. J. T. Bos, *Tetrahedron Lett.*, 1978, 4839.
[183] G. J. Verhoeckx, J. Kroon, A. C. Brouwer, and H. J. T. Bos, *Acta Crystallogr., Sect. B*, 1980, **36**, 484.

have an 80% yield.[184] The keten (109) is observed by p.e. spectroscopy when (110) is heated to 350 °C.[184] Neither thiet has a tendency to open. The thiadiazine (111) can be used as the starting material for the preparation of (112; $n = 0$, 1, or 2).[185]

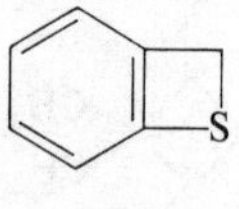

(107) 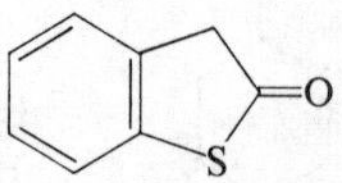   (108) 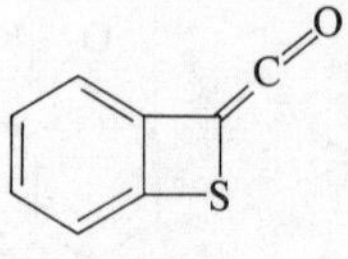   (109) 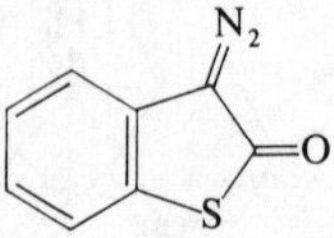   (110)

(111) 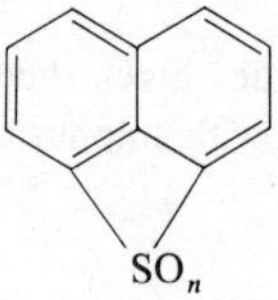   (112) 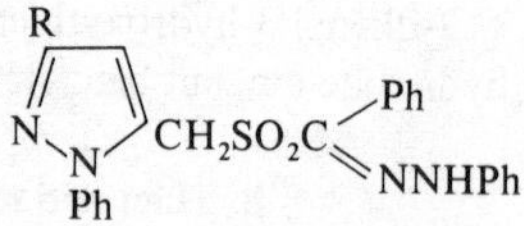   (113)

Thiet sulphones are prepared by elimination reactions of the appropriate thietan sulphones.[144, 168, 175]

Reactions.—Condensation of thiet sulphone with 2 moles of halogeno-hydrazones, RC(Cl)=NNHPh, gives isoimidazoles (113).[175] Cycloaddition of thiet sulphone to cyclopentadiene occurs with an *exo/endo* ratio of 0.25.[172] 2,4-Diphenylthiet dioxide is hydrolysed to dibenzyl sulphone with base, but rearranges to the isomeric sulphinate 3,5-diphenyl-1,2-oxathiolen 2-oxide with concentrated sulphuric acid.[143] Treatment of 4-phenyl-2*H*-thiet 1,1-dioxide with KOH in ethanol gives 3-ethoxy-2-phenylthietan 1,1-dioxide.[144]

A theoretical study concludes that the activation energy for ring-closure of an $\alpha\beta$-unsaturated thione to a thiet should be about 30 kJ mol^{-1} more favourable than the closure of butadiene to cyclobutene.[186] Acid catalysis should increase the difference by another 45 kJ mol^{-1}.

9 Dithietans, Dithiets, and their Derivatives

1,2-Dithietans.—1,2-Dithietan (114) is the first isolated derivative in that class. It arises by dimerization of ethyl sulphine, probably by rearrangement of a first-formed sulphenyl sulphinate[187] (see Chap. 3, Pt. II, p. 161).

1,2-Dithiets.—Two dithiets (115) and (116) have been isolated,[188] and more observed.[189, 190] Benzodithiet is trapped by dimethyl acetylenedicarboxylate to give

[184] R. Schulz and A. Schweig, *Tetrahedron Lett.*, 1980, **21**, 343.
[185] J. Nakayama, T. Fukushima, E. Seki, and M. Hoshino, *J. Am. Chem. Soc.*, 1979, **101**, 7684.
[186] J. P. Snyder, *J. Org. Chem.*, 1980, **45**, 1341.
[187] E. Block, A. A. Bazzi, and L. K. Revelle, *J. Am. Chem. Soc.*, 1980, **102**, 2490.
[188] A. Krebs, H. Colberg, U. Hopfner, H. Kimling, and J. Odenthal, *Heterocycles*, 1979, **12**, 1153.
[189] N. Jacobsen, P. de Mayo, and A. C. Weedon, *Nouv. J. Chim.*, 1978, **2**, 331.
[190] P. de Mayo, A. C. Weedon, and G. S. K. Wong, *J. Org. Chem.*, 1979, **44**, 1977.

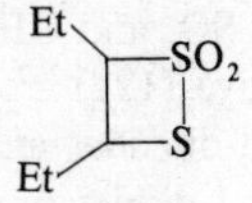

(114)

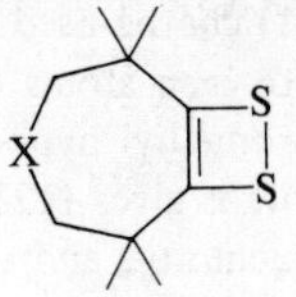

(115) X = S
(116) X = SO₂

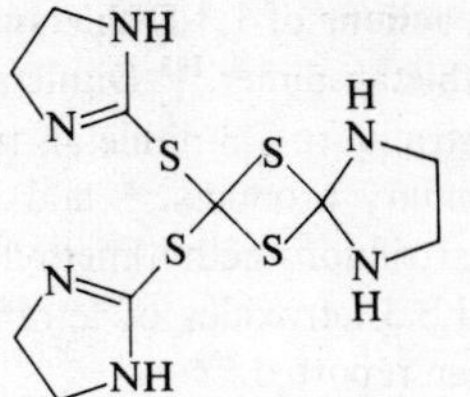

(117)

a benzo-1,4-dithiin.[190] Tervalent phosphorus compounds insert into the S—S bond of bis-(trifluoromethyl)dithiet.[191]

Formation of 1,3-Dithietans.—The dimerization of thiones or potential thiones is a major source of 1,3-dithietans; *e.g.*, treatment of α-chlorosulphenyl chlorides with triphenylphosphine gives derivatives of dithietan.[192] Monothiobenzil gives 2,4-diphenyl-2,4-dibenzoyl-1,3-dithietan on standing in solution, but can be regenerated photochemically.[193] Symmetrically substituted dithietans often accompany the preparation of thiones.[194-196] Spiro-derivatives (117) arise from ethylenethiourea and carbon disulphide in the presence of NaH in THF.[197]

Dithiocarboxylic acids (118) can be converted into 1,3-dithietans (119) by acid chlorides,[198,199] iodine,[199] HCl,[199] DCCI,[200] or upon standing for a long time,[199] and they are formally thioketen dimers. The cycloaddition of two C=S groups yields thioketen dimers[118,151,201-203] and (120) from methyl isothiocyanate.[204] Derivatives (121) are prepared by the reaction of dithiocarboxylic acids (118) with phosgene.[199,205]

For papers on biologically active dithietans and for spectroscopic information see the bibliography on p. 232.

R₂CHCS₂H

(118)

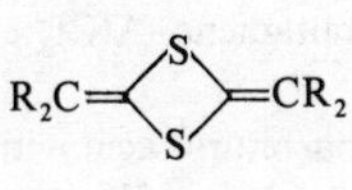

(119)

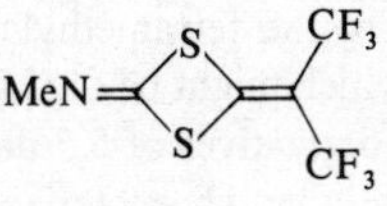

(120)

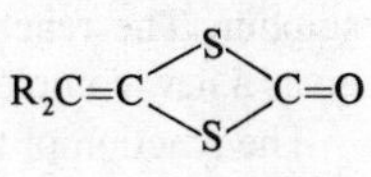

(121)

[191] B. C. Burros, N. J. De'Ath, D. B. Denney, D. Z. Denney, and I. J. Kipnis, *J. Am. Chem. Soc.*, 1978, **100**, 7300.
[192] K. Oka, *J. Org. Chem.*, 1979, **44**, 1736.
[193] C. Bak and K. Praefcke, *Z. Naturforsch., Teil B*, 1980, **35**, 372.
[194] P. Metzner, J. Vialle, and A. Vibet, *Tetrahedron*, 1978, **34**, 2289.
[195] M. M. A. Abdel-Rahman, *Acta Chim. Acad. Sci. Hung.*, 1977, **93**, 425.
[196] B. S. Pedersen, S. Scheibye, N. H. Nilsson, and S. O. Lawesson, *Bull. Soc. Chim. Belg.*, 1978, **87**, 223.
[197] M. Yokoyama, S. Ohtuki, M. Muraoka, and T. Takeshima, *Tetrahedron Lett.*, 1978, 3823.
[198] M. Augustin, C. Groth, H. Kristen, K. Peseke, and C. Wiechmann, *J. Prakt. Chem.*, 1979, **321**, 205.
[199] E. Schaumann and F.-F. Grabley, *Liebigs Ann. Chem.*, 1979, 1715.
[200] T. Takeshima, A. Yano, N. Fukada, Y. Hirose, and M. Muraoka, *J. Chem. Res. (S)*, 1979, 140.
[201] H. Kohn, Y. Gopichand, and P. Charumilind, *J. Org. Chem.*, 1978, **43**, 4955.
[202] H. Boehme and P. Malcherek, *Arch. Pharm. (Weinheim, Ger.)*, 1980, **313**, 81.
[203] E. Schaumann and F.-F. Grabley, *Liebigs Ann. Chem.*, 1979, 1702.
[204] M. S. Raasch, *J. Org. Chem.*, 1978, **43**, 2500.
[205] E. Schaumann, E. Kausch, and E. Rossmanith, *Liebigs Ann. Chem.*, 1978, 1543.

Reactions of 1,3-Dithietans.—Monothiobenzil can be generated by photolysis of its dithietan dimer.[193] Dithietanones (121) can be used as thioketen equivalents.[199, 205] Tetrafluoro-1,3-dithietan reacts with oxygen atoms to give CF_2, SO, and CF_2S as primary products,[206] and with trifluoromethyl hypochlorite to give (122).[207] With tris(trifluoromethyl)methyl hypochlorite it gives (123).[208] The 1,1-dioxides and the 1,1,3,3-tetroxides of tetrafluoro-, tetrachloro-, and tetrabromo-1,3-dithietans have been reported.[209]

Aldehydes can be converted into their hexafluoroisopropylidene derivatives (124) by reaction with triphenylphosphine and 2,2,4,4-tetrakis(trifluoromethyl)-1,3-dithietan.[210]

Selenium Derivatives.—Pyrolysis of 1,2,3-selenadiazoles at 500–600 °C gives 2,4-bis(methylene)-1,3-diselenetans.[211] Treatment of divinyl sulphone with selenium tetrabromide gave a compound that was claimed to be (125).[212]

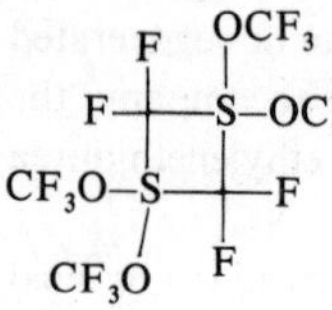

(122) 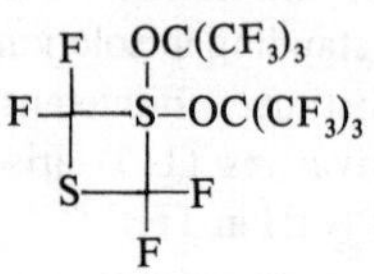   (123) 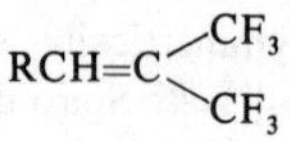   (124) 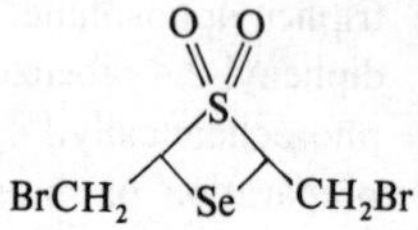   (125)

10 Four-membered Rings containing Sulphur and Oxygen

Calculations concerning the stabilities of 1,2-oxathietan 2-oxide,[128] 1,2-oxathietan-3-one and its enol,[128] and 1,2,3-[128] and 1,3,2-dioxathietans[128, 129] have appeared.

The anhydride (126) has been prepared at 10 K by photochemical addition of SO_2 to keten.[213] It decomposes photochemically to formaldehyde, CO_2, and sulphur. The reaction of the tetramethylcyclobutadiene–$AlCl_3$ complex with SO_2 gives a new complex, which might be (127).[214]

The reaction of two derivatives of 5,5-diallylbarbituric acid with sulphuric acid in acetonitrile is said to give characterizable β-sultones.[215] Addition of SO_3 to hexafluorobutadiene occurs in a 1,2 manner to give (128).[216] The double bond of (128) can be brominated to give another stable β-sultone. The addition is reversible above 20 °C, and (128) reacts with excess SO_3 to give (129) and SO_2.[216]

[206] I. R. Slagle and D. Gutman, *Int. J. Chem. Kinet.*, 1979, **11**, 453.
[207] T. Kitazume and J. M. Shreeve, *Inorg. Chem.*, 1978, **17**, 2173.
[208] Q.-C. Mir, K. A. Laurence, R. W. Shreeve, D. P. Babb, and J. M. Shreeve, *J. Am. Chem. Soc.*, 1979, **101**, 5949.
[209] R. Seelinger and W. Sundermeyer, *Angew. Chem., Int. Ed. Engl.*, 1980, **19**, 203.
[210] D. J. Barton and Y. Inouye, *Tetrahedron Lett.*, 1979, 3397.
[211] A. Holm, C. Berg, C. Bjerre, B. Bak, and H. Svanholt, *J. Chem. Soc., Chem. Commun.*, 1979, 99.
[212] Yu. V. Migalina, V. G. Lend'el, A. S. Koz'min, and N. S. Zefirov, *Khim. Geterotsikl. Soedin.*, 1978, 708.
[213] I. R. Dunkin and J. G. MacDonald, *J. Chem. Soc., Chem. Commun.*, 1978, 1020.
[214] P. B. J. Driessen and H. Hogeveen, *J. Organomet. Chem.*, 1978, **156**, 265.
[215] M. Konieczny and A. Radzicka, *Arch. Immunol. Ther. Exp.*, 1978, **26**, 955.
[216] N. B. Kaz'mina, I. L. Knunyants, G. M. Kuz'yants, E. I. Mysov, and E. P. Lur'e, *Izv. Akad. Nauk SSSR, Ser. Khim.*, 1979, 118.

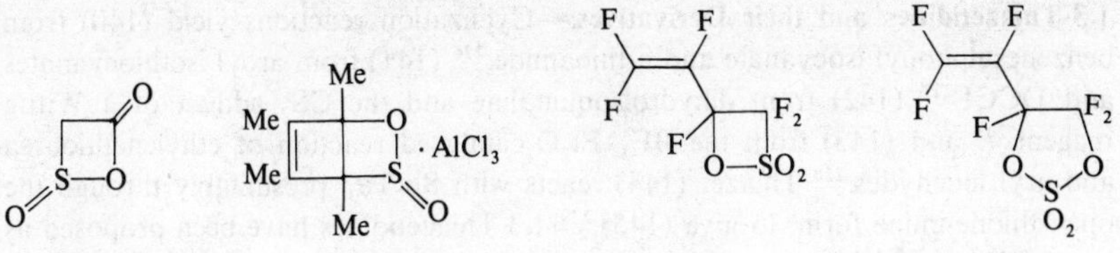

(126)   (127)   (128)   (129)

Oxathietan 1,1-dioxide (130) reacts with alcohols to give (2-fluorosulphonyl)-carboxylic esters, and with BF_3 and Et_3N to give fluorosulphonyl(trifluoro-methyl)keten.[217] The latter reaction is more efficient if the ring is first opened with KF and HF. Chloral cyclo-adds to sulphenes to give β-sultones.[170,218] 4-Tri-chloromethyl-1,2-oxathietan 2,2-dioxide reacts with primary and secondary amines to give β-hydroxy-sulphonamides.[218]

An oxathiet (131) has been observed in the low-temperature photolysis of (132).[190] The thermolysis of diphenylthiiren 1-oxide also yields an oxathiet.[122] The alkylidene sulphate (133) has been reported.[219]

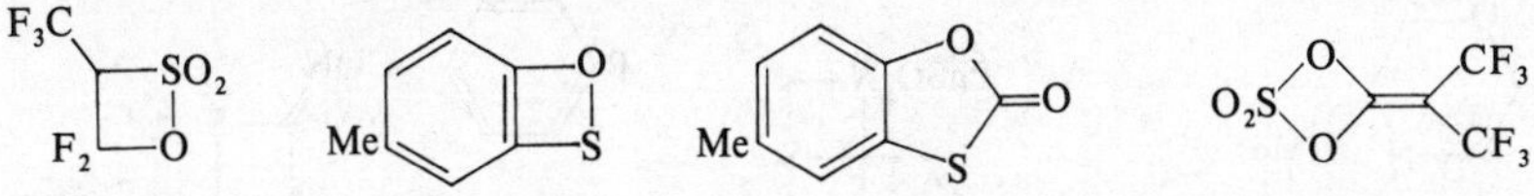

(130)   (131)   (132)   (133)

11 Four-membered Rings containing Sulphur and Nitrogen and/or Phosphorus

1,2-Thiazetidines and their Derivatives.—Carboxamides with an acidic α-hydrogen react with thionyl chloride in the presence of pyridine to give 1,2-thiazetidin-3-one 1-oxides (134).[220] N-Phenyl-1,2-thiazetidin-3-one 1-oxide, prepared from keten and N-sulphinylaniline, is opened by anilines to give (135).[221] Thiofluorenone reacts with (136) to give, amongst other things, a small yield of (137); this is oxidized to its sulphonamide by MCPBA.[222]

N-(Alkylsulphonyl)imides (138), prepared by treating N-alkylsulphamoyl chlorides with Et_3N, react with enamines that bear no hydrogens β to the morpholino-group to give 1,2-thiazetidine 1,1-dioxides (139).[223] The preparations and e.s.r. spectra of two relatively stable 1,2-thiazet-2-yl radicals have been reported.[224,225]

[217] A. F. Eleev, S. I. Pletnev, G. A. Sokol'skii, and I. L. Knunyants, *Zh. Vses. Khim. O-va*, 1978, **23**, 229 [*Mendeleev Chem. J.*, 1978, **23**, No. 6, p. 45].
[218] W. Hanefeld and D. Kluck, *Arch. Pharm. (Weinheim, Ger.)*, 1978, **311**, 698.
[219] A. F. Eleev, G. A. Sokol'skii, and I. L. Knunyants, *Izv. Akad. Nauk SSSR, Ser. Khim.*, 1978, 2360.
[220] L. Capuano, G. Urhahn, and A. Willmes, *Chem. Ber.*, 1979, **112**, 1012.
[221] J. E. Semple and M. M. Joullié, *J. Org. Chem.*, 1978, **43**, 3066.
[222] G. Mazzanti, G. Maccagnani, B. F. Bonini, P. Pedrini, and B. Zwanenburg, *Gazz. Chim. Ital.*, 1980, **110**, 163.
[223] T. Nagai, T. Shingaki, M. Inagaki, and T. Ohshima, *Bull. Chem. Soc. Jpn.*, 1979, **52**, 1102.
[224] R. Mayer, G. Domschke, S. Bleisch, and A. Bartl, *Tetrahedron Lett.*, 1978, 4003.
[225] R. Mayer, S. Bleisch, G. Domschke, A. Tkac, A. Stasko, and A. Bartl, *Org. Magn. Reson.*, 1979, **12**, 532.

1,3-Thiazetidines and their Derivatives.—Cyclization reactions yield (140) from benzenesulphonyl isocyanate and a thioamide,[226] (141) from aroyl isothiocyanates and DCCI,[227] (142) from dihydroisoquinoline and the CS_2 adduct of a Wittig reagent,[203] and (143) from the $BF_3 \cdot Et_2O$-catalysed reaction of ethylenethiourea and aryl aldehydes.[228] Thiazet (144) reacts with Sb_2Te_3, presumably through the open thione-imine form, to give (145).[229] 1,3-Thiazetidines have been proposed as intermediates.[205, 230]

(134) X = CO or SO_2

(135)

(136)

(137)

(138)

(139)

(140)

(141)

(142)

(143)

(144)

(145)

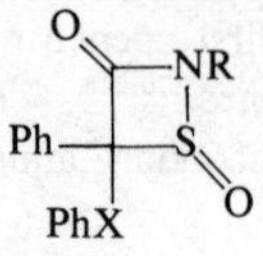

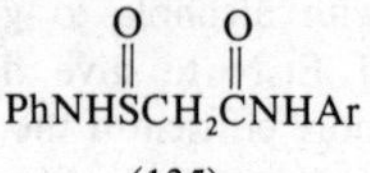

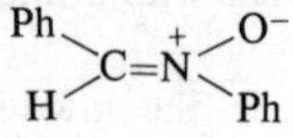

Other Four-membered Rings containing Sulphur and Nitrogen.—The reaction of tris(imino)sulphur derivatives with a slight excess of sulphonyl isocyanates gives eventually (146).[231, 232] Tris(trimethylsilylimino)sulphur reacts with *N*-(sulphinyl)-heptafluoroisopropylamine to give (147), possibly through the intermediacy of (148).[232] The compound (149) arises from the reaction of (150a) and (150b).[232]

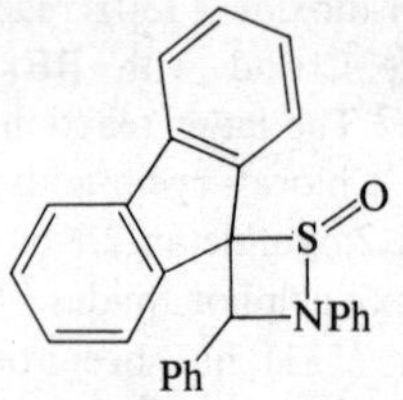

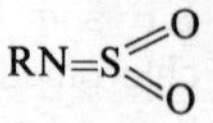

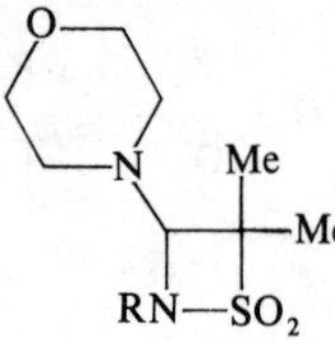

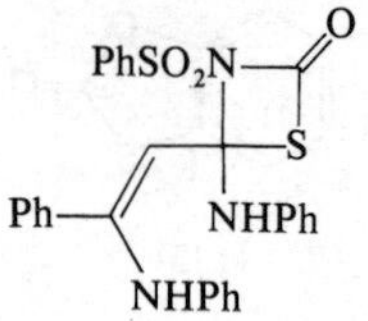

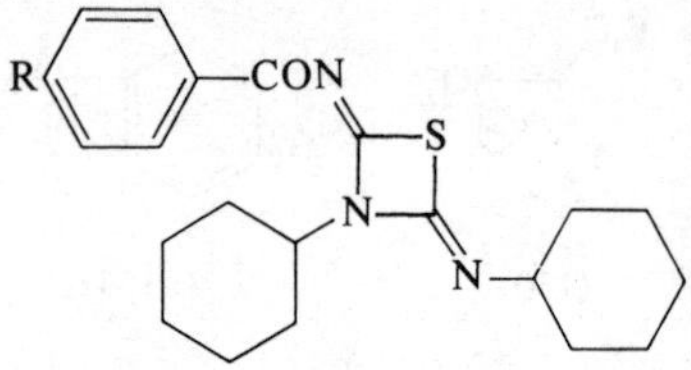

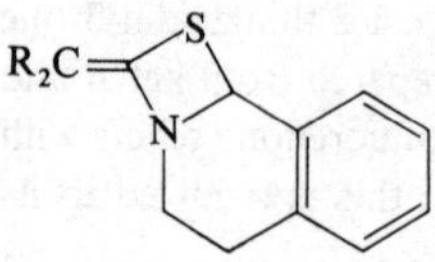

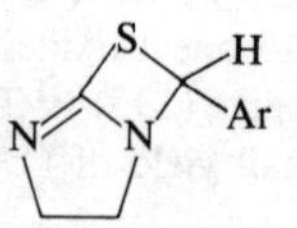

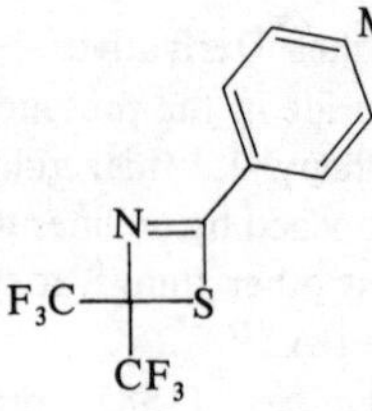

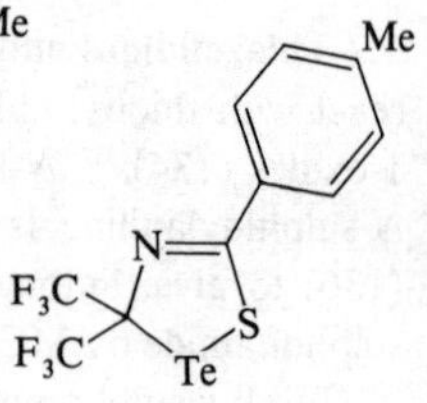

[226] W. Zankowska-Jasinska and H. Borowiec, *Pol. J. Chem.*, 1978, **52**, 1683.
[227] O. Hritzova and P. Kristian, *Collect. Czech. Chem. Commun.*, 1978, **43**, 3258.
[228] M. Yokoyama and H. Monma, *Tetrahedron Lett.*, 1980, **21**, 293.
[229] K. Burger, R. Ottlinger, A. Proksch, and J. Firl, *J. Chem. Soc., Chem. Commun.*, 1979, 80.
[230] E. Schaumann and E. Kausch, *Liebigs Ann. Chem.*, 1978, 1560.
[231] F.-M. Tesky, R. Mews, B. Krebs, and M. R. Udupa, *Angew. Chem., Int. Ed. Engl.*, 1978, **17**, 677.
[232] F.-M. Tesky, R. Mews, and B. Krebs, *Angew. Chem., Int. Ed. Engl.*, 1979, **18**, 235.

Dimer (151) is opened and stabilized by pyridine and S_4N_4.[233] Sulphur trioxide gives (152) with N-(sulphinyl)trifluoromethanesulphonamide.[234] Both (151) and (152) react with benzonitrile to give (153).[233,234] With aryl isocyanates, they give (154).[234] Both 1,2,3,4-dithiadiazetidines[235] and 1,3,2,4-dithiadiazetidines[236] have been proposed ⸍s reaction intermediates.

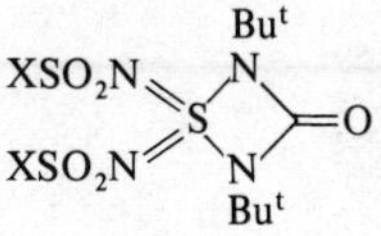

(146)

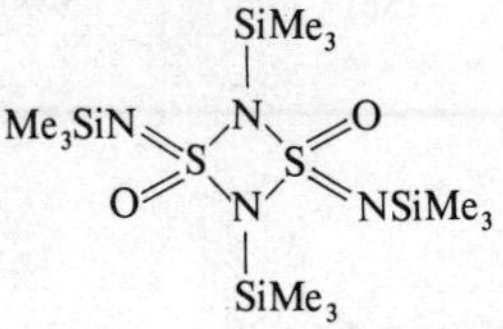

(147)

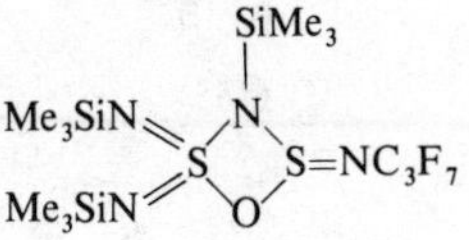

(148)

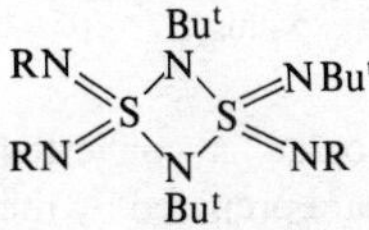

(149) R = CF$_3$, C$_2$F$_5$, or C$_3$F$_7$

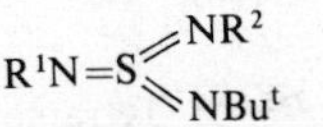

(150) a; R^1 = Bu^t; R^2 = CF$_3$, C$_2$F$_5$, or C$_3$F$_7$
b; R^1 = R^2 = CF$_3$, C$_2$F$_5$, or C$_3$F$_7$

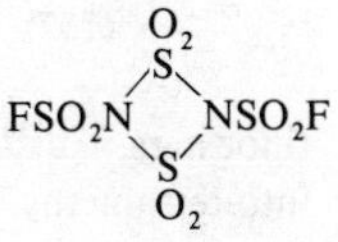

(151)

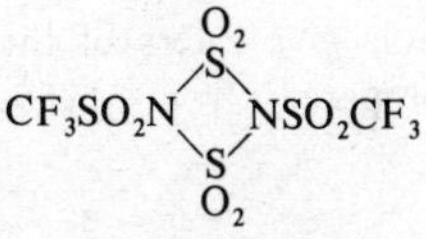

(152)

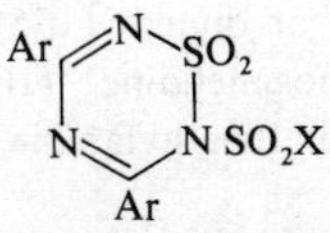

(153) X = F or CF$_3$

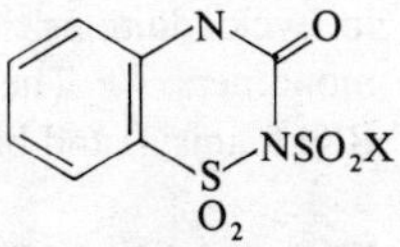

(154)

1,3,2,4-Dithiadiphosphetans.—2,4-Disubstituted 1,3,2,4-dithiadiphosphetan 2,4-disulphides (155) are the anhydrides of dithiophosphonic acids. They have been prepared by the reaction of (156) with H_2S.[237–239] The di-t-butyl derivative (155; R = Bu^t) has been prepared from tetra-t-butylcyclotetraphosphine by heating with excess sulphur,[240] and (155; R = Ph) from bis(trifluoromethylthio)phenylphosphine and sulphur.[241] These anhydrides, particularly (155; R = p-anisyl), have been exploited for the preparation of thiocarbonyl groups from the corresponding carbonyls (replacing the more obnoxious P_2S_5) e.g., thioketones,[196]

[233] H. W. Roesky and M. Aramaki, *Angew. Chem., Int. Ed. Engl.*, 1978, **17**, 129.

[234] H. W. Roesky, M. Aramaki, and L. Schonfelder, *Z. Naturforsch., Teil B*, 1978, **33**, 1072.

[235] H. Kagami and S. Motoki, *Bull. Chem. Soc. Jpn.*, 1979, **52**, 3463.

[236] Y. Inagaki, T. Hosogai, R. Okazaki, and N. Inamoto, *Bull. Chem. Soc. Jpn.*, 1980, **53**, 205.

[237] P. S. Khokhlov, L. I. Markova, G. D. Sokolova, B. Ya. Chvertkin, S. G. Zhemchuzhin, and A. I. Ermakov, *Zh. Obshch. Khim.*, 1978, **48**, 564.

[238] K. D. Dzhundubaev, A. S. Sulaimanov, and B. Batyrkanova, *Zh. Obshch. Khim.*, 1978, **48**, 2037.

[239] O. N. Grishina, N. A. Andreev, and E. I. Babkina, *Zh. Obshch. Khim.*, 1978, **48**, 1321.

[240] M. Baudler, C. Gruner, G. F. Furstenberg, B. Kloth, F. Saykowski, and U. Oezer, *Z. Anorg. Allg. Chem.*, 1978, **446**, 169.

[241] T. Vakratsas, *Chem. Chron.*, 1978, **7**, 125 (*Chem. Abstr.*, 1979, **90**, 137 911).

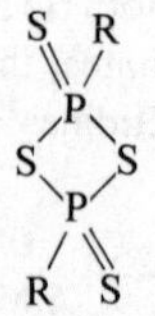

(155)

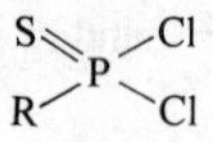

(156)

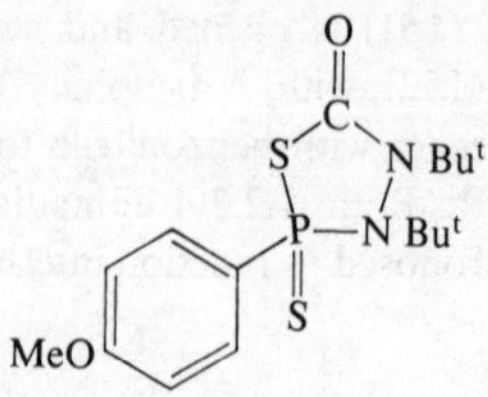

(159)

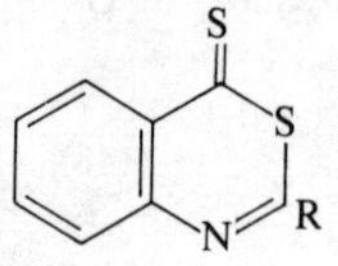

(157)

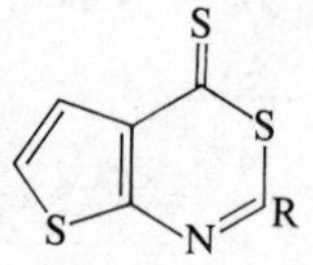

(158)

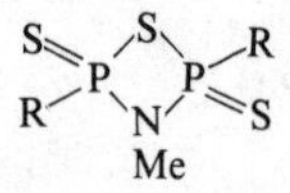

(160) R = *p*-anisyl

thioesters,[242-244] and thioamides.[245-249] Tetramethylene sulphoxide is converted into tetramethylene disulphide.[250] Heterocycles (157) and (158) are prepared by the reaction of the corresponding *o*-acylamino-ethoxycarbonyl derivatives with (155; R = *p*-anisyl).[251,252] The latter reagent also reacts with 1,2-di-t-butyldiaziridinone to give (159).[253] Grignard reagents attack the dimers at phosphorus to give derivatives of dithiophosphinic acid.[254,255] Dimers (155) also serve as donors of the monomer in cycloadditions.[255] Reactions of dimers (155) with alcohols give esters of the monomers.[239] The thiazadiphosphetidine (160) was prepared from (155; R = *p*-anisyl) and bis(trimethylsilyl)methylamine.[256]

Four-membered Rings containing Sulphur, Nitrogen, and Phosphorus.—The compound (160) results from the reaction of (155; R = *p*-anisyl) and $(Me_3Si)_2NMe$.[256] Heterocycle (161) is obtained from isocyanates and (162).[257] Thionyl chloride reacts with (163) to give (164).[258]

[242] B. S. Pedersen, S. Scheibye, K. Clausen, and S.-O. Lawesson, *Bull. Soc. Chim. Belg.*, 1978, **87**, 293.
[243] R. Shabana, S. Scheibye, K. Clausen, S. O. Olesen, and S.-O. Lawesson, *Nouv. J. Chim.*, 1980, **4**, 47.
[244] S. Scheibye, J. Kristensen, and S.-O. Lawesson, *Tetrahedron*, 1979, **35**, 1339.
[245] S. Scheibye, B. S. Pedersen, and S.-O. Lawesson, *Bull. Soc. Chim. Belg.*, 1978, **87**, 229.
[246] S. Scheibye, B. S. Pedersen, and S.-O. Lawesson, *Bull. Soc Chim. Belg.*, 1978, **87**, 299.
[247] H. Fritz, P. Hug, S.-O. Lawesson, E. Logemann, B. S. Pedersen, H. Sauter, S. Scheibye, and T. Winkler, *Bull. Soc. Chim. Belg.*, 1978, **87**, 525.
[248] H. J. Meyer, C. Nolde, I. Thomsen and S.-O. Lawesson, *Bull. Soc. Chim. Belg.*, 1978, **87**, 621.
[249] W. Walter and T. Proll, *Synthesis*, 1979, 941.
[250] J. B. Rasmussen, K. A. Jorgensen, and S.-O. Lawesson, *Bull. Soc. Chim. Belg.*, 1978, **87**, 307.
[251] K. Clausen and S.-O. Lawesson, *Bull. Soc. Chim. Belg.*, 1979, **88**, 305.
[252] K. Clausen and S.-O. Lawesson, *Nouv. J. Chim.*, 1980, **4**, 43.
[253] G. L'abbé, J. Flemal, J.-P. Declercq, G. Germain, and M. Van Meerssche, *Bull. Soc. Chim. Belg.*, 1979, **88**, 737.
[254] (a) W. Kuchen, R. Uppenkamp, and K. Diemert, *Z. Naturforsch., Teil B*, 1979, **34**, 1398; (b) K. Diemert, P. Haas, and W. Kuchen, *Chem. Ber.*, 1978, **111**, 629.
[255] (a) J. Hogel and A. Schmidpeter, *Z. Naturforsch., Teil B*, 1979, **34**, 915; (b) W. Zeiss and A. Schmidpeter, *ibid.*, p. 1042.
[256] W. Zeiss, H. Henjes, D. Lux, W. Schwarz, and H. Hess, *Z. Naturforsch., Teil B*, 1979, **34**, 1334.
[257] A. Schmidpeter and T. von Criegern, *Angew. Chem., Int. Ed. Engl.*, 1978, **17**, 443.
[258] O. J. Scherer, N.-T. Kulbach, and W. Glassel, *Z. Naturforsch., Teil B*, 1978, **33**, 652.

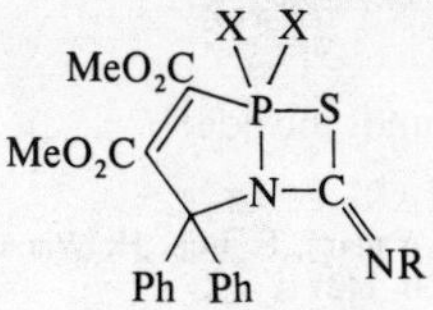

(161) R = Ph or COPh
X = Me or Ph

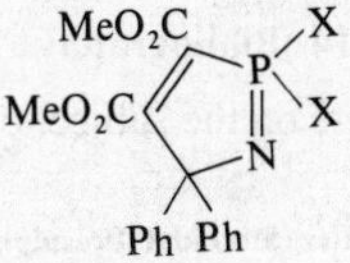

(162) X = Me or Ph

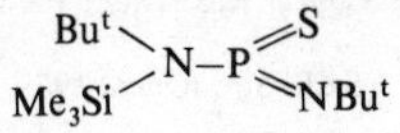

(163)

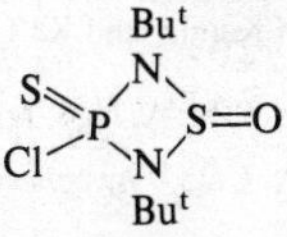

(164)

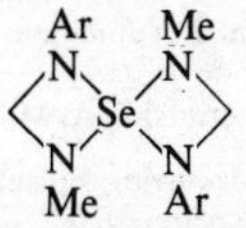

(165) Ar = *m*-CF$_3$C$_6$H$_4$

Selenium Derivatives.—The spiro-compound (165) is produced in 47% yield in the reaction of *m*-CF$_3$C$_6$H$_4$N(SiMe$_3$)CON(Me)SiMe$_3$ with SeOCl$_2$.[259]

12 Four-membered Rings containing Sulphur and Silicon or Germanium

3,3-Diethyl-2,4-dimethyl-3-silathietan is prepared from divinyl sulphide by its reaction with diethylsilane.[260] Tetramethyl-2,4-disila-1,3-dithietan disproportionates with octamethyl-1,4-dithia-2,3,5,6-tetrasilacyclohexane to give hexamethyl-1,3-dithia-2,4,5-trisilacyclopentane.[261]

2,2-Diethyl-2-germathietan has been proposed as an intermediate in the reaction of thiiran with diethylgermylene.[78]

13 Thiadiazaboretidines

Cyclization of bis(dimethylamino)phenylborane with sulphuryl isocyanates gives (166).[262]

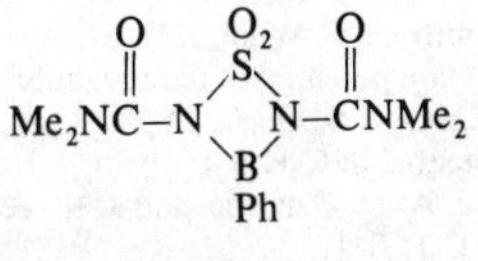

(166)

[259] H. W. Roesky and K. Ambrosius, *Z. Naturforsch., Teil B*, 1978, **33**, 759.
[260] M. G. Voronkov, S. V. Kirpichenko, T. J. Barton, V. V. Keiko, V. A. Pestunovich, and B. A. Trofimov, *Tezisy Dokl. Nauchn. Sess. Khim. Tekhnol. Org. Soedin. Sery Sernistykh Neftei, 14th*, 1975, 147 (*Chem. Abstr.*, 1978, **89**, 43 569).
[261] H. Noth, H. Fußstetter, H. Prommerening, and T. Taeger, *Chem. Ber.*, 1980, **113**, 342.
[262] H. W. Roesky and S. K. Mehrotra, *Angew. Chem., Int. Ed. Engl.*, 1978, **17**, 599.

14 Bibliography

Further papers on the subject of small-ring compounds include:

Biologically Active Steroidal Episulphides.—Y. Muraoka, I. Yahara, F. Itoh, H. Watanabe, and N. Nara, *Oyo Yakuri*, 1978, **16**, 739 (*Chem. Abstr.*, 1979, **90**, 146 142)

F. Kobayashi, H. Furukawa, M. Ogawa, Y. Muraoka, I. Yahara, and Y. Hayashi, *Oyo Yakuri*, 1978, **16**, 779 (*Chem. Abstr.*, 1979, **90**, 146 153)

K. Takeda, *Symp. Pap., IUPAC Int. Symp. Chem. Nat. Prod., 11th*, 1978, **4** (Part 1) 464 (*Chem. Abstr.*, 1980, **92**, R34 512)

T. Hori, T. Miyake, K. Takeda, and J. Kato, *Prog. Cancer Res. Ther.*, 1978, **10**, 159

Y. Nomura, J. Yamagata, H. Kondo, K. Kanda, and K. Takenaka, *Prog. Cancer Res. Ther.*, 1978, **10**, 15

A. V. Kamernitskii, A. M. Ustynyuk, and T. M. A. Ngo, *Izv. Akad. Nauk SSSR, Ser. Khim.*, 1979, 180

Naturally occurring Episulphides.—R. A. Cole, *Phytochemistry*, 1978, **17**, 1563 [in autolysis of turnip, *Brassica campestris*]

L. N. Nixon, E. Wong, C. B. Johnson, and E. J. Birch, *J. Agric. Food Chem.*, 1979, **27**, 355 [aroma of cooking mutton]

M. Sakaguchi and T. Shibamoto, *Agric. Biol. Chem.*, 1979, **43**, 667, and *J. Agric. Food Chem.*, 1978, **26**, 1179 [in model system for browning of food]

Spectroscopy.—W. J. Broer and W. D. Weringa, *Org. Mass Spectrom.*, 1978, **13**, 232 [$C_2H_5S^+$ ions]

J. L. Holmes, A. S. Blair, G. M. Weese, A. D. Osborne, and J. K. Terlouw, *Adv. Mass Spectrom., Sect. B*, 1978, **7**, 1227 [$C_2H_4S^+$ ions]

G. Baykut, K.-P. Wanczek, and H. Hartmann, *Adv. Mass Spectrom., Sect. A*, 1980, **8**, 186 [thiiran spectrum and ion chemistry]

D. W. Berman, V. Anicich, and J. L. Beauchamp, *J. Am. Chem. Soc.*, 1979, **101**, 1239 [stability of $C_2H_5S^+$ by ion cyclotron resonance]

H. J. Mockel, *Fresenius' Z. Anal. Chem.*, 1979, **295**, 241 [CI mass spectrum of methylthiiran]

C. A. Kingsbury, D. L. Durham, and R. Hutton, *J. Org. Chem.*, 1978, **43**, 4696 [$^3J_{CH}$ in thiirans]

R. Gleiter and J. Spanget-Larsen, *Top. Current Chem.*, 1979, **86**, 139 [Photoelectron spectrum of thiiran]

Thiirenium Ions.—R. Destro, T. Pilati, and M. Simonetta, *Nouv. J. Chim.*, 1979, **3**, 533 [X-ray crystal structure shows pyramidal sulphur]

P. Kollman, S. Nelson, and S. Rothenberg, *J. Phys. Chem.*, 1978, **82**, 1403 [theory]

I. G. Csizmadia, V. Lucchini, and G. Modena, *Gazz. Chim. Ital.*, 1978, **108**, 543 [theory]

Biologically Active Thietans and Dithietans.—D. R. Crump, *Tetrahedron Lett.*, 1978, 5233 [smell of stoat]

C. Brinck, R. Gerell, and G. Odham, *Oikos*, 1978, **30**, 68 [pheromone of mink]

M. Sakaguchi and T. Shibamoto, *Agric. Biol. Chem.*, 1979, **43**, 667 [$C_2H_4S_2$ in model system for browning of food]

R. W. McLeod and G. T. Khair, *Ann. Appl. Biol.*, 1978, **88**, 81 [nematocidal activity]

K. G. Choudhari and S. Z. Koli, *Pesticides*, 1978, **12**, 29 [nematocidal activity]

Physical Properties of Thietans and Dithietans.—J. Barrett and F. S. Deghaidy, *Spectrochim. Acta, Part A*, 1979, **35**, 509 [electronic spectra]

P. Klaboe and Z. Smith, *Spectrochim. Acta, Part A*, 1978, **34**, 489 [vibrational spectra]

V. F. Kalasinsky, R. Block, D. E. Powers, and W. C. Harris, *Appl. Spectrosc.*, 1979, **33**, 361 [matrix vibrational spectrum]

M. L. Kaplan, R. C. Haddon, F. B. Bramwell, F. Wudl, J. H. Marshall, D. O. Cowan, and S. Gronowitz, *J. Phys. Chem.*, 1980, **84**, 427 [reduction potential of tetracyanobis(methylene)]

G. S. Wilson, D. D. Swanson, J. T. Klug, R. S. Glass, M. D. Ryan, and W. K. Musker, *J. Am. Chem. Soc.*, 1979, **101**, 1040 [oxidation potential of $C_2H_4S_2$]

J. L. Holmes, A. S. Blair, G. M. Weese, A. D. Osborne, and J. K. Terlouw, *Adv. Mass Spectrom., Sect. B*, 1978, **7**, 1227 [mass spectrum of $C_2H_4S^+$]

H. J. Moeckel, *Fresenius' Z. Anal. Chem.*, 1979, **295**, 241 [CI mass spectrometry of thietan]

A. A. Scala and I. Colon, *J. Heterocycl. Chem.*, 1978, **15**, 421 [mass spectra of thietan and both oxides]

R. Gleiter and J. Spanget-Larsen, *Top. Current Chem.*, 1979, **86**, 139 [photoelectron spectra of thietan and a dithietan]

5
Saturated Cyclic Compounds of Sulphur, Selenium, and Tellurium

BY P. K. CLAUS

1 Introduction

The organization of this chapter follows the lines of previous volumes, with the exception that a subsection on sulphur-containing macrocycles has been omitted.

A number of reviews on several aspects of sulphur chemistry and comprehensive books on general organic chemistry include material pertinent to this Report. Among these are various chapters of recent volumes of a well-known reference book[1] and of a new comprehensive book on Organic Chemistry.[2] A review on ^{13}C n.m.r. data of non-aromatic heterocyclic compounds includes numerous data on saturated sulphur heterocycles.[3] The synthesis of saturated heterocyclic rings has been summarized.[4] Conformations of five-membered rings have been reviewed in detail.[5] New tellurium heterocyclic chemistry has been summarized.[1,6]

2 Thiolans, Thians, Thiepans, Thiocans, and their Oxides, Dioxides, and Imides

Synthesis.—A new general synthesis of α-substituted three- to nine-membered cyclic sulphides (2) has been developed:[7] cyclization, with NaOH, of the pyridinium salts (1), prepared by the reaction of α,ω-dihalogenoalkanes with N-methyl-($1H$)-pyridine-2-thione, afforded the cyclic sulphides (2).

Several 2- and 3-substituted thiolans have been prepared by the bromination of

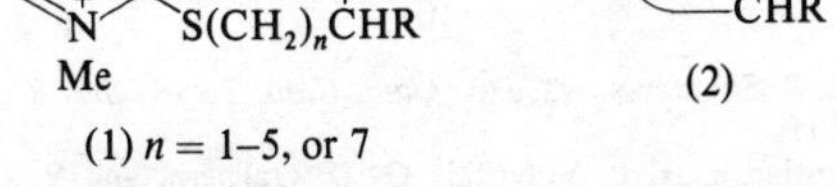

(1) $n = 1-5$, or 7

(2)

[1] 'Rodd's Chemistry of Carbon Compounds', Elsevier, Amsterdam, Vol. IV, Part E (1977), Chap. 21; Part H (1978), Chap. 40; and Part K (1979), Chap. 50, 52, 55, and 56.
[2] 'Comprehensive Organic Chemistry', Vol. 3, ed. D. N. Jones, and Vol. 4, ed. P. G. Sammes, Pergamon Press, London, 1979.
[3] E. L. Eliel and K. M. Pietrusiewicz, *Top. Carbon-13 NMR Spectrosc.*, 1979, **3**, 171.
[4] N. F. Elmore, in 'General and Synthetic Methods' (Specialist Periodical Reports), ed. G. Pattenden, The Chemical Society, London, 1978, Vol. 1, p. 197.
[5] B. Fuchs, *Top. Stereochem.*, 1978, **10**, 1.
[6] K. J. Irgolic, *J. Organomet. Chem.*, 1978, **158**, 235, 267.
[7] K. Sotoya, M. Yamada, T. Takamoto, T. Sakakibara, and R. Sudoh, *Synthesis*, 1977, 884.

δ-methoxy-alkanols and cyclization with Na_2S.[8] Bicyclic S-heterocycles, *e.g.* (3), have been obtained by intramolecular electrophilic addition of thiol groups to double bonds.[9] The thiopyrano-indole (4) was obtained by oxidative cyclization of 3-(indol-3-yl)propanethiol.[10] Ultraviolet irradiation of ethylene sulphide and alkenyl-magnesium bromides gave 4-substituted thians, *e.g.* (5).[11]

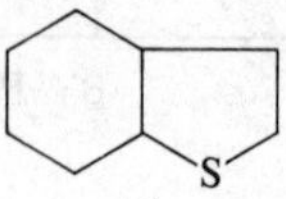

(3) 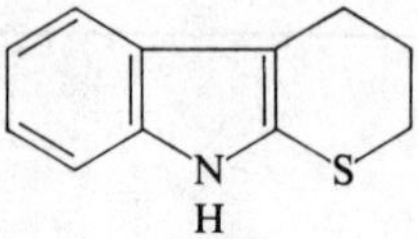　　　　(4) 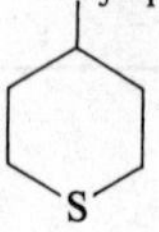　　　　(5)

Ring-opening of episulphides was also involved in the synthesis of substituted thiolans, *e.g.* (6).[12] Methylsulphenylation of but-3-enyl methyl sulphide afforded an intermediate thiiranium salt (7) that suffered intramolecular attack by the methylthio-group to give a diastereoisomeric mixture of the sulphonium salts (8).[13]

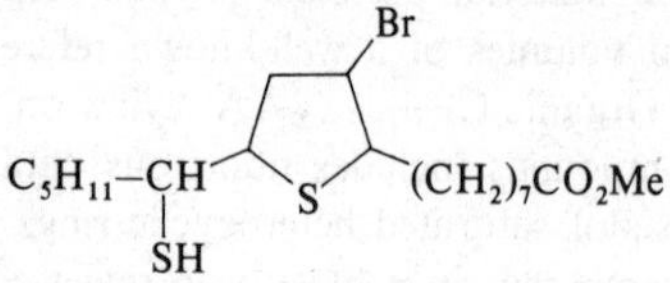

(6) 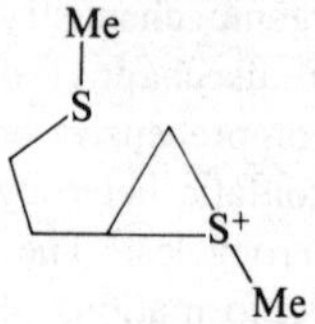　　　　(7) 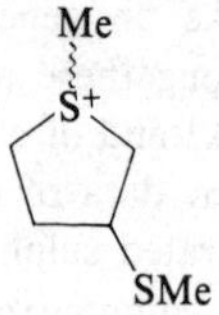　　　　(8)

Photolysis of allyl pent-4-enyl sulphides such as (9) gave mixtures of five- and six-membered sulphur heterocycles *via* unsaturated thiyl radicals.[14] Oxidative decarboxylation of the disulphide (10), and intramolecular trapping of the intermediate carbon radical to give the thiolan (11), indicated a biochemical pathway for the formation of the C(2)—S bond of penicillins.[15]

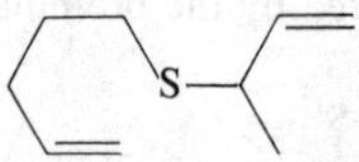

(9) 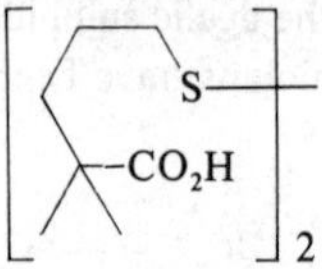　　　　(10) 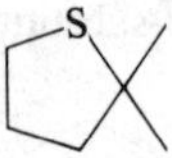　　　　(11)

[8] N. P. Volynskii and L. P. Shcherbakova, *Izv. Akad. Nauk SSSR, Ser. Khim.*, 1979, 1080 (*Chem. Abstr.*, 1979, **91**, 175 105).
[9] A. B. Urin, G. N. Gordadze, N. P. Volynskii, G. D. Gal'pern, and V. G. Zaikin, *Tezisy Dokl. Nauchn. Sess. Khim. Tekhnol. Org. Soedin. Sery Sernistykh Neftei 14th*, 1975, (publ. 1976), 56 (*Chem. Abstr.* 1978, **88**, 190 543).
[10] T. Hino, H. Miura, R. Murata, and M. Nakagawa, *Chem. Pharm. Bull.*, 1978, **26**, 3695.
[11] V. P. Krivonogov, V. I. Dronov, and R. F. Nigmatullina, *Khim. Geterotsikl. Soedin.*, 1977, 1622 (*Chem. Abstr.* 1978, **88**, 152 358).
[12] I. L. Kuranova and E. V. Snetkova, *Zh. Org. Khim.*, 1978, **14**, 2165 (*Chem. Abstr.*, 1979, **90**, 121 320).
[13] M. L. Kline, N. Beutow, J. K. Khim, and M. C. Caserio, *J. Org. Chem.*, 1979, **44**, 1904.
[14] G. Bastien and J. M. Surzur, *Bull. Soc. Chim. Fr.*, Part 2, 1979, 601.
[15] J. E. Baldwin and T. S. Wan, *J. Chem. Soc., Chem. Commun.*, 1979, 249.

Intramolecular Claisen condensation of (12), followed by cyclization, afforded the fused Δ^3-dihydrothiopyran-5-one (13).[16] A new ring-closure reaction was found by heating *N*-acetyl-*S*-(α-keto-alkyl)cysteines with acetic anhydride to give 2,5-dihydrothiophens (14).[17] 3-Methoxycarbonyl-2,5-dihydrothiophen (15), an excellent precursor to 2-methoxycarbonyl-1,3-butadiene, was synthesized by modifications of previously reported procedures.[18] Carbon-13 n.m.r. data for a series of 2,5-dihydrothiophens have been reported.[19]

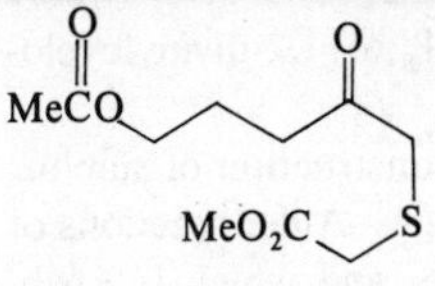

(12) 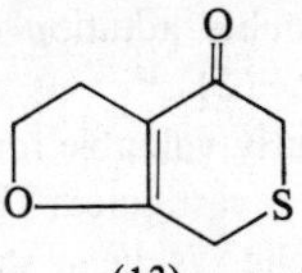   (13) 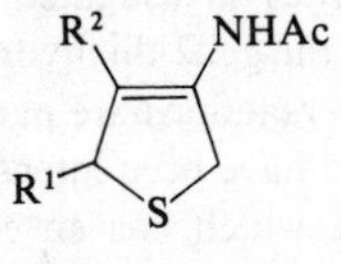   (14) 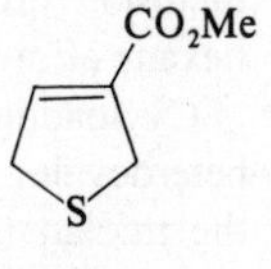   (15)

The studies on cyclization of 2-(methylsulphonylalkyl)cyclohexanones have been continued,[20] and 2-thiahydrindane 2,2-dioxides, *e.g.* (16), and 2-thiadecalin 2,2-dioxides, *e.g.* (17), have been isolated as the main products of such cyclizations. The synthesis of the 2-thiadecalin (18), with an angular hydroxy-group in position 10, has been reported.[21]

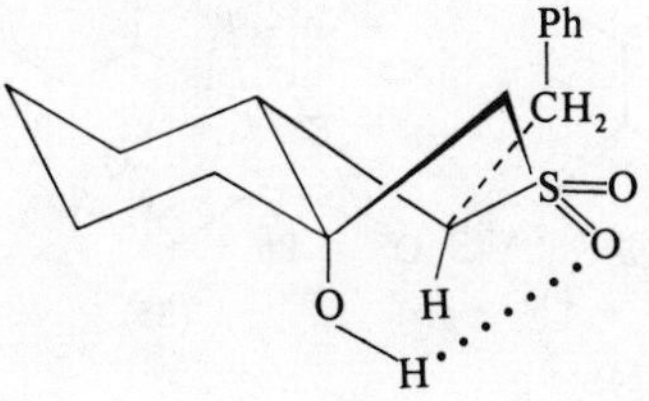

(16) 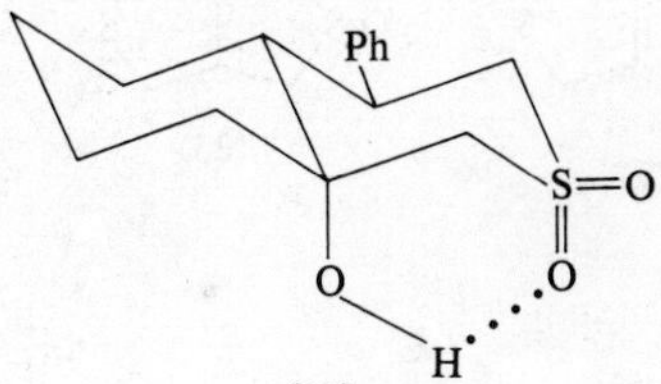   (17)

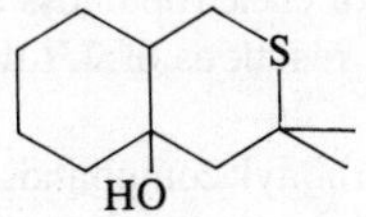

(18)

Nucleophilic addition of activated methylene compounds to distyryl sulphides, sulphoxides, or sulphones stereoselectively gave the thian derivatives (19).[22] Analogues of the known host compound (20) have been synthesized, and their

[16] S. Bernasconi, C. Capellini, A. Corbella, M. Ferrari, P. Gariboldi, G. Jommi, and M. Sisti, *Gazz. Chim. Ital.*, 1979, **109**, 5.
[17] G. F. Field, *J. Org. Chem.* 1979, **44**, 825.
[18] J. M. McIntosh and R. A. Sieler, *J. Org. Chem.*, 1978, **43**, 4431.
[19] J. M. McIntosh, *Can. J. Chem.*, 1979, **57**, 131.
[20] S. Fabrissin, S. Fatutta, and A. Risaliti, *J. Chem. Soc., Perkin Trans. 1*, 1978, 1321.
[21] G. V. Pavel and M. N. Tilichenko, *Zh. Org. Khim.*, 1978, **14**, 2369 (*Chem. Abstr.*, 1979, **90**, 137 626).
[22] O. H. Hartwig and H. Yamamura, *Liebigs Ann. Chem.*, 1977, 1500; H. Yamamura and H. H. Otto, *Arch. Pharm. (Weinheim, Ger.)*, 1978, **311**, 762.

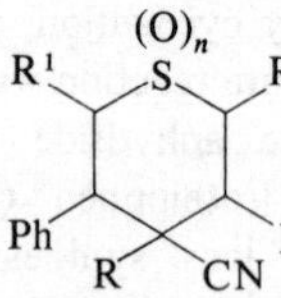

(19) $n = 0$, 1, or 2
R = CN, CO_2R^2, or $CONH_2$

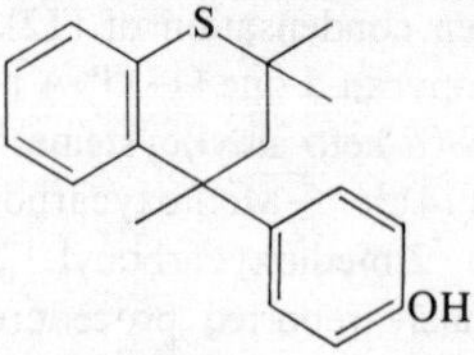

(20)

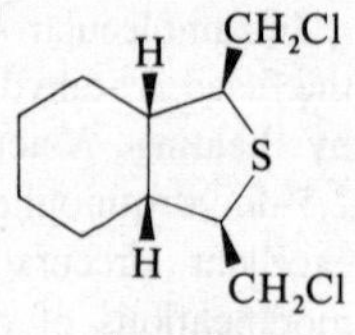

(21)

inclusion properties investigated.[23] Double addition of SCl_2 to 1,2-divinylcyclo-hexane afforded the *cis*-2-thiahydrindane (21).[24]

Cycloaddition reactions are particularly valuable for the construction of sulphur heterocycles, and have been intensively investigated. Thus, Diels–Alder reactions of the thiolan (22), which has an exocyclic *cis*-diene structure, and which is easily prepared from 2,3-bis(bromomethyl)buta-1,3-diene, have been studied.[25] Addition of SO_2 gave the cyclo-adduct (23). A series of dithiabicycloheptan-3-ones have been prepared by [3 + 2]cycloadditions of the cyclic thiocarbonyl ylide (24) to a variety of dipolarophiles; *e.g.*, to formaldehyde to give (25).[26]

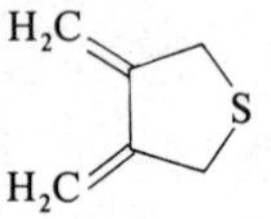

(22)

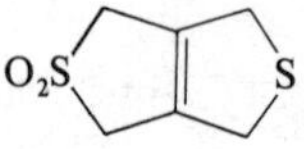

(23)

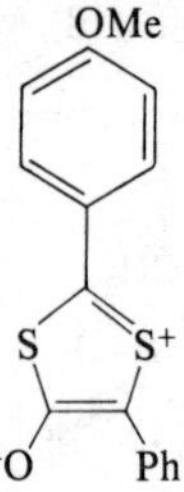

(24)

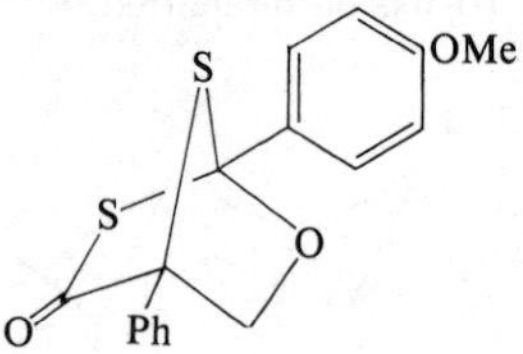

(25)

The double Diels–Alder reaction of tetrachlorothiophen 1,1-dioxide and allyl vinyl sulphide proceeded with cheletropic loss of SO_2 to give the hetero-isotwistene (26).[27] Various Diels–Alder reactions of 2*H*-thiopyran yielded *endo*-adducts, such as (27).[28]

Cycloadditions of thiocarbonyl compounds to dienes, or of vinyl thiocarbonyl derivatives to dienophiles, have received particular attention. Thus, Δ^3-dihydro-thiopyrans, *e.g.* (28), have been obtained by the addition of diphenyl thioketone to 1,3-dienes.[29] Oxidation of (28), followed by a vinylogous Pummerer reaction, gave the 2*H*-thiopyran (29).[29]

1,2-Dithiocarbonyl compounds such as dithio- or tetrathio-oxalates (30) served

[23] A. D. U. Hardy, J. J. McKendrick, and D. D. McNicol, *J. Chem. Soc., Perkin Trans. 2*, 1979, 1072.
[24] N. N. Novitskaya, N. K. Pokoneshchikova, R. G. Kantyukova, E. E. Zaev, L. M. Khalilov, and G. A. Tolstikov, *Tezisy Dokl. Nauchn. Sess. Khim. Tekhnol. Org. Soedin. Sery Sernistykh Neftei 14th*, 1975, (publ. 1976), 102 (*Chem. Abstr.*, 1978, **88**, 190 512).
[25] Y. Gaoni and S. Sadeh, *J. Org. Chem.*, 1980, **45**, 870.
[26] H. Gotthardt and C. M. Weisshuhn, *Chem. Ber.*, 1976, **111**, 3029, 3037, 3171, 3178; H. Gotthardt, C. M. Weisshuhn, and B. Christl, *Liebigs Ann. Chem.*, 1979, 360.
[27] M. S. Raasch, *J. Org. Chem.*, 1980, **45**, 856.
[28] R. H. Fleming and B. M. Murray, *J. Org. Chem.*, 1979, **44**, 2280.
[29] K. Praefcke and C. Weichsel, *Liebigs Ann. Chem.*, 1979, 784; *ibid.*, 1980, 333.

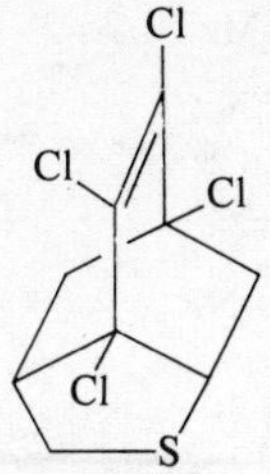

(26) 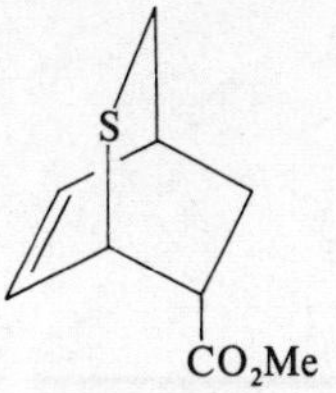   (27) 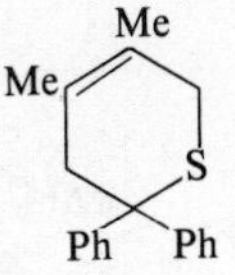   (28) 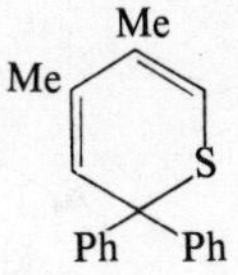   (29)

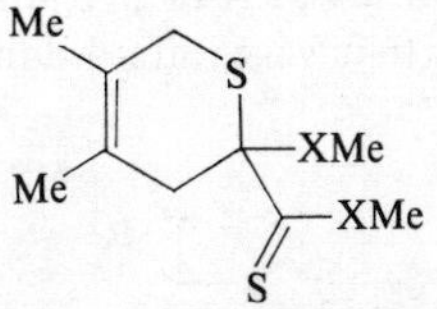

(31) 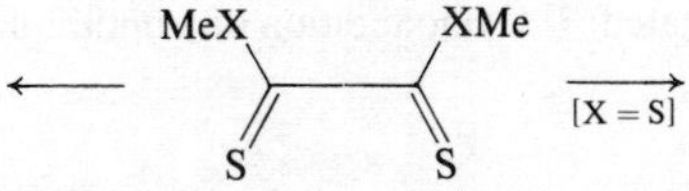   (30) X = O or S 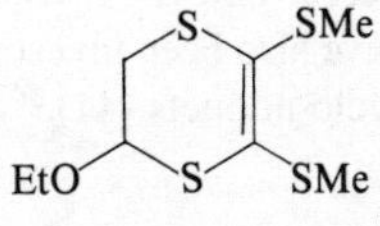   (32)

in cycloadditions both as 2π and 4π components, yielding either Δ^3-dihydro-thiopyrans (31) or dithiin derivatives, *e.g.* (32)[30] (see also Chap. 3, Pt III, p. 196).

2-Acyl-Δ^2-dihydrothiopyrans, *e.g.* (33), have been obtained by Diels–Alder reactions and Cope rearrangements of intermediate 2-thioacyl-Δ^3-dihydropyrans.[31] Certain 3-aryl-2-cyano-thioacrylamides, such as (34), dimerize in a $(4_s + 2_s)$ mode to give a Δ^2-dihydrothiopyran, *e.g.* (35).[32]

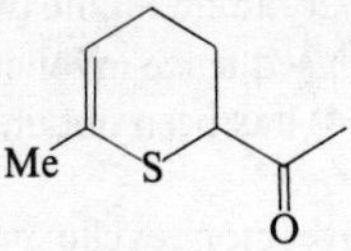

(33) 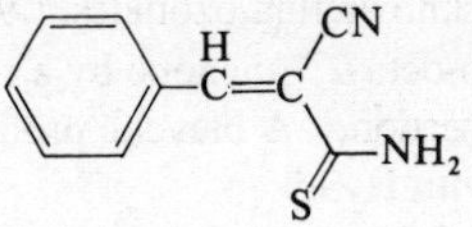   (34) 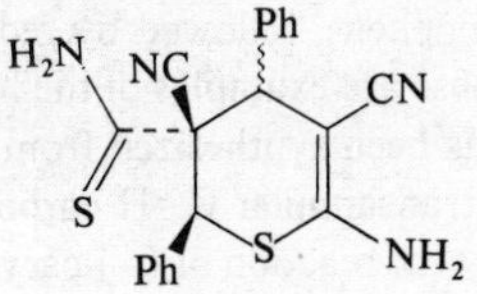   (35)

The $\alpha\beta$-unsaturated thioketone (36), generated by thermolysis of its dimer, reacts with a variety of dienophiles to give cyclo-adducts with the Δ^2-dihydrothiopyran structure, *e.g.* (37).[33] A similar cycloaddition has been used for the preparation of (38).[34]

The *o*-thioquinone methide (39), generated by the photoreaction of benzo-1,2-dithiole-3-thione with olefins, dimerizes to give a head-to-head dimer.[35a] Further

[30] K. Hartke, T. Kissel, J. Quante, and G. Henssen, *Angew. Chem.*, 1978, **90**, 1016.
[31] K. B. Lipkowitz and B. P. Mundy, *Tetrahedron Lett.*, 1977, 3417; K. B. Lipkowitz, S. Scarpone, B. P. Mundy, and W. G. Bornmann, *J. Org. Chem.*, 1979, **44**, 486.
[32] J. S. A. Brunskill, A. De, and D. F. Ewing, *J. Chem. Soc., Perkin Trans. 1*, 1978, 629.
[33] T. Karakasa and S. Motoki, *J. Org. Chem.*, 1978, **43**, 4147; *ibid.*, 1979, **44**, 4151; T. Karakasa, H. Yamaguchi, and S. Motoki, *ibid.*, 1980, **45**, 927.
[34] H. Abdel Reheem Ead, N. Abdel Latif Kassab, H. Koeppel, W. D. Bloedorn, and K. D. Schleinitz, *J. Prakt. Chem.*, 1980, **322**, 155.
[35] (*a*) R. Okazaki, F. Ishii, K. Sunagawa, and N. Inamoto, *Chem. Lett.*, 1978, 51; (*b*) R. Okazaki, K. T. Kang, K. Sunagawa, and N. Inamoto, *ibid.*, p. 55; R. Okazaki, K. Sunagawa, K. T. Kang, and N. Inamoto, *Bull. Chem. Soc. Jpn.*, 1979, **52**, 496.

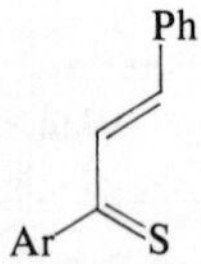

(36)

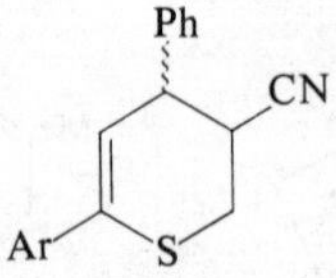

(37)

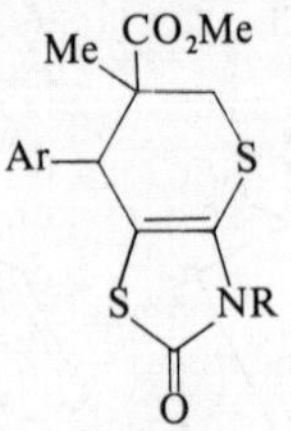

(38)

stereo- and regio-selective [4 + 2]-cycloaddition reactions of (39), *e.g.* to give (40), have also been investigated.[35b] Cycloaddition of tropothione to fulvenes afforded the cyclo-adducts (41).[36]

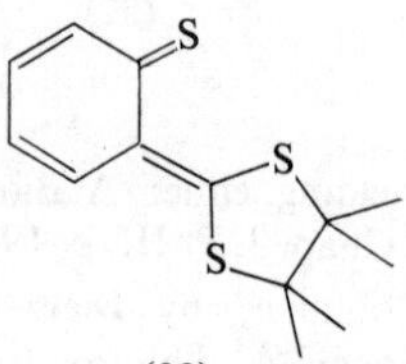

(39)

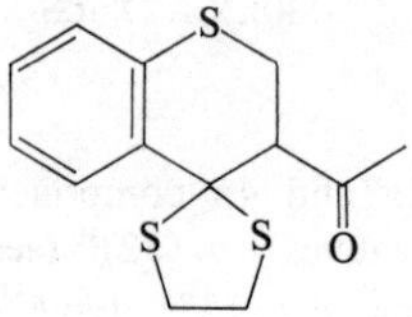

(40)

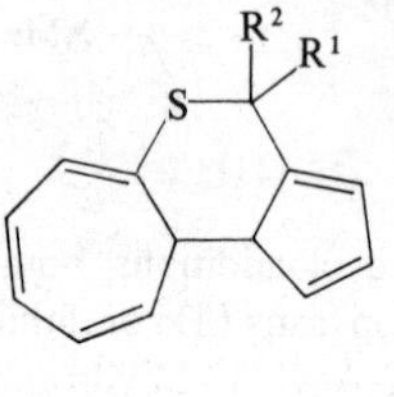

(41)

Various thiolans have been prepared by ionic hydrogenation of thiophens with Et_3SiH in the presence of acids.[37] Addition of singlet oxygen to alkyl-substituted thiophens, followed by reduction with di-imine, gave the bicyclic compounds (42); these are examples of the hardly known thia-ozonides.[38] 9-Thia-noradamantane (43) has been synthesized from cyclo-octa-2,7-dienone by a reaction sequence involving a transannular C—H carbene insertion.[39] A bicyclic product (44) has been obtained by the reaction of (−)-carvone with H_2S.[40]

Photolysis of 9-thiabicyclo[3.3.1]nonane-2,6-dione (45) gave monocyclic products, *e.g.* (46), by α-cleavage and C_α—S bond cleavage.[41] The 2,6-dithia-

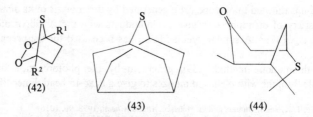

(42)

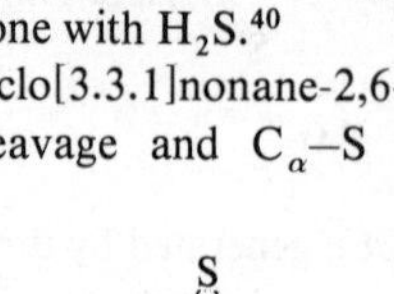

(43)

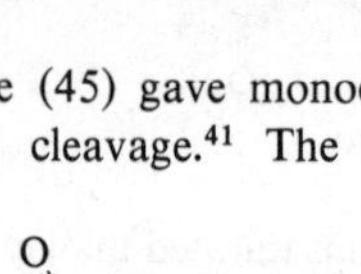

(44)

[36] T. Machiguchi, M. Watanabe, and M. Hoshino, *Koen Yoshishu-Hibenzekei Hokozoku Kagaku Toronkai [oyobi] Kozo Yuki Kagaku Toronkai, 12th*, 1979, 5 (*Chem. Abstr.*, 1980, **92**, 128 670).
[37] Z. N. Parnes, Yu. I. Lyakhovetskii, M. I. Kalinkin, D. N. Kursanov, and L. I. Belen'kii, *Tetrahedron*, 1978, **34**, 1703; A. V. Anisimov, V. F. Ionova, V. S. Babaitsev, V. K. Govorek, and E. A. Viktorova, *Khim. Geterotsikl. Soedin.*, 1979, 1062 (*Chem. Abstr.*, 1980, **92**, 41 801).
[38] W. Adam and H. J. Eggelte, *Angew. Chem.*, 1978, **90**, 811.
[39] T. Sasaki, S. Eguchi, and T. Hioki, *J. Org. Chem.*, 1978, **43**, 3808.
[40] M. Hargreaves, R. McDougall, and L. Rabari, *Z. Naturforsch., Teil B*, 1978, **33**, 1535.
[41] P. H. McCabe and C. R. Nelson, *Tetrahedron Lett.*, 1978, 2819.

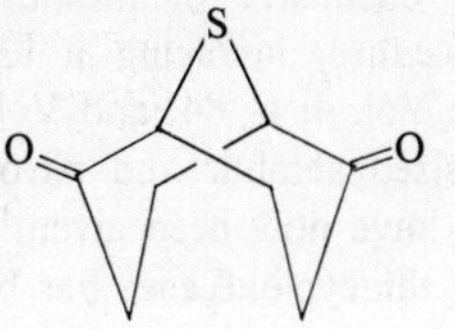

(45) 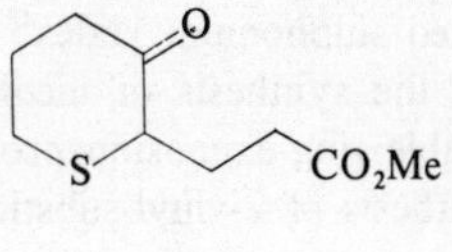   (46) 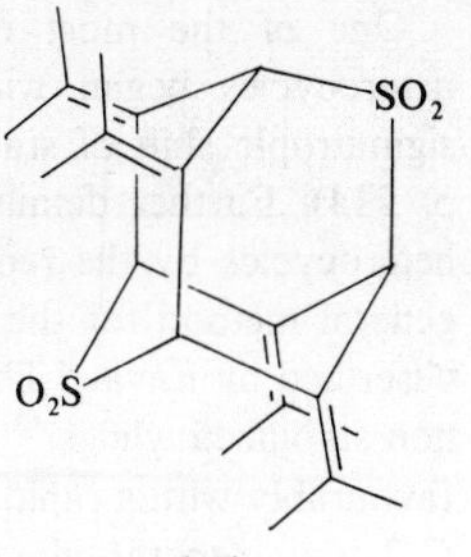   (47)

adamantane (47) has, surprisingly, been formed by the Bu^{n}Li-induced cyclo-dimerization of diallenyl sulphone.[42]

Tetrahydro-1- and -2-benzothiepins have been prepared by ring-expansion of thio- and isothio-chroman-4-ones, using ethyl diazoacetate.[43] Photocycloaddition of cyclohexene to 3-acetoxybenzo[b]thiophen 1,1-dioxide and base-catalysed hydrolytic ring-expansion of the intermediate tricyclic adduct gave (48).[44] Expansion of the ring was also observed in the acid-catalysed rearrangement of the $\alpha\beta$-epoxy-sulphone (49) to the thiepan-3-one (50). A similar ring-expansion of unsaturated sultones afforded β-keto-sultones.[45]

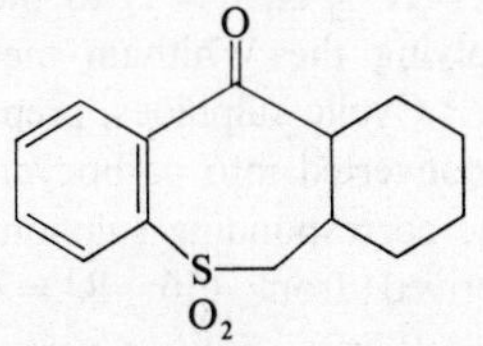

(48) 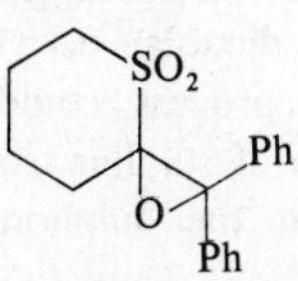   (49) 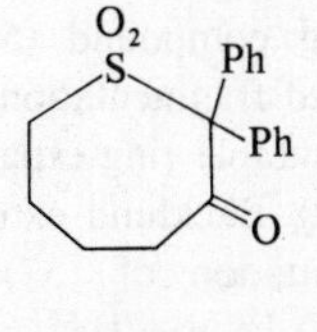   (50)

The bicyclic thia-octadiene (51) and its partially and fully saturated derivatives have been prepared by known methods.[46] The preparation of the dibenzodithiocins (52) and (53) has been reported.[47] The chiral *cis*-isomer (53) could be partially resolved.[47b]

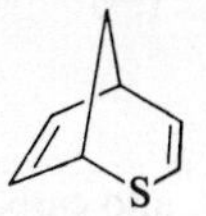

(51) 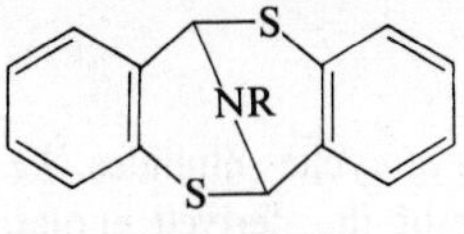   (52)

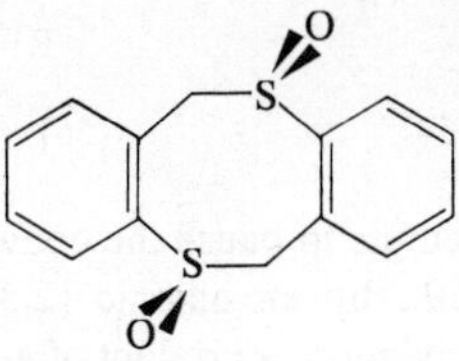

(53)

[42] S. Braverman, D. Reisman, and M. Sprecher, *Tetrahedron Lett.*, 1979, 901.
[43] R. Pellicciari and B. Natalini, *J. Chem. Soc., Perkin Trans. 1*, 1977, 1822.
[44] N. V. Kirby and S. T. Reid, *J. Chem. Soc., Chem. Commun.*, 1980, 150.
[45] T. Durst, K. C. Tin, F. Dereinachhirtzbach, J. M. Decesare, and M. D. Ryan, *Can. J. Chem.*, 1979, 57, 258.
[46] P. Barraclough, S. Bilgic, and D. W. Young, *Tetrahedron*, 1979, 35, 91; P. Barraclough, S. Bilgic, J. B. Pedley, A. J. Rogers, and D. W. Young, *ibid.*, p. 99.
[47] (a) J. Becher, C. Dreier, and O. Simonsen, *Tetrahedron* 1979, 35, 869; (b) M. Nogradi, *J. Chem. Soc., Perkin Trans. 1*, 1979, 1261.

One of the most intriguing developments in the chemistry of mesocyclic heterocycles began with Vedejs' ring-expansion procedure, involving a [2,3]-sigmatropic shift of stabilized sulphonium ylides[48] (see Vol. 4, p. 84, and Vol. 5, p. 214). Further details of the synthesis of medium-sized sulphur and nitrogen heterocycles by the repeatable ring-expansion process have now been given.[49] A general method for the synthesis of 2-vinyl-substituted thiacycloalkanes has been described by Fava.[50] The ring-expansion procedure has been extended to the use of non-stabilized ylides.[50] It can be shown that ring expansion does not compete favourably with a rapid isomerization by reversible deprotonation–reprotonation at C-2, *i.e.*, α to the vinyl group. Methylides (54; $R^1 = R^2 = H$) that are without hydrogen at C-2 ($R^3 = Me$) undergo highly stereoselective but non-stereospecific ring-enlargements: only the five-membered ylide with the carbanionic and the vinyl moieties on the same side of the ring reacted in a [2,3]-sigmatropic fashion, to give a 17:1 mixture of (*Z*)- and (*E*)-5-methylthiacyclo-oct-4-ene [(55; and (56; $R^1 = R^2 = H$, $R^3 = Me$, $n = 1$)], but both isomers of the six-membered ylide (54; $n = 2$) gave exclusively (*E*)-5-methylthiacyclonon-4-ene (56; $n = 2$).[51] The results have been rationalized in terms of conformational effects of the ground state and reactant-like transition states. The DBU-induced rearrangement of the ylide (54; $R^1 = H$, $R^2 = CO_2Et$, $R^3 = H$, $n = 2$) afforded the thiacyclononene (55; $R^1 = R^3 = H$, $R^2 = CO_2Et$, $n = 2$), in 97% isolated yield.[52]

The configurational inversion of (55; $R^1 = R^2 = R^3 = H$, $n = 1$) to the corresponding compound (56) was achieved by applying the Whitham method of concerted fragmentation of a dioxolan derivative.[53] Cyclic sulphides, prepared by the repeatable ring-expansion process, could be converted into carbocycles by a Ramberg–Bäcklund extrusion of sulphur from the corresponding sulphones; *e.g.*, the formation of (57) from the sulphone derived from (56; $R^1 = R^3 = H$, $R^2 = CO_2Et$, $n = 3$).[54]

(54)   (55)   (56)   (57)

Attempts to cause the conversion of cyclic sulphides, *e.g.* (58), into carbocycles, *e.g.* (59), by an anionic [2,3]-shift of the derived enolate anions failed.[55] After *S*-methylation, a product of a sigmatropic shift could be obtained only in the case of the six-membered ring (58), but Stevens 1,2-migrations and fragmentations to

[48] E. Vedejs and J. P. Hagen, *J. Am. Chem. Soc.*, 1975, **97**, 6878.
[49] E. Vedejs, M. J. Arco, D. W. Powell, J. M. Renga, and S. P. Singer, *J. Org. Chem.*, 1978, **43**, 4831.
[50] V. Ceré, C. Paolucci, S. Pollicino, E. Sandri, and A. Fava, *J. Org. Chem.*, 1978, **43**, 4826.
[51] V. Ceré, C. Paolucci, S. Pollicino, E. Sandri, and A. Fava, *J. Org. Chem.*, 1979, **44**, 4128.
[52] E. Vedejs, M. J. Arco, and J. M. Renga, *Tetrahedron Lett.*, 1978, 523.
[53] V. Ceré, A. Guenzi, S. Pollicino, E. Sandri, and A. Fava, *J. Org. Chem.*, 1980, **45**, 261
[54] E. Vedejs and S. P. Singer, *J. Org. Chem.*, 1978, **43**, 4884.
[55] E. Vedejs, M. J. Arnost, and J. P. Hagen, *J. Org. Chem.*, 1979, **44**, 3230.

alkanes also occurred, and these were the exclusive pathways in the case of the eight-membered derivatives. The *trans*-1-thiacyclo-oct-4-ene (60), which incorporates the C-1 to C-7 segment of a macrocyclic antibiotic with the natural stereochemistry at C-2, C-3, and C-6, has been obtained by a remarkably stereospecific cyclization of an intermediate sulphenic acid, followed by DBU-promoted [2,3]-sigmatropic ring-expansion, which transformed the stereochemistry at C-4 of the intermediate thiolanium ylide into the stereochemistry at C-6 of (60).[56]

The construction of saturated sulphur heterocycles was also involved in numerous contributions to the synthesis of natural products or modified natural products. Further syntheses of '*d*'- or '*dl*'-biotin have been reported. An intramolecular [3 + 2]-cycloaddition of the nitrile oxide (61) to give the tricyclic product (62) was the key step in a synthesis of racemic biotin, starting with cycloheptene.[57] A quite different approach to racemic biotin, involving only nine steps, started with a cycloaddition of chlorosulphonyl isocyanate to chromene.[58] In a synthesis of '*d*'-biotin, the monoaldehyde of adipic acid was transformed into the heptanoate (63), which was resolved.[59] A sequence of thirteen steps led from '*d*'-glucosamine to '*d*'-biotin.[60] Several biotin analogues have been prepared,[61] and the [13]C n.m.r. spectrum of biotin has been studied.[62]

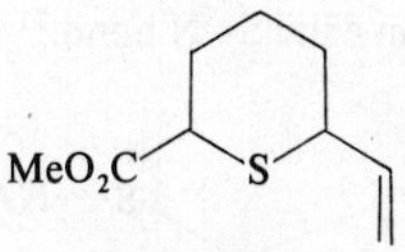

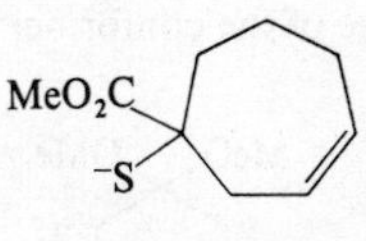

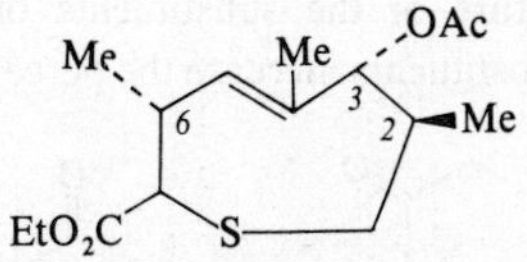

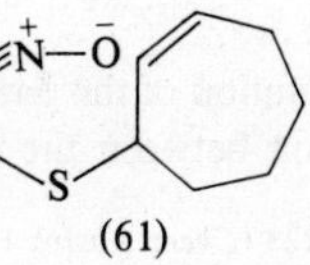

Various approaches to the synthesis of sulphur-containing prostaglandins and prostacyclins, and of thia-steroids with sulphur in ring A, B, C, or D have been

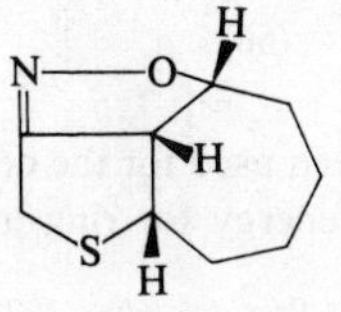

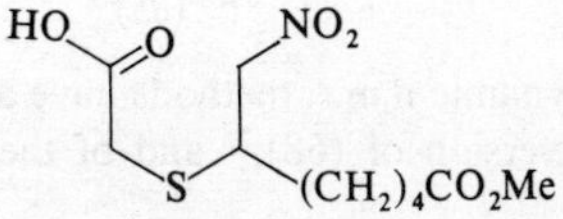

[56] E. Vedejs and M. J. Mullins, *J. Org. Chem.*, 1979, **44**, 2947.
[57] P. N. Confalone, E. D. Lollar, G. Pizzolato, and M. R. Uskokovic, *J. Am. Chem. Soc.*, 1978, **100**, 6291.
[58] A. Fliri and K. Hohenlohe-Oehringen, *Chem. Ber.*, 1980, **113**, 607.
[59] J. Vasilevskis, J. A. Gualteri, S. D. Hutchings, R. C. West, J. W. Scott, D. R. Parrish, F. T. Bizzarro, and G. F. Field, *J. Am. Chem. Soc.*, 1978, **100**, 7423.
[60] H. Ohrui, N. Sueda, and S. Emoto, *Agric. Biol. Chem.*, 1978, **42**, 865 (*Chem. Abstr.*, 1978, **89**, 59 857).
[61] S. D. Mikhno, T. M. Filippova, N. S. Kulachkina, I. G. Suchkova, and V. M. Berezovskii, *Zh. Org. Khim.*, 1978, **14**, 1706 (*Chem. Abstr.*, 1978, **89**, 197 247).
[62] H. J. Bradbury and R. N. Johnson, *J. Magn. Reson.*, 1979, **35**, 217; S. P. Singh, V. I. Stenberg, S. A. Franum, and S. S. Parmar, *Spectrosc. Lett.*, 1978, **11**, 259 (*Chem. Abstr.*, 1978, **89**, 41 785).

reported.[63,64] The synthesis and properties of 6-thiatetracycline have been described,[65] and the X-ray structure of 6-thiatetracycline has been determined.[66]

Properties.—Stereochemical aspects in the chemistry of saturated sulphur heterocycles are still emphasized; ^{1}H and, in particular, ^{13}C n.m.r. spectroscopy have been widely used for conformational and configurational assignments. Thus, n.m.r. studies indicated that thian-3-one, its 1-oxide (64), and its 1,1-dioxide have the chair conformation.[67,68] The conformational equilibria for (64) and for the 4-oxo-isomer (65; R = H, $n = 1$) and its dimethyl acetal were shown to be heavily shifted towards the axial sulphoxide conformer, compared with that for the thian 1-oxide itself.[67] Carbon-13 n.m.r. shift data indicated a transannular interaction between the sulphur and the carbonyl group of (64) and of its 1,1-dioxide.[68] The conformations of a series of 2,6-diphenylthian-4-ones (65; R = Ph, $n = 0$) and of a few thian-4-ols have been investigated by ^{13}C n.m.r. methods.[69] Low-temperature ^{13}C n.m.r. spectroscopy was used for a determination of the barrier to ring inversion of the ketal (66) (42.7 kJ mol^{-1})[70] and for a study of the conformational equilibria of a series of 1-imide derivatives of thian, and of some oxa- and di-thians.[71] In a series of thian 1-N-arylimides (67), which generally prefer the conformation with an equatorial S—N bond, the conformational preference was found to depend on the nature of the substituents on the aromatic ring; strongly electron-withdrawing substituents increase the percentage of the conformer with an axial S—N bond.[71]

(64)   (65)   (66)   (67)

Dynamic n.m.r. methods have also been used for the determination of the barrier to inversion of (68),[72] and of the free energy for ring inversion between the two

[63] J. H. Jones, J. B. Bicking, and E. J. Cragoe, Jr., *Prostaglandins*, 1979, **17**, 223 (*Chem. Abstr.*, 1979, **90**, 203 934); K. C. Nicolaou, R. L. Magolda, and W. E. Barnette, *J. Chem. Soc., Chem. Commun.*, 1978, 375; K. C. Nicolaou, S. P. Seitz, W. J. Sipio, and J. F. Blount, *J. Am. Chem. Soc.*, 1979, **101**, 3884; K. C. Nicolaou, W. E. Barnette, and R. L. Magolda, *ibid.*, 1978, **100**, 2567; K. Shimoji, Y. Arai, and M. Hayashi, *Chem. Lett.*, 1978, 1375.
[64] P. S. Jogdeo and G. V. Bhide, *Heterocycles*, 1977, **7**, 985; S. Kano, K. Tanaka, S. Hibino, and S. Shibuya, *J. Org. Chem.*, 1979, **44**, 1580, 1582; S. R. Ramadas and P. S. Srinivasan, *Steroids*, 1977, **30**, 213; N. Rao and L. D. Quin, *Phosphorus Sulfur*, 1979, **5**, 371; T. Terasawa and T. Okada, *Heterocycles*, 1978, **11**, 171; *ibid.*, 1980, **14**, 99; *J. Chem. Soc., Perkin Trans. 1*, 1978, 576; *ibid.*, 1979, 990.
[65] R. Kirchlechner and W. Rogalski, *Tetrahedron Lett.*, 1980, **21**, 247; K. Irmscher, *Symp. Pap. — IUPAC Intern. Symp. Chem. Nat. Prod. 11th*, 1978, **4** (Part 1), 494 (*Chem. Abstr.*, 1980, **92**, 94 142).
[66] R. Prevo and J. J. Stezowski, *Tetrahedron Lett.*, 1980, **21**, 251.
[67] W. A. Nachtergaele, D. Tavernier, and M. J. O. Anteunis, *Bull. Soc. Chim. Belg.*, 1980, **89**, 33.
[68] J. A. Hirsch and A. A. Jarmas, *J. Org. Chem.*, 1978, **43**, 4106.
[69] K. Ramalingam, K. D. Berlin, R. A. Loghry, D. van der Helm, and N. Satyamurthy, *J. Org. Chem.*, 1979, **44**, 471, 477.
[70] R. Borsdorf, E. Kleinpeter, S. Agurakis, and H. Jancke, *J. Prakt. Chem.*, 1978, **320**, 309.
[71] P. K. Claus, F. W. Vierhapper, and R. L. Willer, *J. Org. Chem.*, 1979, **44**, 2863.
[72] K. Ramarajan and K. D. Berlin, *Proc. Okla. Acad. Sci.*, 1979, **59**, 70 (*Chem. Abstr.* 1980, **92**, 163 468).

half-chair forms of (35), which are about equally populated at $-25\ °C$.[73] Considering contributions from torsional interactions of the substituents, a value of about 37 kJ mol^{-1} has been calculated for unsubstituted dihydrothiopyrans;[73] thus the introduction of a sulphur atom into a six-membered unsaturated ring increases the barrier to inversion by about 15 kJ mol^{-1}. The *cis*- and *trans*-isomers of 2,6-diphenyl-5,6-dihydro-2*H*-thiopyran (69) have been found to exist in half-chair rather than in boat conformations.[74] Two conformational processes, with free-energy barriers of 28.1 and 34.1 kJ mol^{-1}, respectively, and which were assigned to pseudorotation and to ring-inversion processes, were found for 1-thiacyclo-octan-5-one (70), which has been deduced to have an unsymmetrical boat-chair conformation.[75]

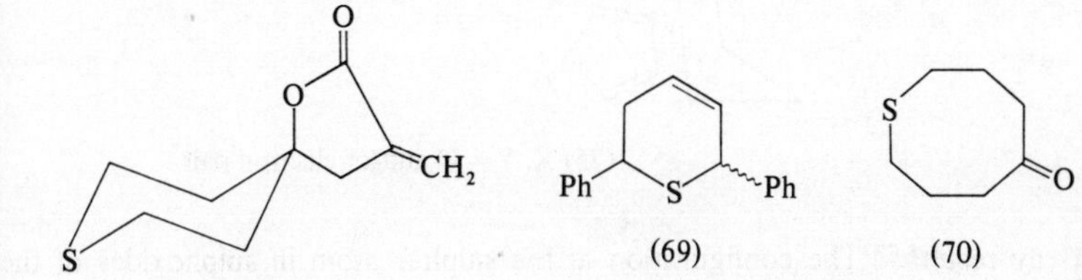

(68) (69) (70)

Carbon-13 n.m.r. spectra of *trans*-1-thiadecalin (71a), of 4-hetera-*trans*-1-thiadecalins (71b), and of the corresponding sulphoxides (α- or β-oriented) and sulphones have been reported.[76] The *trans*- and *cis*-stereoisomers of 1-thionia-bicyclo[4.4.0]decane salts (72) were prepared, and their stereostructures assigned by ^{13}C and ^{1}H n.m.r. spectroscopy and by *X*-ray analysis.[77] Activation parameters for the configurational inversion of (72) and of the derived ylides were determined. The *cis*-ylide has been found to be much more stable than the *trans*-form.[77] Values of substituent-induced ^{13}C n.m.r. chemical shifts in the 9-thiabicyclo[3.3.1]non-2-ene skeleton were evaluated by measuring the spectra of a number of derivatives (73).[78]

An elucidation of the structure, of the conformation, and of the configuration at sulphur of the bicyclic compounds (74) was based on ^{1}H and ^{13}C n.m.r. evidence

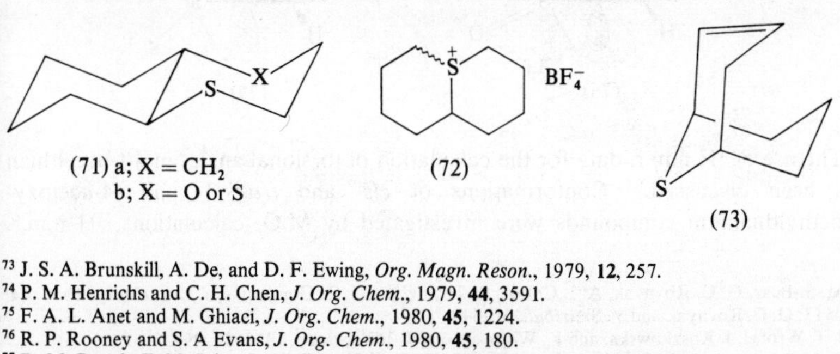

(71) a; X = CH$_2$ (72)
     b; X = O or S

(73)

[73] J. S. A. Brunskill, A. De, and D. F. Ewing, *Org. Magn. Reson.*, 1979, **12**, 257.
[74] P. M. Henrichs and C. H. Chen, *J. Org. Chem.*, 1979, **44**, 3591.
[75] F. A. L. Anet and M. Ghiaci, *J. Org. Chem.*, 1980, **45**, 1224.
[76] R. P. Rooney and S. A Evans, *J. Org. Chem.*, 1980, **45**, 180.
[77] D. M. Roush, E. M. Price, L. K. Templeton, D. H. Templeton, and C. H. Heathcock, *J. Am. Chem. Soc.*, 1979, **101**, 2971.
[78] J. W. de Haan, L. J. M. van de Ven, H. Vlems, M. M. E. Scheffers-Sap, H. Gillissen, and H. M. Buck, *Tetrahedron*, 1980, **36**, 799.

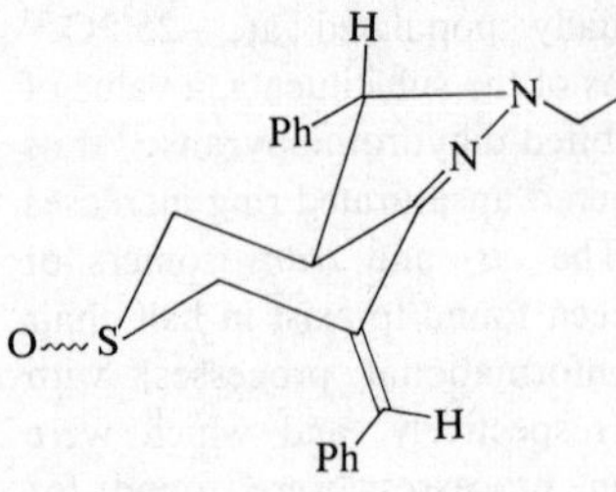

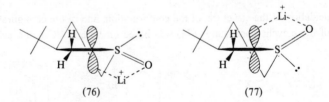

(74)

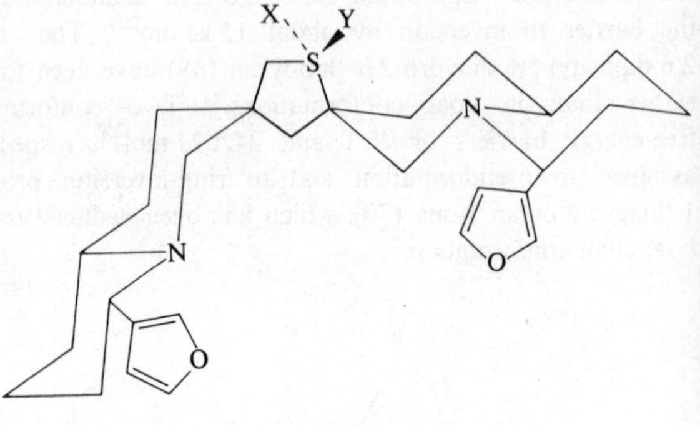

(75) X, Y = O and/or electron pair

and X-ray results.[79] The configuration at the sulphur atom in sulphoxides of the thiaspiran alkaloids (75), extracted from *Nuphar* species, was established by ASIS and LIS methods and by analysis of S—O bond-anisotropy-induced shifts.[80] LIS studies have also been proven to be useful in structural investigations of sulphonium salts.[81] There is strong evidence that a weak complexation occurs between the lanthanide ion and the anion of the sulphonium compound, and not with the electron pair on sulphur.

On the basis of the observed coupling constants, the lithio-derivatives of *cis*- and *trans*-4-t-butylthian 1-oxides (76) and (77) have been proposed to exist in half-chair conformations, with a planar metallated carbon atom and the Li^+ held in a stable co-ordinated structure.[82]

(76)   (77)

The use of ¹H n.m.r. data for the calculation of torsional angles in 4-t-butylthian has been discussed.[83] Conformations of *cis*- and *trans*-3- and -4-acetoxy-1-methylthianium compounds were investigated by M.O. calculations, ¹H n.m.r.

[79] M. S. Puar, G. C. Rovnyak, A. I. Cohen, B. Töplitz, and J. Z. Gougoutas, *J. Org. Chem.*, 1979, **44**, 2513; G. C. Rovnyak and V. Shu, *ibid.*, p. 2518.

[80] J. T. Wrobel, J. Ruszkowska, and K. Wojtasiewicz, *J. Mol. Struct.*, 1978, **50**, 299.

[81] R. L. Caret and A. N. Vennos, *J. Org. Chem.*, 1980, **45**, 361.

[82] G. Chassaing, R. Lett, and A. Marquet, *Tetrahedron Lett.*, 1978, 471; R. Lett and G. Chassaing, *Tetrahedron*, 1978, **34**, 2705.

[83] M. Brigodiot, S. Boileau, M. C. Taieb, and H. Cheradame, *Tetrahedron*, 1979, **35**, 87.

spectroscopy, and X-ray crystallography.[84] The assignment of the configuration and conformation of cyclic dipeptides with sulphur-containing saturated rings, *e.g.* (78),[85] of the two isomeric sulphoxides (79) of N-methyl-L-thiazolidine-4-carboxylic acid hydantoin,[86] of the bicyclic compound (80),[87] and of two isomeric sulphoxides with the 8,10-diaza-4-thiabicyclo[5.3.0]decan-9-one structure, *e.g.* (81),[88] was based on n.m.r. data.

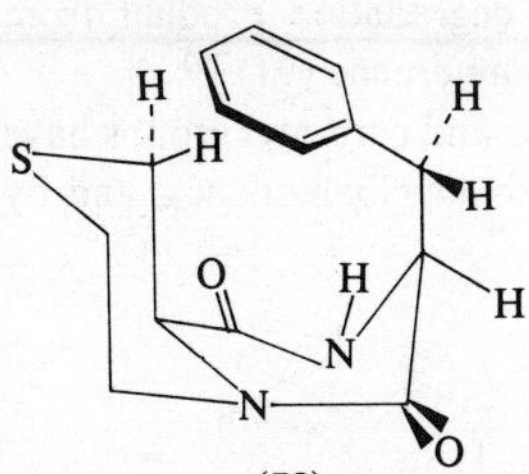

(78) 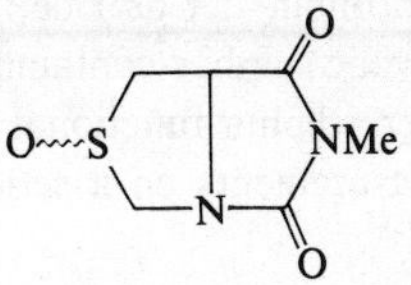   (79) 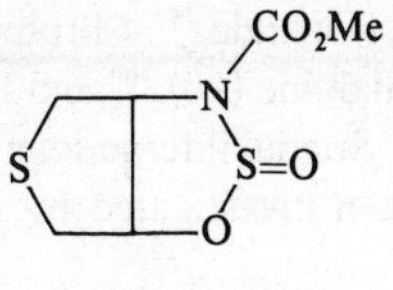   (80)

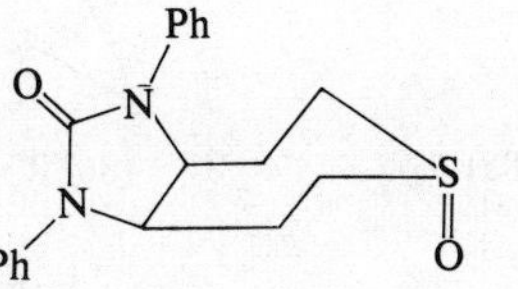

(81)

The conformations of thioxanthene S-oxide (82)[89] and of several thio-xanthenium salts (83)[90] have been investigated. The conformation of (82) was shown to be changed from S—O pseudo-equatorial in weakly interacting solvents to pseudo-axial S—O in associating solvents such as CF_3CO_2H.[89] The solvent-dependent conformational free-energy of the 2-methoxy-group of (84) was determined by BF_3-catalysed equilibration.[91]

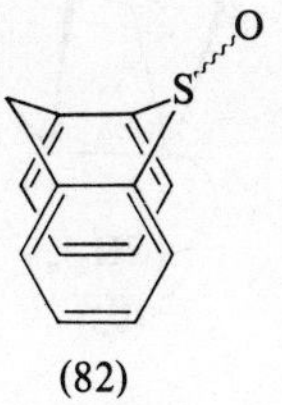

(82) 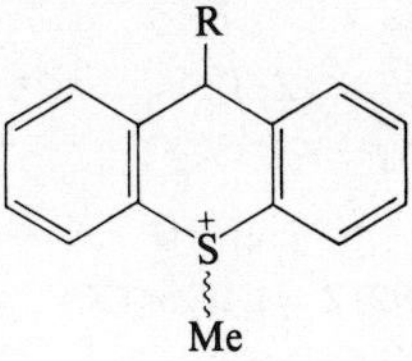   (83) 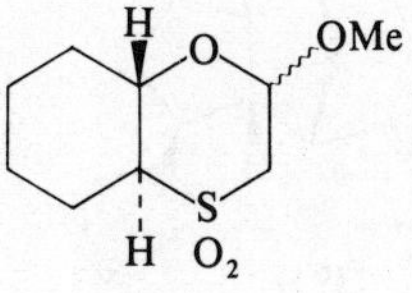   (84)

[84] H. D. Höltje, *Arch. Pharm. (Weinheim, Ger.)*, 1978, **311**, 311 (*Chem. Abstr.*, 1978, **89**, 42 136); H. D. Höltje, B. Jensen, and G. Lambrecht, *Eur. J. Med. Chem.-Chim. Ther.*, 1978, **13**, 453 (*Chem. Abstr.*, 1979, **90**, 161 938); G. Lambrecht, *Arch. Pharm. (Weinheim, Ger.)*, 1978, **311**, 636 (*Chem. Abstr.*, 1978, **89**, 107 866).
[85] M. J. O. Anteunis, R. Schrooten, W. A. Nachtergaele, and I. Neirinckx, *Bull. Soc. Chim. Belg.*, 1979, **88**, 684; M. J. O. Anteunis, C. Bécu, B. Liberek, and A. Kolodziejczyk, *ibid.*, p. 817.
[86] M. J. O. Anteunis, W. A. Nachtergaele, and M. Budesinsky, *Bull. Soc. Chim. Belg.*, 1979, **88**, 233.
[87] T. M. Filippova, S. D. Mikhno, N. S. Kulchkina, I. G. Suchkova, and V. M. Berezovskii, *Vopr. Stereokhim.*, 1978, **7**, 35 (*Chem. Abstr.*, 1979, **91**, 38 783).
[88] L. Lacombe and S. Lavielle, *Org. Magn. Reson.*, 1979, **12**, 39.
[89] A. L. Ternay, Jr., J. Herrmann, and B. A. Hayes, *J. Org. Chem.*, 1980, **45**, 189.
[90] M. Hori, T. Kataoka, H. Shimizu, and S. Ohno, *Heterocycles*, 1979, **12**, 1555.
[91] D. Lee, J. C. Keifer, R. P. Rooney, T. B. Garner, and S. A. Evans, *J. Org. Chem.*, 1979, **44**, 2580.

A partial conformational analysis of a series of substituted thian-4-ols, based on a study of the kinetics of acetylation, has been reported.[92]

The structures of several sulphur heterocycles have been determined using X-ray methods, including the thian-4-ol (85), which still showed a flattened chair form,[92] thian-4-ones, *e.g.* (86),[69] 2-chloro-4-t-butylthian 1-oxides,[93] 1-methylthiolanium iodide,[94] the *trans*-2-thiadecalin derivative (87),[95] *cis*- and *trans*-fused tricyclic compounds (88),[96] the bicyclic dihydrothiopyran (89),[97] thiochroman-4-one 1,1-dioxide,[98] 4-bromothiochroman 1,1-dioxide,[99] the degradation product from thiamine (90),[100] and the tricyclic sulphur-containing sesquiterpene (91).[101]

Spatial interactions between sulphur functional groups and carbonyl groups have been investigated by i.r. measurements on a series of thiacycloalkanones, and by

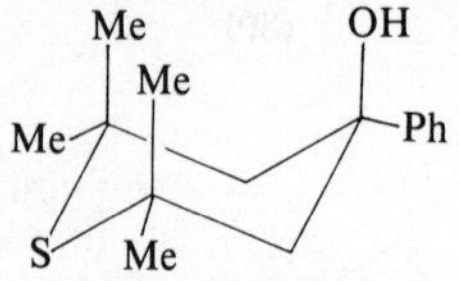

(85)

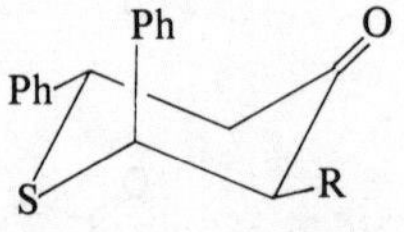

(86) R = Me or Et

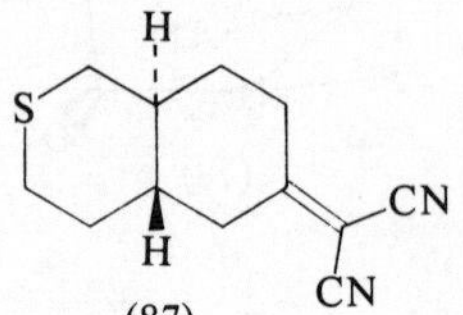

(87)

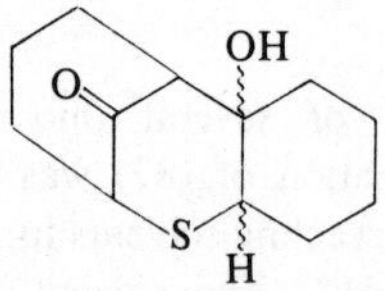

(88)

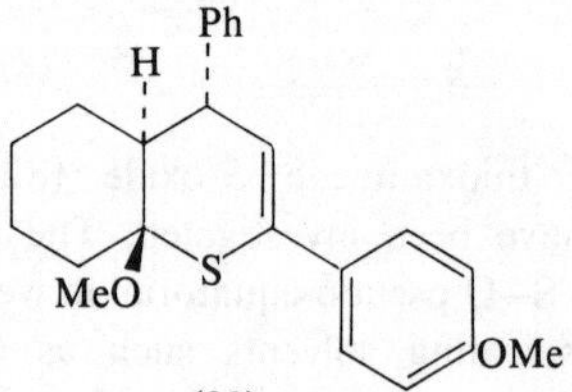

(89)

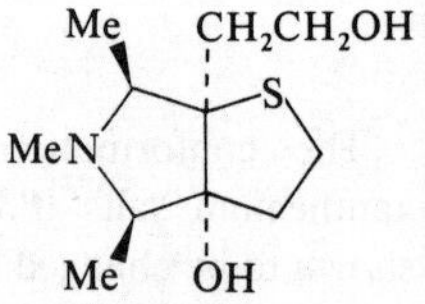

(90)

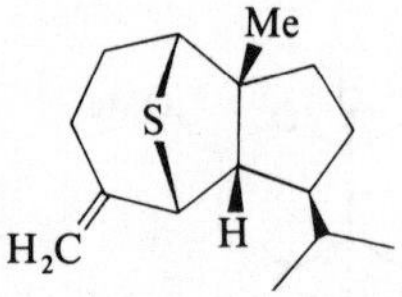

(91)

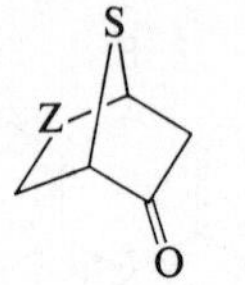

(92) Z = CH₂ or CO

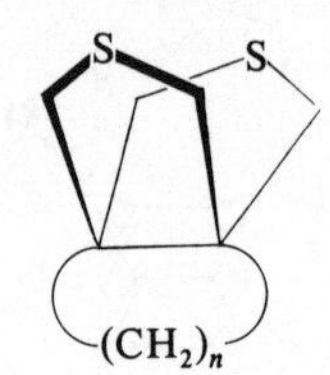

(93)

[92] N. Satyamurthy, R. Sivakumar, K. Ramalingam, K. D. Berlin, R. A. Loghry, and D. van der Helm, *J. Org. Chem.*, 1980, **45**, 349.

[93] F. Robert, *Acta Crystallogr., Sect. B*, 1977, **33**, 3480, 3484.

[94] F. Miyoshi, K. Tokuno, T. Watanabe, M. Matsui, and T. Ohashi, *Yakugaku Zasshi*, 1979, **99**, 924 (*Chem. Abstr.*, 1980, **92**, 41 192).

[95] K. Goubitz and C. H. Stam, *Cryst. Struct. Commun.*, 1978, **7**, 503.

[96] A. A. Usol'tsev, E. S. Karaulov, M. N. Tilichenko, M. Yu. Antipin, S. G. Il'in, and Yu. T. Struchkov, *Khim. Geterotsikl. Soedin.*, 1978, 1044 (*Chem. Abstr.*, 1978, **89**, 214 799).

[97] S. V. Soboleva, O. A. D'yachenko, L. O. Atovmyan, V. G. Kharchenko, and S. K. Klimenko, *Zh. Strukt. Khim.*, 1978, 499 (*Chem. Abstr.*, 1978, **89**, 196 860).

[98] S. E. Ealick, D. van der Helm, and J. R. Baker, *Acta Crystallogr., Sect. B*, 1979, **35**, 495.

[99] G. D. Andreetti, G. Bocelli, and P. Sgarabotto, *Cryst. Struct. Commun.*, 1978, **7**, 543.

[100] K. Kamiya, Y. Wada, and M. Takamoto, *Takeda Kenkyusho Ho*, 1979, **38**, 44 (*Chem. Abstr.*, 1980, **92**, 22 423).

[101] T. Yoshida, S. Muraki, and K. Takahashi, *J. Chem. Soc., Chem. Commun.*, 1979, 512.

studying u.v. spectra of 7-thiabicyclo[2.2.1]heptan-2-one and the 2,5-dione (92).[102] P.e. spectra of dithiapropellanes (93) indicated increasing spatial interaction between the $3p$ lone pairs on the sulphur centres with increasing values of n.[103] The preferred conformation of thiolan-2-yl radicals at low temperatures is probably the envelope form, as indicated by e.s.r. studies.[104] Dipole moments of 1-thiaindans[105] and gas-phase ionization potentials, electrochemical redox potentials, and spectroscopic properties of a series of thiochroman-4-ones and thiochromones[106] have been determined.

Reactions.—The stereoselectivity of the abstraction of diastereotopic protons α to trico-ordinated sulphur and the stereochemistry of electrophilic reactions at the carbanion centres of α-lithio-thian 1-oxides still excite considerable interest. The rate factors for base-catalysed H–D exchange of α-protons in several conformationally biassed or rigid thianium cations have been measured: with increasing rigidity, the reactivity of H_a decreases, while that of H_e remains about constant.[107] It was proposed that the removal of H_a requires a change to a half-chair conformation in order to get the incipient lone pair at C-α aligned *syn* or *anti* to one of the vicinal S–C bonds.

While the methylation of α-lithio-thian 1-oxides usually gave stereoselectively the α-methylated isomer with the *trans* configuration, the alkylation of seven- or eight-membered 1-thiacycloalkane 1-oxides proceeded with inversed stereochemistry, and the more abundant isomer always had the *cis* configuration.[108] Thus the mechanistic explanation (see Vol. 5, p. 91) (based on experiments involving methylations by trimethyl phosphate[82]) which proposed inversion of configuration because of a sterically hindered approach by CH_3I, but retention when the electrophilic species can act as a chelating agent, does not seem to be generally applicable. A *trans* arrangement of the entering methyl and the S-phenyl group was always obtained in low-temperature methylations of six-membered cyclic sulphonium ylides, *e.g.* (94),[109] even if configurational inversion is necessary, while methylation of five-membered ylides was not found to be stereoselective. The Pummerer reactions of both *cis*- and *trans*-4-(p-chlorophenyl)thian S-oxide gave the *trans* product (95), but the *cis* product when an excess of DCC or 2,6-lutidine was present.[110] A detailed mechanism for the α-halogenation of sulphoxides, particularly thian 1-oxides, has been proposed,[111] involving the formation of a positively charged ylide (96). Chlorination of 2,6-diphenylthian-4-one with DMF and $POCl_3$ gave 4-chloro-2,6-diphenyl-3-formyl-$2H$-dihydrothiopyran, whereas

[102] R. Borsdorf and H. Remane, *J. Prakt. Chem.*, 1980, **322**, 152; M. Kobayashi and R. Gleiter, *Z. Naturforsch., Teil B*, 1978, **33**, 1057.
[103] I. Tabushi, Y. Tamaru, and Z. Yoshida, *Bull. Chem. Soc. Jpn.*, 1978, **51**, 1178.
[104] C. Gaze and B. C. Gilbert, *J. Chem. Soc., Perkin Trans. 2*, 1978, 503.
[105] I. U. Numanov, L. M. Kim, and I. M. Nasyrov, *Izv. Akad. Nauk Tadzh. SSR, Otd. Fiz.-Mat. Geol.-Khim. Nauch*, 1978, 120 (*Chem. Abstr.*, 1979, **91**, 192 638).
[106] R. O. Loutfy, I. W. J. Still, M. Thompson, and T. S. Leong, *Can. J. Chem.*, 1979, **57**, 638.
[107] G. Barbarella, G. Dembach, A. Garbesi, F. Bernardi, A. Bottoni, and A. Fava, *J. Am. Chem. Soc.*, 1978, **100**, 200.
[108] V. Cerè, S. Pollicino, E. Sandri, and A. Fava, *Tetrahedron Lett.*, 1978, 5239.
[109] A. Garbesi, *Tetrahedron Lett.*, 1980, **21**, 547.
[110] T. Numata, O. Itoh, and S. Oae, *Tetrahedron Lett.*, 1979, 161.
[111] J. Klein, *Chem. Lett.*, 1979, 359.

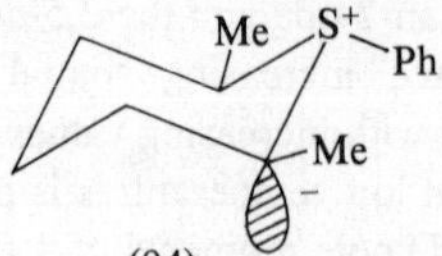

(94) 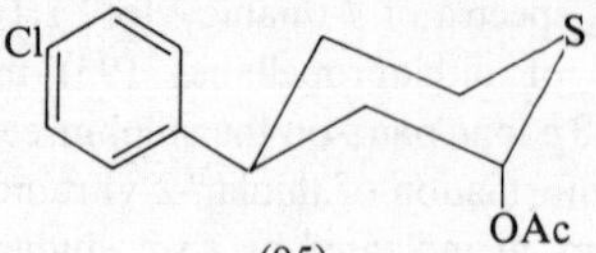   (95)

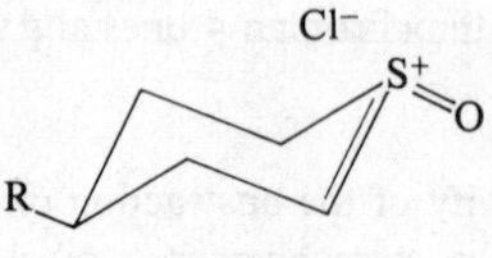

(96)

NCS in pyridine gave 2,6-diphenyldihydro-4*H*-thiopyran-4-one.[112] Full details of the kinetically controlled additions of various photogenerated carbenes and nitrenes to 4-t-butylthian have been published.[113] Only *N*-aryl sulphimides with equatorial S—N bonds were obtained from conformationally mobile or rigid thians and from *cis*- and *trans*-1-thiadecalins.[114]

Sulphimides with axial S—N bonds have been prepared from cyclic sulphoxides with equatorial S—O bonds.[114] [2,3]-Sigmatropic rearrangements of configurationally and conformationally homogeneous sulphimides, *e.g.* (97), proceed with high stereospecificity to give the *cis*-rearranged product *e.g.* (98)[115] (see also Chap. 2, Pt III, p. 135). The reduction of thian-4-ones with hydrides has also been studied.[116] The acceleration of the reaction, compared to the rate observed with cyclohexanone, has been attributed to a transannular 'through-space' effect.[116a]

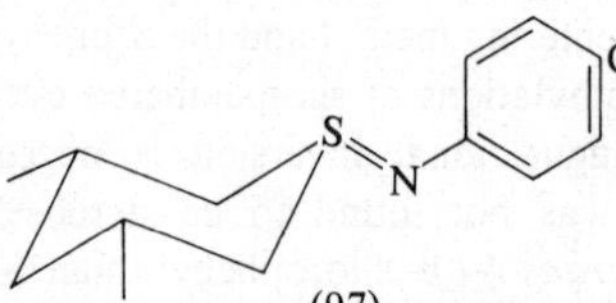

(97) 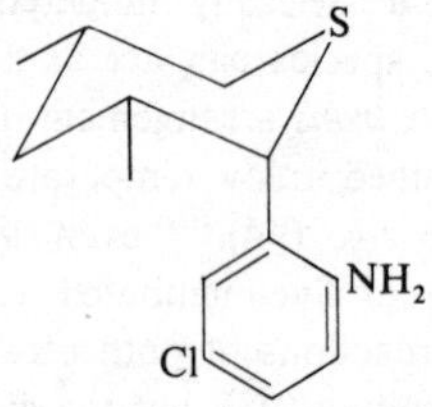   (98)

The influence of the sulphur functional group on the kinetics of various reactions at position 4 of thian derivatives has been studied, and participation of the electron pair on sulphur, steric approach control, and distortion of the chair conformation were considered in order to explain the observed effects.[117] Base-catalysed

[112] J. A. Van Allan, G. A. Reynolds, and C. H. Chen, *J. Heterocycl. Chem.*, 1977, **14**, 1399; C. H. Chen and G. A. Reynolds, *J. Org. Chem.*, 1979, **44**, 3144.

[113] D. C. Appleton, D. C. Bull, J. McKenna, J. M. McKenna, and A. R. Walley, *J. Chem. Soc., Perkin Trans. 2*, 1980, 385.

[114] P. K. Claus, W. Rieder, and F. W. Vierhapper, *Monatsh. Chem.*, 1978, **109**, 609, 631.

[115] P. K. Claus, W. Rieder, F. W. Vierhapper, and J. Bailer, *Tetrahedron*, 1979, **35**, 1373.

[116] (a) P. Geneste, R. Durand, I. Hugon, and C. Reminiac, *J. Org. Chem.*, 1979, **44**, 1971; (b) N. Satyamurthy and K. Ramalingam, *Indian J. Chem., Sect. B*, 1979, **17**, 62 (*Chem. Abstr.*, 1979, **91**, 123 305).

[117] V. Baliah and R. Jeyaraman, *Indian J. Chem., Sect. B*, 1977, **15**, 832, 852 (*Chem. Abstr.*, 1978, **88**, 88 973, 189 872); V. Baliah and K. Pandiarajan, *ibid.*, 1978, **16**, 807 (*Chem. Abstr.*, 1979, **90**, 103 269); V. Baliah, N. Bhavani, and J. Chandrasekharan, *ibid.*, 1979, **18**, 243 (*Chem. Abstr.*, 1980, **92**, 163 452); V. Baliah and N. Bhavani, *ibid.*, 1978, **16**, 776 (*Chem. Abstr.*, 1979, **90**, 186 718).

cyanoethylation of thian-4-ones and condensations of thiolan-3-ones with acrolein have been reported.[118] Participation of the β-sulphur atom in solvolysis reactions with cyclic systems has been studied intensively. The observed reactivity in solvolysis of various 3-thiacycloalkyl derivatives of different ring size was in the order $4 > 6 > 3 \gg 5$.[119] Solvolysis studies with β-substituted five- and six-membered thiacycloalkanes indicated an intermediate episulphonium ion (99).[120] The episulphonium ion (100), formed on solvolysis of a 6-substituted 2-thiabicyclo-[2.2.1]heptane, could be isolated as the perchlorate.[120] The unusually high *exo/endo* rate ratio found in this reaction is the highest observed for participation of β-S.

The 2-tetrahydrothienyl group has been suggested as an excellent protecting group for alcohols, and it is easily introduced by acetal exchange reactions between alcohols and 2-tetrahydrothienyl diphenylacetate (101; $R = OCOCHPh_2$), pre-

(99)   (100)   (101)

pared from 2-chlorothiolan (101; $R = Cl$).[121] Reactions of phosphorus, nitrogen, and carbon nucleophiles with 2-chlorothiolan afforded 2-substituted thiolans, *e.g.* (101; $R = Ph_3P^+$ or R_2^2PO), which are useful as equivalents of acyl anions in subsequent Wittig–Horner reactions, and [101; $R = CH(CO_2Et)_2$]. Quenching of the anion derived from the last compound with MeI resulted in α-methylation and in S-methylation with ring-opening.[121]

3-Tosyl-sulpholens (102) have been prepared as precursors of electron-deficient dienes.[122] The reaction of the bicyclic sulphone (103a) with Grignard reagents was found to produce butadienyl sulphoxides, which are interesting as synthetic intermediates. Cyclopentenones were prepared by thermal cheletropic elimination of SO_2 from (103b).[123] Miscellaneous reactions of di- and tetra-hydrothiophen 1,1-dioxides (alkylations, nitrations, displacement reactions, reactions of 2,3- and 3,4-epoxy-derivatives, Pummerer reactions, and desulphonylations) have been investigated by various groups.[124]

[118] K. Kh. Tokmurzin, *Tezisy Dokl.-Vses. Konf. Stereokhim. Konform. Anal. Org. Neftekhim. Sint. 3rd*, 1976, 83 (*Chem. Abstr.*, 1978, **88**, 169 899); S. D. Mikhno, T. M. Filippova, T. N. Polyanskaya, S. I. Peretokina, I. G. Suchkova, and V. M. Berezovskii, *Khim. Geterotsikl. Soedin.*, 1978, 1620 (*Chem. Abstr.*, 1979, **90**, 121 322).

[119] H. Morita and S. Oae, *Symp. Heterocycl.*, 1977, 95 (*Chem. Abstr.*, 1979, **90**, 5503).

[120] J. Ohishi and S. Ikegami, *Chem. Pharm. Bull.*, 1978, **26**, 3211 (*Chem. Abstr.*, 1979, **90**, 54 321); J. Ohishi, K. Tsuneoka, S. Ikegami, and S. Akaboshi, *J. Org. Chem.*, 1978, **43**, 4013.

[121] C. G. Kruse, E. K. Poels, F. L. Jonkers, and A. van der Gen, *J. Org. Chem.*, 1978, **43**, 3548; C. G. Kruse, E. K. Poels, and A. van der Gen, *ibid.*, 1979, **44**, 2911; C. G. Kruse, A. C. V. Janse, V. Dert, and A. van der Gen, *ibid.*, p. 2916.

[122] K. Inomata, H. Kinoshita, H. Takemoto, Y. Murata, and H. Kotake, *Bull. Chem. Soc. Jpn.*, 1978, **51**, 3341.

[123] Y. Gaoni, *Tetrahedron Lett.*, 1977, 4521.

[124] Only the latest publications of the various groups are given: E. M. Asatryan, A. Ts. Malkhasyan, R. T. Grigoryan, A. P. Engoyan, S. M. Mirakyan, and G. T. Martirosyan, *Arm. Khim. Zh.*, 1979, **32**, 44 (*Chem. Abstr.*, 1980, **92**, 41 672); V. M. Berestovitskaya, M. V. Titova, and V. V. Perekalin, *Zh. Org. Khim.*, 1979, **15**, 877 (*Chem. Abstr.*, 1979, **91**, 20 226); P. G. Dul'nev, *Khim. Geterotsikl. Soedin.*,

(*continued overleaf*)

(102)   (103) a; $R^4 = H$ (104)
               b; $R^4 = Cl$ (105)

The reactivity of cyclic sulphides toward hydrogen atoms[125] and the isomerization and dehydrogenation of thian and other monocyclic sulphides in the presence of acid catalysts[126] have been studied. Among the new reagents for the oxidation of sulphides to sulphoxides was found the oxaziridine (104), which oxidizes 4-t-butylthian to give 64% of the *trans*-sulphoxide and 16% of the *cis*-isomer.[127] Oxidation of thian by the photochemical transfer of oxygen from the selenoxide (105) and by direct photo-oxidation was also observed.[128] P_2S_5 has been used as a new mild reagent for the reduction of sulphoxides to sulphides, including cyclic derivatives.[129] Various base-induced photoreactions (reductive ring-cleavage, or fragmentation with rearrangement of the carbon skeleton) of the 3-(acetyl)thiotetronic acid (106) were reported.[130] The reaction of four- to six-membered cyclic sulphoxides with t-butoxyl radicals gave rise to the formation of acyclic alkyl radicals by a ring-opening reaction.[131] A series of chromium pentacarbonyl complexes of dihydro- and tetrahydro-thiophens and of their sulphoxides, in which chromium is bonded to sulphur, have been prepared.[132] The rates of ring-opening and -closing and the proportion of the sugar with a free SH group have been

(106)   (107)

(*reference 124 continued*)

   1979, 892 (*Chem. Abstr.*, 1979, **91**, 157 536); U. M. Dzhemilev, R. V. Kunakova, F. V. Sharipova, L. V. Spirikhin, L. M. Khalilov, E. V. Vasil'eva, and G. A. Tolstikov, *Izv. Akad. Nauk SSSR, Ser. Khim.*, 1979, 1822 (*Chem. Abstr.*, 1980, **92**, 6341); L. A. Mukhamedova and L. I. Kursheva, *Khim. Geterotsikl. Soedin.*, 1979, 613 (*Chem. Abstr.*, 1979, **91**, 107 852); N. N. Novitskaya, B. V. Flekhter, L. V. Spirikhin, and G. A. Tolstikov, *ibid.*, p. 563 (*Chem. Abstr.*, 1979, **91**, 39 234); J. Polakova, M. Palecek, and M. Prochazka, *Collect. Czech. Chem. Commun.*, 1979, **44**, 3705; F. Borg-Visse, F. A. Dawans, and E. Marechal, *Synthesis*, 1979, 817.
[125] O. Horie, J. Nishino, and A. Amano, *J. Org. Chem.*, 1978, **43**, 2800.
[126] A. K. Yus'kovich, T. A. Danilova, and E. A. Viktorova, *Tezisy Dokl. Nauchn. Sess. Khim. Tekhnol. Org. Soedin. Sery Sernistykh Neftei 14th*, 1975, (publ. 1976), 222 (*Chem. Abstr.*, 1978, **89**, 43 031).
[127] F. A. Davis, R. Jenkins, Jr., and S. G. Yocklovich, *Tetrahedron Lett.*, 1978, 5171.
[128] T. Tezuka, H. Suzuki, and H. Miyazaki, *Tetrahedron Lett.*, 1978, 4885; T. Tezuka, H. Miyazaki, and H. Suzuki, *ibid.*, p. 1959.
[129] I. W. J. Still, S. K. Hasan, and K. Turnbull, *Can. J. Chem.*, 1978, **56**, 1423.
[130] K. Saito and T. Sato, *Chem. Lett.*, 1978, 307.
[131] W. B. Gara and B. P. Roberts, *J. Chem. Soc., Perkin Trans. 2*, 1977, 1708.
[132] J. H. Eekhof, M. Hogeveen, and R. M. Kellogg, *J. Organomet. Chem.*, 1978, **161**, 361.

measured for several thio-sugars, *e.g.* for (107), by following the rates of the reactions with dipyridyl disulphides.[133]

The rate of hydration of the strained 9-thiabicyclononane (108)[134] and the halogenolysis of cyclohexa-1,4-diene monoepisulphide (109), which formed the sulphur-bridged products (110), have been studied.[135]

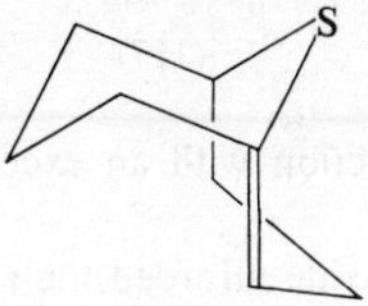

(108)  (109) 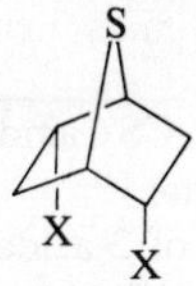 (110)

Several bi- and tri-cyclic thiepan derivatives were obtained by autoxidation of the cyclobutadiene (111) or by various reactions with strained sulphur-containing acetylenes, *e.g.*. the dithiet (112).[136] The ylides derived from *cis*- and *trans*-(72) were alkylated stereoselectively at low temperatures,[77] and they reacted with carbonyl compounds to give ring-opened oxirans (113).[77,137] Reduction of the 6-methoxy-carbonyl derivative of (72) afforded the ten-membered cyclic sulphide (114).[77]

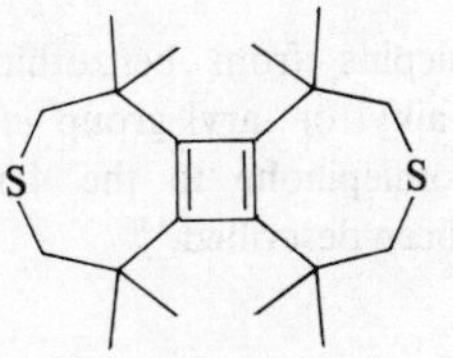

(111) 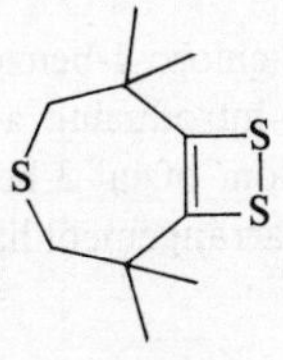 (112) 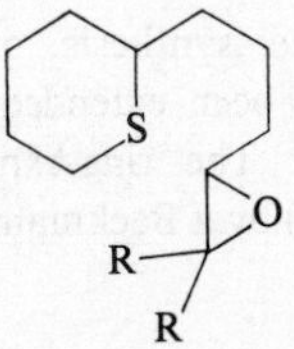 (113) 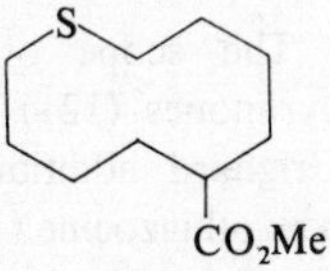 (114)

Kinetic studies on bimolecular substitutions at the benzylic centre of (115) indicated that the cyclic structure of (115) prevents an orthogonal arrangement of nucleophile and leaving group, and thus stabilization of the transition state by conjugation with the aromatic ring.[138] Various reactions of thiochromanones and related compounds have been studied, including the formation of the 1-thioflavone 1-oxide (116) from the corresponding 1-thioflavanone;[139] the photodehydrogen-ation of 1-thiochroman-4-ones;[140] the formation of the cyclic thiol ester (117) from isothiochromane;[141] and the t-butylation of isothiochroman-4-one by lithiation,

[133] C. E. Grimshaw, R. L. Whistler, and W. W. Cleland, *J. Am. Chem. Soc.*, 1979, **101**, 1521.
[134] W. K. Chwang, A. J. Kresge, and J. R. Wiseman, *J. Am. Chem. Soc.*, 1979, **101**, 6972.
[135] P. H. McCabe and A. Stewart, *J. Chem. Soc., Chem. Commun.*, 1980, 100.
[136] A. Krebs, R. Kemper, H. Kimling, K. H. Klaska, and R. Klaska, *Liebigs Ann. Chem.*, 1979, 473; A. Krebs, H. Colberg, U. Höpfner, A. Kimling, and J. Odenthal, *Heterocycles*, 1979, **12**, 1153; H. J. Lindner and B. Klitschke, *Chem. Ber.*, 1978, **111**, 2047.
[137] K. Tokuno, F. Miyoshi, Y. Arata, Y. Itatani, Y. Arakawa, and T. Ohashi, *Yakugaku Zasshi*, 1978, **98**, 1005 (*Chem. Abstr.*, 1979, **90**, 6195).
[138] J. F. King and G. T. Y. Tsang, *J. Chem. Soc., Chem. Commun.*, 1979, 1131.
[139] R. Bognar, J. Balint, and M. Rakosi, *Liebigs Ann. Chem.*, 1977, 1529.
[140] A. Couture, A. Lablache-Combier, and T. Qu Ninh, *Tetrahedron Lett.*, 1977, 2873.
[141] H. Böhme and H.-J. Wilke, *Liebigs Ann. Chem.*, 1978, 1123.

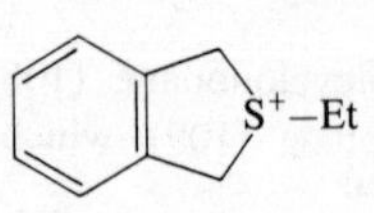

(115) 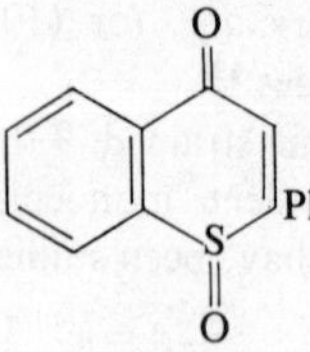   (116) 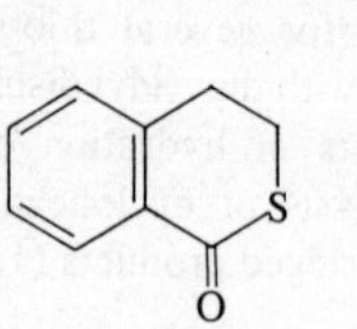   (117)

reaction with CS_2 amd MeI, and subsequent reaction with an excess of lithium dimethylcuprate.[142]

Photolysis of 2-azido-4-thiochromanone 1,1-dioxide afforded the ring-expanded product (118),[143] while (119) and (120) were formed by base-catalysed sulphimide rearrangements.[144]

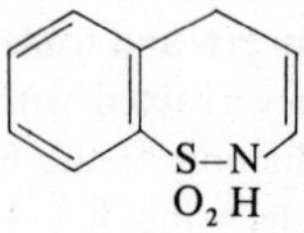

(118) 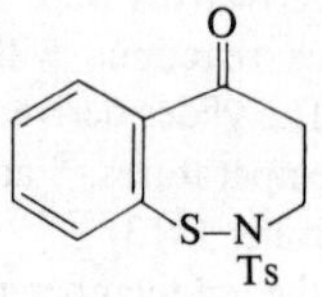   (119) 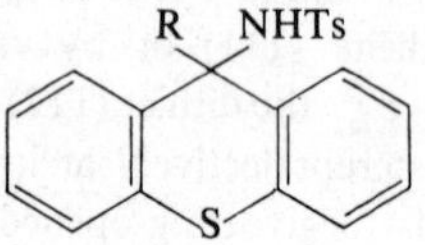   (120)

The scope of the synthesis of 4-chloro-1-benzothiepins from benzothio-pyranones (121) has been extended by introducing a 5-alkyl or -aryl group *via* Grignard addition.[145] The ring-expansion of a 3-benzothiepinone to the 4,1-benzothiazocine (122) by a Beckmann rearrangement has been described.[146]

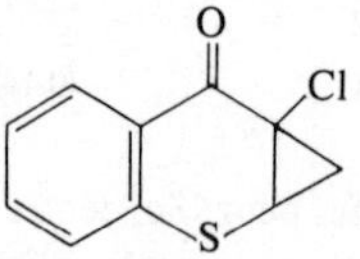

(121) 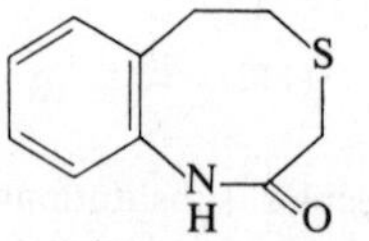   (122) 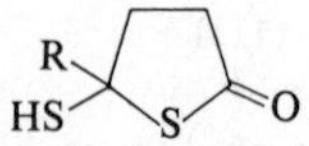   (123)

Several approaches to cyclic thio- and dithio-lactones have been described: thus, the reaction of γ-oxo-esters with H_2S gave 5-mercaptothiolan-2-ones (123)[147] (see Vol. 5, p. 249). A general approach to γ- and δ-thiolactones, *e.g.* (124), involves the reaction of a bis-metallated thioacid with α- or β-chloro-ketones and cyclization of the resulting γ- or δ-chloro-β-hydroxy-thioacids.[148] Reactions of cyclic lactones with the dimeric phenylthionophosphine sulphide (125) gave dithiolactones.[149] γ-Dithiolactones, *e.g.* (126), were obtained from 2-alkenyl alkanedithioates, which

[142] O. H. Johansen and K. Undheim, *Acta Chem. Scand., Ser. B*, 1979, **33**, 460.
[143] I. W. J. Still and T. S. Leong, *Can. J. Chem.*, 1980, **58**, 369.
[144] Y. Tamura, S. M. Bayoni, C. Mukai, and M. Ikeda, *Tetrahedron Lett.*, 1980, **21**, 533; Y. Tamura, Y. Nishikawa, C. Mukai, K. Sumoto, M. Ikeda, and M. Kise, *J. Org. Chem.*, 1979, **44**, 1684.
[145] V. J. Traynelis, J. A. Schield, W. A. Lindley, and D. W. H. McDowell, *J. Org. Chem.*, 1978, **43**, 3379.
[146] H. Böhme and R. Malcherek, *Arch. Pharm. (Weinheim, Ger.)*, 1979, **312**, 648, 653, 714.
[147] J. Brunet, D. Paquer, and P. Rioult, *Phosphorus Sulfur*, 1977, **3**, 377.
[148] A. M. Sarpeshkar, G. J. Gossick, and J. Wemple, *Tetrahedron Lett.*, 1979, 703.
[149] S. Scheibye, J. Kristensen, and S. O. Lawesson, *Tetrahedron* 1979, **35**, 1339.

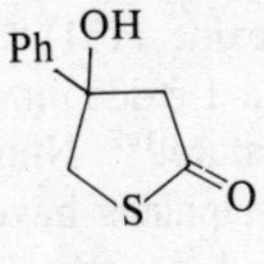

are easily prepared from Grignard reagents, CS_2, and allyl bromides, by a [3,3]-sigmatropic rearrangement and cyclization.[150] Dihydrodithioisocoumarin was easily obtained by a one-pot reaction, involving the $AlCl_3$-promoted reaction of CS_2 with phenylethyl alcohol.[151]

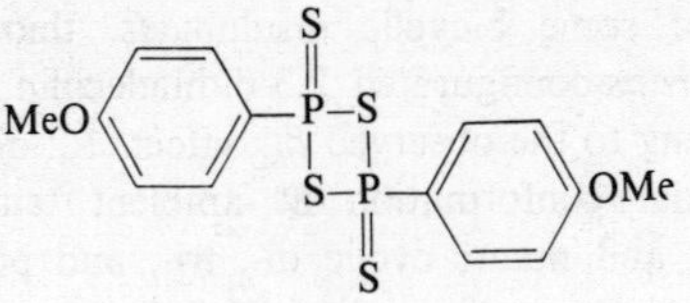

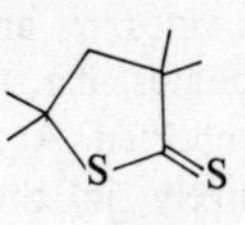

3 Compounds containing Two or More Sulphur Atoms in the Ring

Di- and Poly-sulphides.—A new, simple synthesis of '*dl*'-α-lipoic acid, using a butadiene telomer as the starting material, has been reported.[152] 3,5-Dimethyl-1,2,4-trithiolan (127), a constituent of processed meat, was prepared by a one-step synthesis from α-chloroethanesulphenyl chloride.[153] The isolation of *cis*- and *trans*-3,5-diethyl-1,2,4-trithiolan from a member of the genus *Allium*,[154] and the separation of the *cis*/*trans* isomers of 3,5-dialkyl-1,2,4-trithiolans and their structural investigation by NOE measurements,[155] have been reported. 1,2-Dithiolans (128) (see Vol. 5, p. 306)[156] and cyclic trithiocarbonates have been prepared by various methods.[157] Reactions of α-mercapto-acids with $SOCl_2$ afforded the anhydro-sulphites (129).[158] 1,2-Dithian was formed by treatment of thiolan 1-oxide with the sulphur-transfer reagent (125).[159]

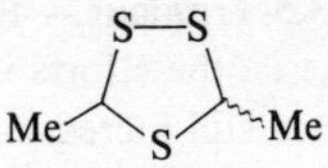

The disulphide (130) was obtained by the reaction of myrcene (7-methyl-3-methylene-octa-1,6-diene) with sulphur.[160] Carbon-13 n.m.r. spectra of 3-methyl-1,2-dithian and of the derived 1- and 2-monoxides, the 1,1- and

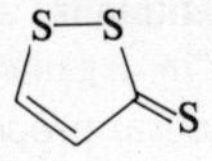

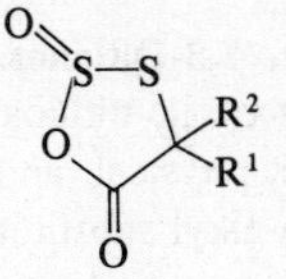

[150] M. Schoufs, J. Meijer, P. Vermeer, and L. Brandsma, *Synthesis*, 1978, 439.
[151] M. Czarniecki and R. Q. Kluttz, *Tetrahedron Lett.*, 1979, 4893.
[152] J. Tsuji, H. Yasuda, and T. Mandai, *J. Org. Chem.*, 1978, **43**, 3606.
[153] P. Dubs and M. Joho, *Helv. Chim. Acta*, 1978, **61**, 1404.
[154] H. Kameoka, Y. Demizu, Y. Iwase, and M. Miyazawa, *Tennen Yuki Kagobutsu Toronkai Koen Yoshishu*, *21st*, 1978, 199 (*Chem. Abstr.*, 1979, **90**, 69 093); H. Kameoka and Y. Demizu, *Phytochemistry*, 1979, **18**, 1397.
[155] H. P. Kruse and H. Anger, *Z. Chem.*, 1980, **20**, 65.
[156] E. Meinetsberger, A. Schöffer, and H. Behringer, *Synthesis*, 1977, 802.
[157] S. Kubota, E. Taniguchi, M. Eto, and K. Maekawa, *J. Fac. Agric., Kyushu Univ.*, 1977, **22**, 1 (*Chem. Abstr.*, 1978, **88**, 74 354).
[158] M. Ali, S. Roy, and B. J. Tighe, *J. Appl. Chem. Biotechnol.*, 1977, **27**, 696 (*Chem. Abstr.*, 1978, **88**, 169 551).
[159] J. B. Rasmussen, K. A. Jörgensen, and S. O. Lawesson, *Bull. Soc. Chim. Belg.*, 1978, **87**, 307.
[160] T. L. Peppard and J. A. Elridge, *Chem. Ind.* (*London*), 1979, 552.

2,2-dioxides,[161] and of some bicyclic disulphides, thiolsulphinates, and thiol-sulphonates, e.g. the trans-configurated 2,3-dithiadecalin 2β-oxide (131),[162] have been reported. According to the observed β_{SO} effect, 1,2-dithian 1-oxide must exist exclusively in the axial conformation at ambient temperature.[162] Numerous preparations of meso- and macro-cyclic di-, tri-, and poly-sulphides have been reported.[163] Thus, a chair conformation has been assigned to the 1,2,4,5-benzotetrathiepin (132).[163b] The conformation of the cyclic L-cysteinyl-L-cysteine (133) has been investigated by ^{1}H n.m.r. conformational analysis and compared with the conformation in the solid state.[163d] Numerous macrocyclic di-, tri-, and tetra-disulphides with m- or p-cyclophane structures[163f] and five- to 26-membered cyclic mono- or bis-trisulphides[163g] have been prepared.

(130)

(131)

(132)

(133)

1,3-Dithiolans, 1,3-Dithians, 1,4-Dithians, and 1,3,5-Trithians.—Because of the significance of cyclic dithioacetals in organic synthesis, the efforts in investigating their chemical, physical, and structural properties are still increasing. Various di-, tri-, and tetra-alkyl-substituted 1,3-dithians have been prepared,[164,165] and their

[161] T. Takata, Y. H. Kim, S. Oae, and K. T. Suzuki, Tetrahedron Lett., 1978, 4303.

[162] S. W. Bass and S. A. Evans, Jr., J. Org. Chem., 1980, 45, 710.

[163] (a) H. B. Bühl, U. Timm and H. Meier, Chem. Ber., 1979, 112, 3728; (b) F. Fehér and B. Engelen, Z. Naturforsch., Teil B, 1979, 34, 426; (c) A. A. Mahmood and J. M. Fayadh, Egypt. J. Chem., 1976, 19, 539 (Chem. Abstr., 1979, 91, 157 712); (d) S. Capasso, L. Mazzarella, and T. Tancredi, Biopolymers, 1979, 18, 1555 (Chem. Abstr., 1979, 91, 57 517); (e) R. T. Parfitt, D. E. Games, R. F. Cookson, A. C. Richards, and N. Lynaugh, Org. Mass. Spectrom., 1978, 13, 341; (f) F. Bottino, S. Foti, S. Pappalardo and N. Bresciani-Pahor, Tetrahedron Lett., 1979, 1171; F. Bottino, S. Foti, S. Pappalardo, P. Finocchiaro, and M. Ferrugia, J. Chem. Soc., Perkin Trans. 1, 1979, 198, 1712; F. Bottino and S. Pappalardo, Heterocycles, 1979, 12, 1331; M. S. Raasch, J. Org. Chem., 1979, 44, 2629; Z. V. Todres, N. G. Furmanova, S. P. Avagyan, Yu. T. Struchkov, and D. N. Kursanov, Phosphorus Sulfur, 1979, 5, 309; (g) A. Kato, Y. Hashimoto, I. Otsuka, and K. Nakatsu, Chem. Lett., 1978, 1219; J. Emsley, D. W. Griffiths, and R. Osborn, J. Chem. Soc., Chem. Commun., 1978, 658; J. Emsley and D. W. Griffiths, J. Chem. Res. (S), 1979, 251; J. Emsley, D. W. Griffiths, and G. J. J. Jayne, J. Chem. Soc., Perkin Trans. 1, 1979, 228; F. Fehér and B. Engelen, Acta Crystallogr., Sect. B, 1979, 35, 1853; F. Fehér and K. Glinka, Z. Naturforsch., Teil B., 1979, 34, 1031; D. N. Harpp and A. Granata, J. Org. Chem., 1979, 44, 4144; H. W. Roesky, H. Zamankhan, J. W. Bats, and H. Fuess, Angew. Chem., 1980, 92, 122.

[164] A. V. Bogatskii, A. I. Gren, T. I. Davidenko, and A. F. Galatin, Vopr. Stereokhim., 1978, 7, 24 (Chem. Abstr., 1979, 91, 19 741); T. I. Davidenko, ibid., 1978, 7, 20 (Chem. Abstr., 1979, 91, 19 284); A. V. Bogatskii, T. I. Davidenko, A. I. Gren, and Yu. Yu. Samitov, Zh. Org. Khim., 1979, 15, 2195 (Chem. Abstr., 1980, 92, 93 792).

[165] R. Boehm, Pharmazie, 1978, 33, 614 (Chem. Abstr., 1979, 90, 6323).

stereochemistry and spectral properties investigated.[164] The synthesis of 1,3-dithian-2- and -5-ones, and of 1,3-dithian-2-thiones, has been reported.[166,167] The reaction of α-diazo-ketones with CS_2 afforded 1,3-dithiolans.[168] The addition of dithiocarboxylic acids to allenes provided an entry to the 4-alkylidene-1,3-dithiolans (134).[169] A variety of 2-substituted 1,3-dithians has been prepared, partially in view of their synthetic applications. Thus, 2-cyano-1,3-dithian was obtained by lithiation of 1,3-dithian, followed by isocyanide–metal exchange.[170] The reaction of the N-tosylimide of 1,3,5-trithian with alcohols and base leads to 2-alkoxy-1,3,5-trithians (135).[171] 2-Trimethylsilyl-1,3-dithian reacts with acid chlorides to give 2-acyl-1,3-dithians.[172] 2-(α-Keto-alkyl)-1,3-dithiolans and 1,3-dithians were obtained by addition of 1,2-dithiols to acetylenic ketones,[173] and by condensation of 1,3-dicarbonyl compounds with propane-1,3-dithiol,[174] respectively. An approach to the γ-(1,3-dithian-2-yl)-β-keto-ester (136) has been found.[175] Full details of the synthetic applications of 2-chloro-1,3-dithian have appeared, including reactions with Grignard reagents and active-methylene compounds, and electrophilic aromatic substitutions.[176]

(134)   (135)   (136)

Reports have appeared on the preparation and on the Wittig–Horner reactions of (1,3-dithian-2-yl)phosphonium salts, e.g. (137).[177] Cyclocondensation of $O(CH_2SSO_3Na)_2$ with thiolesters gave the 1,3,5-oxathian derivatives (138).[178] 1,2-Dithiols and 2-mercaptoethanols were added to CF_3CN to give 1,3-dithiolans, e.g. (139), and 1,3-oxathiolans, respectively.[179] Several 2-ferrocenyl-1,3-dithiolans, -1,3-oxathiolans, and -1,3,5-trithians have been prepared.[180]

[166] S. Satsumabayashi, S. Motoki, and H. Takahashi, *Synthesis*, 1979, 184.
[167] S. Kubota, E. Taniguchi, M. Eto, and K. Maekawa, *Agric. Biol. Chem.*, 1977, **41**, 1621 (*Chem. Abstr.*, 1978, **88**, 22 766).
[168] P. Yates and J. A. Eenkhorn, *Heterocycles*, 1977, **7**, 961.
[169] G. Levesque, A. Mahjoub, and A. Thuillier, *Colloq. Int. CNRS*, 1977, **278**, 365 (*Chem. Abstr.*, 1979, **91**, 174 492).
[170] H. N. Khatri and H. M. Walborsky, *J. Org. Chem.*, 1978, **43**, 734.
[171] H. Gross and H. G. Walther, *Z. Chem.*, 1978, **18**, 62.
[172] P. Jutzi and O. Lorey, *Phosphorus Sulfur*, 1979, **7**, 203.
[173] M. N. Basyouni, M. T. Omar, and E. A. Ghali, *Synthesis*, 1980, 115; V. N. Elokhina, R. V. Karnaukhova, A. S. Nakhmanovich, I. D. Kalikhman, and M. G. Voronkov, *Zh. Org. Khim.*, 1979, **15**, 57 (*Chem. Abstr.*, 1979, **91**, 5133).
[174] I. Stahl, R. Manske, and J. Gosselck, *Chem. Ber.*, 1980, **113**, 800.
[175] E. C. Taylor and J. L. La Mattina, *J. Org. Chem.*, 1978, **43**, 1200.
[176] C. G. Kruse, A. Wiseman, and A. van der Gen, *J. Org. Chem.*, 1979, **44**, 1847.
[177] H. J. Cristau, H. Cristol, and D. Bottaro, *Synthesis*, 1978, 826; H. J. Cristau, J. P. Vors and H. Cristol, *Tetrahedron Lett.*, 1979, 2377; B. Costisella and H. Gross, *J. Prakt. Chem.*, 1978, **320**, 128.
[178] M. G. Gadzhieva, Sh. T. Akhmedov, I. G. Alizade, and A. Z. Gumbatalieva, *Zh. Org. Khim.*, 1978, **14**, 2188 (*Chem. Abstr.*, 1979, **90**, 72 153).
[179] O. Glemser and J. M. Shreeve, *Inorg. Chem.*, 1979, **18**, 2319.
[180] A. Ratajczak and A. Pierko, *Rocz. Chem.*, 1977, **51**, 967 (*Chem. Abstr.*, 1978, **88**, 37 936); A. Ratajczak, *ibid.*, p. 1735 (*Chem. Abstr.*, 1978, **88**, 50 997).

$$Ph_3\overset{+}{P}CH_2 \quad (137) \qquad (138) \qquad (139)$$

Several papers have appeared concerning new preparative pathways to cyclic keten SS-acetals.[181] Both the synthetic exploration of the chemistry of these keten dithioacetals and the synthetic aspects of dithian chemistry in general will not be covered in this Report (see Chap. 2, Pt I, p. 90).

The preparation and several reactions and spectral properties of a variety of tetra- and hexa-thia-adamantanes have been reported.[182]

Retention of configuration was observed in several preparations of bicyclic 1,3-dithiolans from mono- or bi-cyclic precursors that have a *cis* or a *trans* configuration.[183] Thus, the reaction of the *trans*-acetate (140) with KOH gave *trans*-perhydrobenzo-1,3-dithiole (141), probably *via* an intermediate thiiranium ion.

$$(140) \qquad (141)$$

Tricyclic 1,3-dithiolanone derivatives, *e.g.* (142), were obtained by the cyclo-addition of meso-ionic 1,3-dithiolones (143), which were prepared *in situ*.[184] 1,2-Dithiole-3-thione reacts with norbornadiene and other strained olefins to give heterodienes such as (144).[185] 1,3-Dithiolan-2-thione, though thermally unreactive towards acetylenes, does react photochemically to give dithioles (145).[186]

The chemistry of 2-alkoxy-1,3-benzodithioles has been reviewed.[187] A new route for the reduction of carboxylic acid derivatives is provided by their conversion into 1,3-benzodithiolium salts (146), which can be reduced in two steps to give the corresponding hydrocarbons.[188] The reaction of α-mercapto-ketones with am-

[181] M. Mikolajczyk, S. Grzejszczak, A. Zatorski, M. Mlotkowska, H. Gross, and B. Costisella, *Tetrahedron*, 1978, **34**, 3081; M. Augustin and Ch. Groth, *J. Prakt. Chem.*, 1979, **321**, 215; T. Harada, Y. Tamaru, and Z. Yoshida, *Tetrahedron Lett.*, 1979, 3525; Y. Nagao, K. Seno, and E. Fujita, *ibid.*, p. 4403; M. J. Taschner and G. A. Kraus, *J. Org. Chem.*, 1978, **43**, 4235.

[182] G. Levesque, A. Mahjoub, and A. Thuillier, *Tetrahedron Lett.*, 1978, 3847; P. Borg, K. Olsson, and Y. Shahab, *Chem. Scr.*, 1977, **11**, 192 (*Chem. Abstr.*, 1979, **90**, 103 253); V. I. Khvostenko, E. G. Galkin, E. M. Vyrypaev, B. M. Lerman, and G. A. Tolstikov, *Khim. Geterotsikl. Soedin.*, 1978, 748 (*Chem. Abstr.*, 1979, **90**, 21 729); B. M. Lerman, L. I. Umanskaya, Z. Ya. Aref'eva and G. A. Tolstikov, *Tezisy Dokl. Nauchn. Sess. Khim. Tekhnol. Org. Soedin. Sery Sernistykh Neftei 14th* 1975 (publ. 1976), 186 (*Chem. Abstr.*, 1978, **88**, 190 759).

[183] R. C. Forster and L. N. Owen, *J. Chem. Soc., Perkin Trans 1*, 1978, 822, 1208.

[184] H. Gotthardt, O. M. Huss, and C. M. Weisshuhn, *Chem. Ber.*, 1979, **112**, 1650.

[185] V. N. Drozd and G. S. Bogomolova, *Zh. Org. Khim.*, 1977, **13**, 2012 (*Chem. Abstr.*, 1978, **88**, 37 668).

[186] M. Ohashi, N. Mino, N. Nakayama, and S. Yamada, *Chem. Lett.*, 1978, 1189.

[187] J. Nakayama, K. Ueda, M. Hoshino, and T. Takemasa, *Synthesis*, 1977, 770.

[188] I. Degani and R. Fochi, *J. Chem. Soc., Perkin Trans. 1*, 1978, 1133.

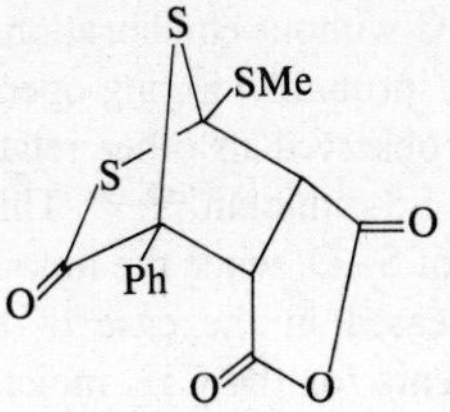

(142)

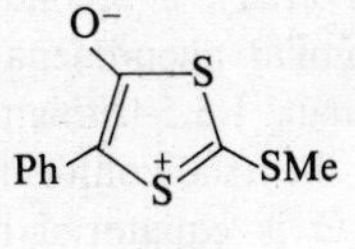

(143)

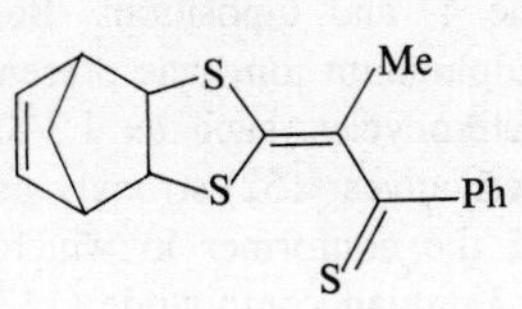

(144)

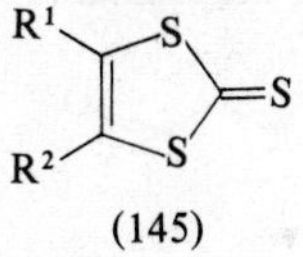

(145)

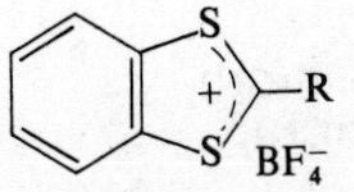

(146)

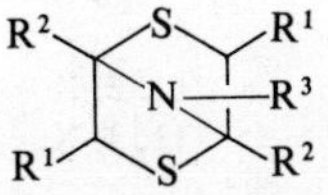

(147)

monia, primary amines, diamines, hydroxylamines, or substituted hydrazines provides a versatile synthesis of 2,5-epimino-1,4-dithians (147).[189]

The 1,4-dithian derivatives (148) and (149) have been isolated after the reaction of ethane-1,2-thiol with 2-chlorocyclohexanone[190] and α-bromoacetaldehyde diethyl acetal,[191] respectively. A novel cleavage reaction, with the formation of complexes such as (150), was observed when 1,4-dithian reacted with $[Fe_2(CO)_9]$.[192]

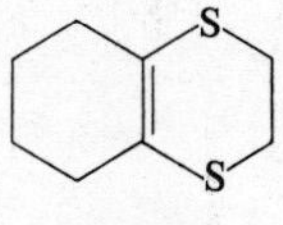

(148)

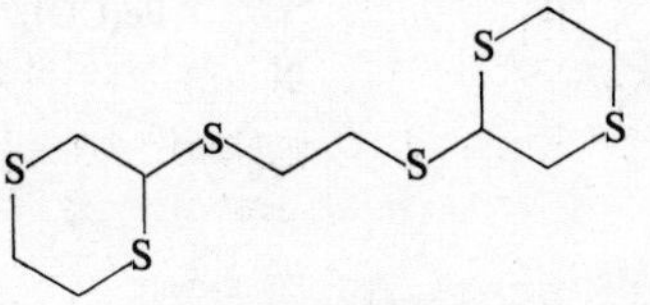

(149)

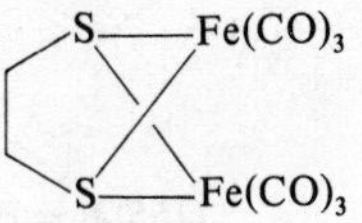

(150)

The protonation of 1,3-dithian in FSO_3H has been thoroughly studied.[193,194] At slow S—H exchange, the monoprotonated species (151) appeared to be predominantly equatorial. The formation of a hydrogen-bonded complex between 1,3-dithian and FSO_3H has been suggested in order to explain the observation that

[189] F. Asinger, A. Saus, and M. Bähr-Wirtz, *Liebigs Ann. Chem.*, 1979, 708.
[190] I. G. Mursakulov, F. F. Kerimov, N. K. Kasumov, E. A. Ramazanov, and N. S. Zefirov, *Azerb. Khim. Zh.*, 1979, 93 (*Chem. Abstr.*, 1979, **91**, 56 925).
[191] G. Giusti, M. Ambrosio, R. Faure, G. Schembri, E. J. Vincent, and C. Feugeas, *C.R. Hebd. Seances Acad. Sci., Ser. C*, 1979, **288**, 441.
[192] A. Shaver, P. J. Fitzpatrick, K. Steliou, and I. S. Butler, *J. Am. Chem. Soc.*, 1979, **101**, 1313.
[193] J. B. Lambert, E. Vulgaris, S. I. Featherman, and M. Majchrzak, *J. Am. Chem. Soc.*, 1978, **100**, 3269.
[194] J. B. Lambert, M. W. Majchrzak, and D. Stec, III, *J. Org. Chem.*, 1979, **44**, 4689.

the axial and equatorial 2-protons are equivalent at 60 °C without equilibration of the 4- and 6-positions. Besides (151), a second form, probably a ring-opened sulphonium ion, was present. Similar phenomena were observed in other related heterocycles, such as 1,3-diselenan, 1,3,5-trithian, and 1,3-dithiolan.[193,194] Thian derivatives (152) strongly prefer the axial conformation of S—O, while the amount of the conformer in which S—O is equatorial is increased in the case of the 1,3-dithian compounds (153).[195] Similar steric requirements for the CH_2 moieties and sulphur atoms have been concluded on the basis of the similar barriers to phenyl rotation that are observed for the dithian (154) and phenylcyclohexane.[196]

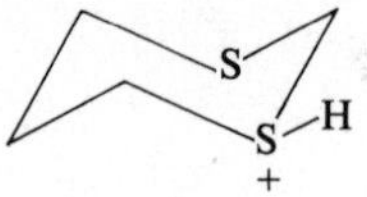

(151)

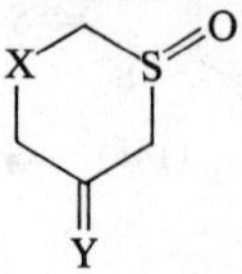

(152) X = CH_2, Y = O or $C(CN)_2$
(153) X = S, Y = O or $C(CN)_2$

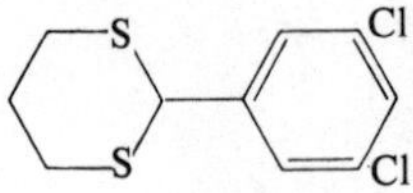

(154)

An axial preference of the aryl moiety of spirocyclic metacyclophanes (155) has been deduced from [1]H n.m.r. data.[197] The energy barriers to ring inversion were determined for the complexes (156)[198] and (157).[199] The coalescence phenomenon observed in the spectra of the dihalogen complexes (157) was found to be due not to reversal of the 1,4-dithian ring but rather to inversion of the configuration at the sulphur atom.[199]

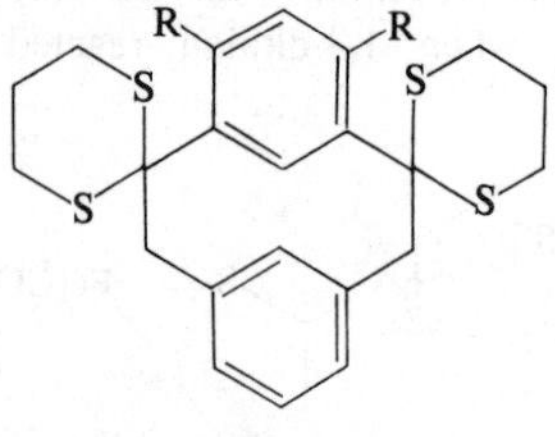

(156)

(157)

(155) R = H or Me

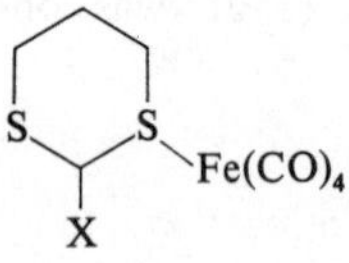

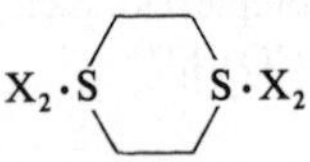

The existence of a repulsive S/S or S/O *gauche* effect has been demonstrated by *cis/trans* equilibration studies with 1,3-dithians (158).[200] Carbon-13 n.m.r. data have been reported for 2-X-substituted (X = Me, $SiMe_3$, $GeMe_3$, $SnMe_3$, or $PbMe_3$) and a series of methyl-substituted 1,3-dithians.[198,201] Incremental shifts of the carbon atoms of the ring have been used for assignments of the stereochemistry of S—O of anancomeric compounds (159).[202] Other spectral studies have dealt with

[195] K. Bergesen, B. M. Carden, and M. J. Cook, *J. Chem. Soc., Perkin Trans. 2*, 1978, 1001.
[196] T. Schaefer, W. Niemczura, and W. Danchura, *Can. J. Chem.*, 1979, **57**, 355.
[197] E. Langer, H. Lehner, and M. J. O. Anteunis, *Monatsh. Chem.*, 1978, **109**, 719.
[198] D. J. Cane, W. A. G. Graham, and L. Vancea, *Can. J. Chem.*, 1978, **56**, 1538.
[199] G. Hunter, R. F. Jameson, and M. Shiralian, *J. Chem. Soc., Perkin Trans. 2*, 1978, 712.
[200] E. L. Eliel and E. Juaristi, *J. Am. Chem. Soc.*, 1978, **100**, 6114.
[201] K. Pihlaja and B. Bjoerkvist, *Org. Magn. Reson.*, 1977, **9**, 533.
[202] M. J. O. Anteunis and L. Van Acker, *Acta Cienc. Indica, Ser. Chem.*, 1979, **5**, 57 (*Chem. Abstr.*, 1980, **92**, 146 145).

(158) X = O or S   (159) Y = CH₂, O, or S

the i.r. spectra of 1,3,5-trithian and of its 1-oxide,[203] and with the electron-impact fragmentation of 1,3,5-trithians.[204]

The lower-melting *cis*-1,3-dioxide of 2,2-diphenyl-1,3-dithian (160) exists in the crystal in the conformation with both sulphoxide oxygens *syn*-diaxial.[205] Other *X*-ray data have been reported on the structure of the trimethyl-1,3,5-trithian (161)[206] and on several 1,4-dithian derivatives.[207]

(160)   (161)

The high stereoselectivity exhibited by the 2-carbanions of 1,3-dithians in reactions with electrophiles evoked further investigations on the structure of the lithiated species. Proton and ^{13}C n.m.r. spectra of the 2-lithio- and 2-potassio-derivatives of 2-phenyl-1,3-dithian suggested that neither lithium–sulphur inter-actions nor the concomitant intimate ion-pairing, but rather an intrinsically preferred equatorial orientation of the lone pair in carbanions derived from 1,3-dithians, may be responsible for such stereoselectivity.[208] An *X*-ray analysis of the TMEDA complex of 2-lithio-2-methyl-1,3-dithian showed the complex to have the dimeric structure (162), with the two lithium atoms in equatorial positions[209] (see Chap. 2, Pt I, p. 89). The barriers to phenyl rotation in 2-lithio-2-phenyl-4,6-dimethyl-1,3-dithian have been discussed in terms of the formation of tight and solvent-separated ion pairs.[210] Exchange studies with 2,2-dimethyl-1,3-dithian, as well as *ab initio* SCF calculations, gave evidence that the polarizability of the C—S α-bond at the carbanionic site accounts for the regiochemistry of C—H acidification by sulphur.[211] The synthetic scope of 2-lithio-2-stannyl-1,3-dithians (163) has been explored.[212] Tritiation of the 2-anion of 2-nonyl-1,3-dithian was involved in the synthesis of (*R*)-[1-^{2}H, 1-^{3}H]decane

[203] M. Asai and K. Noda, *Spectrochim. Acta, Part A*, 1978, **34**, 695.
[204] P. Wolkoff and J. L. Holmes, *Org. Mass Spectrom.*, 1978, **13**, 338.
[205] R. F. Bryan, F. A. Carey, and R. W. Miller, *J. Org. Chem.*, 1979, **44**, 1540.
[206] K. Sekido, H. Ono, T. Noguchi, and S. Hirokawa, *Bull. Chem. Soc. Jpn.*, 1977, **50**, 3149.
[207] K. H. Linke and H. G. Kalker, *Z. Anorg. Allg. Chem.*, 1977, **434**, 157; M. Bukowska-Strzyzewska, W. Dobrowolska, and B. Pniewska, *Pol. J. Chem.*, 1978, **52**, 1843 (*Chem. Abstr.*, 1979, **90**, 86 606); J. Kaiser, R. Richter, L. Mögel, and W. Schroth, *Tetrahedron*, 1979, **35**, 505.
[208] A. G. Abatjoglou, E. L. Eliel, and L. F. Kuyper, *J. Am. Chem. Soc.*, 1977, **99**, 8262.
[209] R. Amstutz, D. Seebach, P. Seiler, B. Schweizer, and J. D. Dunitz, *Angew. Chem.*, 1980, **92**, 59.
[210] L. F. Kuyper and E. L. Eliel, *J. Organomet. Chem.*, 1978, **156**, 245.
[211] W. T. Borden, E. R. Davidson, N. H. Andersen, A. D. Denniston, and N. D. Epiotis, *J. Am. Chem. Soc.*, 1978, **100**, 1604.
[212] D. Seebach, I. Willert, A. K. Beck, and B. T. Gröbel, *Helv. Chim. Acta*, 1978, **61**, 2510.

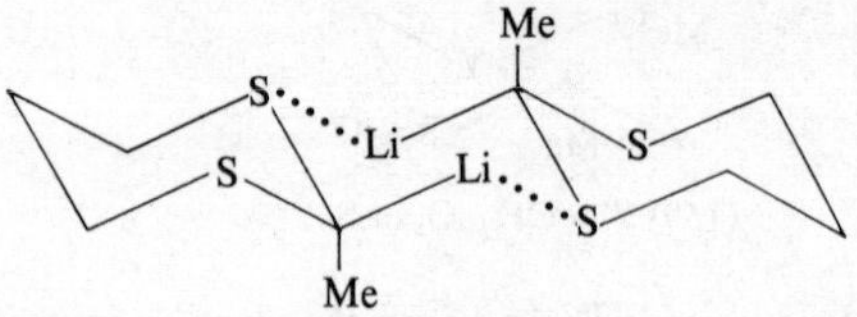

(162) 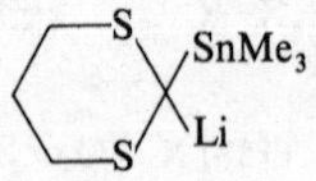   (163) 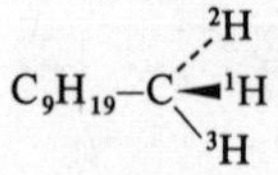   (164)

(164).[213] A novel synthesis of α-diketones from aldehydes proceeds by the formation of the anion of the ethylene dithioacetal of the aldehyde, reaction with $Fe(CO)_5$, alkylation, and hydrolysis.[214]

In contrast to 2-phenyl-1,3-dioxan-5-one, where nucleophiles add, regardless of their size, from the axial side of the carbonyl group, the same nucleophiles add to 2-phenyl-1,3-dithian-5-one exclusively from the equatorial side, to yield (165).[215] A ring-expansion reaction has been observed when 2-methyl-2-ethyl-1,3-dithiolan was brominated, giving the dithiin (166).[216]

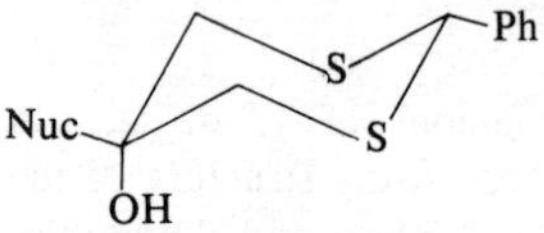

(165) 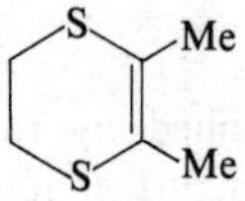   (166) 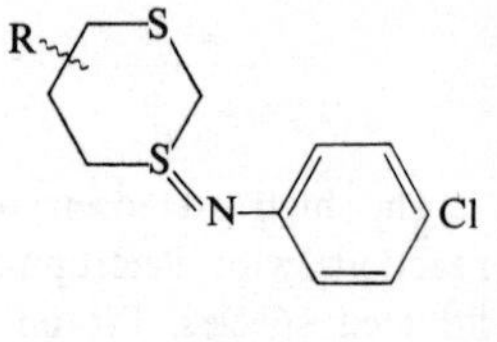   (167)

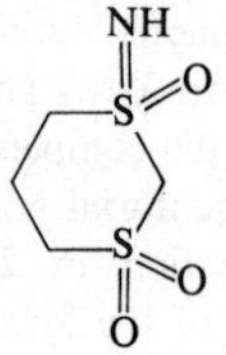

(168) 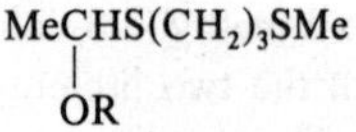   (169) 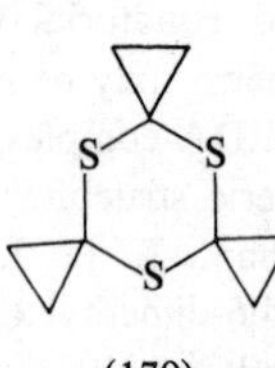   (170)

The liquid-phase oxidation of 1,3-dithian, and oxidation of alcoholic functional groups in the presence of the 1,3-dithian moiety by the Pfitzner–Moffatt reagent, has been studied.[217,218] Both *cis*- and *trans*-1,3-dithian 1-*N*-arylimides (167) were prepared and they rearranged with high stereospecificity.[219] 1,3-Dithian 1-imide 1,3,3-trioxide (168) has been obtained, in excellent yield, by the cleavage of the corresponding *N*-tosyl compound with Na and NH_3.[220]

[213] B. T. Golding, P. V. Ioannou, and I. F. Eckhard, *J. Chem. Soc., Perkin Trans. 1*, 1978, 774.
[214] M. Yamashita and R. Suemitsu, *J. Chem. Soc., Chem. Commun.*, 1977, 691.
[215] Y. M. Kobayashi, J. Lambrecht, J. C. Jochims, and U. Burkert, *Chem. Ber.*, 1978, **111**, 3442.
[216] G. Giusti and G. Schembri, *C.R. Hebd. Seances Acad. Sci., Ser. C*, 1978, **287**, 213.
[217] S. A. Agisheva, F. N. Latypova, S. S. Zlotskii, V. S. Martem'yanov, A. I. Gren, and D. L. Rakhmankulov, *Tezisy Dokl. Nauchn. Sess. Khim. Tekhnol. Org. Soedin. Sery Sernistykh Neftei 14th*, 1975, (publ. 1976), 176 (*Chem. Abstr.*, 1978, **88**, 189 680).
[218] I. Dyong, R. Hermann, and G. von Kiedrowski, *Synthesis*, 1979, 526.
[219] J. Bailer, P. K. Claus, and F. W. Vierhapper, *Tetrahedron*, 1980, **36**, 901.
[220] R. B. Greenwald and D. H. Evans, *Synthesis*, 1977, 650.

Ring cleavage of 1,3-dithiolans and 1,3-dithians by methylthiolation,[221] and of s-trithian by treatment with bromine or with Na in liquid NH_3,[222] has been reported. New methods for dethioacetalization of cyclic and acyclic SS- and OS-acetals included reactions with isoamyl nitrite, $Tl(NO_3)_2$, $Hg(ClO_4)_2$, nitrosonium or nitronium salts, bromodimethylsulphonium bromide, benzeneseleninic anhydride, electrochemical oxidation, or treatment of S-oxides with triethyloxonium salts and hydrolysis.[223] The reaction of 1,3-dithianonium salts with alcohols to give hemithioacetals (169) has been tested as a new method for the protection of alcohols.[224] The pyrolysis of 2,4,6-trimethyl-1,3,5-trithian, affording vinyl mercaptan,[225] and of trimeric cyclopropanethione (170), as a precursor of allene episulphide,[226] has been investigated. The abstraction of hydrogen at C-2 of 1,3-dithians, 1,3-dithiolans, and 1,3-oxathiolans by hydroxyl or t-butoxyl radicals has been studied by e.s.r. methods.[227]

Medium-sized Dithiacycloalkanes.—A convenient procedure for the preparation of 1,4-dithiepan has been given recently,[228] and the synthesis of various dithiepan derivatives, *i.e.* 6-benzoyl-1,4-dithiepan,[229] 4,7-dihydro-1,3-dithiepin,[230] and 2-substituted benzothiepins (171),[231] has been described. Dynamic ^1H and ^{13}C n.m.r. spectroscopy revealed the most stable conformation of the seven-membered ring of (171) to be a chair, on which the 2-methyl group adopts an equatorial position while the 2-methoxy-group takes the axial position.[231b] Several dithiepan derivatives were prepared by ring-expansion methods: dehydration of 1,3-dithian-2-ylmethanol gave

(171)　　　　(172)　　　　(173)

[221] J. K. Kim, J. K. Pau, and M. C. Caserio, *J. Org. Chem.*, 1979, **44**, 1544.
[222] E. Weissflog and M. Schmidt, *Phosphorus Sulfur*, 1978, **4**, 383; *ibid.*, 1979, **6**, 453.
[223] K. Juji, K. Ichikawa, and E. Fujita, *Tetrahedron Lett.*, 1978, 3561; E. Fujita, Y. Nagao, and K. Kaneko, *Chem. Pharm. Bull.*, 1978, **26**, 3743 (*Chem. Abstr.*, 1979, **90**, 203 915); R. A. J. Smith and D. J. Hannah, *Synth. Commun.*, 1979, **9**, 301; G. A. Olah, S. C. Narang, G. F. Salem, and B. G. B. Gupta, *Synthesis*, 1979, 273; G. A. Olah, Y. D. Vankar, M. Arvanaghi, and G. K. S. Prakash, *ibid.*, 1979, 720; D. H. R. Barton, N. J. Cussans, and S. V. Ley, *J. Chem. Soc., Chem. Commun.*, 1977, 751; Q. N. Porter and J. H. P. Utley, *J. Chem. Soc., Chem. Commun.*, 1978, 255; M. Platen and E. Steckhan, *Tetrahedron Lett.*, 1980, 511; I. Stahl, J. Apel, R. Manske, and J. Gosselck, *Angew. Chem.*, 1979, **91**, 179.
[224] T. A. Hase and R. Kivikari, *Synth. Commun.*, 1979, **9**, 107.
[225] E. N. Deryagina, M. A. Kuznetsova, O. B. Bannikova, and M. G. Voronkov, *Zh. Org. Khim.*, 1978, **14**, 2012 (*Chem. Abstr.*, 1979, **90**, 22 713).
[226] E. Block, R. E. Penn, M. D. Ennis, T. A. Owens, and S. L. Yu, *J. Am. Chem. Soc.*, 1978, **100**, 7436.
[227] V. V. Zorin, S. S. Zlotskii, V. F. Shuvalov, A. P. Moravskii, and D. L. Rakhmankulov, *Zh. Org. Khim.*, 1979, **15**, 178 (*Chem. Abstr.*, 1979, **90**, 186 034); Ch. Gaze and B. C. Gilbert, *J. Chem. Soc., Perkin Trans. 2*, 1979, 763.
[228] E. Weissflog, *Phosphorus Sulfur*, 1979, **6**, 489.
[229] V. I. Dronov, R. F. Nigmatullina, L. M. Khalilov, and Yu. E. Nikitin, *Zh. Org. Khim.*, 1979, **15**, 1709 (*Chem. Abstr.*, 1980, **92**, 5662).
[230] D. N. Harpp, K. Steliou, and B. T. Friedlander, *Org. Prep. Proced. Int.*, 1978, **10**, 133.
[231] (*a*) H. Gross, I. Keitel, and B. Costisella, *J. Prakt. Chem.*, 1978, **320**, 255; (*b*) F. Sauriol-Lord and M. St. Jacques, *Can. J. Chem.*, 1979, **57**, 3221.

6,7-dihydro-5H-1,4-dithiepins (172),[232] while the reaction of keten SS-dithio-acetals, derived from 1,3-dithian, with Pb(OAc)$_4$ afforded 3-alkyl-1,4-dithiepan-2-ones (173).[233]

Bicyclic 1,4-dithiepans, e.g. (174), have been obtained from tropone and dithiols.[234] A tetrathiapropellane[235] and numerous spirocyclic 1,4-dithiepans, e.g. (175), have been prepared.[236]

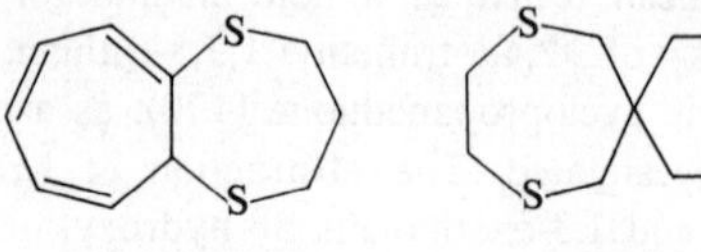

(174)   (175)

Oxidation of mesocyclic dithioethers[237] gave cyclic dithioether cations, e.g. (176), and cationic cyclic sulphur radicals, e.g. (177); these are the first long-lived, non-aromatic sulphur-containing radicals to be reported, characterized by a new S—S bond that is established by the interaction of three electrons.[237b] Similarly, a long-lived N—S-bonded cation radical and the dication (178) were obtained from 5-methyl-1-thia-5-azacyclo-octane (see also Chap. 2, Pt III, p. 142). Kinetic investigations on the reduction of mesocyclic sulphoxides, e.g. (179), indicated that there is anchimeric assistance by the transannular thioether group, with the formation of intermediate dithioether cations.[237]

Participation of transannular sulphur atoms has also been suggested to operate in the facile ring-opening of a dibenzodithiecin by brominolysis.[238] The crystal-structure determination of (180) revealed a double boat conformation, with all sulphur atoms nearly coplanar.[239]

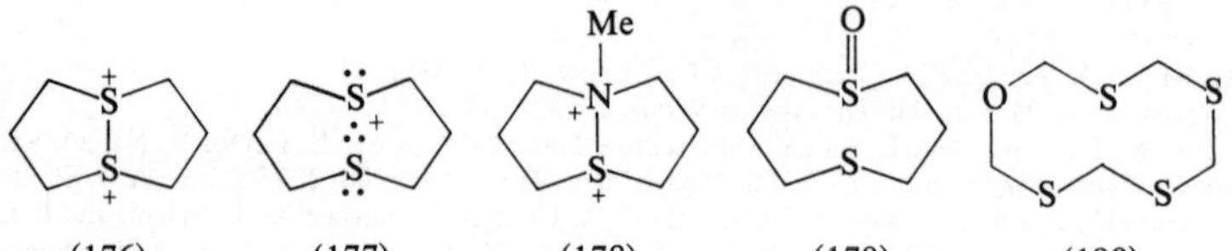

(176)   (177)   (178)   (179)   (180)

[232] P. M. Weintraub, J. Heterocycl. Chem., 1979, 16, 1081.

[233] K. Hiroi and S. Sato, Chem. Lett., 1979, 923.

[234] M. Cavazza, G. Morganti, and F. Pietra, Tetrahedron Lett., 1978, 2135.

[235] J. Jamrozik, J. Prakt. Chem., 1979, 321, 437.

[236] S. Smolinski, J. Mokrosz, M. Jamrozik, and M. Kubaszek, Monatsh. Chem., 1977, 108, 885; S. Smolinski and J. Mokrosz, Zesz. Nauk. Uniw. Jagiellon, Pr. Chem., 1978, 23, 7, 19 (Chem. Abstr., 1979, 90, 87 416, 87 417).

[237] (a) P. B. Roush and W. K. Musker, J. Org. Chem., 1978, 43, 4295; J. T. Doi and W. K. Musker, J. Am. Chem. Soc., 1978, 100, 3533; W. K. Musker, T. L. Wolford, and P. B. Roush, ibid., p. 6416; W. K. Musker, A. S. Hirschon, and J. T. Doi, ibid., p. 7754; G. S. Wilson, D. D. Swanson, J. T. Klug, R. S. Glass, M. D. Ryan, and W. K. Musker, ibid., 1979, 101, 1040; (b) K. D. Asmus, Colloq. Int. CNRS, 1977, 278, 305 (Chem. Abstr., 1979, 91, 210 573).

[238] M. K. Au, T. C. W. Mak, and T. L. Chan, J. Chem. Soc., Perkin Trans. 1, 1979, 1475.

[239] G. Valle and G. Zanotti, Cryst. Struct. Commun., 1977, 6, 651.

4 Sulphur- and Oxygen-containing Rings

Sultenes, Sultines, and Sultones.—Further reports have appeared on the formation of sultones from SO_3 and dienes[240] and from the cycloaddition of α-amino-methylene-ketones and sulphene.[241] Unusual sultone formations occurred during the reaction of a ketose derivative with mesyl chloride[242] and during the re-arrangements of bicyclic ketones that were induced by sulphuric acid, respec-tively.[243] The kinetics of reactions of sultones (181) with nucleophiles[244] and the enthalpy changes accompanying their alkaline hydrolysis,[245] which allow an estimation of differences in ring strain between the five- and six-membered sultones, have been studied.

A further report on a $[\pi + \pi\pi]$-cycloaddition of a diene and SO_2 has appeared:[246] the reaction of *o*-quinodimethane with SO_2 afforded mainly the cyclic sulphinate (182), and only small amounts of the $[n + \pi\pi]$ product were formed.

(181) $n = 1$ or 2   (182)

In an SbF_5-promoted cycloaddition of SO_2 to cyclohexa-1,3-diene, the sultine (183) was formed, probably *via* a [1,2]-H shift in an initially generated allylic intermediate.[247] γ-Sultines have been shown to extrude SO_2 thermally to give cyclopropanes when there is an aryl group at C-5.[248] An earlier report on a synthesis of 1,2-oxathiolan-5-one (184a) was shown to have been erroneous,[249] yet the first stable members of this class of mixed carboxylic–sulphinic acid anhydrides have been reported: thus treatment of a four-membered thiolactone with MCPBA gave the tetraphenyl derivative (184b), probably by a ring-expanding isomerization

(183)   (184) a; R = H   (185)
                     b; R = Ph

[240] A. V. Semenovskii, E. V. Polunin, I. M. Zaks, and A. M. Moiseenkov, *Izv. Akad. Nauk SSSR, Ser. Khim.*, 1979, 1327 (*Chem. Abstr.*, 1979, **91**, 123 689); T. Akiyama, M. Sugihara, T. Imagawa, and M. Kawanisi, *Bull. Chem. Soc. Jpn.*, 1978, **51**, 1251; F. van der Griendt and H. Cerfontein, *Tetrahedron Lett.*, 1978, 3263.

[241] L. Mosti, G. Menozzi, G. Bignardi, and P. Schenone, *Farmaco, Ed. Sci.*, 1977, **32**, 794 (*Chem. Abstr.*, 1978, **88**, 62 262); L. Mosti, P. Schenone, and G. Menozzi, *J. Heterocycl. Chem.*, 1979, **16**, 177; F. Evangelisti, P. Schenone, and A. Bargagna, *ibid.*, p. 217.

[242] M. B. Yunker and B. Fraser-Reid, *J. Chem. Soc., Chem. Commun.*, 1978, 325.

[243] D. S. Brown, H. Heaney, S. V. Ley, K. G. Mason, and P. Singh, *Tetrahedron Lett.*, 1978, 3937.

[244] T. Deacon, C. R. Farrar, B. J. Sikkel, and A. Williams, *J. Am. Chem. Soc.*, 1978, **100**, 2525.

[245] E. Izbacka and D. W. Bolen, *J. Am. Chem. Soc.*, 1978, **100**, 7625.

[246] T. Durst and L. Tetreault-Ryan, *Tetrahedron Lett.*, 1978, 2353.

[247] K. S. Fongers and H. Hogeveen, *Tetrahedron Lett.* 1979, 275.

[248] T. Durst, J. D. Finlay, and D. J. H. Smith, *J. Chem. Soc., Perkin Trans. 1*, 1979, 950.

[249] J. G. Kasperek and G. J. Kasperek, *J. Org. Chem.*, 1978, **43**, 3393.

of the intermediate α-keto-sulphoxide.[250] Heating of 2-(benzylsulphinyl)isophthalic acid in a mixture of acetic acid and acetic anhydride afforded the mixed anhydride (185).[251]

Several sultams have been prepared by cycloaddition reactions similar to those applied to the synthesis of sultones.[252,253] Thus, the generation of sulphene in the presence of pyridine gave the sultam (186).[252] Activated dienes reacted with simple N-sulphonylamines, *e.g.* MeN=SO$_2$, to give unsaturated sultams such as (187),[254] and thione S-imides reacted smoothly as 1,3-dipoles to give cyclic sulphenamides (188), or as dienophiles, with the formation of Δ^3-dihydrothiopyran 1-imides.[255]

(186)   (187)   (188)

1,3-Oxathiolans and 1,3- and 1,4-Oxathians.—The attractive and repulsive *gauche* effects, including effects observed in 2-substituted 1,4-oxathians and 5-substituted 1,3-dithians, have been reviewed.[256]

A series of methyl-substituted 1,3-oxathians has been prepared by conventional procedures.[257] Details of a great number of preparative procedures for the synthesis of a variety of oxathiolan and oxathian derivatives have been published, including the synthesis of 1,3-benzoxathioles and 1,3-oxathiolans by cyclocondensation of o-mercaptophenol and β-mercaptoethanol, respectively, with acetylenic ketones;[258] the condensation of benzaldehyde with α-monothioglycerol to give a mixture of the *cis*- and *trans*-isomers of the 1,3-oxathian (189) and of the 1,3-oxathiolan (190);[259] the free-radical addition of thioglycolic acid to appropriate olefins as a key step for the preparation of 1,4-oxathian- and 1,4-oxathiepan-2-ones, *e.g.* of (191);[260] cyclocondensation of orthoesters with α-mercapto-carboxylic acids to give 1,3-oxathiolan-5-ones;[261] the synthesis of 1,4-oxathian-2-ones (192) by the reaction of chalcone epoxides with thioglycolic acid;[262] the electrophilic addition of SCl$_2$ to vinylic ethers, to afford mixtures of the *cis*- and *trans*-isomers of 2,6-dichloro-

[250] H. Kohn, P. Charumilind, and S. H. Simonsen, *J. Am. Chem. Soc.*, 1979, **101**, 5431.
[251] W. Walter, B. Krische, G. Adiwidjaja, and J. Voss, *Chem. Ber.*, 1978, **111**, 1685.
[252] J. St. Grossert, M. M. Bharadwaj, R. F. Langler, T. St. Cameron, and R. E. Cordes, *Can. J. Chem.*, 1978, **56**, 1183.
[253] W. E. Truce and J. P. Shepherd, *J. Am. Chem. Soc.*, 1977, **99**, 6453; T. Iwasaki, S. Kajigaeshi, and S. Kanemasa, *Bull. Chem. Soc. Jpn.*, 1978, **51**, 229.
[254] J. A. Kloek and K. L. Leschinsky, *J. Org. Chem.*, 1979, **44**, 305.
[255] T. Saito and S. Motoki, *J. Org. Chem.*, 1979, **44**, 2493.
[256] E. Juaristi, *J. Chem. Educ.*, 1979, **56**, 438.
[257] K. Pihlaja, P. Pasanen, and J. Wähäsilta, *Org. Magn. Reson.*, 1979, **12**, 331.
[258] V. N. Elokhina, A. S. Nakhmanovich, I. D. Kalikhman, and M. G. Voronkov, *Tezisy Dokl.-Vses. Konf. Khim. Atsetilena, 5th*, 1975, 298 (*Chem. Abstr.*, 1978, **88**, 170 015).
[259] R. Böhm, *Pharmazie*, 1978, **33**, 465.
[260] D. I. Davies, L. Hughes, Y. D. Vankar, and J. E. Baldwin, *J. Chem. Soc., Perkin Trans. 1*, 1977, 2476.
[261] Yu. A. Davidovich, L. A. Pavlova, and S. V. Rogoshin, *Zh. Org. Khim.*, 1978, **14**, 664 (*Chem. Abstr.*, 1978, **88**, 190 647).
[262] A. A. Hamed, A. Essawy, and M. A. Salem, *Indian J. Chem., Sect. B*, 1978, **16**, 693 (*Chem. Abstr.*, 1979, **90**, 121 517).

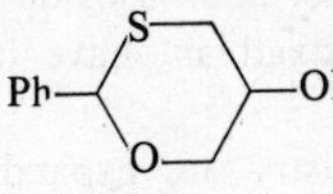

(189)

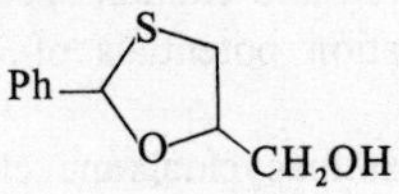

(190)

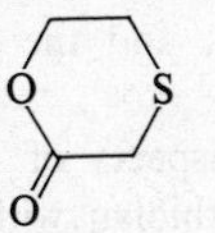

(191)

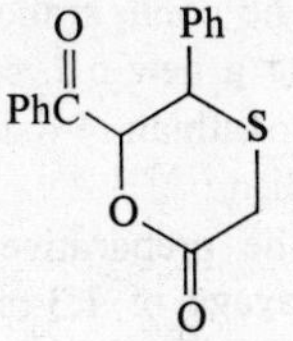

(192)

1,4-oxathians;[263] the formation of cyclic trithio-orthocarbonates (193) by the thermolysis of bis-(dithiocarbonates);[264] the preparation of 1,3-oxathiol-2-ones and of 2-substituted 1,3-oxathioles by treatment of αα-dibromo-ketones with potassium ethyl xanthate;[265] a novel synthesis of 2H-1,3-benzoxathiol-2-ones (194) by the HI- or pyridine-hydrochloride-induced cyclization of thiocarbamates;[266] and the base-catalysed cyclization and Smiles rearrangement of 3-(2,4,6-trinitrophenyl-thio)propanol to give the benzoxathiepin (195).[267]

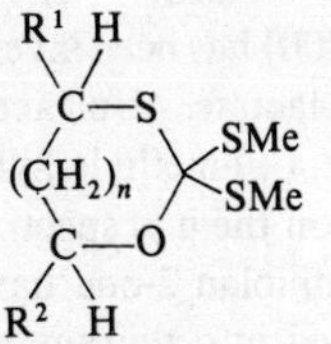

(193) *n* = 0 or 1

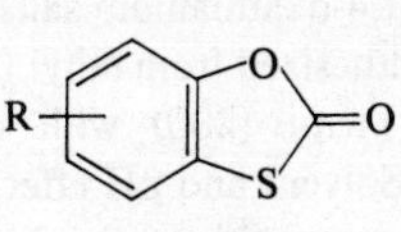

(194)

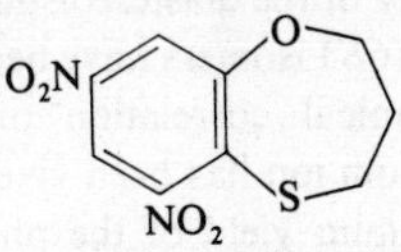

(195)

The formation of the benzoxathianone (196) by an intramolecular Pummerer reaction has been studied by a combination of ^{18}O labelling and stereoselective deuterium labelling and the use of optically active compounds.[268]

Conformations and chair-twist equilibria of a series of methyl-substituted 1,3-oxathians have been studied by ^1H n.m.r. spectroscopy.[257] Similarly, the conformations of various other 1,3- and 1,4-oxathians have been investigated.[269] Phenoxathiin sulphoxide (197) has been found to exist in the pseudo-axial conformation in both the solid state and in solution.[270]

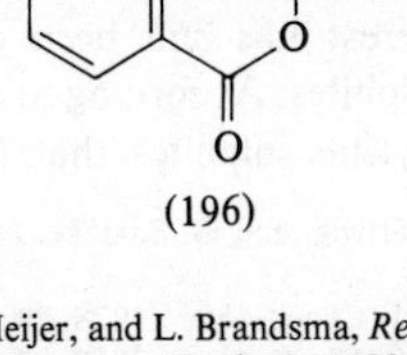

(196)

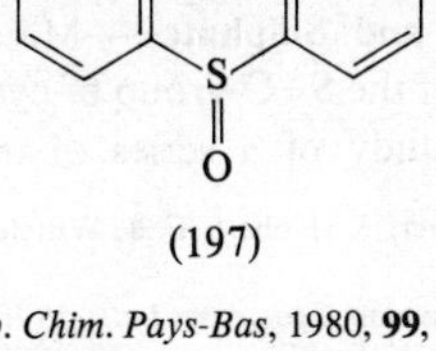

(197)

[263] M. Schoufs, J. Meijer, and L. Brandsma, *Recl. Trav. Chim. Pays-Bas*, 1980, **99**, 12.
[264] A. Faure and G. Descotes, *Synthesis*, 1978, 286.
[265] C. Bak, G. Höhne, and K. Praefcke, *Chem.-Ztg.*, 1978, **102**, 66.
[266] J. T. Traxler, *J. Org. Chem.*, 1979, **44**, 4971.
[267] V. N. Knyazev, V. N. Drozd, and V. M. Minov, *Zh. Org. Khim.*, 1978, **14**, 105 (*Chem. Abstr.*, 1978, **89**, 43 366).
[268] S. Wolfe, P. M. Kazmaier, and H. Auksi, *Can. J. Chem.*, 1979, **57**, 2404.
[269] T. Spoormaker and M. J. A. de Bie, *Recl. Trav. Chim. Pays-Bas*, 1980, **99**, 15; A. I. Gren, A. M. Turyanskaya, V. I. Sidorov, and A. Weigt, *Vopr. Stereokhim.*, 1977, **6**, 87 (*Chem. Abstr.*, 1979, **90**, 186 259); A. M. Turyanskaya, *ibid.*, 1978, **7**, 68 (*Chem. Abstr.*, 1979, **91**, 38 786).
[270] J. S. Chen, W. H. Watson, D. Austin, and A. L. Ternay, Jr., *J. Org. Chem.*, 1979, **44**, 1989.

The configurations of several steroid 1,3-oxathiolans have been established by using a new c.d. sector rule.[271] Infrared and Raman spectra of thian 1-oxide and 1,4-oxathian 4-oxide, and the ionization potentials of 1,4-oxathian, have been studied.[272,273]

The preparative aspects of 1,3-oxathiacycloalkane chemistry are expanding. Cleavage of 1,3-oxathiolan with Me_3SiI afforded the thioether $Me_3SiO(CH_2)_2S\text{-}CH_2I$, which has been found to be useful as an efficient alkylating agent.[274] 4,4-Dimethyl-1,3-oxathiolan 3,3-dioxide (198) offers a new equivalent of the formyl anion which readily demasks under thermal conditions. Alkylation and pyrolysis gave aldehydes in excellent yields. The reaction of the 2-lithio-derivative of (198) with ketones, followed by pyrolysis, afforded either α-hydroxy-aldehydes or ring-expanded acyloins.[275] 2-Substituted 1,3-benzoxathiolium salts (199), readily obtainable from esters, react with alcohols to furnish good yields of esters after Hg^{2+}-promoted hydrolysis. Thus, the compounds (199) are masked esters from which the protecting group can be easily removed.[276]

Alkylations of 2-lithio-1,3-oxathians have been briefly studied.[277] The stereo-chemistry of the diastereoisomeric 1,4-oxathianium salts (200) has been investigated and the (6S)-isomers have been synthesized from ethyl (S)-lactate. A full account of the chemical correlation of (6S)-trans-(200) with the (+)-methylethylpropyl-sulphonium ion has been given.[278] Solvent and pH effects on the u.v. spectrum and the quantum yield of the photodecomposition of 1,3-oxathiolan-2-one have been investigated.[279] 1,3-Oxathiolan-5-ones (201) were converted into thiirans by flash vacuum pyrolysis, *via* the intermediate thiocarbonyl ylides.[280] The reaction proceeds by clean inversion of configuration and is assumed to involve disrotatory ring-opening followed by conrotatory ring-closure. The liquid complex of borane and 1,4-oxathian has been suggested as a new, convenient hydroborating agent.[281]

(198)   (199)   (201)

(200)

Cyclic Sulphites and Sulphates.—Major interest has still been devoted to the stereochemistry of the S=O group of cyclic sulphites. According to a systematic ^1H and ^{13}C n.m.r. study of a series of trimethylene sulphites that have phenyl or

[271] P. Welzel, I. Müther, K. Kobert, F. J. Witteler, T. Hartwig, and G. Snatzke, *Liebigs Ann. Chem.*, 1978, 1333.

[272] Y. Hase and Y. Kawano, *Spectrosc. Lett.*, 1978, **11**, 161 (*Chem. Abstr.*, 1978, **89**, 5595).

[273] J. Jalonen, *Ann. Acad. Sci. Finn.*, Ser. A2, 1979, **189**, 88 (*Chem. Abstr.*, 1979, **91**, 174 694).

[274] G. E. Keyser, J. D. Bryant, and J. R. Barrio, *Tetrahedron Lett.*, 1979, 3263.

[275] G. W. Gokel and H. M. Gerdes, *Tetrahedron Lett.*, 1979, 3375, 3379.

[276] G. Aimo, I. Degani, and R. Fochi, *Synthesis*, 1979, 223.

[277] K. Juji, M. Ueda, and E. Fujita, *J. Chem. Soc., Chem. Commun.*, 1977, 814.

[278] E. Kelstrup, *J. Chem. Soc., Perkin Trans. 1*, 1979, 1029, 1037.

[279] H. Chandra and K. S. Sidhu, *Indian J. Chem.*, Sect. B, 1977, **158**, 823 (*Chem. Abstr.*, 1978, **88**, 169 301).

[280] T. B. Cameron and H. W. Pinnick, *J. Am. Chem. Soc.*, 1979, **101**, 4755; *ibid.*, 1980, **102**, 744.

[281] H. C. Brown and A. K. Mandal, *Synthesis*, 1980, 153.

t-butyl groups at C-4 or at both C-4 and C-6 which are potentially *syn*-diaxial to S=O, it appears that the ring inverts to a conformation with both S=O and the bulky substituent equatorial except for *trans*-4,6-disubstituted systems, which are reluctant to adopt a chair conformation with a *syn*-diaxial orientation of the bulky substituent and sulphoxide oxygen. Twist geometries have been suggested for these latter compounds.[282] Both isomers of the sulphite (202) have been prepared, and the chair conformations established.[282b] X-Ray structures have been determined for several trimethylene sulphites:[283] the unusual equatorial position of the S=O group of (203) has been confirmed,[283a] and a twist conformation has been established for (204).[283c]

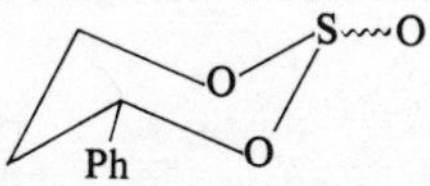

(202) 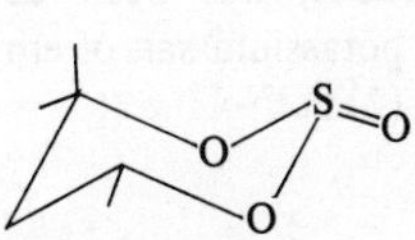   (203)

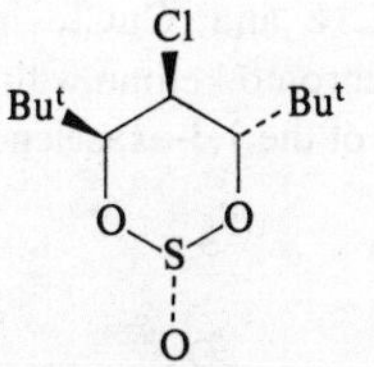

(204)

Cyclic sulphinamates (205), like cyclic sulphites, also prefer a chair conformation with an axial S=O group.[284] Substituents R on the nitrogen atom are preferentially axial. The 3-t-butyl-*cis*-4-methyl derivative (206) has been shown to prefer (in solution and in the solid state) a slightly distorted chair conformation, with nitrogen at the top of a very flattened pyramid, with axial S=O and methyl, and with the t-butyl group displaced towards the axial direction. Unlike cyclic sulphites, the chair form with equatorial S=O is rarely involved, and twist forms with a 3,6- or 1,4-axis were observed.[284] Carbon-13 n.m.r. spectra of the five-membered analogues (207) have also been related to the conformational situation.[285]

Unsaturated analogues are available by various cyclization procedures, *e.g.* by condensation of SOCl₂ with a 2-acyl-acetamide to give (208).[286] Other cyclizations

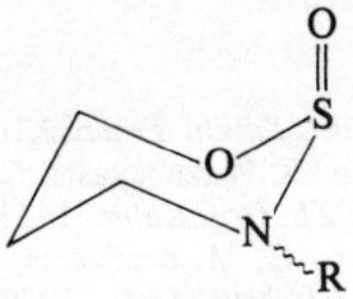

(205) 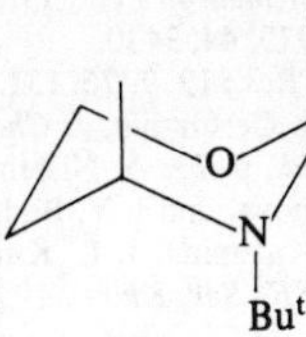   (206) 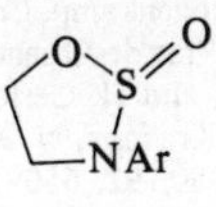   (207) 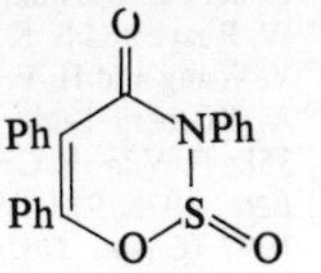   (208)

[282] (a) G. W. Buchanan, C. M. E. Cousineau, and T. C. Mundell, *Can. J. Chem.*, 1978, **56**, 2019; (b) *Tetrahedron Lett.*, 1978, 2775.

[283] (a) G. Petit, A. T. H. Lenstra, W. Van de Mierop, H. J. Geise, and D. G. Hellier, *Recl. Trav. Chim. Pays-Bas*, 1978, **97**, 202; (b) G. Petit, A. T. H. Lenstra, and H. J. Geise, *Bull. Soc. Chim. Belg.*, 1978, **87**, 659; (c) A. C. Carbonelle, Y. Jeannin, and F. Robert, *Acta Crystallogr., Sect. B*, 1978, **34**, 1631; (d) G. Petit, A. T. H. Lenstra, H. J. Geise, and P. Swepston, *Cryst. Struct. Commun.*, 1980, **9**, 187.

[284] P. Tisnes, P. Maroni, and L. Cazaux, *Org. Magn. Reson.*, 1979, **12**, 481, 490; L. Cazaux, P. Tisnes, and J. Jaud, *J. Chem. Res. (S)*, 1980, 10.

[285] T. Nishiyama, T. Mizuno, and F. Yamada, *Bull. Chem. Soc. Jpn.*, 1978, **51**, 323.

[286] L. Capuano, G. Urhahn, and A. Willmes, *Chem. Ber.*, 1979, **112**, 1012.

with $SOCl_2$ led to the formation of eight- to sixteen-membered cyclic sulphites.[287] The structures of cyclic sulphites of sugars have been established by ^1H n.m.r. methods,[288] and the preparation of cyclic sulphite esters has been used to establish the stereochemistry, *meso* or DL, of chiral pinacols.[289]

Several cyclic sulphates have been obtained, by various sulphonation procedures.[290]

Selenium- and Tellurium-containing Rings.—The synthesis of selenacyclohexane from n-pentane and SeO_2 was improved by using a zeolite catalyst containing rare-earth elements.[291] A direct formation of selenacycloalkanes, *e.g.* (209) and (210), by intramolecular oxyselenation of diallyl ether and other olefins with KSeCN and $CuCl_2$ in MeOH, has been described.[292] The reaction of an $\alpha\alpha$-dibromo-ketone with the potassium salt of ethyl diselenocarbonate gave a 61% yield of the 1,3-oxaselenolan (211).[293]

(209)   (210)   (211)

In contrast to the less reactive 1-(vinylthio)alkynes, their sulphoxides or sulphones reacted readily with Na_2Se to give 1,4-thiaselenins (212).[294] Other studies dealt with the formation of 5,6-dihydro-2*H*-selenopyran,[295] the synthesis of optically active 8-thia- and 8-selena-derivatives of 1,1'-binaphthyls, *e.g.* (213),[296] and the formation of a tetracyclic selenium heterocycle on treatment of bicyclic selenoketals with Bu^nLi.[297] A product that was obtained after oxidation of dimedone with SeO_2 has been identified as a tricyclic 1,3-oxaselenole.[298] Selena- and

[287] T. Nishiyama, K. Ido, and F. Yamada, *J. Heterocycl. Chem.*, 1979, **16**, 597; A. C. Guimaraes, J. B. Robert, L. Cazaux, C. Picard, and P. Tisnes, *Tetrahedron Lett.*, 1980, **21**, 1039.

[288] W. Reeve and S. K. Davidsen, *J. Org. Chem.*, 1979, **44**, 3430.

[289] Y. Wang and H. P. C. Hogenkamp, *Carbohydr. Res.*, 1979, **76**, 131.

[290] A. Koeberg-Telder, F. Van de Griendt, and H. Cerfontain, *J. Chem. Soc., Perkin Trans. 2*, 1980, 358; F. Van de Griendt and H. Cerfontain, *ibid.*, p. 23; A. Nishinaga and S. Wakabayashi, *Chem. Lett.*, 1978, 913; V. L. Krasnov, N. K. Tulegenova, and I. V. Bodrikov, *Zh. Org. Khim.*, 1979, **15**, 1997 (*Chem. Abstr.*, 1980, **92**, 6509); N. B. Kaz'mina, I. L. Knunyants, G. M. Kuz'yants, E. I. Mysov, and E. P. Lur'e, *Izv. Akad. Nauk SSSR, Ser. Khim.*, 1979, 118 (*Chem. Abstr.*, 1979, **90**, 137 784).

[291] E. Sh. Mamedov, S. B. Kurbanov, R. D. Mishiev, and T. N. Shakhtakhtinski, *Zh. Org. Khim.*, 1979, **15**, 1554 (*Chem. Abstr.*, 1980, **92**, 163 231).

[292] A. Toshimitsu, S. Uemura, and M. Okano, *J. Chem. Soc., Perkin Trans. 1*, 1979, 1206.

[293] C. Bak, K. Praefcke, and L. Hendriksen, *Chem.-Ztg.*, 1978, **102**, 361.

[294] W. Verboom, R. S. Sukhai, and J. Meijer, *Synthesis*, 1979, 47.

[295] C. H. Chen, G. A. Reynolds, N. Zumbulyadis, and J. A. Van Allan, *J. Heterocycl. Chem.*, 1978, **15**, 289.

[296] M. M. Harris and P. K. Patel, *J. Chem. Soc., Perkin Trans. 2*, 1978, 304.

[297] H. M. J. Gillissen, P. Schipper. P. J. J. M. Van Ool, and H. M. Buck, *J. Org. Chem.*, 1980, **45**, 319.

[298] T. Laitalainen, T. Simonen, M. Klinga, and R. Kivekaes, *Finn. Chem. Lett.*, 1979, 145 (*Chem. Abstr.*, 1980, **92**, 163 902); T. Laitalainen and T. Simonen, *ibid.*, p. 147 (*Chem. Abstr.*, 1980, **92**, 58 273).

[299] J. Bergman and L. Engman, *Org. Prep. Proced. Int.*, 1978, **10**, 289.

(212)

(213)

(214) X = Se or Te

tellura-phthalic anhydrides (214) were obtained by treating $NaBH_4$ with selenium or tellurium, followed by phthaloyl chloride.[299]

The thermolysis of 2,2-dimethyl-2,3-dihydrobenzo[*b*]selenophen oxide (215) proceeded by *syn*-elimination of selenoxide and intramolecular addition of the intermediate benzeneselenenic acid.[300] Cleavage of *s*-triselenan by its reaction with bromine offers a preparative pathway to bis-(bromomethyl) selenide.[301] A series of naphthalene and tetracene dichalcogenides (216; X, Y = S, Se, or Te) were

(215)

(216)

prepared, and their *X*-ray structures, magnetic susceptibilities, Raman spectra, half-wave oxidation potentials, and their conducting complexes with inorganic anions were investigated.[302] *X*-Ray structures and the determination of the activation energy for an intramolecular ligand-reorganization process of selenuranes have been reported.[303] Spectral studies on saturated selenium and tellurium compounds included studies on infrared and Raman spectra and investigations of ASIS effects.[304] Several five- and six-membered 1,1-dihalogeno-1-tellura-compounds have been prepared, and their properties studied.[305]

[300] H. J. Reich, S. Wollowitz, J. E. Trend, F. Chow, and D. F. Wendelborn, *J. Org. Chem.*, 1978, **43**, 1697.

[301] E. Weissflog, *Phosphorus Sulfur*, 1980, **8**, 87.

[302] J. Meinwald, D. Dauplaise, and J. Clardy, *J. Am. Chem. Soc.*, 1977, **99**, 7743; D. Dauplaise, J. Meinwald, J. C. Scott, H. Temkin, and J. Clardy, *Ann. N.Y. Acad. Sci.*, 1978, **313**, 382 (*Chem. Abstr.*, 1979, **91**, 157 665); O. N. Eremenko, S. P. Zolotukhin, A. I. Kotov, M. L. Khidekel, and E. B. Yagudskii, *Izv. Akad. Nauk SSSR, Ser. Khim.*, 1979, 1507 (*Chem. Abstr.*, 1979, **91**, 175 269).

[303] B. Dahlen and B. Lindgren, *Acta Chem. Scand., Ser. A*, 1979, **33**, 403; D. B. Denney, D. Z. Denney, and J. F. Hsu, *J. Am. Chem. Soc.*, 1978, **100**, 5982.

[304] M. G. Giorgini, G. Paliani, and R. Cataliotti, *Spectrochim. Acta, Part A*, 1977, **33**, 1083 (*Chem. Abstr.*, 1978, **89**, 59 342); J. R. Durig and W. J. Natter, *J. Chem. Phys.*, 1978, **69**, 3714; A. L. Esteban and E. Diez, *J. Magn. Reson.*, 1979, **36**, 113; A. L. Esteban and E. Diez, *J. Org. Chem.*, 1979, **44**, 3425.

[305] C. Knobler and R. F. Ziolo, *J. Organomet. Chem.*, 1979, **178**, 423; Yu. V. Migalina, I. M. Balog, V. G. Lendel, A. S. Koz'min, and N. S. Zefirov, *Khim. Geterotsikl. Soedin.*, 1978, 1212 (*Chem. Abstr.*, 1979, **90**, 6360); T. N. Srivastava, R. C. Srivastava, and M. Singh, *J. Organomet. Chem.*, 1978, **157**, 405; T. N. Srivastava, R. C. Srivastava, and H. B Singh, *Indian J. Chem., Sect. A*, 1979, **18**, 71 (*Chem. Abstr.*, 1979, **91**, 193 143); T. N. Srivastava, R. C Srivastava, and A. Bhargava, *ibid.*, p. 236 (*Chem. Abstr.*, 1980, **92**, 163 801); T. N. Srivastava, R. C. Srivastava, H. B. Singh, and M. Singh, *ibid.*, p. 367 (*Chem. Abstr.*, 1980, **92**, 110 772); R. F. Ziolo and W. H. H. Guenther, *J. Organomet. Chem.*, 1978, **146**, 245.

Treatment of sodium phenylacetylide with tellurium in DMSO gave the ditelluroles (217).[306] Thermolysis of the easily available spirocyclic tellurans (218) afforded cyclic bis(disulphides) in good yields.[307] The first heterocycles to contain sulphur, nitrogen, and tellurium, *e.g.* (219), have been obtained by the reaction of the intermediate 1,3-thiazabuta-1,3-dienes with Sb_2Te_3.[308] Proton and [13]C n.m.r. spectra of telluran-3,5-diones have been reported.[309]

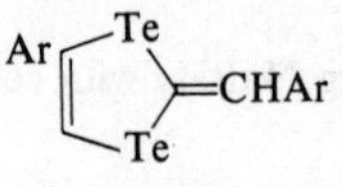

(217) 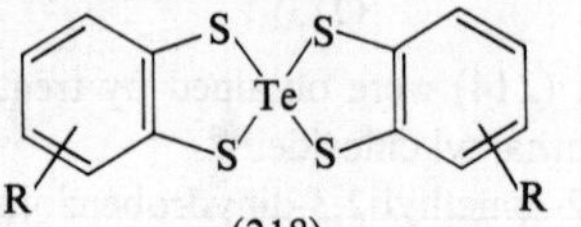   (218) 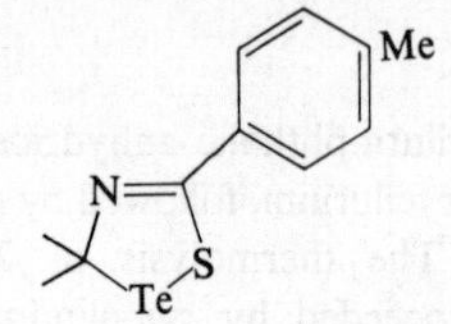   (219)

Acknowledgement. A conventional literature search has been assisted by computer, through the Lockheed search system, and a CA Current Awareness Search has been provided by the Computer Informationsdienst Graz. This work was supported by the Bundesministerium für Wissenschaft und Forschung.

[306] M. L. Petrov, V. Z. Laishev, and A. A. Petrov, *Zh. Org. Khim.*, 1979, **15**, 2596 (*Chem. Abstr.*, 1980, **92**, 163 912).

[307] B. Nakhdjavan and G. Klar, *Liebigs Ann. Chem.*, 1977, 1683.

[308] J. C. Dewan, W. B. Jennings, J. Silver, and M. S. Tolley, *Org. Magn. Res.*, 1978, **11**, 449.

[309] K. Burger, R. Ottlinger, A. Proksch, and J. Firl, *J. Chem. Soc., Chem. Commun.*, 1979, 80.

6
Heteroaromatic Compounds of Sulphur, Selenium, and Tellurium

BY M. DAVIS

1 Introduction

The organization of this Chapter is a radical change from the arrangement of the corresponding parts of Volume 5. This Reporter has tried to review, from the standpoint of sulphur chemistry rather than of heterocyclic chemistry, the highlights of current work which in the previous Volume comprised Chapters 6, 10, 11, 12, and 13, and parts of Chapters 7 and 8. It is not, of course, comprehensive; the main criteria used in selection were novelty, utility, and theoretical interest.

Two relevant reviews, on meso-ionic ring systems[1] and on selenium–nitrogen heterocycles,[2] have been published; and a three-part volume on 'Thiazole and Its Derivatives' has appeared.[3] Parts 1 and 2 of this volume are concerned with the chemistry of thiazole and substituted thiazoles, and the third part contains chapters on meso-ionic compounds, on cyanine dyes, and on selenazoles.

2 Theoretical and Spectroscopic Studies

Molecular Orbital Calculations.—The agreement between *ab initio* M.O. calculations and experimental data, usually spectroscopic, for heteroaromatic sulphur compounds is often not good. There are difficulties in deciding which M.O. terms should be included. Nevertheless, this is an active area, and some interesting results have been reported. For example, the conformations, stabilities, and charge distributions in 2- and 3-monosubstituted thiophens have been studied; the calculations show, in accord with experience, that the thienyl group prefers to act, relative to phenyl, as a π-electron donor or σ-electron acceptor, and that such electronic effects are greater at the 2- rather than at the 3-position.[4] The π-electron distributions in thiophen, selenophen, and tellurophen (1; X = S, Se, and Te, respectively) have been calculated;[5] and similar calculations have been used to correlate the observed magnetic circular dichroism (m.c.d.) of these rings.[6] The m.c.d. of thiophen is only slightly perturbed by substituents, unlike the different and

[1] K. T. Potts, *Lect. Heterocycl. Chem.*, 1978, **4**, 35.
[2] I. Lalezari and A. Shafiee, *Adv. Heterocycl. Chem.*, 1979, **24**, 109.
[3] 'The Chemistry of Heterocyclic Compounds', Vol. 34, ed. J. V. Metzger, Wiley–Interscience, New York, 1979.
[4] J. Kao and L. Radom, *J. Am. Chem. Soc.*, 1979, **101**, 311.
[5] J. Fabian, *Z. Phys. Chem. (Leipzig)*, 1979, **260**, 81.
[6] B. Norden, R. Hakansson, P. B. Pedersen, and E. W. Thulstrup, *Chem. Phys.*, 1978, **33**, 355.

much weaker m.c.d.'s of pyrrole and furan (1; X = NH and O, respectively), which change shape considerably when the molecules are substituted.

The theoreticians have been working their way steadily towards more complex molecules, and recently thiazoles, isothiazoles, and thiadiazoles have received attention in connection with n.q.r.[7] or p.e.[8] spectroscopy. Photoelectron spectroscopy has also been used to examine the energy levels of benzo-fused derivatives of these heterocycles;[9] certain cyclophanes (2; X, Y = O or S) have also been examined, and the extent of through-space and through-bond interaction between the heterocyclic rings has thereby been determined.[10]

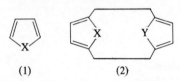

(1)   (2)

Nuclear Magnetic Resonance.—Nitrogen-14, carbon-13, and proton n.m.r. has been used to distinguish isomeric benzazoles[11] and thiadiazoles.[12] There is a large difference in nitrogen shielding, generally more than 20 p.p.m., between isomeric benzenoid (3) and *ortho*-quinonoid (4) structures, and similar differences occur in analogous sulphur compounds.[11] Natural-abundance ^{15}N n.m.r. spectra of such nitrogen–sulphur and nitrogen–selenium rings as 2,1,3-benzothiadiazole (5; X = S) and 2,1,3-benzoselenadiazole (5; X = Se) have been obtained, and it has been suggested that the high-field ^{15}N resonance of the benzothiadiazole can be explained by a significant resonance contribution from a sulphur di-imide form (6).[13]

(3)   (4)   (5)   (6)

Carbon-13 n.m.r. spectra of numerous thiazoles,[14] 1,2,3-thiadiazoles,[15] and benzothiazoles[16] have been reported and correlated with the electronic effects of substituents. Replacement of one or more of the phenyl groups in the cyclo-

[7] M. Redshaw, M. H. Palmer, and R. H. Findlay, *Z. Naturforsch., Teil A*, 1979, **34**, 220.
[8] G. Salmona, R. Faure, E. J. Vincent, C. Guimon, and G. Pfister-Guillouzo, *J. Mol. Struct.*, 1978, **48**, 205.
[9] C. Guimon, G. Pfister-Guillouzo, G. Salmona, and E. J. Vincent, *J. Chim. Phys. Phys.-Chim. Biol.*, 1978, **75**, 859; M. H. Palmer and S. M. F. Kennedy, *J. Mol. Struct.*, 1978, **43**, 33.
[10] F. Bernardi, A. Bottoni, F. P. Colonna, G. Distefano, U. Folli, and P. Vivarelli, *Z. Naturforsch., Teil A*, 1978, **33**, 959.
[11] L. Stefaniak, *Org. Magn. Reson.*, 1978, **11**, 385.
[12] M. Witanowski, L. Stefaniak, A. Grabowska, and G. A. Webb, *Spectrochim. Acta, Part A*, 1978, **34**, 877.
[13] I. Yavari, R. E. Botto, and J. D. Roberts, *J. Org. Chem.*, 1978, **43**, 2542; K. L. Williamson and J. D. Roberts, *Heterocycles*, 1978, **11**, 121.
[14] R. Faure, J.-P. Galy, E.-J. Vincent, and J. Elguero, *Can. J. Chem.*, 1978, **56**, 46.
[15] J. H. Looker, N. A. Khatri, R. B. Patterson, and C. A. Kingsbury, *J. Heterocycl. Chem.*, 1978, **15**, 1383.
[16] S. N. Sawhney and D. W. Boykin, *J. Org. Chem.*, 1979, **44**, 1136.

propenium perchlorate (7) by 2-thienyl groups causes a shift of the absorption bands of the cyclopropenium ions to longer wavelengths, showing that 2-thienyl is more effective than phenyl in delocalizing positive charge; n.m.r. spectra and pK values support this view.[17]

Mössbauer spectra have been reported for some tellurium heterocycles; *e.g.*, (8).[18] The complete structure of 1,2,5-thiadiazole (9) has been determined by double-resonance modulation (DRM) microwave spectroscopy.[19]

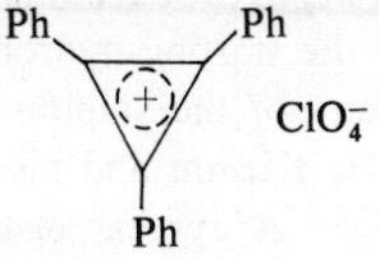

(7) 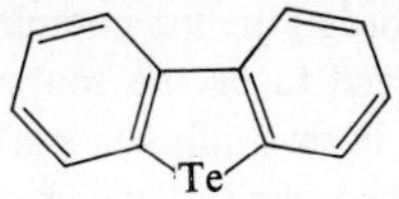   (8)    (9)

3 Thiophens, Selenophens, and Tellurophens

General.—Recent publications include a chapter on thiophens in a comprehensive treatise;[20] an article on recent advances in the synthesis of benzo[*b*]thiophens;[21] a published plenary lecture on condensed thiophens, especially thienothiophens, selenolothiophens, and related systems;[22] and a review on the synthesis of polycyclic thiophens from the direct insertion of heterosulphur bridges into vinylarenes, biaryls, and angularly condensed arenes.[23]

Ring Synthesis and Destruction.—One method of preparing thiophens is by inserting a sulphur atom into an acyclic precursor. The 'sulphur-transfer reagent' may be a simple inorganic compound, as illustrated by the formation of (11) from ethyl phenyl ketone (10) on treatment with pyridine and thionyl chloride.[24] The simplest sulphur-transfer reagent is elemental sulphur, as in the production of the thiophen (13; X = S) from 1,4-diphenyl-2,3-dibenzylbutadiene (12). The reaction of (12) with selenium dioxide affords the corresponding selenophen (13; X = Se).[25]

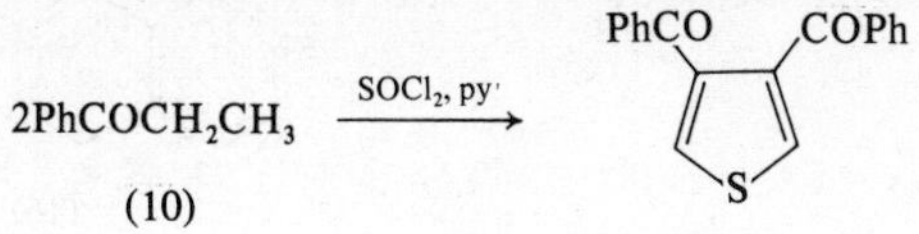

2PhCOCH$_2$CH$_3$ $\xrightarrow{\text{SOCl}_2,\ \text{py}}$

(10)

(11)

[17] G. Martelli, P. Spagnolo, L. Testaferri, and M. Tiecco, *Tetrahedron Lett.*, 1979, 281; *cf.* ref. 4.
[18] N. S. Dance and C. H. W. Jones, *Can. J. Chem.*, 1978, **56**, 1746.
[19] O. L. Stiefvater, *Z. Naturforsch., Teil A*, 1978, **33**, 1511; *ibid.*, p. 1518.
[20] O. Meth-Cohn in 'Comprehensive Organic Chemistry', Vol. 4, ed. P. G. Sammes, Pergamon Press, Oxford, 1979, p. 789
[21] B. Iddon, *Stud. Org. Chem. (Amsterdam)*, 1979, **3** (New Trends Heterocycl. Chem.), 250.
[22] V. P. Litvinov, in 'Topics in Organic Sulphur Chemistry', ed. M. Tišler, University Press, Ljubljana, 1978, p. 157
[23] L. H. Klemm, J. J. Karchesy, and D. R. McCoy, *Phosphorus Sulfur*, 1979, **7**, 9.
[24] K. Oka, *Heterocycles*, 1979, **12**, 461; see also C. M. Bonnin, P. A. Cadby, C. G. Freeman, and D. A. Ward, *Aust. J. Chem.*, 1979, **32**, 833.
[25] L. Lepage and Y. Lepage, *J. Heterocycl. Chem.*, 1978, **15**, 1185; similar sulphur-insertion reactions have been described by D. Nasipuri, A. Sarkar, P. K. Chakraborty, and I. De Dalal, *J. Indian Chem. Soc.*, 1978, **55**, 1232, and by K. Adachi and J. Tanaka, *Nippon Kagaku Kaishi*, 1978, 1666 (*Chem. Abstr.*, 1978, **90**, 137 605).

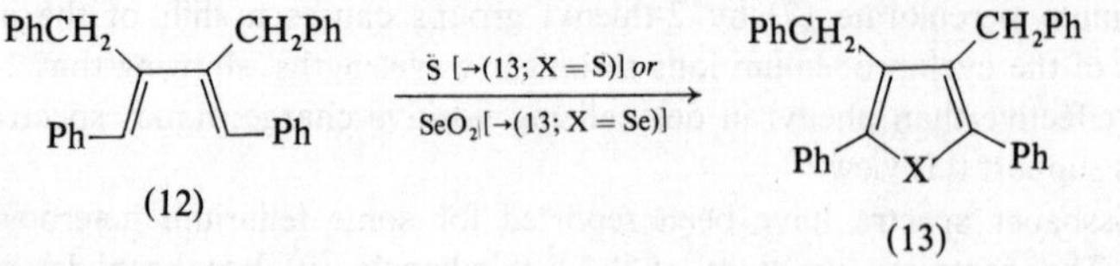

Hydrogen sulphide reacts reversibly with many unsaturated or aromatic hydrocarbons, in the presence of mixed oxide catalysts, to produce thiophens; these may be polycyclic, according to the complexity of the starting hydrocarbon. This reaction, which is believed to be the source of many of the sulphur compounds found in crude oil, has been studied in particular by Klemm and his co-workers, who have now published their detailed results.[23, 26] A typical example is the formation of the isomeric benzonaphthothiophens (15) and (16) from the reaction of 2-phenylnaphthalene (14) with hydrogen sulphide over a sulphided cobaltous oxide–molybdic oxide–alumina catalyst at 450—630 °C in a flow system. Thiophens may also be obtained from aldehydes in like manner.[27] The kinetics and mechanism of these sulphurization reactions, and especially of the industrially important reverse reactions (hydrodesulphurization), have been examined in some detail; thiols may be intermediates.[26, 28]

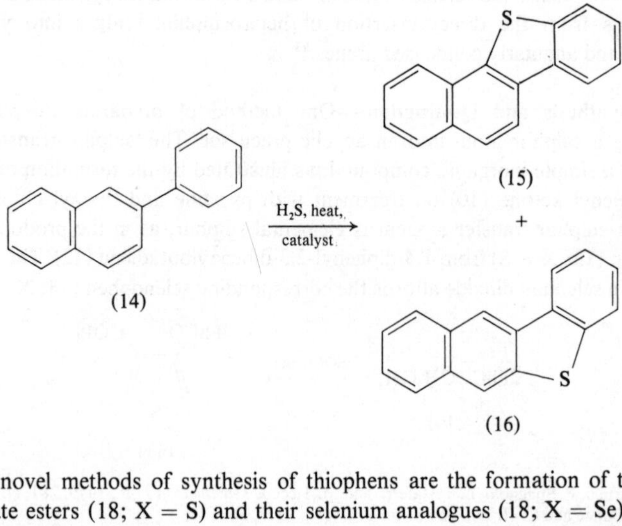

Other novel methods of synthesis of thiophens are the formation of thiophen 3-malonate esters (18; X = S) and their selenium analogues (18; X = Se) from an acyclic precursor (17) by treatment with sodium sulphide or selenide;[29] the synthesis of benzo[b]thiophens by the photocyclization of 2-(arylthio)aceto

[26] L. H. Klemm and J. J. Karchesy, *J. Heterocycl. Chem.*, 1978, **15**, 65; *ibid.*, p. 281; L. H. Klemm, J. J. Karchesy, and R. F. Lawrence, *ibid.*, p. 417; L. H. Klemm and J. J. Karchesy, *ibid.*, p. 561; L. H. Klemm, J. J. Karchesy, and R. F. Lawrence, *ibid.*, p. 773; L. H. Klemm and R. F. Lawrence, *ibid.* 1979, **16**, 599.

[27] J. Barrault, M. Guisnet, J. Lucien, and R. Maurel, *J. Chem. Res. (S)*, 1978, 207.

[28] N. K. Nag, A. V. Sapre, D. H. Broderick, and B. C. Gates, *J. Catal.*, 1979, **57**, 509; J. M. Pazos and P. Andréu, *Can. J. Chem.*, 1980, **58**, 479.

[29] J. P. Clayton, A. W. Guest, A. W. Taylor, and R. Ramage, *J. Chem. Soc., Chem. Commun.*, 1979, 500.

acetates;[30] the ring-contraction of 3-bromothiocoumarin (19) by amines, which yields benzo[b]thiophen-2-carboxamides (20);[31] or the remarkable easy preparation of 2-phenylbenzo[b]thiophen-3-amines (21) from 2-nitrobenzonitrile and toluene-α-thiol.[32]

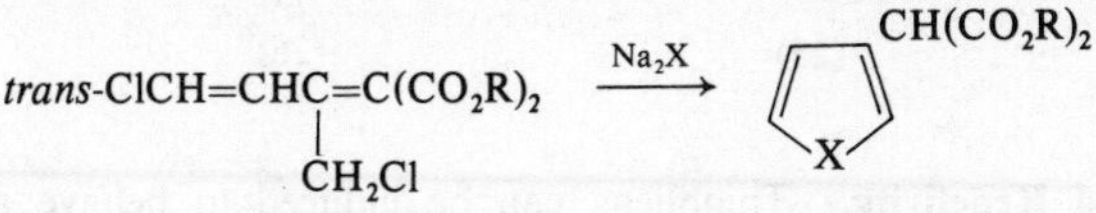

$$trans\text{-ClCH=CHC=C(CO_2R)_2} \xrightarrow{Na_2X}$$

with CH_2Cl substituent

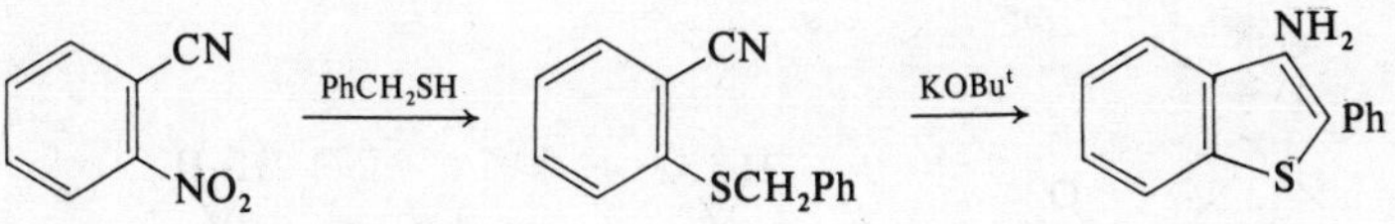

(17) → (18) product $CH(CO_2R)_2$ on ring X

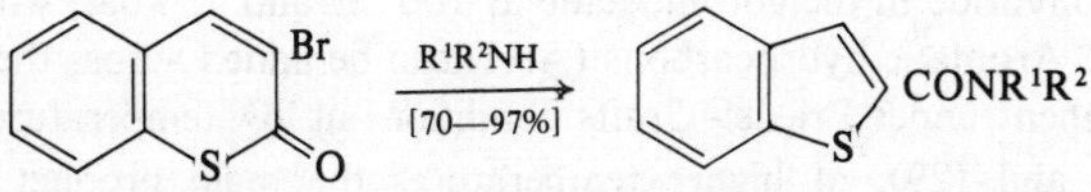

(19) $\xrightarrow[\text{[70—97%]}]{R^1R^2NH}$ (20) $CONR^1R^2$

(with CN, NO₂) $\xrightarrow{PhCH_2SH}$ (with CN, SCH₂Ph) $\xrightarrow{KOBu^t}$ (21) NH_2, Ph

Apart from hydrodesulphurization, reactions causing transformation or destruction of the thiophen ring are rather rare. Three interesting examples have been reported. In the first,[33] photolysis of 3-cyanothiophen (22) in furan affords the adduct (23) as the major product. 2-Cyanothiophen behaves in a similar way, and an analysis of permutation patterns has indicated that there is a mechanism involving first the 2,5-bridged compound (24) followed by a 'walk' of the sulphur atom. Anils (25) are converted into pyrroles (26) by triethyl phosphite;[34] a similar ring-opening by intramolecular insertion of nitrene has also been shown to occur with o-nitrophenyl-di-(2-thienyl)methanes.[35]

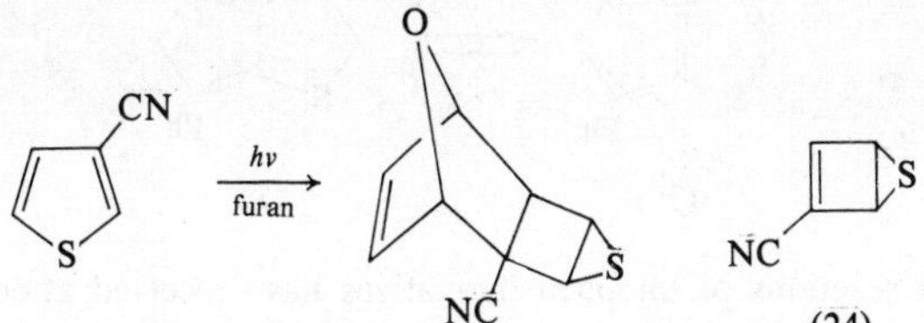

(22) $\xrightarrow[\text{furan}]{h\nu}$ (23) → (24)

[30] T. Sasaki and K. Hayakawa, *Tetrahedron Lett.*, 1980, **21**, 1525.
[31] V. L. Savel'ev, T. G. Afanas'eva, and V. A. Zagorevskii, *Khim. Geterotsikl. Soedin.*, 1978, 1340 (*Chem. Abstr.*, 1979, **90**, 38 740).
[32] J. R. Beck, *J. Heterocycl. Chem.*, 1978, **15**, 513.
[33] J. A. Barltrop, A. D. Day, and E. Irving, *J. Chem. Soc., Chem. Commun.*, 1979, 881.
[34] V. M. Colburn, B. Iddon, H. Suschitzky, and P. T. Gallagher, *J. Chem. Soc., Chem. Commun.*, 1978, 453.
[35] G. Jones, C. Keates, I. Kladko, and P. Radley, *Tetrahedron Lett.*, 1979, 1445.

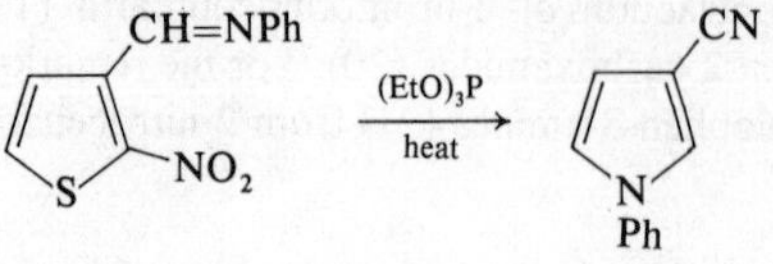

(25)　　　　　　　　　　　　　　　(26)

Reactions and Reactivity.—Thiophens can be induced to behave as dienes in Diels–Alder reactions at high pressure, as in the reaction between thiophen itself and maleic anhydride in dichloromethane at 100 °C and 15 kbar, which gives the adduct (27).[36] Aromatic hydrocarbons (ArH) can be added across the 2,3-bond of benzo[b]thiophens under Friedel–Crafts conditions at low temperatures, giving the adducts (28) and (29); at higher temperatures the main product is a 2-aryl-benzo[b]thiophen.[37]

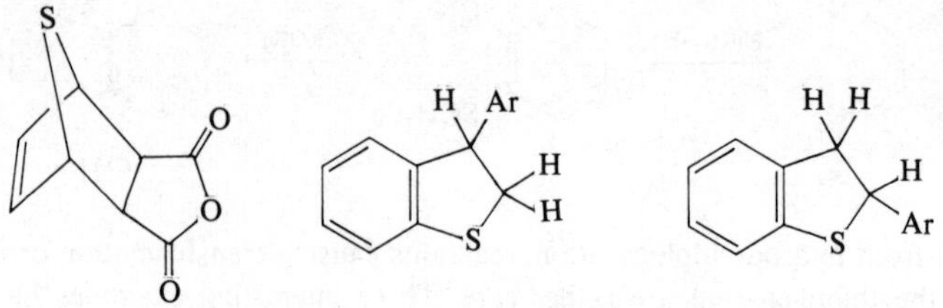

(27)　　　　　　　　　　(28)　　　　　　　　　　(29)

An unexpected finding is that 4,6-diphenyl-4H-cyclopenta[b]thiophen (30) exists in equilibrium with its 5H-isomer (31); the latter compound is fluorescent and possesses high Diels–Alder reactivity as a diene; surprisingly, (30) itself will react with dienophiles under mild conditions.[38]

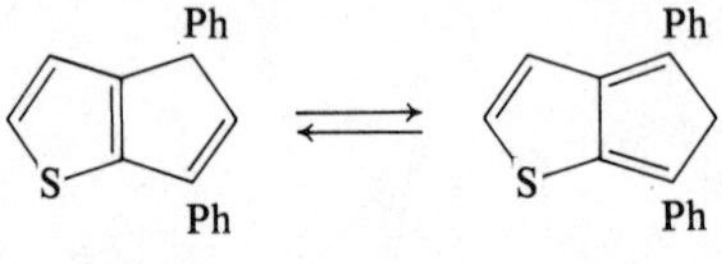

(30)　　　　　　　　　　　　　(31)

Substitution reactions of thiophen derivatives have received attention recently, particularly of nitro-thiophens, in which both *ipso*- and *cine*-substitutions may occur[39] or in which radical chain reactions may be involved.[40] Isomeric bromo-

[36] H. Kotsuki, H. Nishizawa, S. Kitagawa, M. Ochi, N. Yamasaki, K. Matsuoka, and T. Tokoroyama, *Bull. Chem. Soc. Jpn.*, 1979, **52**, 544.

[37] P. D. Clark, K. Clarke, D. F. Ewing, and R. M. Scrowston, *J. Chem. Soc., Perkin Trans. 1*, 1980, 677.

[38] J. Skramstad and T. Midthaug, *Acta Chem. Scand., Ser. B*, 1978, **32**, 413.

[39] P. Cogolli, F. Maiolo, L. Testaferri, M. Tiecco, and M. Tingoli, *J. Heterocycl. Chem.*, 1979, **16**, 1495; M. Novi, G. Guanti, F. Sancassan, and C. Dell'Erba, *J. Chem. Soc., Perkin Trans. 1*, 1978, 1140.

[40] P. J. Newcombe and R. K. Norris, *Aust. J. Chem.*, 1978, **31**, 2463; *ibid.*, 1979, **32**, 2647.

iodo-thiophens give the same mixture of products with sodium methoxide – the so-called 'halogen-dance'.[41]

Rate Data, Substituent Effects, and Hammett Relationships.— Infrared spectroscopic measurements on the isomeric acids (32) and (33) show a good correlation between the data and the σ_I constants of the substituents X; with n.m.r. spectroscopy there was no good single-parameter correlation. The Hammett ρ for the series (32) was 1.5 times that for the corresponding series (33);[42] this Reporter suggests that the presence of only *one* hydrogen atom adjacent to the carboxyl group in (32) might permit greater coplanarity of that group, and thus greater sensitivity to electronic effects that are transmitted through the ring. Reliable rate data for the cyclization of ω-(2-thienyl)alkanoic acids (34) have been obtained for ring sizes up to 21 (*i.e.*, $n = 17$). Below a ring size of nine nuclei, the macrocyclization is exceedingly slow.[43]

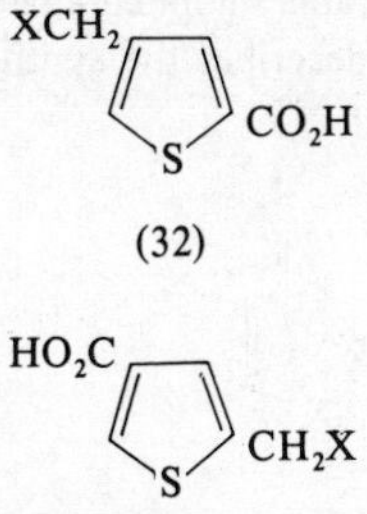

Two groups have examined the rate and equilibrium constants for the formation of Meisenheimer complexes, *e.g.* (35), by methoxide ion with nitro-thiophens and nitro-selenophens, and they have compared the results with those of the corresponding benzene compounds.[44]

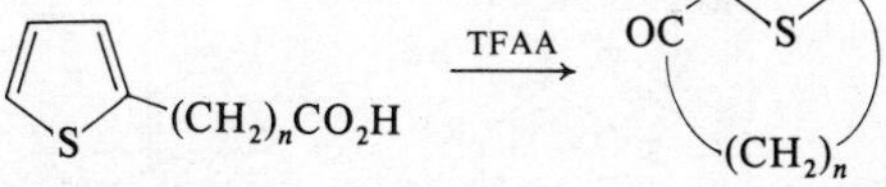

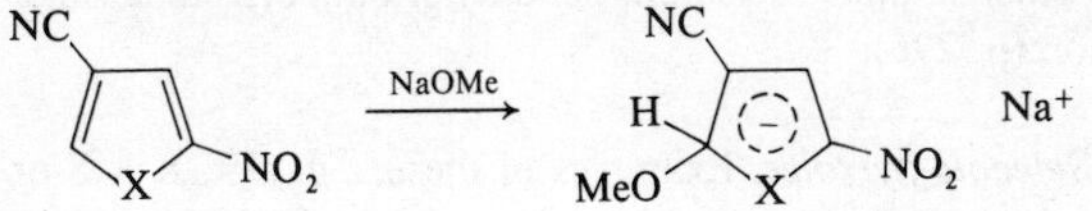

(35) X = S or Se

Synthetic Uses of Thiophen Derivatives.—The reaction between dimethyl diazo-malonate and thiophen is catalysed by rhodium salts; the product is a stable ylide (36) which can be used as an equivalent of bis(methoxycarbonyl)carbene. Thus the heating of (36) with cyclo-octene under reflux affords the bicyclononane (37) in

[41] S. Gronowitz, A. Hallberg, and C. Glennow, *J. Heterocycl. Chem.*, 1980, **17**, 171.
[42] A. Perjéssy, M. Janda, and D. W. Boykin, *J. Org. Chem.*, 1980, **45**, 1366.
[43] C. Galli, G. Illuminati, and L. Mandolini, *J. Org. Chem.*, 1980, **45**, 311.
[44] D. Spinelli, G. Consiglio, and R. Noto, *J. Chem. Res. (S)*, 1978, 242; F. Terrier, A.-P. Chatrousse, and C. Paulmier, *J. Org. Chem.*, 1979, **44**, 1634.

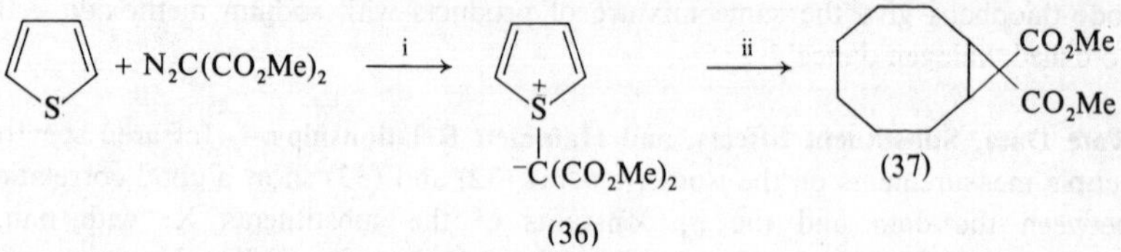

(36)

Reagents: i, Rhodium salts; ii, cyclo-octene, heat

Scheme 1

86% yield, as shown in Scheme 1.[45] Other thiophens, including benzo[*b*]thiophen, react similarly, except where electron-withdrawing substituents are present. Simple diazoketones generally produce 2-substituted thiophens rather than ylides.

Naphthothiophens, especially Thiapseudophenalenones.—Neidlein (see Vol. 5, p. 291) has continued his interesting work on these remarkable thiophen analogues of phenalene and phenalenium cations; a series of papers describes the synthesis and properties of compounds such as (38).[46]

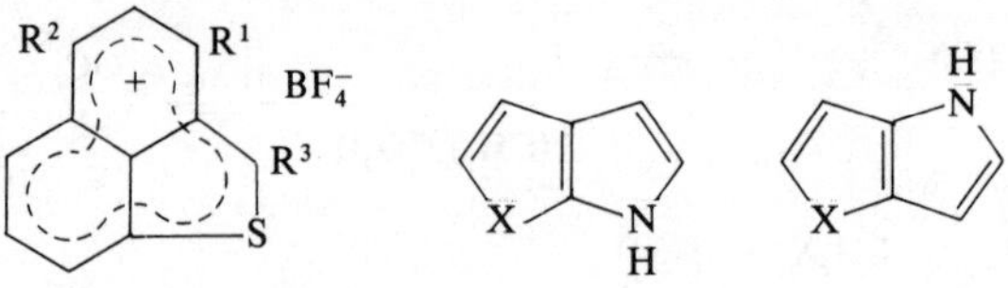

(38) R^1 = H, R^2 = OEt   (39)   (40)
     R^1 = OEt, R^2 = H
     R^3 = Me, Et, or Ph

Thiophen Rings Fused to other Heteroaromatic Systems that do not contain Sulphur, Selenium, or Tellurium.—Very many such annelated systems have been described, and in this section only those of particular note are mentioned. Annelation to other sulphur-, selenium-, and tellurium-containing rings is considered in a later section (p. 296).

Thieno- and Selenolo-pyrroles. Examples of these, *e.g.* (39; X = S or Se) and (40; X = S or Se), have been prepared by the reaction of ethyl azidoacetate with the monoaldehydes of thiophen or selenophen, followed by cyclization, hydrolysis, and decarboxylation of the product.[47]

Thienopyridines. These have received much attention in the past few years, particularly by Klemm, who has continued his examination of all six possible

[45] J. Cuffe, R. J. Gillespie, and A. E. A. Porter, *J. Chem. Soc., Chem. Commun.*, 1978, 641; R. J. Gillespie, J. Murray-Rust, P. Murray-Rust, and A. E. A. Porter, *ibid.*, 1979, 366; R. J. Gillespie and A. E. A. Porter, *J. Chem. Soc., Perkin Trans. 1*, 1979, 2624.
[46] R. Neidlein and H. Seel, *Arch. Pharm. (Weinheim, Ger.)*, 1978, **311**, 324; R. Neidlein and K. F. Cepera, *Liebigs Ann. Chem.*, 1978, 627; R. Neidlein and L. Seguil-Camargo, *Arch. Pharm. (Weinheim, Ger.)*, 1978, **311**, 710; R. Neidlein, K. F. Cepera, and A. Hotzel, *ibid.*, 1978, **311**, 861; R. Neidlein and L. Seguil-Camargo, *Liebigs Ann. Chem.*, 1979, 965.
[47] S. Soth, M. Farnier, and C. Paulmier, *Can. J. Chem.*, 1978, **56**, 1429.

isomers.[48] Other workers have studied, especially, the preparation and properties of derivatives of the [2,3-*b*] isomer (41; $R^1 = R^2 = H$).[49] One versatile synthetic procedure is the Vilsmeier formylation of acetamido-thiophens, as in the preparation of the halogenated compound (41; $R^1 = Br$, $R^2 = Cl$).[50] Thieno-[2,3-*b*]pyridinium ylide (42), prepared by *N*-amination of thieno[2,3-*b*]pyridine with hydroxylamine *O*-mesitylenesulphonate and base, undergoes ring-expansion on irradiation, producing the previously unknown thieno-1*H*-1,2-diazepines (43).[51] Thieno[2,3-*c*]pyridines, on similar treatment, afford mixtures of the analogous thieno-1*H*-1,3- and -3*H*-2,3-diazepines.[52]

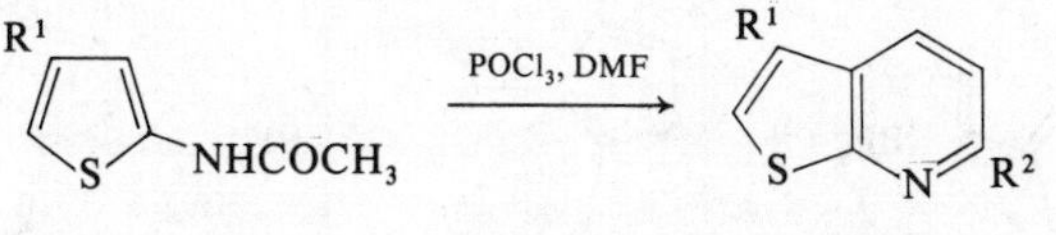

(41)

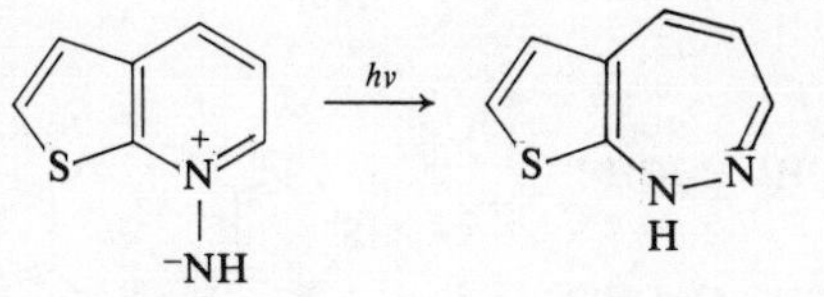

(42)   (43)

Thieno[3,2-*b*]pyridines[53] and thieno[3,2-*c*]pyridines[54] have also received attention.

A simple synthesis of thieno[3,2-*c*:4,5-*c'*]dipyridine (44) by photolysis of 4,4'-thiodipyridine has been described.[55]

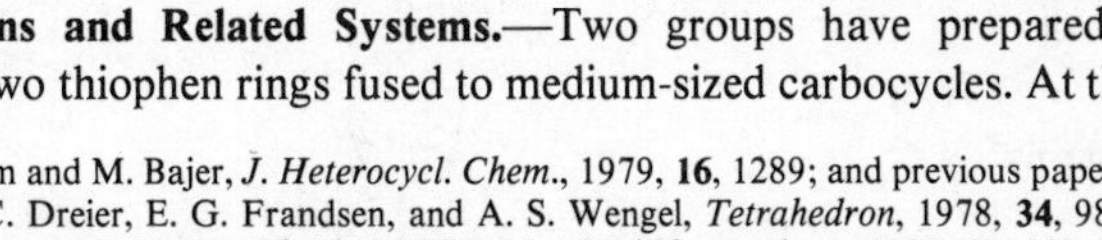

(44)

Polythiophens and Related Systems.—Two groups have prepared compounds containing two thiophen rings fused to medium-sized carbocycles. At the University

[48] L. H. Klemm and M. Bajer, *J. Heterocycl. Chem.*, 1979, **16**, 1289; and previous papers of the series.

[49] J. Becher, C. Dreier, E. G. Frandsen, and A. S. Wengel, *Tetrahedron*, 1978, **34**, 989; C. Corral, R. Madronero, and N. Ulecia, *Afinidad*, 1978, **35**, 129 (*Chem. Abstr.*, 1978, **89**, 129 477); K. Gewald, H. Schaefer, and K. Sattler, *Monatsh. Chem.*, 1979, **110**, 1189; P. M. Gilis, A. Haemers, and W. Bollaert, *Eur. J. Med. Chem. — Chim. Ther.*, 1978, **13**, 265 (*Chem. Abstr.*, 1978, **89**, 129 437); I. Lalezari, *J. Heterocycl. Chem.*, 1979, **16**, 603; V. I. Shvedov, T. P. Sycheva, and T. V. Sakovich, *Khim. Geterotsikl. Soedin.*, 1979, 1331 (*Chem. Abstr.*, 1979, **92**, 76 361); *ibid.*, 1979, 1336 (*Chem. Abstr.*, 1979, **92**, 94 271); *ibid.*, 1979, 1340 (*Chem. Abstr.*, 1979, **92**, 146 648).

[50] O. Meth-Cohn and B. Narine, *Tetrahedron Lett.*, 1978, 2045.

[51] T. Tsuchiya, M. Enkaku, J. Kurita, and H. Sawanishi, *Chem. Pharm. Bull.*, 1979, **27**, 2183.

[52] T. Tsuchiya, M. Enkaku, and H. Sawanishi, *Heterocycles*, 1979, **12**, 1471.

[53] J. M. Barker, P. R. Huddleston, and A. W. Jones, *J. Chem. Res. (S)*, 1978, 393.

[54] J. P. Maffrand and R. Boigegrain, *Heterocycles*, 1979, **12**, 1479.

[55] J. E. Rockley and L. A. Summers, *Chem. Ind. (London)*, 1979, 666.

of Lund the dithienotropylium salts (45) and (46) have been made by standard, if lengthy, syntheses from substituted thiophens;[56] and a group of workers at University College London have prepared the related compound (49) by the base-catalysed rearrangement and dimerization of the diyne (48).[57] Derivatives of the closely related dithienothiepin (47) and its isomers have been made independently by workers at the Universities of Lund and of Dijon (France).[58,59] If the substituents are bulky, as in the diacid (47; R = CO_2H), enantiomers are produced, but the barrier to inversion is not large.[58]

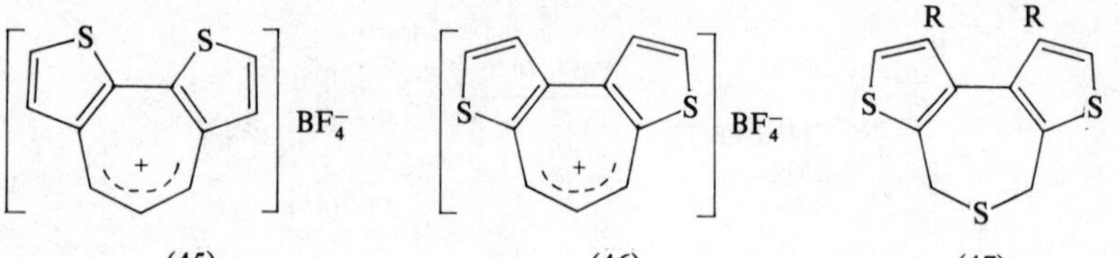

(45)   (46)   (47)

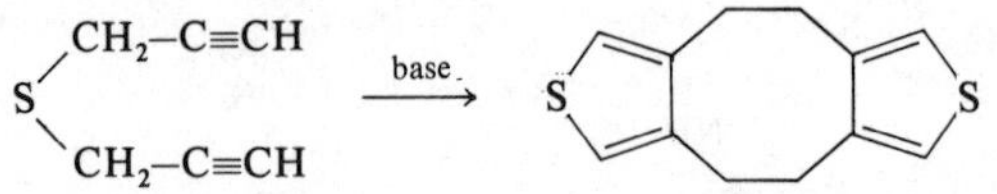

(48)   (49)

Benzo[1,2-*c*:3,4-*c'*:5,6-*c''*]trithiophen (51), a 'hexaradialene', or exocyclic benzene, has been prepared by the oxidation of the trisulphide (50), as shown in Scheme 2. It is unreactive towards dienophiles and appears to be a reasonably stable aromatic system.[60] It would be interesting to determine the lengths of the C–C bonds of the central ring, if only to decide whether this compound (51) could be described like a 'donut' ('a hole surrounded by jam and dough'). A related molecule,

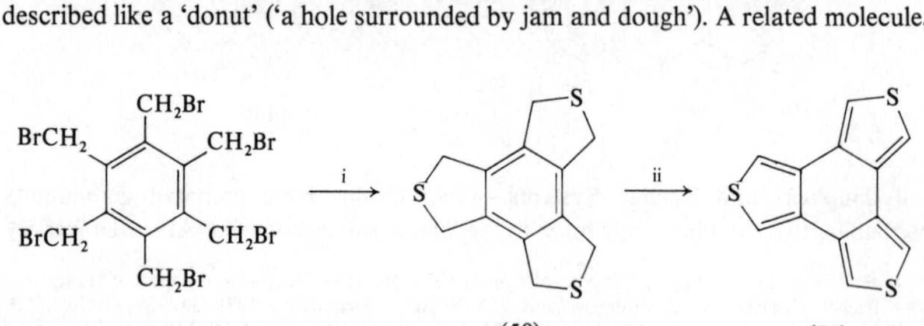

Reagents: i, Na_2S; ii, DDQ or *o*-chloranil   (50)   (51)

Scheme 2

[56] S. Gronowitz and P. Pedaja, *Tetrahedron*, 1978, **34**, 587.
[57] Y. S. P. Cheng, E. Dominguez, P. J. Garratt, and S. B. Neoh, *Tetrahedron Lett.*, 1978, 691; P. J. Garratt and S. B. Neoh, *J. Org. Chem.*, 1979, **44**, 2667.
[58] A. Almqvist and R. Hakansson, *Chem. Scr.*, 1977, **11**, 180; *ibid.*, p. 186.
[59] P. Meunier, *J. Heterocycl. Chem.*, 1978, **15**, 593.
[60] H. Hart and M. Sasaoka, *J. Am. Chem. Soc.*, 1978, **100**, 4326.

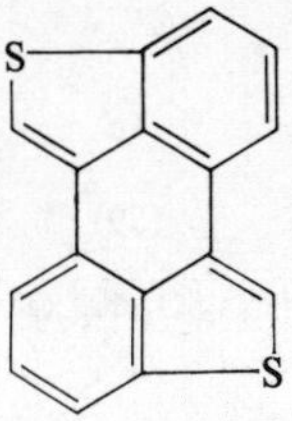

(52) 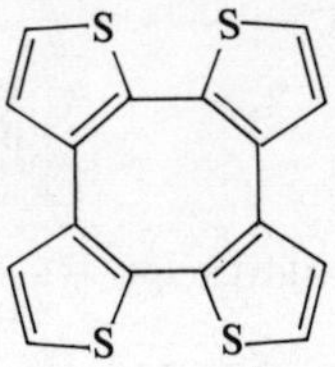   (53) 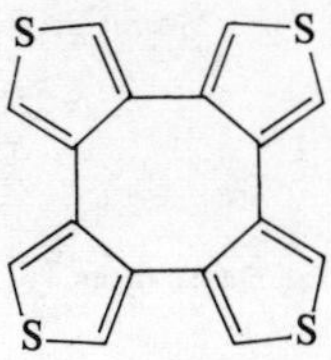   (54)

3,4:3′,4′-bibenzo[*b*]thiophen (52), has been made. It is isoelectronic with perylene and, indeed, has similar physical and chemical properties.[61]

The tetrathiophens (53) and (54) have been synthesized, in moderate yield, by the reaction of dibromobithiophens with butyl-lithium, followed by oxidative coupling with ferric or cupric chloride.[62]

Non-classical Thiophens.—Thiophens which formally require quadricovalent bonding about one or more sulphur atoms are usually called 'non-classical'. Numerous examples have been described during the past decade (see Vol. 5, pp. 282–3). Tetraphenylthieno[3,4-*c*]thiophen (55) and its radical-cation and -anion have now been subjected to a detailed spectroscopic examination, and the results have been rationalized by M.O. models.[63] Non-classical thiophens usually need several aromatic substituents for reasonable stability; examples in which the stability is conferred, at least in part, by fusion to polycyclic nitrogen heterocycles, *e.g.* (56) and (57), have been described, but the degree of stability is not great; (57), for example, has only a transient existence.[64]

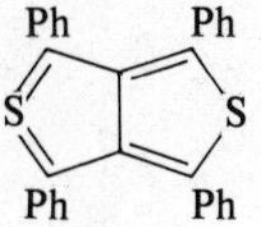

(55) 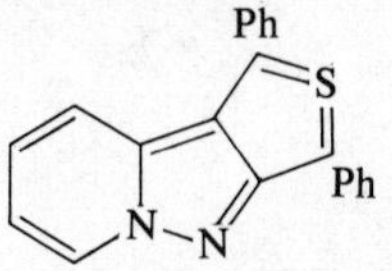   (56)

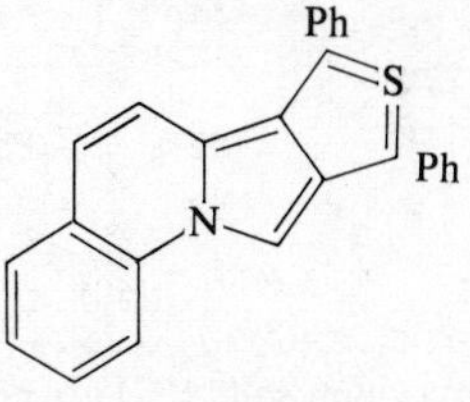

(57)

Selenophens and Tellurophens.—The mass spectrometric fragmentations of selenophen and tellurophen have been elucidated; they are analogous to that of thiophen.[65] An improved synthesis (Scheme 3) of tellurophen (59) has been described, starting from elemental tellurium and using 1,4-bis(trimethylsilyl)buta-

[61] F. Wudl, R. C. Haddon, E. T. Zellers, and F. B. Bramwell, *J. Org. Chem.*, 1979, **44**, 2491.
[62] T. Kauffmann, B. Greving, R. Kriegesmann, A. Mitschker, and A. Woltermann, *Chem. Ber.*, 1978, **111**, 1330.
[63] R. Gleiter, R. Bartetzko, G. Braehler, and H. Bock, *J. Org. Chem.*, 1978, **43**, 3893; P. Fuerderer, F. Gerson, M. P. Cava, and M. V. Lakshmikantham, *Heterocycles*, 1978, **11**, 93.
[64] K. T. Potts and S. Yao, *J. Org. Chem.*, 1979, **44**, 977; K. T. Potts, H. P. Youzwak, and S. J. Zurawel, Jr., *J. Org. Chem.*, 1980, **45**, 90.
[65] F. Fringuelli and A. Taticchi, *J. Heterocycl. Chem.*, 1978, **15**, 137.

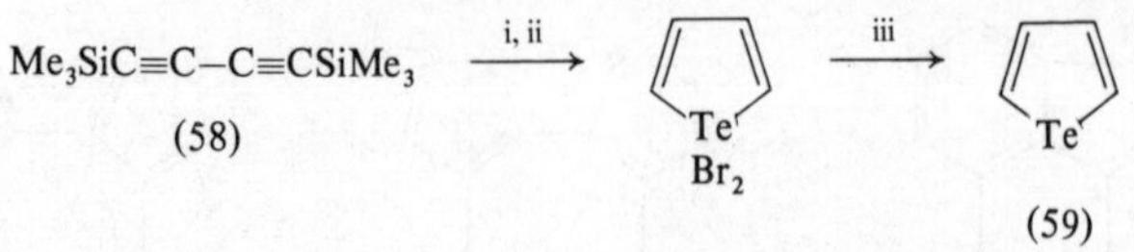

$$Me_3SiC\equiv C-C\equiv CSiMe_3 \quad \xrightarrow{i,\,ii} \quad \text{(structure with Te, Br}_2\text{)} \quad \xrightarrow{iii} \quad \text{(structure with Te)}$$

(58)   (59)

Reagents: i, Na$_2$Te (from Te + NaOH + HOCH$_2$SO$_2$Na, in water); ii, Br$_2$; iii, Na$_2$SO$_3$, K$_2$CO$_3$, H$_2$O

Scheme 3

1,3-diyne (58) as the organic moiety.[66] Compounds in which selenophen rings are fused on to benzene or naphthalene,[67] on to benzofurans,[68] or on to quinolines or carbazoles[69] have been prepared. The last-named compounds are the selenophen analogues of known carcinogens.

4 Thiopyrylium and Selenopyrylium Salts

Cyclic voltammetry has been used to measure redox potentials for two series of pyrylium and thiopyrylium salts in order to compare the electron-donating properties of oxygen and sulphur in a quantitative manner. Ionic and radical systems are more stable in the sulphur analogue in each case. The wavelength of the intramolecular charge-transfer band, and the electrochemical reduction potentials, indicate that the thiopyrylium group is more electron-withdrawing than the pyrylium one.[70] A similar comparison has been carried out, using the methyl-substituted salts (60; X = O or S) and measuring the reactivity of the methyl group towards carbonyl compounds.[71]

The benzoseleninium salt (61) has been made by the reaction of selenoflavone with phenylmagnesium bromide, followed by perchloric acid.[72]

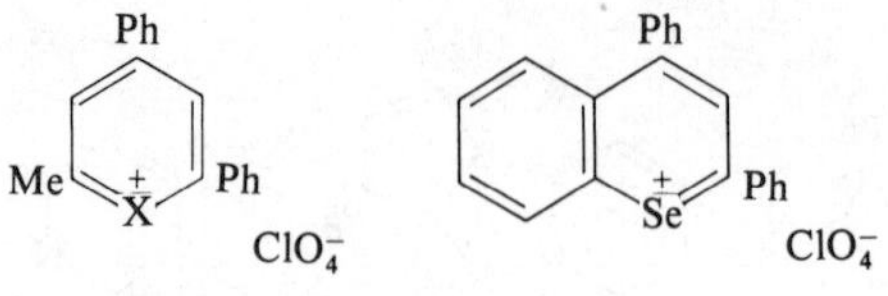

(60)   (61)

Thiopyrylium salts (62) are converted into diazo-esters (63) by the lithium salt of ethyl diazoacetate (see Scheme 4); further treatment with palladium(II) complexes produces thiepins (64), the stability of which depends on R. Remarkably, the 2-propyl compound (64; R = CHMe$_2$) extrudes sulphur at -70 °C, yet the t-butyl analogue (64; R = CMe$_3$) has a half-life of some 7 hours at 131 °C.[73]

[66] W. Lohner and K. Praefcke, *Chem. Ber.*, 1978, **111**, 3745.
[67] P. Cagniant, G. Kirsch, and L. Christiaens, *C.R. Hebd. Seances Acad. Sci., Ser. C*, 1978, **287**, 333 (*Chem. Abstr.*, 1978, **90**, 87 187).
[68] A. Shafiee and E. Behnam, *J. Heterocycl. Chem.*, 1978, **15**, 589.
[69] G. Marechal, L. Christiaens, M. Renson, and P. Jacquignon, *Collect. Czech. Chem. Commun.*, 1978, **43**, 2916.
[70] F. D. Saeva and G. R. Olin, *J. Am. Chem. Soc.*, 1980, **102**, 299.
[71] R. Neidlein and I. Koerber, *Arch. Pharm. (Weinheim, Ger.)*, 1978, **311**, 236.
[72] R. Neidlein and I. Koerber, *Arch. Pharm. (Weinheim, Ger.)*, 1978, **311**, 170.
[73] S. Yano, K. Nishino, K. Nakasuji, and I. Murata, *Chem. Lett.*, 1978, 723; K. Nishino, S. Yano, Y. Kohashi, K. Yamamoto, and I. Murata, *J. Am. Chem. Soc.*, 1979, **101**, 5059.

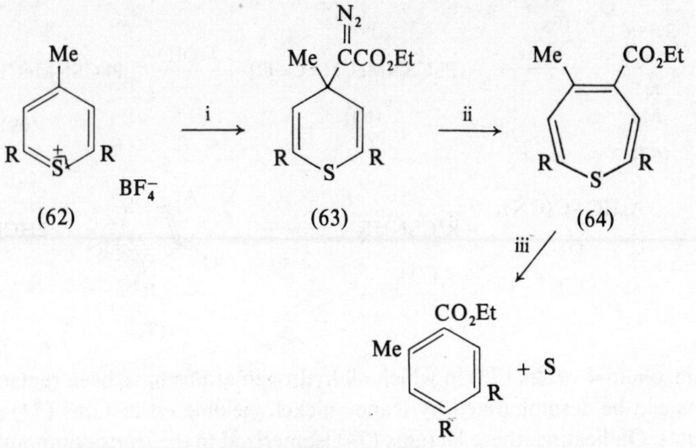

Reagents: i, $N_2C(Li)CO_2Et$; ii, palladium(II) complexes; iii, heat

Scheme 4

5 Thiazoles and Selenazoles

General.—The three-part book[3] on thiazoles, which includes a chapter on selenazoles, has been mentioned earlier.

Mass Spectrometry.—Several groups have examined in detail the mass spectra of thiazoles and isothiazoles,[74] of benzothiazoles,[75,76] and of benzoselenazoles.[75] That thiazole and isothiazole isomerize to a common structure before fragmentation has been shown by an analysis of the abundances of metastable ions and the releases of kinetic energy.[74]

Synthesis and Reactions.—Oxazolium-5-olates (65) add carbon disulphide with high regioselectivity, and the adducts lose carbon dioxide to afford thiazolium-5-thiolates (66) in excellent yield.[77] The thiazolium-5-olate (67) undergoes photochemical ring-opening to give the keten (68), which can be trapped by ethanol, producing the ester (69); a competing fragmentation, yielding carbonyl sulphide, also occurs.[78] The reaction of *gem*-dicyano-epoxides (70) with thioamides (71) yields the corresponding thiazolium-4-olates (72)[79] (see Chap. 3, Pt III, p. 186).

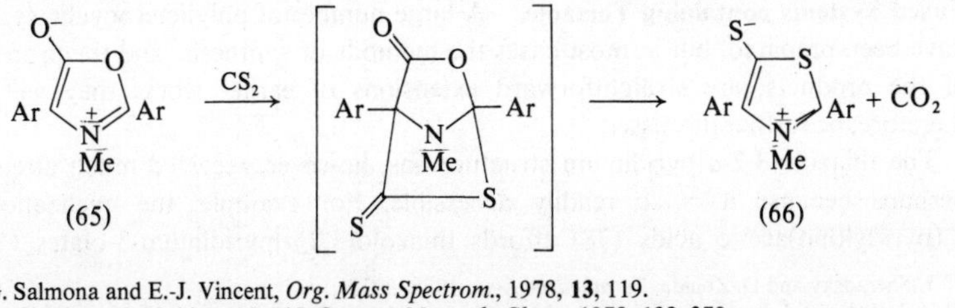

[74] G. Salmona and E.-J. Vincent, *Org. Mass Spectrom.*, 1978, **13**, 119.
[75] G. Ciurdaru, Z. Moldovan, and I. Oprean, *Monatsh. Chem.*, 1978, **109**, 379.
[76] S. Claude, R. Tabacchi, L. Duc, R. Fuchs, and K.-J. Boosen, *Helv. Chim. Acta*, 1980, **63**, 682.
[77] R. Huisgen and T. Schmidt, *Liebigs Ann. Chem.*, 1978, 29.
[78] N. H. Toubro, B. Hansen, N. Harrit, A. Holm, and K. T. Potts, *Tetrahedron*, 1979, **35**, 229.
[79] M. Baudy, A. Robert, and A. Foucaud, *J. Org. Chem.*, 1978, **43**, 3732.

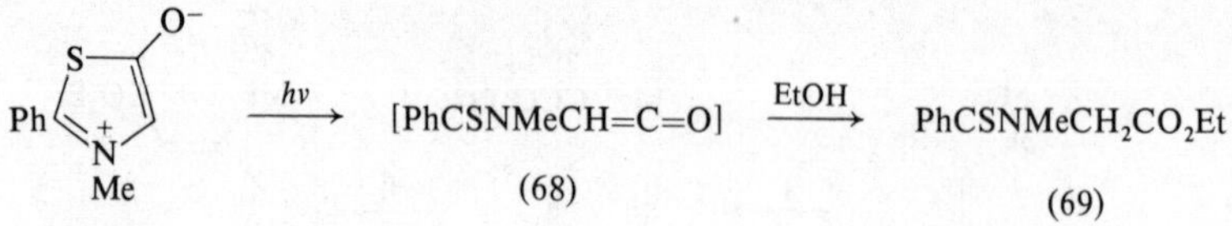

(67) → (68) → (69)

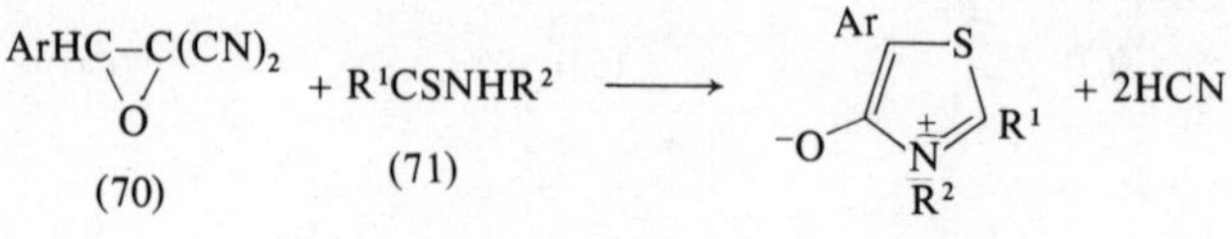

(70) + (71) → (72)

Thiazolium-4-olates (73) in which all hydrogen atoms have been replaced by aryl groups can be desulphurized by Raney nickel, yielding *cis*-lactams (74) as the sole products. On heating, these lactams (74) isomerized to the *trans*-compounds (75).[80]

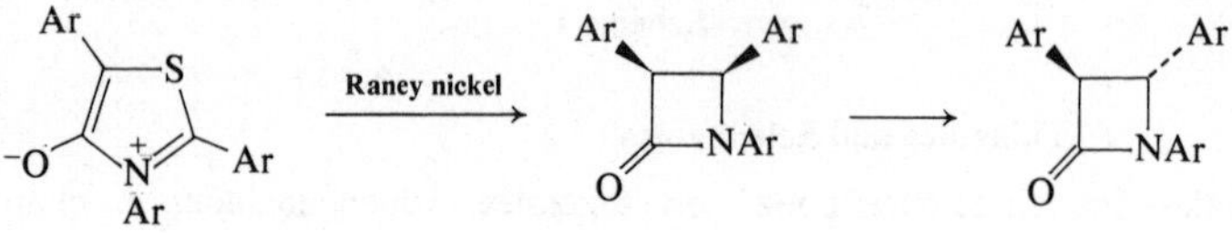

(73) → (74) → (75)

Quaternary thiazolium salts that possess a 2-chloroethyl group, *e.g.* (76), are rapidly rearranged by aqueous base to thietan derivatives, *e.g.* (77), presumably *via* an intermediate bicyclic sulphonium salt.[81]

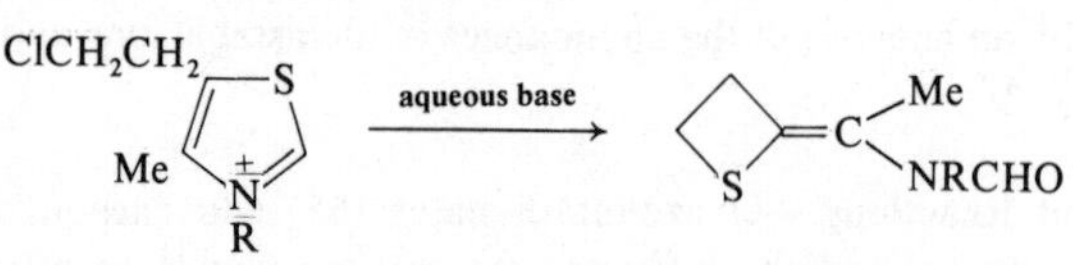

(76) → (77)

Benzothiazoles can be regarded for synthetic purposes as carbonyl equivalents, and the sequence shown in Scheme 5 illustrates the synthesis of aldehydes. Yields are almost quantitative throughout.[82]

Fused Systems containing Thiazole.—A large number of polyheterocyclic systems have been prepared, but in most cases the methods of synthesis, and the properties of the products, are straightforward extensions of earlier work; they will not, therefore, be further discussed.

The thiazolo[3,2-*a*]pyridinium structure has, however, received much attention, perhaps because it is so readily accessible. For example, the cyclization of 2-(pyridylthio)acetic acids (78) affords thiazolo[3,2-*a*]pyridinium-3-olates (79).[83]

[80] T. Sheradsky and D. Zbaida, *Tetrahedron Lett.*, 1978, 2037.
[81] H. J. Federsel, J. Bergman, and U. Stenhede, *Heterocycles*, 1979, **12**, 751.
[82] E. J. Corey and D. L. Boger, *Tetrahedron Lett.*, 1978, 5.
[83] L. T. Gorb, N. N. Romanov, E. D. Sych, and A. I. Tolmachev, *Dopov. Akad. Nauk Ukr. RSR, Ser. B*, 1978, 894 (*Chem. Abstr.*, 1978, **90**, 38 831); L. T. Gorb, N. N. Romanov, and A. I. Tolmachev, *Khim. Geterotsikl. Soedin.*, 1979, 1343 (*Chem. Abstr.*, 1979, **92**, 128 789).

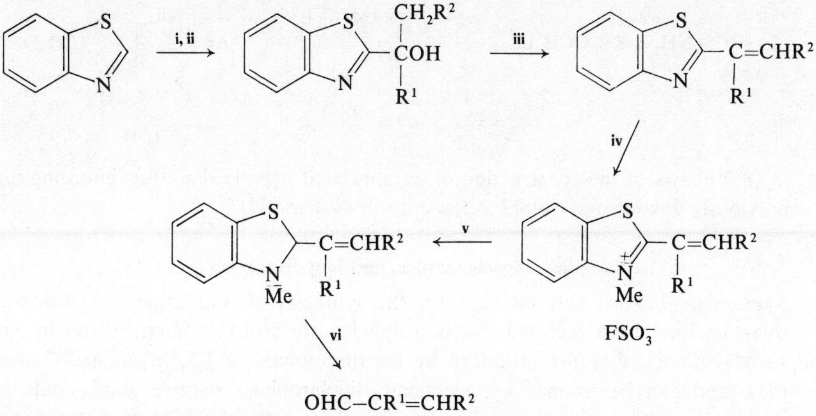

Reagents: i, BuLi; ii, $R^1COCH_2R^2$; iii, P_2O_5, $MeSO_3H$; iv, FSO_3Me; v, $NaBH_4$; vi, $AgNO_3$, H_2O

Scheme 5

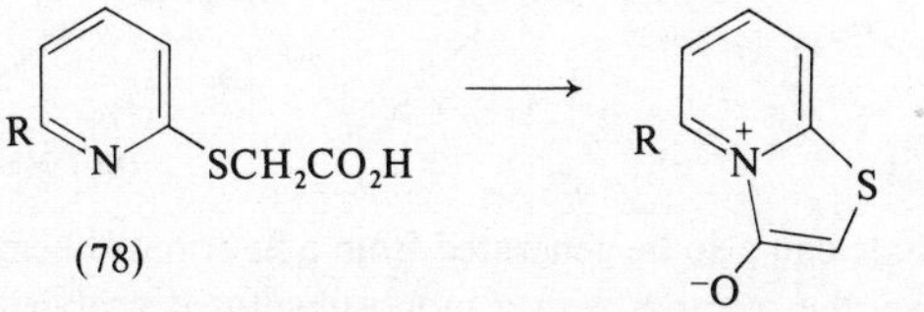

The nucleus C-2 of the thiazole ring in these compounds is appreciably nucleophilic, and it will react readily with electrophiles, including diazonium salts and aldehydes.

In much the same way, 3-hydroxypyridine-2-thione (80) reacts with 1,1,2,2-tetrabromoethane to afford the 8-olate (81); the remaining bromine atom of (81) allows selective deuteriation on C-2 and C-3 of the thiazole ring.[84] Other halogen compounds react with (80) in an analogous fashion;[85] quinoline-2-thione (82) can also be used.[86]

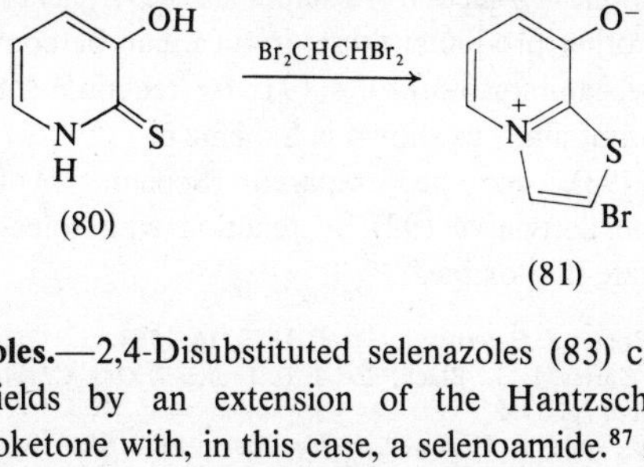

Selenazoles.—2,4-Disubstituted selenazoles (83) can be prepared in moderate to good yields by an extension of the Hantzsch reaction, the reaction of a halogenoketone with, in this case, a selenoamide.[87] Appropriate manipulation of R^2

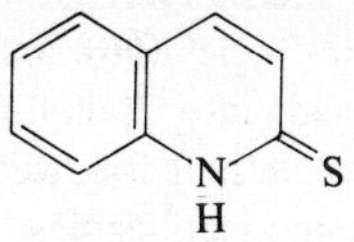

[84] T. Laerum, G. A. Ulsaker, and K. Undheim, *Acta Chem. Scand., Ser. B*, 1978, **32**, 651.
[85] G. A. Ulsaker, T. Laerum, and K. Undheim, *Acta Chem. Scand., Ser. B*, 1978, **32**, 460.
[86] K. T. Potts and D. R. Choudhury, *J. Org. Chem.*, 1978, **43**, 2700.
[87] V. I. Cohen, *Synthesis*, 1979, 66.

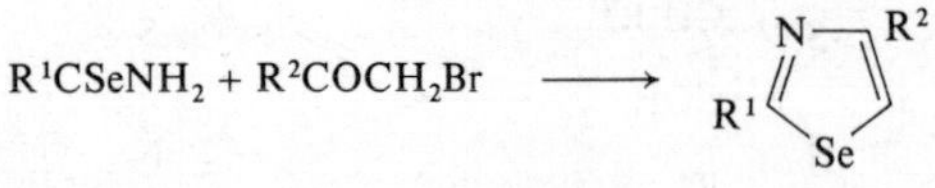

(83) 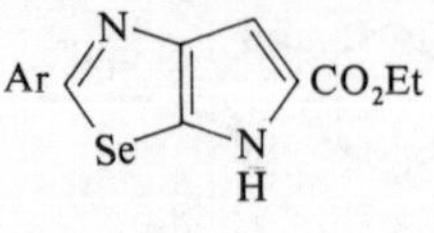   (84)

in (83) allows of the construction of an annelated pyrrole ring, thus affording the previously unknown pyrrolo[3,2-*d*]selenazole system (84).[88]

6 Isothiazoles, Isoselenazoles, and Isotellurazoles

Synthesis.—Several new methods for the synthesis of isothiazoles and benzisothiazoles have been reported. Nitrile sulphides are probable intermediates in one such synthesis; they are produced by the thermolysis of 1,3,4-oxathiazol-2-ones (85), and can be trapped by acetylenic dipolarophiles to give good yields of isothiazoles (86); unsymmetrical acetylenes afford a mixture of isomeric products. Isothiazolines are obtained in reasonable yields if olefinic dipolarophiles are used.[89]

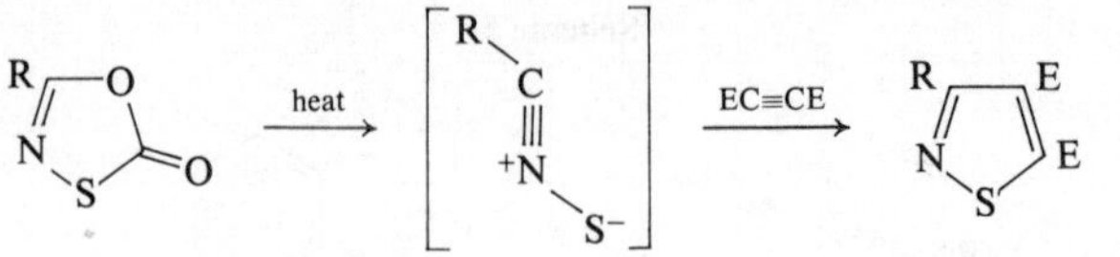

(85)   (86) E = CO_2Me

The nitrile sulphide can also be generated from a fluorinated benzylamine derivative (87); in this case, the reaction with a monosubstituted acetylene yields predominantly the 4-substituted isothiazole (88).[90] These 4-substituted isothiazoles (88) have also been prepared by an alternative synthesis from 3-aminocinnamate esters.[91]

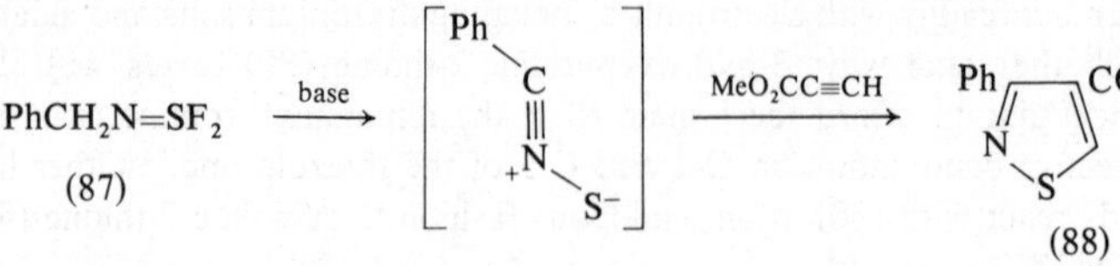

(87)   (88)

The thermolysis of azetidinone (β-lactam) disulphides (89) yields the isothiazolin-3-ones (90), amongst other products; these, in turn, can be converted (in high yield) into 3-alkyl-(or aryl-)aminoisothiazoles (91) by reaction with methyl fluorosulphonate followed by ammonia,[92] as shown in Scheme 6.

4-Amino-isothiazoles, *e.g.* (93), can be prepared (Scheme 7) from the corresponding *O*-hydroxylamine derivative (92) by reaction with toluene-α-thiol and cyclization of the intermediate thio-oxime.[93]

[88] A. Shafiee, A. Mazloumi, and V. I. Cohen, *J. Heterocycl. Chem.*, 1979, **16**, 1563.

[89] R. K. Howe, T. A. Gruner, L. G. Carter, L. L. Black, and J. E. Franz, *J. Org. Chem.*, 1978, **43**, 3736; R. K. Howe and J. E. Franz, *ibid.*, p. 3742.

[90] M. J. Sanders, S. L. Dye, A. G. Miller, and J. R. Grunwell, *J. Org. Chem.*, 1979, **44**, 510.

[91] R. K. Howe, T. A. Gruner, L. G. Carter, and J. E. Franz, *J. Heterocycl. Chem.*, 1978, **15**, 1001.

[92] M. Sako and Y. Maki, *Chem. Pharm. Bull.*, 1978, **26**, 1236; J. Rokach and P. Hamel, *J. Heterocycl. Chem.*, 1978, **15**, 695; J. Rokach, P. Hamel, Y. Girard, and G. Reader, *J. Org. Chem.*, 1979, **44**, 1118.

[93] K. Gewald and P. Bellmann, *Liebigs Ann. Chem.*, 1979, 1534.

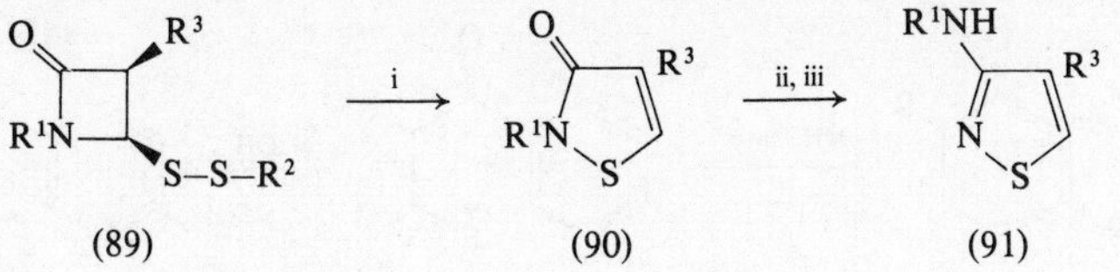

(89)   (90)   (91)

Reagents: i, heat; ii, MeSO$_3$F; iii, NH$_3$

Scheme 6

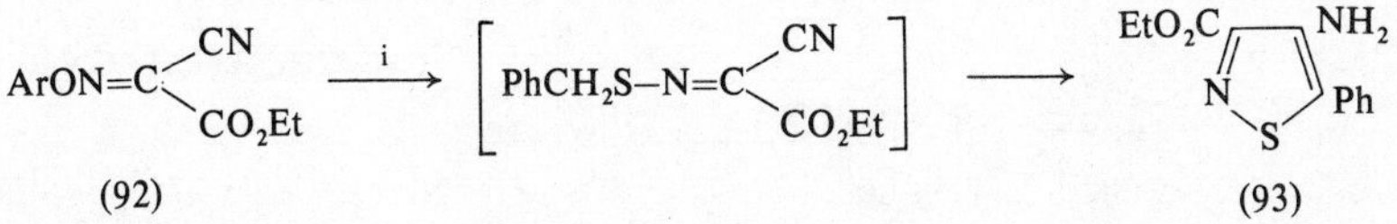

(92)   (93)

Reagents: i, PhCH$_2$SH

Scheme 7

Isothiazolium-4-olates (94) are obtained from 1,2-dithiolium-4-olates (p. 291) by successive treatment with ammonia, alkylating agents, and base,[94] as shown in Scheme 8.

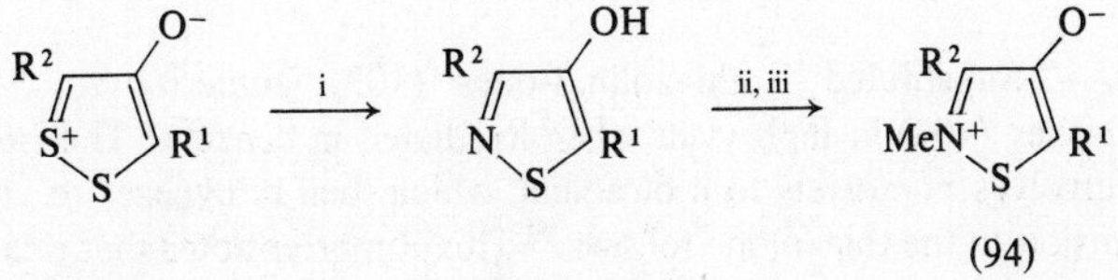

(94)

Reagents: i, NH$_3$; ii, Me$_2$SO$_4$; iii, NaOH

Scheme 8

The thiochromanones (95; R = Me or Ph) react with hydroxylamine O-mesitylenesulphonate to yield sulphilimines (96); these, with base, rearrange to 1,2-benzisothiazoles (97).[95] A remarkable preparation of a 1,2-benzisothiazole (99; X = S) and of a 1,2-benzisoselenazole (99; X = Se), by the reaction of 2,1,3-benzothia-(or selena-)diazole (98; X = S or Se) with benzyne, has been reported.[96] Another interesting synthesis, this time of a 2,1-benzisothiazole (102), is by the reaction of the fluorosilylated amine derivative (100) with bis(trimethylsilyl)sulphur di-imide (101).[97]

[94] D. Barillier, *Phosphorus Sulfur*, 1978, **5**, 251; *Bull. Soc. Chim. Fr.*, Part 2, 1979, 26; *Phosphorus Sulfur*, 1980, **8**, 79.

[95] Y. Tamura, S. M. Bayomi, C. Mukai, M. Ikeda, M. Murase, and M. Kise, *Tetrahedron Lett.*, 1980, **21**, 533.

[96] C. D. Campbell, C. W. Rees, M. R. Bryce, M. D. Cooke, P. Hanson, and J. M. Vernon, *J. Chem. Soc., Perkin Trans. 1*, 1978, 1006.

[97] U. Klingebiel and D. Bentmann, *Z. Naturforsch., Teil B*, 1979, **34**, 123; see also M. Davis, *ibid.*, 1980, **35**, 405.

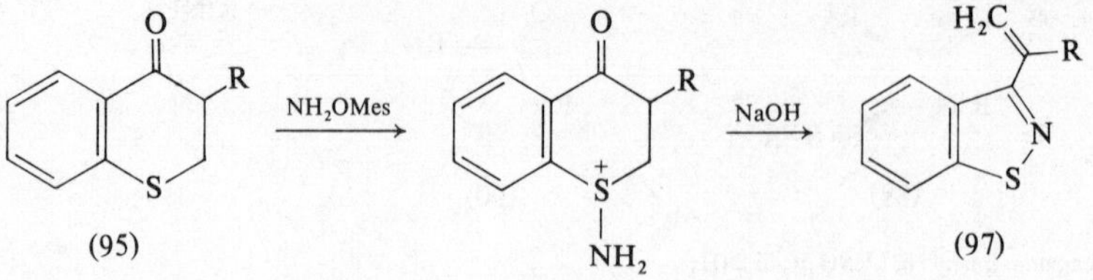

(95)   (96)   (97)

(Mes = mesitylenesulphonate)

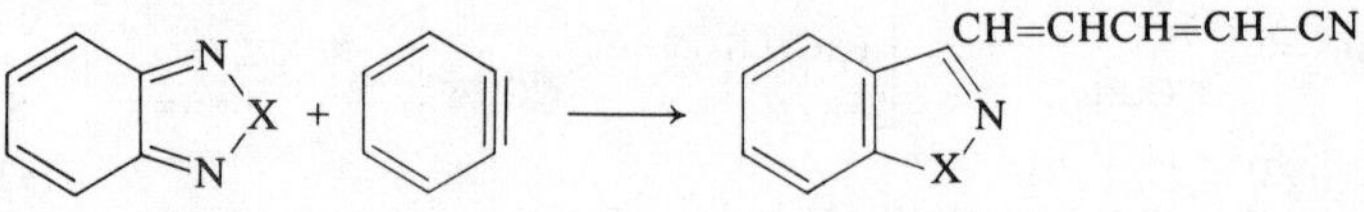

(98)   (99)

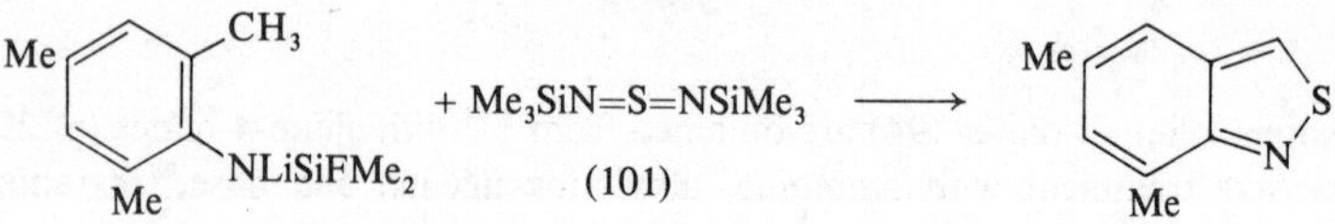

(100)   (101)   (102)

Reactions.—2-Substituted isothiazolin-3-ones (103) isomerize to 3-substituted thiazolin-2-ones (104) in high yield when irradiated in benzene. The isomerization probably involves homolysis to a biradical, which then re-cyclizes to an α-lactam. Ring-expansion to the thiazolone follows.[98] This Reporter notes that this reaction is almost the exact converse of the thermolysis of β-lactam disulphides that was mentioned earlier.[92] Thermolysis of isothiazoles (105; R^1, R^2 = H or Me) in the gas phase at 590 °C yields thioketens (RCH=C=S), and kinetic experiments showed that this decomposition is not a radical process.[99]

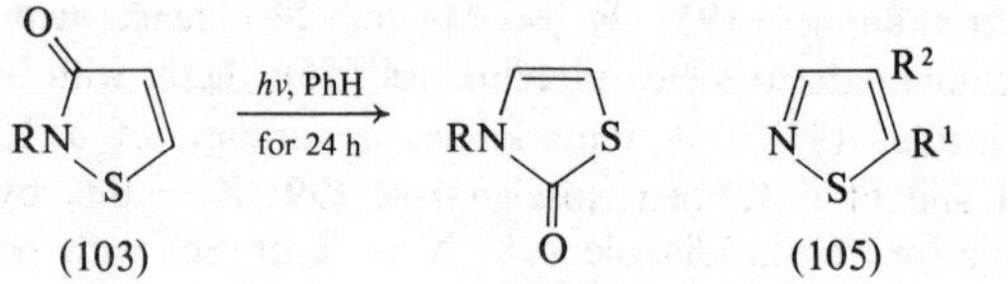

(103)   (104)   (105)

Isothiazolin-3-one 1-oxides and 1,1-dioxides (106; n = 1 or 2) react with dienes to give adducts, *e.g.* (107; n = 1 or 2), sometimes in nearly quantitative yield.[100]

Isothiazolium chlorides (108) undergo ring-expansion when treated with cyanide ion; the product is a 1,3-thiazine (109), illustrating once again the lability of the N—S bond in isothiazolium salts.[101]

[98] J. Rokach and P. Hamel, *J. Chem. Soc., Chem. Commun.*, 1979, 786.
[99] G. E. Castillo and H. E. Bertorello, *J. Chem. Soc., Perkin Trans. 1*, 1978, 325.
[100] E. D. Weiler and J. J. Brennan, *J. Heterocycl. Chem.*, 1978, **15**, 1299.
[101] J. Rokach, P. Hamel, Y. Girard, and G. Reader, *Tetrahedron Lett.*, 1979, 1281.

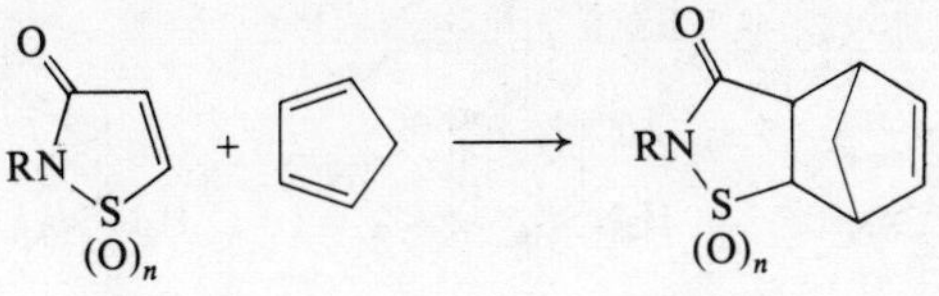

(106) + (107)

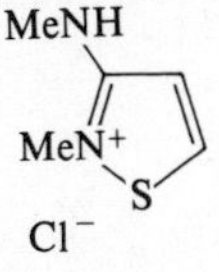

(108) 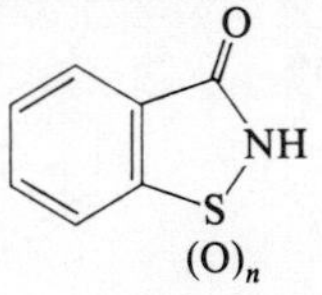   (110) 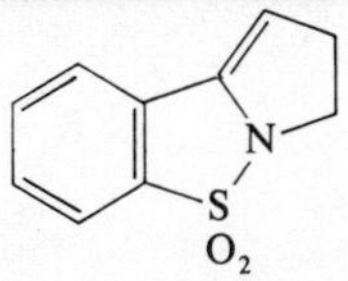   (112)

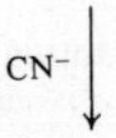

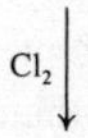

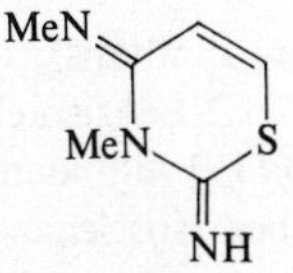

(109) 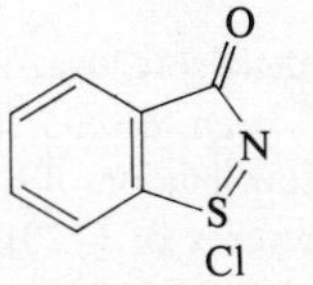   (111) 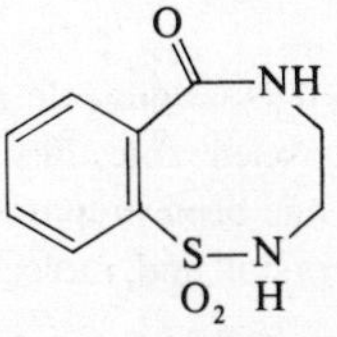   (113)

3-Chloro-1,2-(and -2,1-)benzisothiazoles are useful synthetic intermediates, since the chlorine atom is readily displaced by nucleophiles (see Vol. 5, pp. 352, 355), and recently reported work[102] has extended the range of products thereby obtained. 1-Chloro-1,2-benzisothiazol-3-one (111), obtained by the chlorination of 1,2-benzisothiazolin-3-one (110; $n = 0$), may also prove a useful intermediate.[103]

Saccharin (110; $n = 2$) and its derivatives continue to be the subject of numerous papers; little has changed in regard to the carcinogenicity controversy (see Vol. 5, p. 353), but a typical example of a 'chemical' paper is one in which it is shown how saccharin may be transformed into the imidazo-fused derivative (112) or the thiadiazocine (113).[104]

Isothiazoles Fused to other Nitrogen-containing Heterocycles.—In view of the ease with which, in the laboratory at least, an isothiazole ring may be fused to 'biological' nitrogen heterocycles, it is, perhaps, surprising that no naturally occurring isothiazole has yet been found. Typical recent work includes the efficient synthesis of isothiazolo[3,4-d]pyrimidines (114; X = NHR or SMe),[105] of isothiazolo[4,5-b]-

[102] H. Boeshagen and W. Geiger, *Synthesis*, 1979, 442; *Chem. Ber.*, 1979, **112**, 3286; A. H. Albert, D. E. O'Brien, and R. K. Robins, *J. Heterocycl. Chem.*, 1978, **15**, 529.

[103] E. S. Levchenko and T. N. Dubinina, *Zh. Org. Khim.*, 1978, **14**, 862 (*Chem. Abstr.*, 1978, **89**, 109 201).

[104] J. Ashby, D. Griffiths, and D. Paton, *J. Heterocycl. Chem.*, 1978, **15**, 1009.

[105] H. Takahashi, N. Nimura, and H. Ogura, *Chem. Pharm. Bull.*, 1979, **27**, 1147; H. Okuda, Y. Tominaga, Y. Matsuda, and G. Kobayashi, *Yakugaku Zasshi*, 1979, **99**, 989 (*Chem. Abstr.*, 1979, **92**, 128 845).

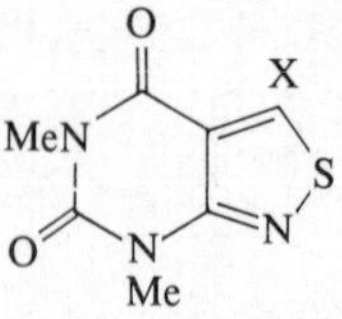

(114) 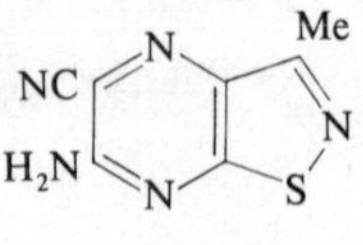   (115) 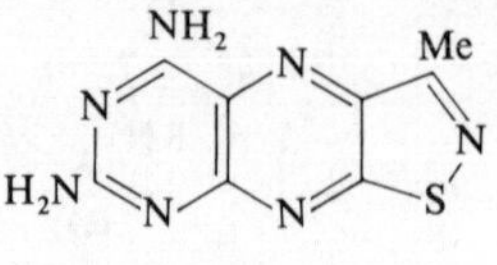   (116)

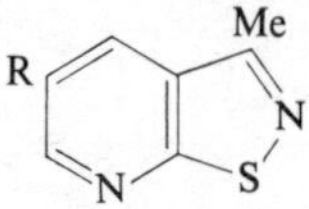

(117) 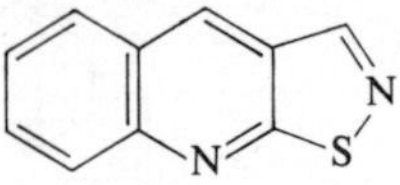   (118)

pyrazines (115) and isothiazolo[4,5-g]pteridines (116),[106] of isothiazolo[5,4-b]-pyridines (117; R = alkyl),[107] and of isothiazolo[5,4-b]quinoline (118).[108]

1,2-Benzisoselenazole and 1,2-Benzisotellurazole.—One new synthesis[96] of a 1,2-benzisoselenazole has already been discussed (p. 287). 1,2-Benzisotellurazole (120) has been prepared by the cyclization of the aldehyde (119) with ammonia.[109] The crystal and molecular structures of (120) and of 1,2-benzisoselenazole have

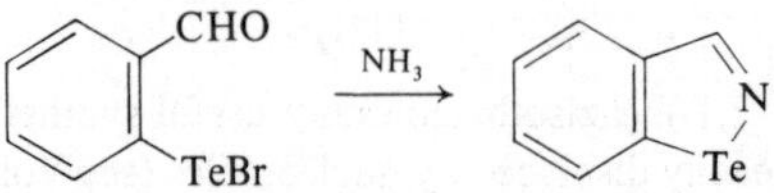

(119)   (120)

been determined; the tellurium compound has anomalous physical properties (solubility, m.pt, *etc.*), and such anomalies have been attributed to the very short intermolecular Te—N bonds (2.4 Å) in the crystal; 1,2-benzisoselenazole does not have similar short Se—N bonds.[110]

7 Oxathiolium and Dithiolium Salts

General.—It was pointed out in the previous volume (see Vol. 5, p. 306) that much of the interest in 1,3-dithioles is a result of the discovery of electrical conductivity in tetrathiafulvalenes ('organic metals'). The charge-transfer complexes formed between tetrathiafulvalenes and bromine, iodine, or other electron sinks (TCNE, TCNQ, *etc.*) may involve the production of 1,3-dithiolium ions, but discussion of this aspect is beyond the scope of this Chapter, and no attempt has been made to review the voluminous literature on this topic.

[106] E. C. Taylor and E. Wachsen, *J. Org. Chem.*, 1978, **43**, 4154.
[107] C. Skoetsch and E. Breitmaier, *Synthesis*, 1979, 370.
[108] I. Iijima and K. C. Rice, *J. Heterocycl. Chem.*, 1978, **15**, 1527.
[109] R. Weber, J. L. Piette, and M. Renson, *J. Heterocycl. Chem.*, 1978, **15**, 865.
[110] H. Campsteyn, L. Dupont, J. Lamotte-Brasseur, and M. Vermeire, *J. Heterocycl. Chem.*, 1978, **15**, 745.

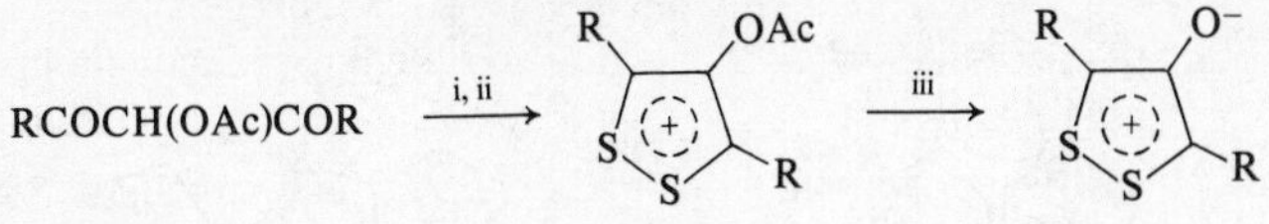

$$\text{RCOCH(OAc)COR} \xrightarrow{\text{i, ii}} (121) \xrightarrow{\text{iii}} (122)$$

Reagents: i, P_4S_{10}, CS_2 or xylene; ii, $HClO_4$; iii, py

Scheme 9

Synthesis.—3,5-Diaryl- and 3,5-dialkyl-1,2-dithiolium salts (121; R = Ar or Bu^t) can be prepared in good yield by the sulphurization of diketones (Scheme 9). Treatment of these salts (121) with pyridine affords the corresponding 4-olates (122), which can be *O*-alkylated with iodomethane and similar compounds. With ammonia, ring scission and re-cyclization occurs, and the corresponding iso-thiazolium-4-olates are formed.[94] An unusual synthesis of 1,2-dithiolium salts (124) is by the photolysis of 1,2-dithiole-3-thiones (123) in benzene or acetonitrile, in the presence of aryl or alkenyl bromides or iodides.[111]

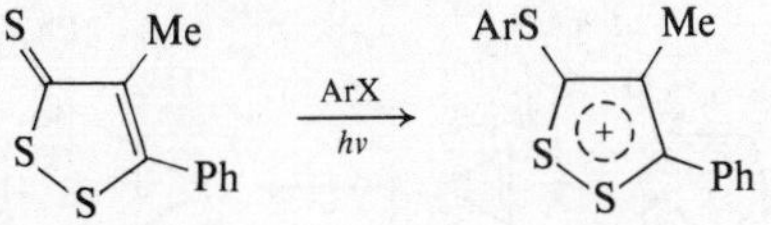

$$(123) \xrightarrow[hv]{\text{ArX}} (124)$$

1,3-Dithiolium salts, *e.g.* (126), can be made from the dithioester (125) by reaction with acetic anhydride; in polar solvents the meso-ionic 1,3-dithiolium-4-olate (127) is readily produced. If dithiocarbamates are used as starting materials, then 2-amino-1,3-dithiolium-4-olates are formed.[112]

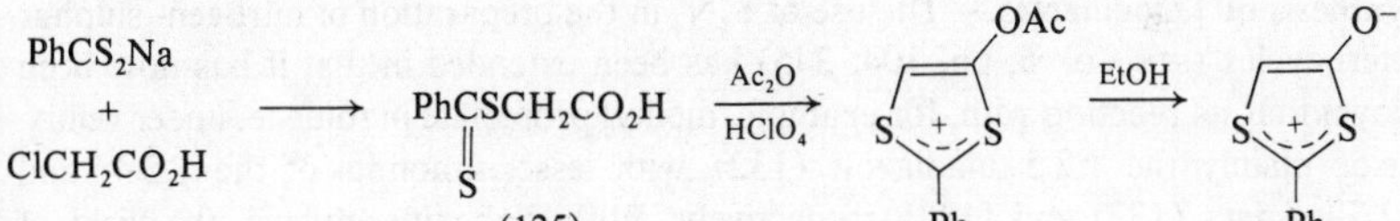

$$\begin{array}{c} \text{PhCS}_2\text{Na} \\ + \\ \text{ClCH}_2\text{CO}_2\text{H} \end{array} \longrightarrow \underset{(125)}{\text{PhCSCH}_2\text{CO}_2\text{H}} \xrightarrow[\text{HClO}_4]{\text{Ac}_2\text{O}} (126) \xrightarrow{\text{EtOH}} (127)$$

Physical and Chemical Properties.—It has been shown that charge-transfer complexes of 1,2-dithiolium cations with TCNQ have conductivities comparable with, but smaller than, those of analogous 1,3-dithiolium salts.[113]

The 1,9-dithiaphenalenyl radical (128) is a stable, monomeric, coplanar species; its e.s.r. spectrum and other spectroscopic properties have been described.[114]

[111] V. N. Drozd, G. S. Bogomolova, and Yu. M. Udachin, *Zh. Org. Khim.*, 1979, **15**, 1069 (*Chem. Abstr.*, 1979, **91**, 74 509).
[112] H. Gotthardt and C. M. Weisshuhn, *Chem. Ber.*, 1978, **111**, 2021; *ibid.*, p. 3178.
[113] G. Le Coustumer, J. Amzil, and Y. Mollier, *J. Chem. Soc., Chem. Commun.*, 1979, 353.
[114] R. C. Haddon, F. Wudl, M. L. Kaplan, J. H. Marshall, and F. B. Bramwell, *J. Chem. Soc., Chem. Commun.*, 1978, 429; R. C. Haddon, F. Wudl, M. L. Kaplan, J. H. Marshall, R. E. Cais, and F. B. Bramwell, *J. Am. Chem. Soc.*, 1978, **100**, 7629.

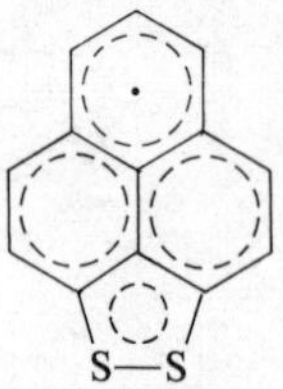

(128)

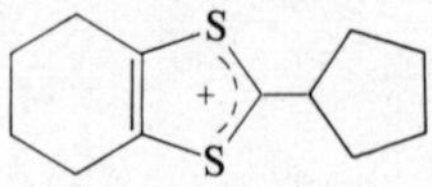

(129)

It has been pointed out that ^{13}C n.m.r. spectroscopy indicates clearly that 1,3-dithiolium cations, *e.g.* (129), are carbenium ions that are stabilized by heteroatoms, and are not delocalized systems that involve the double bond.[115] The same is probably true of 1,3-dioxolium and 1,3-diselenolium ions and analogous mixed systems.

1,3-Dithiolium-4-olates (130) undergo cycloaddition reactions with cyclic olefins and with benzoquinone; the stereochemistry of the adducts, *e.g.* (131), depends on the particular olefin.[116]

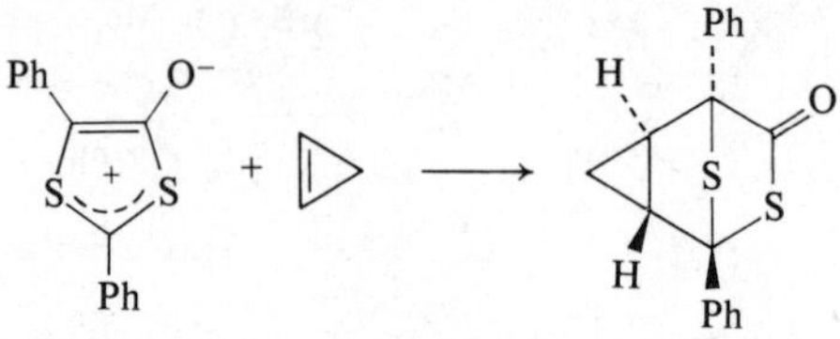

(130)   (131)

8 Thiadiazoles and Related Compounds

Synthesis of Thiadiazoles.—The use of S_4N_4 in the preparation of nitrogen–sulphur heterocycles (see Vol. 5, pp. 104, 345) has been extended in that it has now been shown that its reaction with, for example, methyl propiolate in toluene, under reflux, yields mainly the 1,2,5-thiadiazole (132), with lesser amounts of the 1,2,4- and 1,2,3-isomers (133) and (134), respectively. With diphenylacetylene, the yield of 3,4-diphenyl-1,2,5-thiadiazole is a remarkable 87%.[117] Thionyl chloride is another useful sulphur reagent, and its reaction with substituted hydrazones (135; R^1, R^2 = alkyl, aryl, or H; R^3 = COMe, CO_2Et, SO_2Tol, or $CONH_2$) gives good yields of 1,2,3-thiadiazoles (136).[118]

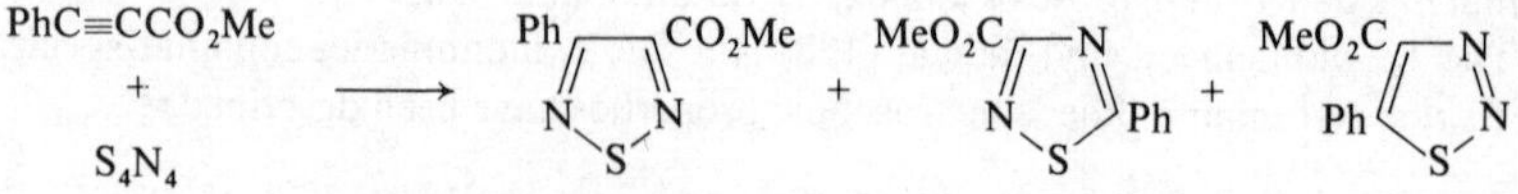

(132)   (133)   (134)

[115] U. Timm, U. Pluecken, H. Petersen, and H. Meier, *J. Heterocycl. Chem.*, 1979, **16**, 1303.
[116] H. Gotthardt, C. M. Weisshuhn, and B. Christl, *Chem. Ber.*, 1978, **111**, 3037.
[117] S. Mataka, K. Takahashi, Y. Yamada, and M. Tashiro, *J. Heterocycl. Chem.*, 1979, **16**, 1009.
[118] H. Meier, G. Trickes, E. Laping, and U. Merkle, *Chem. Ber.*, 1980, **113**, 183.

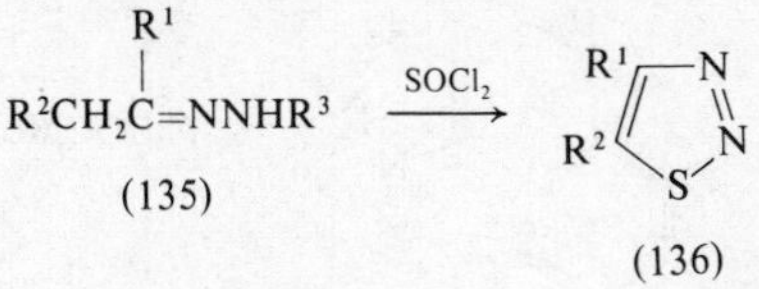

(135) (136)

Meso-ionic 1,2,5-thiadiazolium-4-olates (138) are obtained by the action of sulphur monochloride on the *N*-methylated amino-acid amide (137);[119] see Scheme 10. Several other meso-ionic thiadiazole derivatives have been reported, especially 1,2,3-thiadiazolium-4-(and -5-)olates and -thiolates.[120]

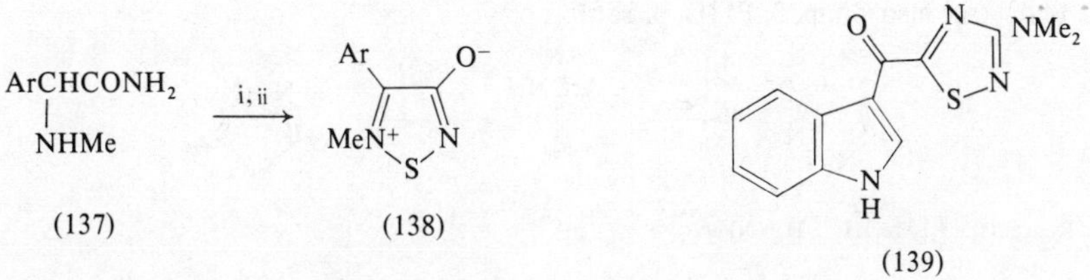

(137) (138)

(139)

Reagents: i, S₂Cl₂, DMF; ii, NaHCO₃ (aq.)

Scheme 10

The First S^II–N Bond in a Natural Product?—Dendrodoine, of which 98 mg was isolated from 2.2 tonnes of a marine tunicate, *Dendrodoa grossularia* (Styelidae), has been shown by *X*-ray and spectral data to be an indole derivative of 1,2,4-thiadiazole (139).[121] A noteworthy feature is that, as far as this Reporter is aware, this is the first natural product with a sulphur(II)-nitrogen bond. All the other fourteen possible combinations of the five common biological elements (C, H, N, O, and bivalent S) are elaborated by living cells; only S—N has, so far, been missing.

Synthesis of Selenadiazoles.—One important selenium-transfer reagent is selenium dioxide, which reacts with hydrazones such as (140) to form 1,2,3-selenadiazoles (141); steroid-like compounds (142; R¹,R² = H, Me, or OMe) can be made in this way,[122] and a similar reaction has been described in which $H_2{}^{75}SeO_3$ was used to insert a radioactive label into a potential adrenal-imaging agent (143).[123]

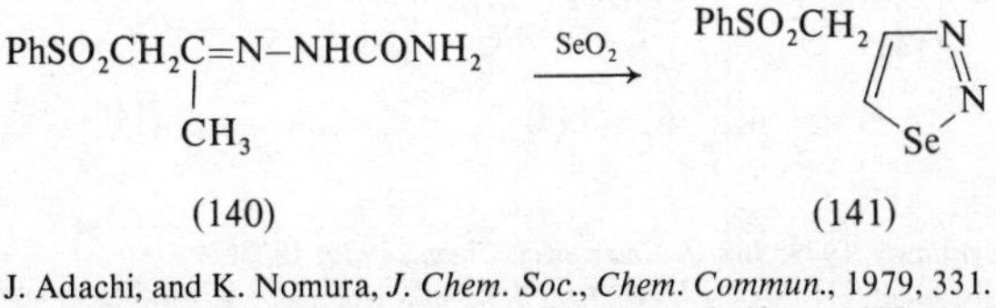

(140) (141)

[119] K. Masuda, J. Adachi, and K. Nomura, *J. Chem. Soc., Chem. Commun.*, 1979, 331.
[120] P. Demaree, M. C. Doria, and J. M. Muchowski, *J. Heterocycl. Chem.*, 1978, **15**, 1295; K. Masuda, J. Adachi, and K. Nomura, *J. Chem. Soc., Perkin Trans. 1*, 1979, 956; *ibid.*, p. 2349.
[121] S. Heitz, M. Durgeat, M. Guyot, C. Brassy, and B. Bachet, *Tetrahedron Lett.*, 1980, **21**, 1457.
[122] I. Lalezari, A. Shafiee, J. Khorrami, and A. Soltani, *J. Pharm. Sci.*, 1978, **67**, 1336; I. Lalezari, A. Shafiee, and S. Sadeghi-Milani, *J. Heterocycl. Chem.*, 1979, **16**, 1405; I. Lalezari and S. Sadeghi-Milani, *ibid.*, 1978, **15**, 501.
[123] R. N. Hanson and M. A. Davis, *J. Labelled Compd. Radiopharm.*, 1979, **16**, 31 (*Chem. Abstr.*, 1979, **92**, 22 436).

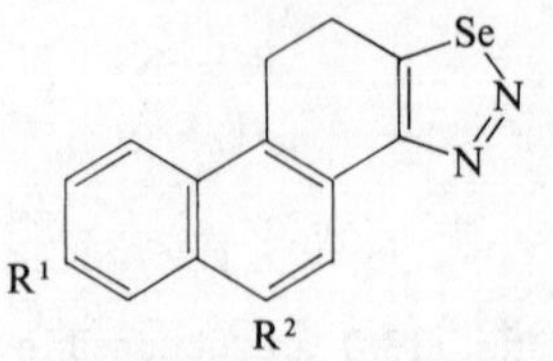

(142) 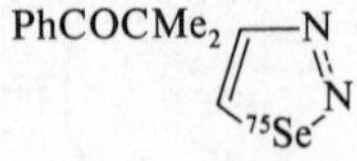   (143)

Aryl selenoamides react with iodine to give 3,5-diaryl-1,2,4-selenadiazoles; with hydrazine, however, the corresponding 1,3,4-selenadiazoles are formed (Scheme 11)[124] (see also Chap. 3, Pt III, p. 185).

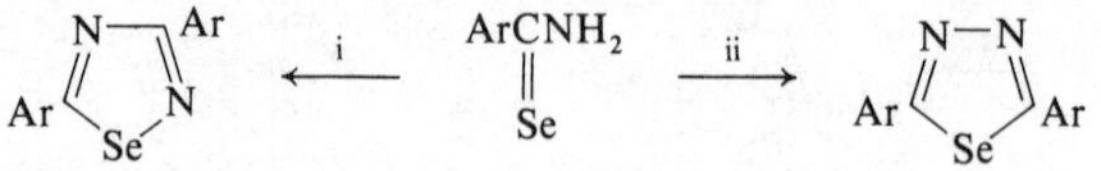

Reagents: i, I₂, MeOH; ii, H₂NNH₂

Scheme 11

Photolysis and Thermolysis of Thiadiazoles and Selenadiazoles.—A remarkably large number of papers on the photolysis of 1,2,3-thiadiazoles, and the likely production of thiirens thereby, have appeared (see Vol. 5, pp. 197, 432).[125] This subject is discussed in Chapter 4 (p. 216), but it may be noted here that two groups have now isolated thiirens and recorded their spectra by low-temperature matrix techniques.[126]

Thermolysis of 1,2,3-selenadiazole affords only selenoketen (144) and acetylene; above 800 °C, acetylene is the only organic product. Thermolysis of 1,2,3-benzoselenadiazole (145) yields the fulvaselone (146), which was observed as an unstable red deposit on a cold finger at −196 °C; it was identified by p.e. spectroscopy.[127]

When 1,2,5-selenadiazoles (147) are photolysed, nitrile selenides (148) are the

$$H_2C=C=Se$$

(144)

(145)   (146)

[124] V. I. Cohen, *Synthesis*, 1978, 768; *J. Heterocycl. Chem.*, 1979, **16**, 365.
[125] U. Timm, H. Buehl, and H. Meier, *J. Heterocycl. Chem.*, 1978, **15**, 697; U. Timm and H. Meier, *ibid.*, 1979, **16**, 1295; H. Buehl, U. Timm, and H. Meier, *Chem. Ber.*, 1979, **112**, 3728; H. Meier, S. Graw, U. Timm, and T. Echter, *Nouv. J. Chim.*, 1979, **3**, 715; R. C. White, J. Scoby, and T. D. Roberts, *Tetrahedron Lett.*, 1979, 2785; E. Schaumann, J. Ehlers, and H. Mrotzek, *Liebigs Ann. Chem.*, 1979, 1734.
[126] A. Krantz and J. Laureni, *Ber. Bunsenges. Phys. Chem.*, 1978, **82**, 13; M. Torres, A. Clement, J. E. Bertie, H. E. Gunning, and O. P. Strausz, *J. Org. Chem.*, 1978, **43**, 2490; M. Torres, I. Safarik, A. Clement, J. E. Bertie, and O. P. Strausz, *Nouv. J. Chim.*, 1979, **3**, 365.
[127] R. Schulz and A. Schweig, *Angew. Chem., Int. Ed. Engl.*, 1980, **19**, 69.

initial products.[128] If 2,1,3-benzothiadiazole 1-oxide (150) is photolysed, one reaction that occurs is a reversible isomerization to 2-thionitroso-nitrosobenzene (151); irreversible isomerization to the *S*-oxide (149) is a competing reaction.[129]

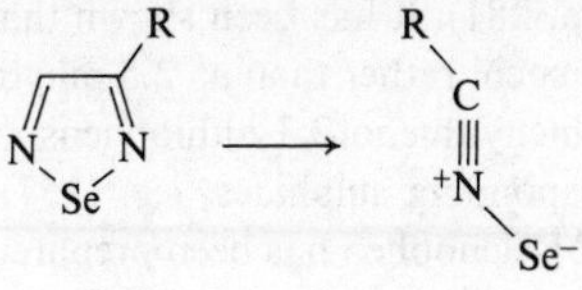

(147)   (148)

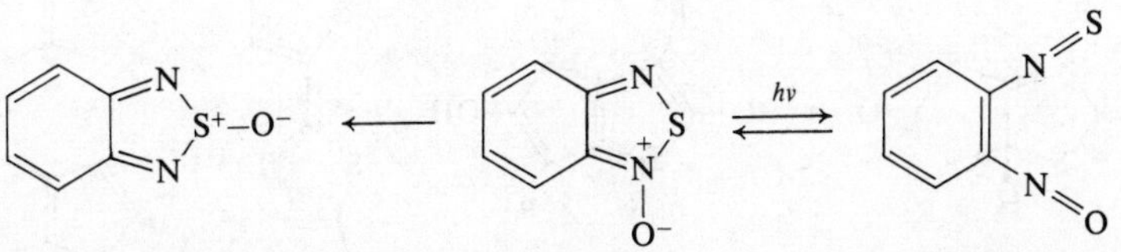

(149)   (150)   (151)

9 Benzodithiazolium Salts and their Selenium Analogues

A careful comparison of the rates of reaction of 1,2,3-benzodithiazolium salts (152; X = Y = S) and their selenium analogues (152; X, Y = S or Se) with *p*-toluidine in acetic acid at 20 °C has shown that substitution of selenium for sulphur at X increases the rate about four-fold; substitution at Y causes only a very slight increase. This observed order of reactivity is in accord with conclusions on charge densities based on M.O. calculations and n.m.r. data.[130] Buffered hydrolysis of some of these salts, *e.g.* (152; X = Y = S, or X = Se, Y = S; R = 6-Cl), yields the corresponding *S*- or *Se*-oxide (153), in which the heterocyclic ring is no longer aromatic. Two stereoisomers of each of these oxides (153), having the N—H and X=O bonds on the same side or on opposite sides of the ring, were observed. These stereoisomers can be interconverted in acidic media.[131]

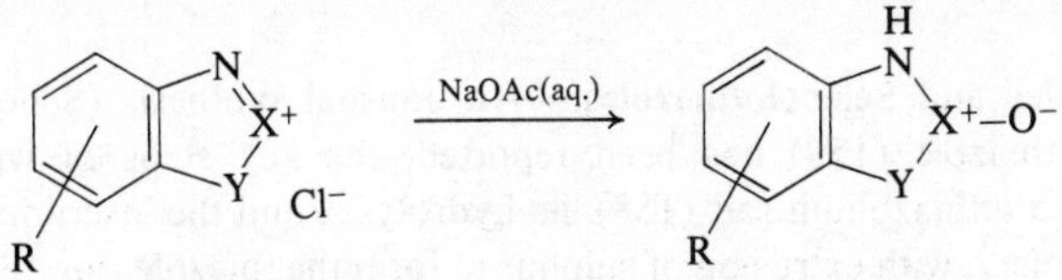

(152)   (153)

[128] C. L. Pedersen, N. Harrit, M. Poliakoff, and I. Dunkin, *Acta Chem. Scand., Ser. B*, 1977, **31**, 848; C. L. Pedersen and N. Hacker, *Tetrahedron Lett.*, 1977, 3982.

[129] C. L. Pedersen, C. Lohse, and M. Poliakoff, *Acta Chem. Scand., Ser. B*, 1978, **32**, 625.

[130] Yu. I. Akulin, M. M. Gel'mont, B. Kh. Strelets, and L. S. Efros, *Khim. Geterotsikl. Soedin.*, 1978, 912 (*Chem. Abstr.*, 1978, **90**, 54 275).

[131] B. Kh. Strelets, M. M. Gel'mont, Yu. I. Akulin, and L. S. Efros, *Khim. Geterotsikl. Soedin.*, 1979, 1205 (*Chem. Abstr.*, 1979, **92**, 128 024).

10 Compounds with Two Fused Aromatic Rings, each containing Sulphur or Selenium Atoms

Thienothiophens and their Selenium Analogues.—Thieno[3,4-c]thiophens have already been mentioned (p. 281). It has been shown that the potentially tautomeric compounds (154) exist as such, rather than as 2,5-dihydroxythieno[3,2-b]thiophens (155).[132] A number of thienylthieno[2,3-b]thiophens (156) and bisthieno[2,3-b]-thiophens and their corresponding sulphides, *e.g.* (157), have been synthesized,[133] and 3-bromoselenolo[2,3-b]selenophen has been prepared (Scheme 12).[134]

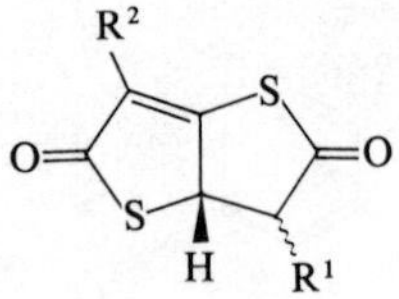

(154) 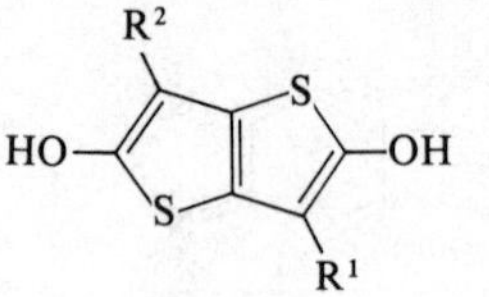   (155) 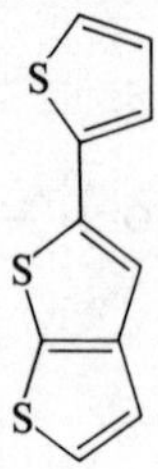   (156) 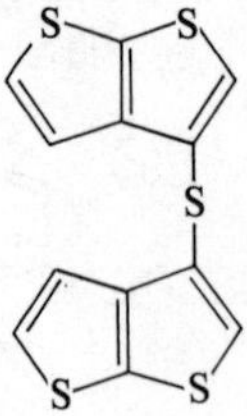   (157)

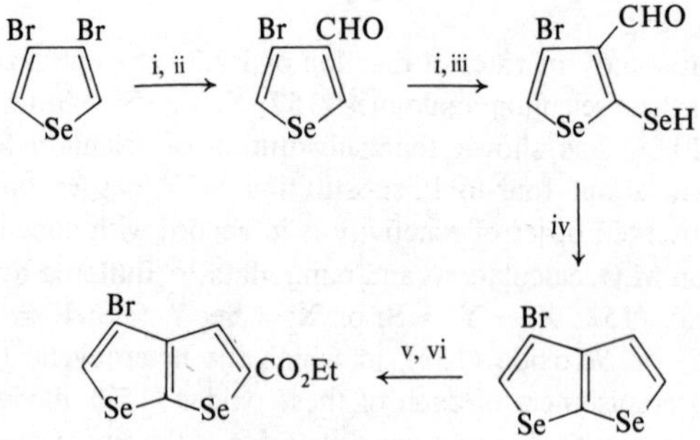

Reagents: i, BuLi; ii, DMF; iii, Se; iv, BrCH$_2$CO$_2$Et, NaOEt; v, H$^+$, H$_2$O; vi, heat

Scheme 12

Thienothiazoles and Selenolothiazoles.—An unusual synthesis (Scheme 13) of a thieno[2,3-d]thiazole (159) has been reported, the key steps of which are the formation of a dithiazolium salt (158), its hydrolysis, and the insertion of a carbon disulphide moiety, with extrusion of sulphur to form the thiazole ring.[135]

2,6-Disubstituted thieno[3,4-d]thiazoles (161; X = S) and selenolo[3,4-d]-thiazoles (161; X = Se) can be prepared from the readily available thiazolyl

[132] L. Testaferri, M. Tiecco, P. Zanirato, and G. Martelli, *J. Org. Chem.*, 1978, **43**, 2197.
[133] P. Meunier, *Bull. Soc. Chim. Belg.*, 1979, **88**, 325.
[134] Ya. L. Gol'dfarb, V. P. Litvinov, and I. P. Konyaeva, *Khim. Geterotsikl. Soedin.*, 1979, 1072 (*Chem. Abstr.*, 1979, **92**, 41 802).
[135] P. I. Abramenko, T. K. Ponomareva, and G. I. Priklonskikh, *Khim. Geterotsikl. Soedin.*, 1979, 477 (*Chem. Abstr.*, 1979, **91**, 74 518).

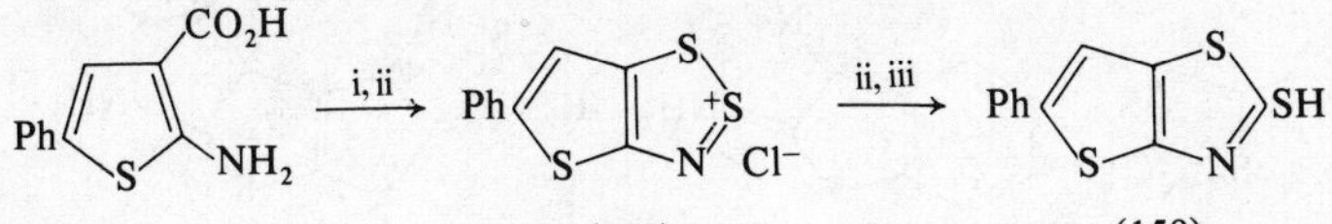

(158) (159)

Reagents: i, S_2Cl_2; ii, H_2O; iii, CS_2, NaOH, H_2O

Scheme 13

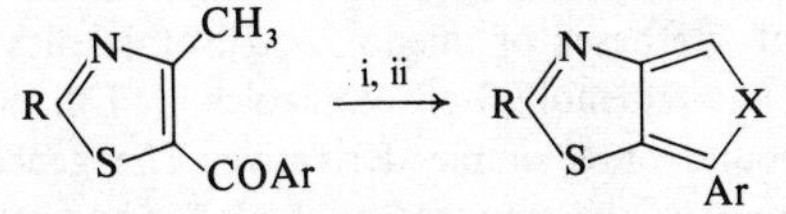

(160) (161)

(R = H or Ph)

Reagents: i, *N*-Bromosuccinimide; ii, $MeCSNH_2$ or $EtCSeNEt_2$

Scheme 14

ketones (160) as shown in Scheme 14; a thioamide or selenoamide is used as the sulphur- or selenium-transfer reagent.[136]

Thieno[3,2-*d*]thiazoles (162; X = S), thieno[3,2-*d*]selenazoles (162; X = Se), and thieno[3,2-*d*]-1,2,3-thiadiazoles (164) are all easily made from the sodium salt of 3-aminothiophen-2-thiol (163; X = S) or its selenol analogue.[137] Benzo[*b*]-selenolo[3,2-*d*]-1,2,3-selenadiazole (166) can be prepared by the cyclocondensation of the semicarbazone (165) with selenium dioxide.[138]

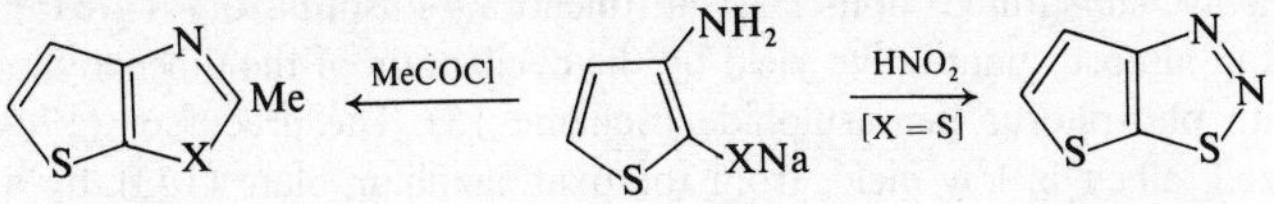

(162) (163) (164)

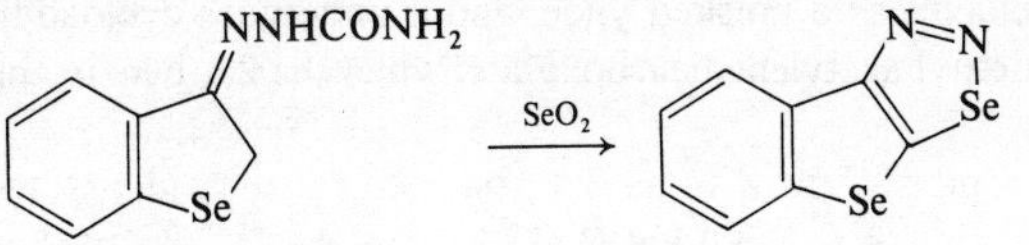

(165) (166)

Thienoisothiazoles.—This seems to be an area of research in which there has suddenly been an increase in activity, judging from the number of papers published in the period under review. Examples of five of the six possible isomers have now been reported, the sole exception being thieno[2,1-*e*]isothiazole (167), which is one of the two possible 'non-classical' isomers.

[136] A. Shafiee and A. Mazloumi, *J. Heterocycl. Chem.*, 1978, **15**, 1455.
[137] C. Paulmier, *Bull. Soc. Chim. Fr.*, Part 2, 1980, 151.
[138] V. P. Litvinov and I. A. Dzhumaev, *Izv. Akad. Nauk SSSR, Ser. Khim.*, 1979, 478 (*Chem. Abstr.*, 1979, **90**, 203 977).

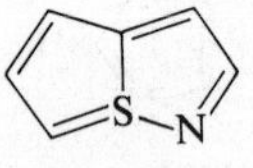

(167) 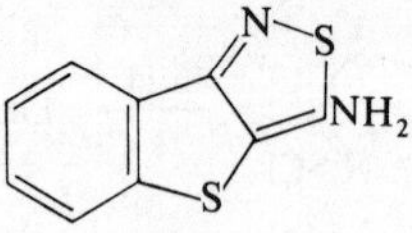   (168) 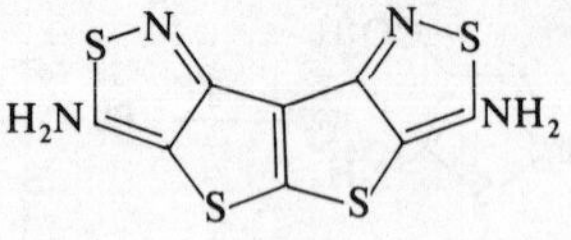   (169)

Thieno[3,2-c]isothiazoles, *e.g.* (168) or (169), have been made, the isothiazole rings being constructed by the well-tried oxidation of *o*-amino-thioamides with a peroxide.[139] In a chemical *tour-de-force*, a group at the University of Hull has devised elegant and efficient syntheses of thieno[2,3-d]isothiazoles (170), thieno-[3,2-d]isothiazoles (171), and thieno[2,3-c]isothiazoles (172), in most cases obtaining the parent compound or a simple derivative. The general procedures employed include not only most of the standard methods for annelating isothiazole rings, in this case on to thiophen precursors, but also the more difficult construction of isothiazoles with substituents capable of forming a thiophen ring.[140] A surprising alternative synthesis of a thieno[2,3-c]isothiazole (172; R = CO_2Et) by the reaction between phosphorus pentasulphide and ethyl 3-cyano-5,5-diethoxy-2-oxo-pentanoate, $(EtO)_2CHCH_2CH(CN)COCO_2Et$, has been briefly reported; the structure of the product was confirmed by an alternative synthesis from an isothiazole-3-thiol.[141]

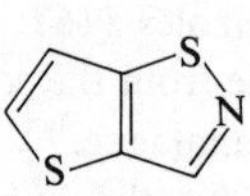

(170) 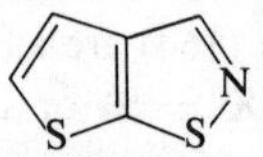   (171) 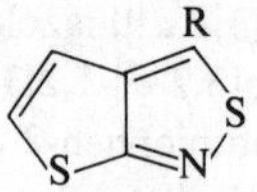   (172)

A heavily substituted non-classical thieno[3,4-c]isothiazole (175) has been prepared in almost quantitative yield by the cyclization of the dibenzoylisothiazole (174) with phosphorus pentasulphide (Scheme 15). The precursor (174) is itself synthesized, albeit in low yield, from the oxathiazolium-olate (173), by a process akin to that described earlier (p. 286).[142] The thiophen ring (but not the isothiazole ring) in (175) behaves as a masked ylide, and it undergoes cycloaddition reactions such as with dimethyl acetylenedicarboxylate, when the 2,1-benzisothiazole (178) is

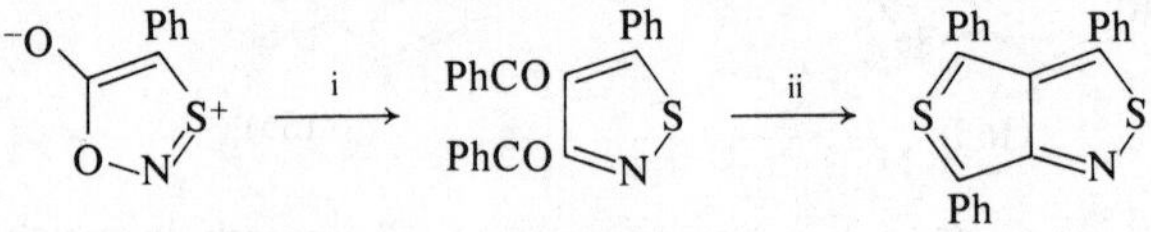

(173)   (174)   (175)

Reagents: i, PhCOC≡CCOPh; ii, P_4S_{10}

Scheme 15

[139] B. Tornetta, M. A. Siracusa, G. Ronsisvalle, and F. Guerrera, *Gazz. Chim. Ital.*, 1978, **108**, 57; G. Seybold and H. Eilingsfeld, *Liebigs Ann. Chem.*, 1979, 1271.
[140] K. Clarke, W. R. Fox, and R. M. Scrowston, *J. Chem. Soc., Perkin Trans. 1*, 1980, 1029.
[141] F. C. James and D. Krebs, *Phosphorus Sulfur*, 1979, **6**, 143.
[142] H. Gotthardt, F. Reiter, R. Gleiter, and R. Bartetzko, *Chem. Ber.*, 1979, **112**, 260.

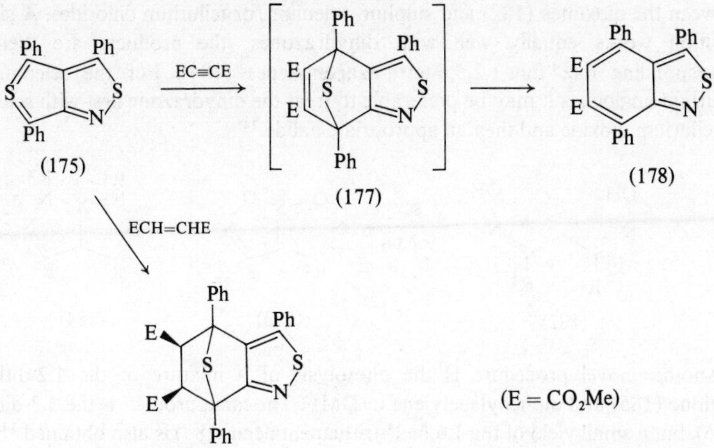

(175)

(177)

(178)

(176)

$(E = CO_2Me)$

formed; the intermediate adduct (177) spontaneously extrudes sulphur. Dimethyl maleate and other olefinic dipolarophiles react stereospecifically, without elimination of sulphur, to give the corresponding adducts, *e.g.* (176), as *exo–endo* isomer pairs.[143]

Meso-ionic Thiazolo[2,3-*b*]thiazoles.—Thiazoline-2-thione (179) reacts readily with 2-bromo-2-phenylacetyl chloride in the presence of base (Scheme 16) to produce 2-phenylthiazolo[2,3-*b*]thiazolium-3-olate (180), which undergoes cycloaddition reactions with acetylenic and olefinic dipolarophiles in much the same way as (175). With dimethyl acetylenedicarboxylate, for example, the thiazolo[3,2-*a*]pyridin-5-one (181) is formed.[144]

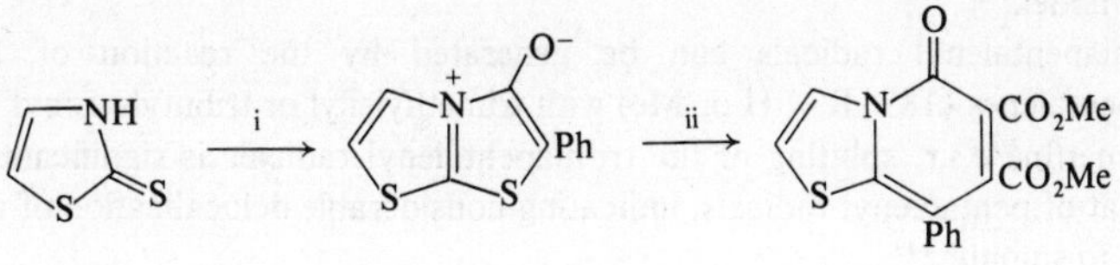

(179)

(180)

(181)

Reagents: i, PhCHBrCOCl, NEt$_3$; ii, MeO$_2$CC≡CCO$_2$Me

Scheme 16

11 1,6-Dihetera-6aλ^4-thiapentalenes and Related Systems

Synthesis.—A remarkably simple synthesis of 1,6-dioxa-6aλ^4-thia-2,5-diaza-pentalenes (183) and their selenium and tellurium analogues is by the reaction

[143] H. Gotthardt and F. Reiter, *Chem. Ber.*, 1979, **112**, 266.
[144] K. T. Potts and D. R. Choudhury, *J. Org. Chem.*, 1978, **43**, 2697; K. T. Potts and S. Kanemasa, *ibid.*, 1979, **44**, 3808.

between the dioximes (182) and sulphur, selenium, or tellurium chlorides. A similar reaction works equally well with dihydrazones; the products are then the corresponding $6a\lambda^4$-thia-1,2,5,6-tetra-azapentalenes (184). For the selenium or tellurium analogues it may be preferable to treat the dihydrazone first with selenium or tellurium dioxide and then an appropriate halide.[145]

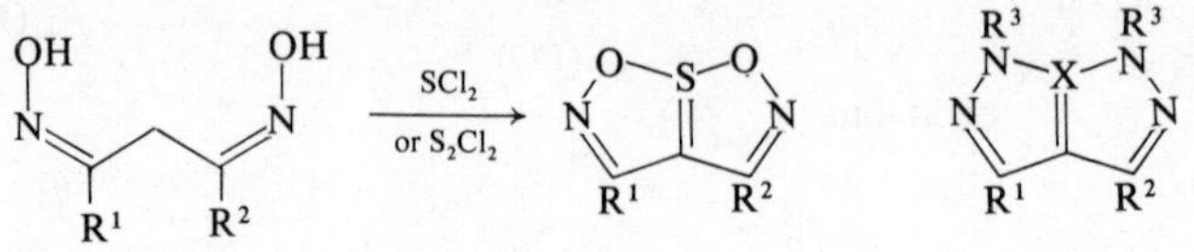

Another novel procedure is the photolysis of a mixture of the 1,2-dithiole-3-thione (185) and diphenylacetylene in DMF. The main product is the 1,3-dithiole (186), but a small yield of the $1,6,6a\lambda^4$-trithiapentalene (187) is also obtained.[146]

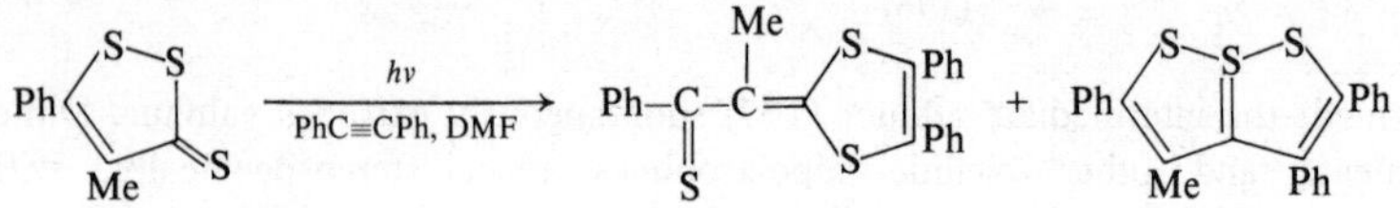

Structural and Spectroscopic Studies.—An *X*-ray crystallographic structure determination of the $1,6a\lambda^4$-dithia-6-azapentalene (188) has been completed;[147] gas-phase p.e. spectra of $1,6$-dioxa-$6a\lambda^4$-thia-2,5-diazapentalene (183; $R^1 = R^2 = H$) have been reported, and discussed in regard to the effect of changes of the 6a-atom on all other non-hydrogen atoms in the molecule. The observed vibrational broadening of the oxygen 1s lines is related to the observed bond lengths, within a simple model.[148]

Trithiapentalenyl radicals can be generated by the reaction of $1,6,6a\lambda^4$-trithiapentalenes (189; R = H or Me) with trimethylsilyl or tributylstannyl radicals. The hyperfine e.s.r. splitting of the trithiapentalenyl radicals is significantly lower than that of pentadienyl radicals, indicating considerable delocalization of unpaired spin on to sulphur.[149]

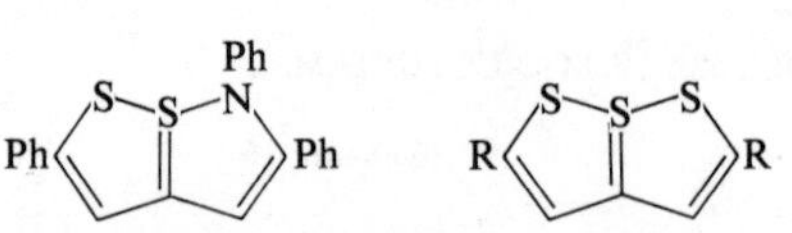

[145] M. Perrier and J. Vialle, *Bull. Soc. Chim. Fr.*, Part 2, 1979, 199; *ibid.*, p. 205.
[146] V. N. Drozd, G. S. Bogomolova, and Yu. M. Udachin, *Zh. Org. Khim.*, 1978, **14**, 2459 (*Chem. Abstr.*, 1978, **90**, 152 052).
[147] M. M. Borel, A. Leclaire, G. Le Coustumer, and Y. Mollier, *J. Mol. Struct.*, 1978, **48**, 227.
[148] L. J. Saethre, N. Martensson, S. Svensson, P. A. Malmquist, U. Gelius, and K. Siegbahn, *J. Am. Chem. Soc.*, 1980, **102**, 1783.
[149] K. U. Ingold, D. H. Reid, and J. C. Walton, *J. Chem. Soc., Chem. Commun.*, 1979, 371.

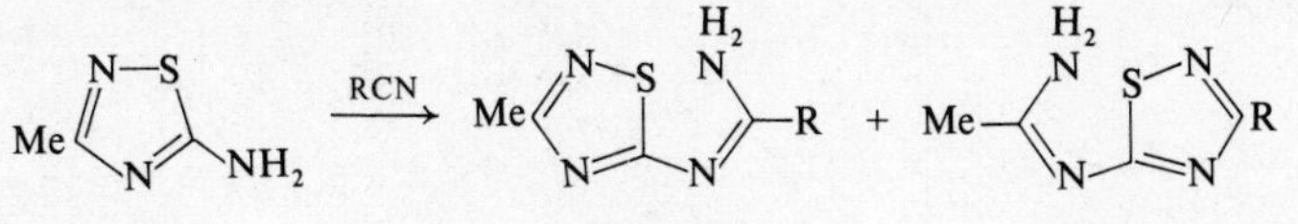

(190)   (191)   (192)

'Bond Switch' at π-Hypervalent Sulphur.—Condensation of 5-amino-3-methyl-1,2,4-thiadiazole (190) with alkyl or aryl cyanides affords an equilibrium mixture of products (191) and (192). A similar bond switch is observed when the 5-imino-1,2,4-thiadiazoline (193) is treated with the imino-ester MeC(=NH)OEt, as shown in Scheme 17. An *X*-ray analysis of the product (194) shows that the S----N and S—N bond lengths are 2.500 and 1.668 Å, respectively.[150] Protonation of (194) causes a reverse switch, the thermodynamically more stable salt (195) being formed.[151] In each case, the probable intermediates are species containing a quadricovalent sulphur atom.

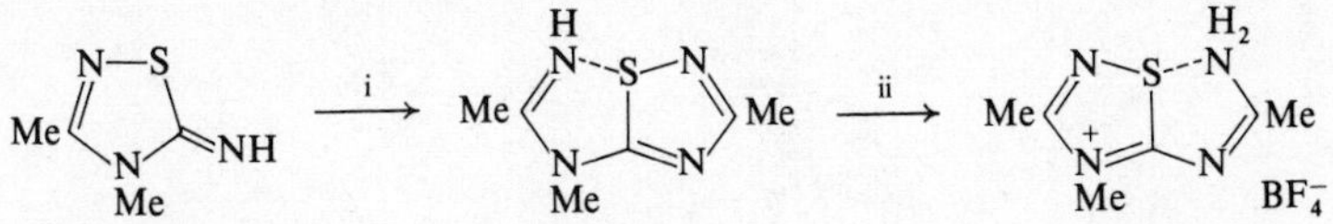

(193)   (194)   (195)

Reagents: i, MeC(=NH)OEt, at 60—80 °C; ii, HBF₄

Scheme 17

[150] K. Akiba, T. Kobayashi, and S. Arai, *J. Am. Chem. Soc.*, 1979, **101**, 5857; K. Akiba, S. Arai, T. Tsuchiya, Y. Yamamoto, and F. Iwasaki, *Angew. Chem.*, 1979, **91**, 176.
[151] K. Akiba, S. Arai, and F. Iwasaki, *Tetrahedron Lett.*, 1978, 4117.

Author Index

303

Author Index

327

Sundermeyer, W., 163, 226
Surzur, J. M., 26, 234
Suschitzky, H., 149, 275
Sutrinso, R., 66, 151
Suzuki, A., 158
Suzuki, F., 50, 103
Suzuki, H., 51, 52, 65, 70, 250
Suzuki, K., 29, 54, 67, 68, 201
Suzuki, K. T., 6, 254
Suzuki, N., 70
Suzuki, R., 68
Suzuki, S., 131, 154
Svanholt, H., 158, 226
Svata, V., 38
Svensson, S., 300
Svoronos, P., 135
Swaminathan, S., 85
Swank, D. W., 142
Swanson, D. D., 232, 262
Swanson, S., 8
Swedo, R. J., 74
Swenton, J. C., 45, 195
Swepston, P., 133, 267
Swern, D., 53, 85, 86, 134, 143
Swindell, C. S., 73
Sych, E. D., 284
Sycheva, T. P., 279
Sydnes, L. K., 188
Sylvia, C. J., 4
Symeonides, K., 220
Symons, M. C. R., 13
Szafranek, J., 197
Szalaiko, U., 70
Szargan, R., 168
Szeja, W., 72
Szilagyi, S., 49, 143
Szmant, H. H., 54
Szymanowski, J., 16

Tabacchi, R., 283
Taber, D. G., 104
Tabushi, I., 247
Tachikawa, A., 133
Taeger, T., 231
Tagaki, W., 76
Tagawa, J., 14
Tagiev, B. A., 3
Taguchi, T., 32, 204
Taguchi, Y., 211
Taieb, M. C., 244
Taira, Z., 199
Tait, T. A., 44, 95, 195
Takagi, M., 44
Takagi, S., 32
Takahaski, F., 84, 144
Takahashi, H., 90, 165, 174, 255, 289
Takahashi, K., 131, 246, 292
Takahashi, M., 93, 101
Takahashi, S., 102
Takahashi, T., 30, 100, 101
Takahashi, Y., 57
Takai, Y., 22
Takaki, K., 41

Takamiya, N., 12
Takamoto, M., 246
Takamoto, T., 233
Takanashi, H., 32
Takano, S., 90, 93
Takase, T., 32
Takata, M., 26, 81
Takata, T., 6, 67, 78, 254
Takayanagi, H., 81, 106
Takeda, K., 42, 138, 165, 232
Takeda, N., 205
Takeda, R., 22, 68
Takeda, T., 85
Takeda, Y., 5
Takei, H., 21, 33, 38, 104
Takei, K., 27
Takemasa, T., 256
Takemoto, H., 249
Takenaka, K., 232
Takeshima, T., 98, 192, 194, 197, 225
Takeuchi, T., 5
Takido, T., 9, 183
Takigawa, T., 208
Takikawa, Y., 63
Takizawa, S., 63
Talaikyte, Z., 26
Tam, C. C., 47, 91
Tam, J. P., 41
Tamagaki, S., 112, 153
Tamao, K., 65
Tamaru, Y., 59, 68, 90, 184, 185, 247, 256
Tamas, J., 6
Tamba, M., 13
Tamura, Y., 60, 65, 138, 140, 147, 166, 179, 252, 287
Tanabe, T., 40
Tanaka, H., 22, 26, 100, 178, 185, 199, 200
Tanaka, J., 273
Tanaka, K., 11, 38, 43, 46, 100, 133, 205, 208, 242
Tanaka, S., 66, 199
Tanaka, T., 5
Tanaskov, M. M., 58
Tancredi, T., 254
Tang, H.-N., 76
Tang, P.-W., 50, 56, 104
Tangerman, A., 163
Tanida, H., 39
Tanigawa, Y., 33
Taniguchi, E., 253, 255
Tanikaga, R., 11, 43, 44, 52, 100, 208
Tanimoto, M., 4
Tanimoto, S., 44, 50, 80, 82
Taniyasu, S., 40
Tantasheva, F. R., 19
Tantawy, A., 165
Tantry, K. N., 157, 199, 204
Tarantelli, T., 7
Tarasenko, A. I., 141
Tarasova, O. A., 32

Tartakovskii, V. A., 81
Taschner, M. J., 93, 256
Tashiro, M., 131, 156, 207, 292
Taticchi, A., 150, 281
Tavernier, D., 242
Taylor, A. J., 70
Taylor, A. W., 274
Taylor, D. R., 174
Taylor, E. C., 255, 290
Taylor, J. B., 144
Tazaki, M., 44
Tebby, J. C., 6
Tember-Kovaleva, T. A., 178
Temkin, H., 269
Templeton, D. H., 81, 243
Templeton, L. K., 81, 243
Terada, H., 174
Teranishi, A. Y., 116
Terasawa, T., 216, 242
Terashime, K., 193
Terayama, Y., 96
Ter-Gabriélyan, E. G., 128
Terlouw, J. K., 232
Ternary, L. L., jun., 245, 265
Terpinski, J., 178
Terrier, F., 277
Tesky, F. M., 133, 228
Teso Vilar, E., 72
Testaferri, L., 17, 273, 276, 296
Testoedeva, S. I., 65
Teterina, L. F., 32
Tetreault-Ryan, L., 263
Tezuka, T., 51, 52, 209, 250
Thaker, K. A., 57
Thamnusan, P., 54, 104
Thea, S., 163
Thi, N. P., 38
Thiel, W., 32, 191
Thielemann, C., 172, 177
Thijs, L., 163
Thind, S. S., 64
Thomas, E. J., 18, 83
Thomas, M. T., 40, 81
Thomas, P. J., 106
Thompson, M., 247
Thomsen, I., 230
Thong, P. D., 172, 177
Thuillier, A., 9, 44, 95, 191, 193, 194, 255, 256
Thulstrup, E. W., 271
Ticozzi, C., 81, 85
Tidwell, T. T., 19
Tiecco, M., 17, 273, 276, 296
Tiensripojamam, A., 54
Tighe, B. J., 253
Tilak, B. D., 74
Tilhard, H.-J., 116
Tilichenko, M. N., 235, 246
Tilley, J. W., 165
Timm, U., 158, 216, 254, 292, 294
Timmerman, A., 164
Timofeeva, G. N., 40
Tin, K.-C., 54, 104, 239